Sustainable Plant Nutrition in a Changing World

This book series aims to provide new insight and improve understanding of plant nutrition, a crucial component in plant production and food security. Plant nutrition encompasses the study of plant growth and metabolism, external supply of necessary chemical compounds and the biochemistry of various nutrients in plants. This field of study has grown in light of recent scientific advances and climate change and has expanded to include the impact of climate change, pollution, nanomaterials, biofortification and genomics on the nutrition of plants.

The component volumes are written by leading scholars from around the globe and contain the latest research and applications. Each volume is fully illustrated with in-depth references for further reading. This series will be of interest to all scientists, academics, students and professionals in the plant sciences and agriculture, as well as biotechnology, ecology and biochemistry.

Renato de Mello Prado
Hassan Etesami • Anoop Kumar Srivastava
Editors

Silicon Advances for Sustainable Agriculture and Human Health

Increased Nutrition and Disease Prevention

Springer

Editors
Renato de Mello Prado
Department of Soil Science
São Paulo State University
Jaboticabal, São Paulo, Brazil

Hassan Etesami
Department of Soil Science
University of Tehran
Tehran, Iran

Anoop Kumar Srivastava
Indian Council of Agricultural Research-
Central Citrus Research Institute
Nagpur, Maharashtra, India

ISSN 2662-2394 ISSN 2662-2408 (electronic)
Sustainable Plant Nutrition in a Changing World
ISBN 978-3-031-69875-0 ISBN 978-3-031-69876-7 (eBook)
https://doi.org/10.1007/978-3-031-69876-7

This Springer imprint is published by the registered company Springer Nature Switzerland AG
The registered company address is: Gewerbestrasse 11, 6330 Cham, Switzerland

Foreword

I started work on silicon (Si) in plants on 1st December 1980. I was based in Bangor, North Wales, and my boss was Dr. Dafydd Wynn Parry, who was one of the relatively few people interested in plant Si and phytoliths at the time. Much has changed in the over 43 years that I have been working in this area. When I went to the library in Bangor in the early 1980s it was a major event when I found a single paper on any aspect of the topic. Now it is impossible to keep up with rapid developments across the whole field! The first work I was involved in at Bangor was an investigation into the possibility that phytoliths could cause esophageal cancer in some parts of the world. We worked together with Dr. Charles O'Neill of the Imperial Cancer Research Fund in London. It was interesting work, but it is fair to say that this idea was never conclusively proved. However, it was indicative of the kind of work that I would be involved in during my career investigating plant Si. I am principally a plant physiologist and electron microscopist, but I have found myself collaborating with a whole range of scientists from other disciplines: agriculture, archaeology, biogeochemistry, cancer research, chemistry, climate science, food science, human nutrition, paleoecology, and soil science.

I was very fortunate to get to know Prof. Emanuel Epstein, one of the world's leading experts on plant mineral relations. Late in his career, Emanuel turned to plant Si, and he wrote two major reviews of the topic that became classics and were responsible for generating much interest in the area. In 1994 he wrote: "A greater awareness of the importance of Si in plants, especially on the part of experimental plant biologists, is bound to have beneficial synergistic effects beyond plant biology *per se*" (*Proc Natl Acad Sci USA* 91:11–17). Emanuel was quite correct, and I have frequently found that my publications in plant science have been used by people working in totally different disciplines. One of the major features of research on plant Si is its multidisciplinary nature.

In 2023, Dr. Renato de Mello Prado was the editor for *Benefits of Silicon in the Nutrition of Plants*, another volume published by Springer Nature. In that book the main focus was very definitely on the roles of Si in plant nutrition. Many of the chapters concentrated on the interactions between Si and other elements. I myself contributed a chapter on the mitigation of aluminum toxicity by Si. Now, Renato has joined Dr. Hassan Etesami and Dr. Anoop Kumar Srivastava to edit a further work on plant Si, including many diverse areas that were not covered in his first volume, making it a truly multidisciplinary book. So, it is a great pleasure to write the foreword for *Silicon Advances for Sustainable Agriculture and Human Health: Increased Nutrition and Disease Prevention*.

In this new volume there are chapters on many important areas. Several chapters look at aspects of soil science and chemistry. These are foundational to our understanding of plant/soil Si relationships. A key topic that has recently received a lot of attention is carbon sequestration by phytoliths, an area I have worked on myself. Clearly, if we could find ways to increase carbon sequestration in the soil and draw down carbon from the atmosphere, that would be of huge benefit to humanity in dealing with climate change. Several chapters consider the roles of Si in reducing abiotic and biotic stresses. There are a number of contributions looking at Si nanoparticles, their uptake by plants, and their importance in plant nutrition. This is a relatively new area of research, but has great potential for the future. The final three chapters focus on various aspects of the role of Si in human nutrition and health. Of course, that Si comes from plants! So, researchers that are interested in almost any area of plant Si research will find something that will be useful to them. Highly recommended reading.

Department of Biological and Medical Sciences
Oxford Brookes University,
Oxford, UK

Martin J. Hodson

Contents

About the Editors

Renato de Mello Prado PhD, is currently working as a Professor of Plant Nutrition at São Paulo State University, São Paulo, Brazil. He also coordinates a research group (Genplant) focused on the nutrition of tropical plants. He has published over 500 articles in peer-reviewed journals and has authored/coauthored about 30 academic books and supervised more than one hundred students (master's/doctorate/post-doctorate).

Hassan Etesami is a research scientist with 15 years of experience in the field of soil biology and biotechnology. He obtained his doctor's degree from the Department of Soil Science, University College of Agriculture & Natural Resources, University of Tehran, Iran, where he is currently a member of the faculty. Dr. Etesami has a special interest in developing biofertilizers and biocontrol agents that meet farmers' demands. He has coauthored over 150 publications (research papers, review papers, and book chapters) in various areas including biofertilizers and biocontrol. He is also reviewer for the Journal of 150 journals. Dr. Etesami's research areas include stressed agricultural management by silicon, microbial ecology, biofertilizers, soil pollution, integrated management of abiotic (salinity, drought, heavy metals, and nutritional imbalance) and biotic (fungal pathogens) stresses, plant-microbe-interactions, environmental microbiology, and bioremediation.

Anoop Kumar Srivastava (MSc and PhD in Soil Science from world famous Banaras Hindu University, India) is currently perched as Principal Scientist (Soil Science) at Central Citrus Research Institute (ICAR), Maharashtra, India. He is specialized in nutrient constraints analysis and mapping by diagnosis and recommendation integrated system (DRIS)-based soil-plant nutrient diagnostics, orchard efficiency modeling, targeted yield-based site-specific nutrient management, rhizosphere specific microbial consortium, Integrated Nutrient Management (INM) concept, rhizosphere hybridization, fertigation partitioning, substrate dynamic, and soil fertility map as decision support for fertilizer recommendation. He is the life member of 33 academic societies, honorary fellow of 3 academic societies, fellow of 13 academic societies, and received lifetime achievement awards from 2 academic societies.

About the Editors

Chapter 1
Challenges of Plant Nutrition and Climate Change: Focus on Fruit Crops

Anoop Kumar Srivastava

Abstract Climate change is a global event, characterized by often uncertain crop responses, often affecting the production and nutritional quality of crops, including fruit. Plant nutrition in this context is seen as strong via-medium in imparting an additional biochemical preparedness in reverting any adverse effects of climate change, including the frequent events. High carbon locking (carbon sequestration within woody framework) ability of fruit crops provides an edge by creating a parallel nutrient sink (fertilization effect of elevated CO_2) to magnify the impact of plant nutrition compared to any other horticultural crops. On the other hand, parameters such as salinity, drought, and high temperature are accompanied with increased accumulation of secondary metabolites and antioxidants (grapes and citrus) coupled with early maturity, thereby creating a niche for underutilized fruits to occupy a place of prominence. Unfortunately, most of the crop responses are not studied in response to the integrated effect of climate change and mark the absence of any yield predictive models (all C3 fruit crops are considered highly responsive to increased CO_2 concentration). Introducing C4 photosynthetic cycle into fruit crops, upscaling 4R (coupled with 4W) principles in climate change scenario, plant nutrition in the interaction of nutrients at various levels, and the use of microbially modified crops associated with the use of potential stress mitigators such as silicon are some of the suggested scientific interventions.

Keywords Climate change · Customized fertilizers · Fruit crops · Fruit quality · Integrated soil fertility module · Nutrient density · Plant nutrition · Soil properties

A. K. Srivastava (✉)
Indian Council of Agricultural Research-Central Citrus Research Institute, Nagpur, Maharashtra, India

R. de Mello Prado et al. (eds.), *Silicon Advances for Sustainable Agriculture and Human Health*, Sustainable Plant Nutrition in a Changing World,
https://doi.org/10.1007/978-3-031-69876-7_1

1.1 Introduction

Climate change is one of the major challenges influencing the production sustainability of fruit crops with wide ranging direct and indirect implications from productive life of trees to postharvest life, somethings very often researchers conveniently overlook. Though climate change is a slow process involving relatively small changes in temperature and precipitation (coupled with extreme changes) over long period of time, these slow changes in climate influence the various soil processes, particularly those related to microbiome-mediated soil nutrient pool (Srivastava 2020; Srivastava and Hota 2020). Visibly or invisibly, climate impacts have already made heavy inroads spatio-temporarily across major fruit belts (temperate or tropical, more so in tropical fruit crops) leading to deterioration in nutrient-depleted suboptimum fruit quality (coupled with suboptimum production outcomes) according to Kavi Kishore et al. (2023). These outcomes adversely affected the human health-related issues of late.

According to an estimate of the World Health Organization (2023), staggering 828 million people are suffering from malnutrition and over 2 billion people are experiencing deficiencies in micronutrients (iron and zinc). These figures are expected further to touch newly 150 million zinc-deficient and 140 million protein--deficient people by 2050, providing strong clues and how our food system is currently engripped in clutches of suboptimum supply level of soil fertility vis-à-vis plant nutrition, posing a great potential threat to the survival of healthy fruit industry.

Climate change beyond any doubt is invariably causing far dramatic alterations in the global environment with respect to fruit crops (as a source of multiple nutrients), with no exception marked by dynamic changes in plant nutrient supply vis-à--vis either fruit yield or fruit quality adversely affecting taste, shape, and nutritional power, more so in temperate than in tropical/subtropical fruit crops (Reddy et al. 2017). Therefore, climate change is perhaps not always so scary looking at the physiological diversity of fruit crops. Parameters such as salinity, drought, and high temperature are often considered favorable for the elevated accumulation of secondary metabolites and antioxidants (grapes and citrus), besides turning some of the unsuitable fruit regions by triggering an early physiological maturity, thus offering a better adaptability to those unconventional fruits (dragon fruits and apple plum). Mounting global evidence suggesting adverse nutritional effects of climate change on fruit crops (carbon-rich-nutrient poor footprint) warrant strong warning signal to undertake periodical stock of the agronomic practices defining quality production and develop "climate-neutral fruit culture" in order to minimize in waging influence of climate change (Bhargava et al. 2011; Datta 2013; Hazarika 2013; Reddy et al. 2017; Dong et al. 2024). And, more importantly, we also need to pick those climate change-related issues that are advantageous to the success of fruit crops growing through balanced carbon farming without any compromise on nutrient density (Bhattacharjee et al. 2022).

Of late, continuously depleting multiple nutrient mining due to increasing annual gap between the amount of nutrients added and the amount of nutrient exported out of the field through edible plant parts has mounted increasing pressure to ensure nutrient density of fruits (e.g., a sharp decline rate was observed in calcium, 57% in lemons, 59% in pineapples, and 65% in tangerines from 1975 to 2001) against any adversity including climate change, thus offering a bigger daunting task (Soares et al. 2019; Bhardwaj et al. 2024). Darmon et al. (2006) suggested nutrient density of fruits and vegetables as nutrients per calorie or nutrients per unit cost calculated per weight (nutrient adequacy score), per calorie (nutrient density score), or per unit cost (nutrient-to-price ratio). Amid all these claims, what are the current evidences to establish decline in the nutrient composition of fruits? Davis (2009) suggested three kinds of evidence that points toward decline in some nutrients in fruits and vegetables comprising the following: (i) early studies of inverse relationships between crop yield and mineral concentrations, widely cited as the "dilution effect"; (ii) three recent studies of apparent historical food composition data median declines of 5–40% or more in some minerals in groups of vegetables and perhaps fruits; and (iii) recent side-by-side plantings of low- and high-yield cultivars of broccoli and grains found consistently negative correlations between yield and concentrations of minerals and protein, a newly recognized "genetic dilution effect." Manjoor et al. (2024) while highlighting the impact of climate change on nutritional quality and cracking incidence of fruit crops highlighted the distinct role of transcriptional factors (underscoring the genetic underpinning behind the stress released phenotypic outcomes) in mediating the plant responses to climate change.

Amid this background, we are looking for some plausible answers to some of the hard intriguing questions such as the following: (i) increase in air carbon dioxide brings any discernible changes in rhizosphere ecology vis-à-vis changes in nutrient bioavailability; (ii) an increase in atmospheric carbon dioxide is associated with consequent increase in soil fertility enrichments and making fruit crops more nutrient dense; (iii) any nutrient imbalance is turning fruit crops into carbon rich but nutrient poor with inferior fruit quality; (iv) we have to come out of the conventional conceptual framework of proteins, vitamins, and minerals; (v) physiological behavior of fruit crops in soil fertility depletion versus fertile soil; (vi) changes in postharvest life of fruit crops in relation to temperature/water deficit/salinity; (vii) climate-smart fruit production system; (viii) carbon storage capacity of fruit crops vis-à-vis changes in preferential accumulation of different nutrients; and lastly (ix) the role of silicon as a nonessential nutrient playing an important role in mitigating multiple stresses under the influence of climate conundrum. All these burning issues are yet to be answered while dealing with fruit nutrition versus climate change, besides consistently not performing mitigation technologies. These core issues will be highlighted in this entire presentation.

1.2 Climate Change as a Global Event

Climate change in the time domain is accountable to (i) natural internal processes, (ii) external forces, (iii) persistent anthropogenic changes in atmospheric composition, and (iv) dynamic land uses. The increased concentration of greenhouse gases leads to an increased temperature which in turn has an impact on the global climate, leading to a phenomena known as climate change. The earth's climate system constantly adjusts to maintain a balance between the energy reaching from the sun and the energy going from the earth back to the space. This means that even a small rise in temperature could bring significant changes in cloud cover and wind patterns. Some of these changes may enhance the warming, while others may counteract. Climate change per se is not necessarily harmful, but the problems emerge from frequent extreme events, very often so difficult to predict, such as more erratic rainfall pattern and unpredictable heat spells that jeopardize our past good efforts on raising the production opportunities. Climate change, like the rest of the world, Asia will also likely to face the brunt of on-going process (Lal 2004).

The projected increases in temperature, variability in precipitation patterns, and increase in frequency of extreme weather events would severely affect the production of fruit crops via changes in rhizosphere structural, genetic and functional microbial diversity of the rhizosphere. These in turn affect different crop developmental phases in varying intensities, which would ultimately determine the productivity and quality on a long-term basis to develop any contingency plan. The emission of carbon dioxide due to anthropogenic activities has enriched the atmosphere influencing the carbon fixation and productivity of fruit crops by creating a parallel nutrient sink (Srivastava et al. 2021). The primary effects of the increased concentration of CO_2 include higher photosynthetic rate, increased light-use efficiency, reduction in transpiration and stomatal conductance, and improved water--use efficiency (Drake et al. 1997). The studies suggest a positive effect of increase in atmospheric carbon dioxide in fruit crops having C_3 photosynthetic pathway (e.g., papaya, mango, guava, citrus, and banana), promoting their growth and productivity. The studies also indicated that the increase in water-use efficiency of fruit crops due to reduction in transpiration offsets the positive effects of elevated CO_2 (Sage and Zhou 2022). The interplay of all these factors associated with climate change would subsequently determine the extent of impact on different crops in different agroecological regions (Srivastava 2012). Therefore, a balanced plant nutrition is expected to moderate the impact of both high and low temperature stresses triggered by climate change (Waraich et al. 2012).

Amidst these facts, the role of silicon is reported to be highly beneficial in fruit crops such as apple (Swierczynski et al. 2023), sapota (Lalithya et al. 2014), mango (More et al. 2015), banana (Hanumanthaiah et al. 2015), and pomegranate (El-Rhman 2010) with regard to growth, yield, and fruit quality parameters, in addition to postharvest fruit quality (Laane 2017; Patil et al. 2017; Abidi et al. 2023). However, studies highlighting silicon nutrition in relation to climate change in fruit crops are still the least understood area of intensive research (Etesami and Jeong 2020).

1.3 Climate Change and Focus on Soil Health

Elaborating climate change impacts on soil fertility and plant nutrition is like opening a pandora box. However, as a common feature, the reduction in nutrient density of fruit crops has the potential to exacerbate nutrient malnutrition. Soil health in relation to different climatic factors that impact nutrient density in fruit crops is an important area of current research (Soares et al. 2019), besides integrating climate change to human health. In this regard, Montgomery and Bikle (2021) while analyzing the soil health and nutrient density, beyond organic and conventional farming, reported that part of the long-running controversy over differences in nutrient concentration between two contrasting systems of farming emerged from variable definitions of a nutrient, while soil health adds a much-needed dimension in the form of provisioning of micronutrients and phytochemicals to support human health.

Climate change (natural or anthropogenic or both) is a global phenomenon occurring continuously since the earth came into existence, often marked by cold and hot cycles in the history of earth's climate. However, these changes became relatively rapid in the last 150–200 years around the world (Fauchereau et al. 2003), making climate change an issue to reckon within the current geopolitical scenario. The fourth assessment report of the Intergovernmental Panel on Climate Change (IPCC) suggests a global average temperature increase of 0.74 °C over the last 100 years and projected an increase of about 1.8–4.0 °C by 2100. The drivers of climate change such as moisture, temperature, and CO_2 are expected to have variable effects on various soil processes and properties that have relevance in soil fertility and productivity as a consequential response (Singh et al. 1999; Kaur and Sandhu 2020). Further, all these effects will be highly region specific, depending on the magnitude of the climate change, soil properties, and climatic conditions. Since climate change is a reality, it will have direct and indirect impacts on soil developmental processes and properties related to fruit production and appropriate mitigation strategies (Srivastava and Singh 2002). These responses of climate change factors cannot be viewed separately, being one factor influencing the other, and the resultant effect would be complex (Table 1.1).

An increase in atmospheric CO_2 concentration is likely to improve soil physical conditions through elevated supply of organic matter by increasing photosynthetic rates and the water-use efficiency of crops (Srivastava and Srivastava 1993a; Srivastava 2009a). Soil chemical properties play a pivotal role in the availability of plant nutrients and the survival of fruit crops, i.e., tropical or temperate fruits (Srivastava and Singh 2002). An increase in soil temperatures accelerates decomposition of organic matter due to increased microbiological activity leading to the rapid release of nutrients. Pendall et al. (2004) suggested an increased CO_2 may not exert a significant direct effect on N-mineralization, but associated warming may cause an increased N-mineralization, leading to increased solution-phase N. Besides, other microbial-mediated processes related to nutrient availability such as biological nitrogen fixation, phosphorus solubilization, nitrification, denitrification, sulfur oxidation, and reduction are highly dependent on soil temperature and soil moisture

Table 1.1 Response of climate change-related parameters on soil properties

Climate change parameters	Changes in soil properties
Increase in temperature variation	Loss of soil organic matter Disturbance in microbiome and nutrient pool Reduction in soil moisture content Increase in mineralization rate Loss of soil structural stability (aggregate sizes) Increase in soil respiration rate
Increase in CO_2 concentration	Increase in soil organic matter Increase in water-use efficiency Greater availability of carbon to soil microorganisms Accelerated nutrient cycling
Increase in rainfall intensity	Increase in soil moisture or soil wetness Enhanced surface runoff and erosion intensity Increase in soil organic matter and nutrient leaching Increase in reduction of iron and nitrates Increased volatilization loss of nitrogen
Reduction in rainfall intensity	Reduction in soil organic matter Increase in soil salinization Reduction in nutrient availability

(Srivastava and Srivastava 1991a; Dijkstra et al. 2010; Srivastava and Malhotra 2017). The elevated atmospheric CO_2-induced climatic changes in soil properties affect the biological storage and stoichiometry of micronutrients (Natali et al. 2009). Soil moisture and temperature are therefore primary determinants of root growth and development affecting nutrient-use efficiency through direct impacts on root surface area and metabolic functions (Srivastava and Srivastava 1993b).

Oxidation and redox conditions of soils are important in biological and abiotic transformations of plant nutrients in soil (Srivastava 2015). Such biologically mediated nutrient transformation processes are influenced by soil redox conditions including oxidation and reduction of nitrogen, phosphorus, and sulfur during mineralization, nitrification, and denitrification (Srivastava and Srivastava 1992, 1995). Soil moisture and the availability of decomposable organic matter are the main factors for controlling the oxidation-reduction state of soil (Srivastava and Srivastava 1991b). The elevated CO_2 levels also have indirect impacts on soil redox conditions by altering plant growth and supply of decomposable organic residues for soil microbial health. Interestingly, an increase in atmospheric CO_2 concentrations is likely to intensify soil acidity due to enhanced soil respiration under elevated CO_2 conditions (Srivastava and Srivastava 1993a, c). Hence, the formation of more carbonic acid (H_2CO_3) in soil water will accelerate leaching of base cations from the soil, thereby causing soil acidification (Oh and Richter 2004) and will continue to do so in the high CO_2 atmosphere in the future. A long dry spell in waterlogged soils would lead to conversion of reduced sulfur compound to sulfate consequently causing reduction in soil pH (Srivastava and Srivastava 1992; Warren et al. 2001). Salinization is another problem that has long been associated with the successful growth of different fruit crops, with an exception of arid fruits (Srivastava and Srivastava 1994; Srivastava and Kohli 1999; Zhang et al. 2014). In addition, the changes in temperature and precipitation patterns expected from climate change are likely to further affect the salt-water balance of soil ecosystems, and fruit crops are no exception in this regard.

Soil contains myriad of invisible microbes interacting through microbe-microbe and microbe-plant interaction and microbe-soil inorganic constituents forming a complex system of biological activity (Gelybo et al. 2018; Roussos 2024) as main drivers of organic matter decomposition and resultantly the nutrient cycling, organic matter turnover, and degradation of xenobiotic and soil structure. Soil organic matter is central to different soil biological properties, being major source of nutrient supply to soil microorganisms. Climate change variables have impact on soil organic matter directly through temperature and precipitation by altering land use and the rate of addition as well as decomposition of organic residues (Kirshbaum et al. 1996). An enhanced decomposition of soil organic matter (particularly the labile fractions, important for nutrient supply) with an increasing proportion of relatively inert carbon fractions at higher temperatures (Srivastava et al. 2002) and reduced soil organic matter decomposition at higher precipitation are commonly observed (Kirschbaum 1995), thus offering a possibility of interchangeable responses on soil properties with equally varied crop responses, unless supplemented with robust field studies (Srivastava 2023) taking a specific fruit crop. The response of soil microorganisms to changes in plant production under elevated CO_2 is highly variable due to the different patterns of plant C-allocation in different plant–soil systems. The increased CO_2 concentrations, in the one hand, could limit the availability of plant nutrients on account of their sequestration in plant biomass and soil organic matter pools (Luo et al. 2004) but, on the other hand, will have a potential negative effect on the carbon sequestration in soils due to alteration in the release of root-derived compounds. The change in soil organic matter turnover rates is expected to cause a decline in soil organic nitrogen on a long-term basis, although in the short term, increased returns of carbon to soils would maintain soil organic nitrogen contents (Srivastava 2020).

Soil microorganisms are often carbon-limited and therefore, increased carbon availability stimulates microbial growth and activity (Srivastava and Bora 2023). The CO_2-induced increases in soil C availability are reported to increase fungal biomass more than bacterial biomass, signifying their roles in organic matter degradation, nutrient cycling, plant nutrition, and soil aggregate formation, as part of soil functioning (Ngullie et al. 2015). Furthermore, lower N availability at elevated CO_2 may, in part, explain these increases in fungi, as fungi tend to have a higher C/N ratio than bacteria and have a lower demand for N than bacteria (Hu et al. 2001). The soil biological activities are usually doubled with every 10 °C increase in temperature. Several studies showed that increased temperatures accelerate rates of soil respiration. In a 10-year soil warming experiment, Melillo et al. (2002) showed a 28% increase in CO_2 flux in the first 6 years of warming compared to control soils, followed by considerable decreases in CO_2 released in subsequent years, and no significant response to warming in the final year of the experiment. The exact microbial processes that cause such long-term responses to warming conditions are not clearly known, but explanations such as increased temperatures that caused microbes to undergo physiological changes that result in reduced carbon-use efficiency (Allison et al. 2010) or acclimatization of soil microbes to higher soil temperatures by adapting their metabolism and eventually return to normal decomposition rates have been proposed (Joseph et al. 2015). Thus, these studies provide a greater variability on the response of soil community function to climate change and the potential effects of these responses at the ecosystem level. The effects of elevated atmospheric CO_2 concentration on soil microbial community structure are often characterized by an increased mycorrhizal colonization (Liu et al. 2017; Ortas 2020). Many studies confirmed that CO_2 enrichment increased mycorrhizal efficiency by improvising the physiological and biochemical preparedness of inoculated fruit crops (Wu et al. 2013; Zou et al. 2014; Gao et al. 2020; Liu et al. 2024) via increased P-uptake (Ortas et al. 2017; Huang et al. 2014; Shao et al. 2021), improved carbon allocation (Wu et al. 2012; Liu et al. 2014; Zou et al. 2016), rhizosphere health traits (Wu et al. 2014, 2015; Zhang et al. 2015; Wang et al. 2014), root morphology (Wu and Srivastava 2012; Shao et al. 2021), and fruit quality attributes (Wu et al. 2017; Li et al. 2023), giving wholesome clues about utility of microorganisms in quality production of fruit crops by sustaining any possible detrimental effects of climate change.

1.4 Challenges of Plant Nutrition

Historically, economic development has been faster in those world regions where fertilizer use and crop yields rose in parallel. Despite bundles of scientific literature that have been poured in the recent past dealing with a detailed description of one or few scientific goals such as molecular genetics, plant mineral nutrients related to molecular biology, physiology processes related to plant nutrition, fertilizer application strategies, and beneficial plant–microbe interactions, the success of plant nutrition in ensuring the quality production of fruit crops is registered par below efficacy and more candidly so in the climate change scenario (Srivastava 2013). In such scenario of climate change, there are multiple expectations (from biochemical bases underlying nutritional processes described down to molecular level, involving differential selectivity of membrane transporters for nutrients or signaling functions of mineral nutrients in plant development to serving large parts of the society from farmers to policy makers, allowing further enhancements in crop yield and quality, while increasing nutrient efficiency and reducing nutrient losses from crop production systems) from plant nutrition, thereby aiming to accomplish varied objectives (Srivastava and Singh 2006). Hence, the current understanding of plant nutrition is largely focused on monocultural situations, which needs to be augmented by the interactions that occur in more complex systems. Therefore, mineral nutrients in soils and crops are important, and predicting positive as well as negative interactions with global climate change is still difficult (Soares et al. 2019), although negative impacts of climate change appear to outweigh the positive ones (St Clair and Lynch 2010).

Plant nutrition stands on two legs: one leg is anchored in plant physiology explaining the fundamental processes underlying and regulating the acquisition, allocation, and utilization of mineral elements in plants, while the other leg is deeply rooted in an applied, problem-oriented context, explaining the use and action of fertilizers and the cycling of mineral elements in agricultural plant production (Wiren 2011). These dimensions of plant nutrition perch different challenges to different perspectives. Noordwijk and Cadisch (2020) previously pointed out the main challenge of plant nutrition as to increase the width of the domain between the access and excess frontiers, rather than to define the crop nutrient response as a single "economic optimum" point, involving two approaches: (i) the technical paradigm of precision farming and (ii) the ecological analogue approach based on filter functions and complementarity of components in mixed plant systems. Dobermann et al. (2022) proposed a new paradigm for responsible plant nutrition following a food systems and circular economy approach to achieve multiple socioeconomic, environmental, and health objectives involving critical actions such as (i) sustainability-driven nutrient roadmaps, (ii) digital crop nutrition solutions, (iii) nutritious crops, (iv) nutrient recovery and recycling, (v) climate-smart fertilizers, and (vi) accelerated innovation. Ironically, plant responses to deficiencies of different nutrients have been studied mainly as a separate event, and only a few studies discussed the molecular basis of biological interaction among the nutrients, comprehensively overlooking the interaction between macro- and micronutrients at two--by-two or multi-level nutrient interactions involving transcription factor or phosphate starvation response, the master regulator of multiple nutrient homeostasis (Kumar et al. 2021) using an integrative study of multiple nutrient signaling cross talks in plants (Fan et al. 2021) for ensuring better biological significance narrating higher yield and nutrient-use efficiency (Rietra et al. 2017).

According to Brown et al. (2022a), the current definitions of essential or beneficial elements for plant growth rely on narrowly defined criteria that do not fully represent a new vision for plant nutrition and often compromise with fertilizer regulation and practice. An effective plant nutrition is supposed to ensure the level of fruit nutrition, much beyond simple minerals and vitamins and more inclined toward freshness (taste quality) as a matter of extended storability. The increased CO_2 concentration in the environment reduced the overall quantity of 25 minerals in plants, including calcium, potassium, zinc, sulfur, copper, and iron (8% on average), and also increased the ratio of carbohydrates to minerals in food plants (Loladze 2014). A new analysis of long-term trends of the mineral content of fruits and vegetables (1940, 1991, and 2019) showed all elements except P declined in concentrations between 1940 and 2019; the greatest overall reductions during this 80-year period were Na (52%), Fe (50%), Cu (49%), and Mg (10%). Many reasons such as reductions, including changes in crop varieties and agronomic factors associated with the industrialization of agriculture, and increases in carbon dioxide could be ascribed (Mayer et al. 2021). Studies of historical fruit nutrient composition data are inherently limited, and most of the other methods focus on single crops of any kind and can include any nutrient of interest and can be carefully controlled (Davis 2009). Therefore, the proposed new definition of mineral plant nutrient is an element that is essential or beneficial for plant growth and development or for the quality attributes of the plant or harvested product, of a given plant species, grown in its natural or cultivated environment (Brown et al. 2022b).

While protecting plant nutrition from the effects of climate change, there is a renewed strong emphasis on enlarging the proportion of fruits, vegetables, and grains in the diet, likely to increase the intake of a wide range of minerals and phytonutrients thought to have specific nutritional benefit. However, the effect of climate change on the accumulation of these phytochemicals and their biosynthetic pathways remains poorly understood. The recent developments in metabolomic analysis methods are of great help in obtaining novel insights into qualitative and quantitative changes in the composition of plant phytonutrients under different climate change scenarios (Shahzad and Rouached 2022) in addition to linking the carbon sequestration ability of crops with plant nutrition (Elbasiouny et al. 2022).

1.5 Impact of Climate Change on Performance of Fruit Crops

The impact of climate change is variable among countries creating higher food security dangers in low-income groups. The forecast of population increase to about 9.9 billion by 2050 makes food security a formidable social problem due to an unequal distribution of food resources, increase in risks related to soil degradation, biodiversity loss, and mass extinction. Many climate indices are used as diagnostic tools to understand the potential impacts and evolution in both space and time of climate change in fruit crops. It is avidly claimed that crops indigenous to a given agroclimate will first undergo plant nutrition-regulated reduction in yield performance under the climate change scenario (Heckathorn et al. 2020).

The extreme weather events of hot and cold wave conditions are extensively reported to cause considerable damage to many of the high-value fruit crops by causing negative effects on flowering and fruit set dynamics. Freezing and extended periods of high temperature have significantly reduced yield and initiated alternate bearing and production of a heavy "on crop" (high yield) followed by a light "off crop" (low yield) in citrus, mango, avocado, pistachio, and olive (Garcia et al. 2021; Medda et al. 2022). Heat stress can be a concern in many regions of the tropics and subtropics, since high temperature is expected to cause significant damages such as sunburns on leaves, branches, and stems, anticipated leaf senescence and abscission, shoot and root growth inhibition, and fruit discoloration (De Almeida and Valle 2007; Higuchi et al. 1998; Wahid et al. 2007; Yamada and Xu 2000) including reproductive processes (Wahid et al. 2007) with negative effects on number of inflorescences, flowers, and fruits (Ledesma et al. 2008). The plant's response to high-temperature stress was observed to be cultivar-dependent. In perennial crops such as mango and guava, temperature is reported to have an influence on flowering (Krishna et al. 2011). The percentage of hermaphrodite flowers was greater in late emerging panicles, which coincided with higher temperatures. Mango cultivars exposed to different day/night temperatures (15/10 °C, 20/15 °C, 25/20 °C, and 30/25 °C) for 20 weeks showed that vegetative growth increased with increasing temperatures (Geetha et al. 2016). Like mango, higher temperatures in citrus also enhance vegetative growth, but after a certain limit, it retards the shoot elongation. Higher temperature (38/28 °C; day/night temperature) for 10 weeks in sour orange, Troyer citrange, and Valencia oranges showed that seedlings with short internodes and leaves were markedly shorter as compared to normal ambient temperature (28/22 °C) (Krishan et al. 2011). In most of the fruit crops, generally higher temperature decreased the day interval required for flowering and cooler temperature though required more days for flowering, but the number of flowers produced increased proportionally at this temperature. Though grape originated in temperate regions, modifications in production system, taking up two pruning and one crop, have enabled it to adapt to tropical conditions. Under climate change conditions, there would be changes in availability of growing degree-days leading to hastening of the phenological processes. Extreme temperatures may endanger berry quality and economic returns (Schultz et al. 2013), besides direct injuries of high temperature including denaturation of proteins coupled with aggregation and increased fluidity of membrane lipids.

Amidst these physiological responses of fruit crops while responding to climate change, the rise in temperature will hasten nutrient mineralization in soils and decrease fertilizer-use efficiency (Elbasiouny et al. 2022). Increase in temperature at fruit maturity is reported to induce fruit cracking and burning in litchi in addition to premature ripening of mango (Hazarila 2013) as a function of depleted nutrient accumulation by developing fruits as a nutrient sink. High temperature (>38 °C) associated with bright sunshine causes sunburn in the banana leaf and peduncle and top hands of the bunch may also get affected by the sunburn (Reddy et al. 2017). Many slow-growing fruit crops require heavy investment on establishment of orchards. Quick alteration/shifting of fruit species or varieties would be difficult, and painful loss-bearing exercise under the impact of climate change may discourage the onward expansion of the fruit industry. Recent studies indicated that in Kullu district of Himachal Pradesh (India), farmers shifted from apple to vegetables, while in Shimla district at relatively higher altitude, orchards were replaced from high-chilling requiring apple cultivars (Royal Delicious) to low-chilling requiring cultivars and other fruit crops such as kiwi, pear, peach, plum, and vegetables (Bhargava et al. 2011) as an example of impending impacts of climate change. A linear reduction in per cent acid with increasing effective heat units was evident in Valencia and Navel orange varieties indicating the negative relationship of temperature with acid/brix ratios (Datta 2013). Most of these responses are addressed toward different climate change constants in isolation rather than studying their combined response and hardly any attempt to compare the prediction of yield under normal versus climate change scenario (Ram et al. 2007; Srivastava and Singh 2005a). A recent review by Rasera et al. (2023) analyzing climate change and citriculture puts forward two cardinal gaps, viz., (i) lack of studies with combined stresses (biotic plus abiotic) instead of isolated studies and (ii) lack of predictive models of citrus production under different climate change scenarios.

Shift in varietal choice may become necessary in case of grapevines, banana, mango, and other important fruit crops. Amidst these crop phenological responses, a bigger question emerges, how are they connected to plant nutrition? The rise in temperature will hasten nutrient mineralization in soils, decrease fertilizer-use efficiency, and increase evapotranspiration with reduced water-use efficiency, altogether a different dynamic of nutrients and water under climate change than conventional areas where such impacts are least realized in field (Srivastava and Malhotra 2014). However, CO_2 enrichment does not appear to compensate for the detrimental effects of higher temperature on yield. Most importantly, the quality of produce of these fruit crops is likely to be severely affected in addition to nutrient load of edible fruit parts under such plant defense comprising situations plagued by climate change.

1.6 Climate Change vis-à-vis Phenology and Fruit Quality Traits

Fruit crops compared to other annual horticultural or field crops are better tuned to moderate any possible adversities arising out of climate change due to their long juvenile period (vegetative development period), distinct physiologically defined growth stages, perennial nature of woody framework, and climacteric/non--climacteric nature, collectively placing them at an advantage. If these are advantages associated with fruit crops, they perhaps require equally well-guarded care to sustain quality production for an extended period and more so in a climate change scenario (Gora et al. 2019; Nawaz et al. 2019). Undoubtedly, like any other crop, fruit crops too are significantly impacted by climate, frequently characterized by poor fruit quality (often unmarketable) coupled with inconsistent and low quantum handicapped by poor vigor, canopy development, and reproductive traits (fruiting ability and reduction in fruit size) and quality attributes (less and irregular color development, low juice content, misshapen fruits) coupled with reduced shelf life (Malhotra 2017). However, such changes are more frequent in temperate than tropical or subtropical fruit crops due to their grater vulnerability to climate change (Reddy et al. 2017). A systematic review and meta-analysis of 343 peer-reviewed publications by Baranski et al. (2014) reported that concentrations of antioxidants (polyphenols, flavanones, stilbenes, flavones, flavonols, and anthocyanins) were 19–51% higher in organic crops.

In a 9-year comprehensive study featuring the effect of climate change on citrus, fruit quality, high rainfall (>220 mm), and low cumulative temperature (<3150 °C) played a promotional role in titratable acidity (>1.8%) in young fruits (less than 50 mm in diameter). But as the fruits further expanded, size (from 50 to 65 mm), excessive rainfall (300–400 mm), lower cumulative temperature (<2400 °C), and reduced diurnal temperature range (<10 °C) played a negative role, causing a reduction in titratable acidity. Conversely, low rainfall (<220 mm), high cumulative temperature (>3150 °C), and high diurnal temperature range (>14 °C) promoted the accumulation of soluble solids in young fruits (9%) at 120 days after flowering (DAF). Furthermore, low rainfall (<100 mm) favored the accumulation of soluble solids (1.5%) during fruit expansion (195–225 days after fruiting) using regression models (Dong et al. 2024).

Climate change-mediated shift in crop phenology is one of the well-known impacts of climate change, thereby producing a complex cascading response on the entire plant physiology, vegetative or reproductive in nature (Fischer et al. 2016); however, the precise responses are difficult to predict due to the involvement of multidimensional physiological processes. Flowering is a critical stage in fruit development with strong consequential effect on production with standard fruit quality. A large number of reports highlighted this physiological response in a variety of fruit crops as partial abscission in flower primordial in pome fruits producing smaller flower bud clusters similar to leafy spurs in mild winters (Datta 2013), incomplete blossoms and poor fruit set in apple (Pan and Shu 2007), advancement in flowering at the rate of 3.15–3.39 days/°C in citrus (Peng et al. 2000), and delays in mango flowering due to high temperature over the ideal day/night temperature range coupled with cloudy weather (Clonan et al. 2021) are some of the glaring examples about strong interference of climate change-related parameters in fruit physiology warranting further investigation to develop climate-neutral combat techniques. However, these parameters individually or collectively have different perspectives on the performance of fruit crops.

Increased CO_2 concentration increases the net photosynthetic rate in grapes, while decreased stomatal conductance in response to increased CO_2 improves innate water-use efficiency. Hence, raising the atmospheric CO_2 level enhanced the yield, regardless of any positive or negative impacts on grape maturity (Ahmed et al. 2011). Sour orange trees (5–8 years) were particularly sensitive to the higher CO_2 levels even though they were well into the middle-age reproductive maturity (17 years) with higher yield substantiated by higher root biomass, leaf biomass, branch biomass, fruit biomass, and total biomass with CO_2 enrichment (350–650 ppm CO_2) than the trees grown under ambient conditions (Kimball et al. 2007). Thus, the heat produced by CO_2-linked global warming in open environments and soil water evaporation tended to supersede these benefits under enriched CO_2, otherwise leading to shorter growth periods, decreased yield, and higher yield variance. On the other hand, temperature regulates the physiological functions, while any imbalance in normal temperature causes heat stress, inadequate chilling for temperate fruit crops, and disruption in pollination in tropical/subtropical fruits, besides direct association with chilling accumulation in stone fruits and transformation of flower bud into vegetative one (Egea et al. 2022) under warm night conditions. These climates regulated phenological responses that act predisposing to optimum fruit quality development. The production of mango, banana, papaya, and sapota was observed to be inversely correlated with temperature and rainfall over a 10-year period (Kumar 2019).

Table 1.2 Effects of climate change-related parameters on different nutritional quality traits of fruit crops

Fruit crops	Quality traits under influence	Reference
	Increase in temperature	
Aonla (*Emblica officinalis* Gaertn)	Irregular bearing coupled with heavy drop	Bhargava et al. (2011)
Apple (Pyrus *malus* L.)	Decrease in skin and flesh disorders	Marlow and Loescher (1984), Ferguson et al. (1998)
Bael (*Aegle marmelos* L. Correa)	Increase in fruit cracking incidence	Saini et al. (2004)
Avocado (*Persea americana* Mill.)	Decrease in skin disorders	Peet and Wolfe (2000)
Grape (*Vitis vinifera* L.)	Decrease in skin disorder, taste, and wine quality	De Ordunu (2000), Jones et al. (2005)
Lime (Citrus *aurantifolia* swingle)	Decrease in flesh disorders	Davenport and Campbell (1977)
Mango (*Mangifera indica* L.)	Increase in spongy tissue disease	Katrodia et al. (1985)
Pineapple (*Ananas comosus* L. Merr.)	Decrease in flesh disorders	Chen and Paull (2001)
Increase in carbon dioxide concentration		
Grape (*Vitis vinifera* L.)	Inconsistent response on sugars, acids, and wine quality	Bindi et al. (2001), Schultz (2000)
Pear (*Pyrus communis* L.)	Increase in size, skin color, and taste	Han et al. (2012)
Sour orange (*Citrus aurantium* L.)	Increase in vitamin C	Idoso et al. (2002)

Growing fruit crops under elevated carbon dioxide is widely reported to increase photosynthesis, thereby providing an increased fruit yield, flavor, and nutritional content (not limited to vitamins C and E and provitamin A alone) under greenhouse/controlled environmental conditions (Doddrell et al. 2023), despite some reports of higher fruit yield coupled with 22% increase in Brix in Nishi pear under CO_2--enriched environment (700 ppm) that were nullified by reduction in fruit firmness, which is a major negative impact on fruit quality (Han et al. 2012). Response of climate change parameters on nutritional quality traits of different fruit crops is severely affected by and large (Table 1.2).

1.7 Developing Climate-Smart Fruit Production System

The combination of these three elements (exposure, susceptibility, and adaptive capacity) frames the discussion of the best adaptive measures to be implemented to reduce crop vulnerability to climate change. The measures can be part of short- or long-term strategies, depending on the harms posed by climate change. Short-term strategies may include changes to irrigation, soil management, and conservation and tree pruning. Long-term strategies may include introduction of new (early or late maturing) varieties, introduction of new (drought tolerant) crops, or farm relocation (Pechan et al. 2023). Brown and Uriu (1996) estimated that, by 2050, horticulture and plantations in tropical countries have the potential to capture as much as 16.4 Gt C compared to agroforestry with the potential to capture only 6.3 Gt C. Many horticultural crops especially perennial fruit trees, spices, and plantation crops have a major role to play in carbon sequestration, containing greenhouse gases and mitigating climate change, since these fruit crops contain 25–100 times more biomass C than agricultural lands (Kumar et al. 2010; Sharma et al. 2021; Gelaye and Getahun 2024). Kumar et al. (2011) suggested that cultivation of temperate fruit crops species such as apricot, apple, and walnut could be exploited more efficiently for carbon sequestration (40,000 metric tons dry weight sequestering 22,300 metric tons CO_2) in mitigating global CO_2 levels and restoring the degraded land in cold arid areas by improving soil organic carbon contents. Perennial crops, such as citrus, apple (orchards), kiwifruit, and grapes (vineyards), are significantly important for converting more atmospheric carbon compared to annual crops such as flowers and field crops due to the net primary production prevalent in the accumulation of C cycle or distributed into the permanent structures of these fruit trees (Sharma et al. 2021). However, the carbon emission derived from various management practices would not be compensated through C storage in apple trees before reaching the mature stage, considering peak C sequestration capability reaching at the age of 18 years old (14–32 Tg C and C storage in biomass from 230 to 475 Tg C in a 10-year period) followed by a decline in capability later with age in the terrestrial ecosystems of Chinese apple (Wu et al. 2012). Hence, there is an inbuilt mechanism genetically embedded in fruit crops to be scientifically transformed into being climate-smart in nature with the candid support of abling cultural practices.

The carbon fixing ability of fruit crops also offers a possibility of creating a carbon market as a green technology. Studies on carbon sequestration capability of fruit-bearing trees in Nubra valley of Himachal Pradesh (India) covering temperature fruit orchards (apricot, apple, and walnut) were observed to sequester roughly 22,300 MT of carbon, with the apricot orchards sequestering maximum carbon (18,000 MT). These findings throw some exciting prospects that fruit crop species (temperate or tropical in nature) could be effectively utilized for developing carbon markets by storing more recalcitrant carbon into their perennial woody framework instead of sequestering more vulnerable carbon into the soil (Kumar et al. 2010). Annual primary production in Navelina orange tree components showed fruit yield as the major sink of carbon fixation in trees, with 10.5 kg C/tree/year accounting for about 40% of the total C sequestration in citrus trees. Structural organs (twigs, branches, and stems) were another important sink for C partitioning with values of around 6 kg C/tree/year (Perez-Piqueres et al. 2020). A summarized carbon sequestering ability of different fruit crops (Fig. 1.1) further establishes such conviction of carbon sequestration vis-à-vis parallel nutrient sink coupled with additional ecosystem service and somewhere could be a vibrant source of many of the unknown solutions of plant nutrition (elevated carbon sequestration by fruit crops is likely to

Carbon sequestration ability as net primary production in major temperate and tropical fruits crops

Fruit crops	C - sequestration as NPP (Mg C/ha/ year)	Reference
Citrus spp	**6.8-11.4**	Quinones et al.(2013) , Yasin et al. (2021)
Kiwifruit (*Actinidia chinensis Planch*)	**8.0**	Holmes et al. (2014)
Banana (*Musa paradisiaca* L.)	**16.0-80.0**	Zhao et al. (2014)
Apple (*Pyrus malus L.*)	**5.2**	Zanotelli et al. (2015)
Walnut (*Juglans regia* L.)	**30.0**	Schindler et al. (2023)
Plum (*Prunus domestica* L.)	**13.1**	Velazquez-Marti and Cazco-Logrono (2018)
Peach (*Prunus persica* L.)	**5.0**	Montanaro et al. (2016)
Citrus (*Citrus clementina* hort.ex Tanaka)	**15.4**	Iglesias et al. (2013)
Guava (Psidium guajava L.)	**26.7**	Gelaye (2024)
Olive (*Olea europaea* L.)	**9.4**	Nardino et al. (2013)
Coconut (*Cocos nucifera* L.)	**16.1**	Ranasinghe and Thimothias et al. (2012)
Apricot (*Prunus armeniaca* L.)	**9.0**	Fiore et al. (2018)
Pomegranate (Punica granatum L.)	**12.0**	Gelaye and Getahun (2024)
Litchi (*Litchi chinensis* Sonn.)	**5.7**	Naik et al. (2018)
Grape (*Vitis vinifera* L.)	**14.0**	Scandellari et al. (2016)
Mango (*Magnifera indica* L.)	**75.6-401.4**	Ganeshamurthy et al. (2019)
Sapota (*Manilkara zapota* L.)	**35.5**	Ramchandra et al. (2022)
	FOREST ECOSYSTEM	
Peatland	**1.4**	Schulze et al.(2015)
Tropical rain	**9.2**	Soepadmo (1993)
Moist tropical	**2.7**	Djomo et al. (2011)

'Net primary production is the amount of CO_2 removed from the atmosphere through net C exchange (NPP = GPP + Ra) as amount of C retained by plant after assimilation through photosynthesis (GPP) and autotrophic respiration (Ra)

Fig. 1.1 Snapshot of carbon sequestration capacity of different fruit crops comparing typical forest-based ecosystems (the carbon fixation capacity is subject to dynamic changes depending upon planting density, tree age and species, soil, agroclimate, orchard gradients, single crop or intercropped, and cultural practices)

transfer at least 8–10% of canopy-accumulated carbon into the rhizosphere, thereby recruiting new microbes and facilitating better development of both microbial diversity and microbial functions to finally maintain high nutrient influx within rhizosphere of fruit crops), yet they are unprioritized research agenda.

Extreme climatic conditions will also negatively impact soil fertility and increase the speed of soil erosion (Srivastava 2020) bringing formable challenges to effective plant nutrition of fruit crops. Thus, an additional fertilizer application or improved nutrient-use efficiency of crops will be needed to maintain productivity or harness the potential for enhanced crop growth due to increased atmospheric CO_2 (Srivastava et al. 2021). Modifying fertilizer application to enhance nutrient availability and usage of soil amendments to improve soil fertility and enhance nutrient uptake (Srivastava et al. 2015a, b) would be another exercise to seriously look at, besides developing 4W (stewardship of irrigation water use like fertilizers), orchard floor management, and crop regulation.

It is also possible that the long-time horizon of perennial crops creates situations such as suitable areas that may become unsuitable during the life of a fruit orchard. The choice of a variety is further complicated by the risk that the best variety for the current climate may be poorly suited for future climates (Srivastava and Pongener 2023). Thus, while adaptations such as planting new varieties and shifting to new areas may reduce impacts in the long term, short-term losses may be largely unavoidable. In wine grape, each grape variety grows in a range of temperatures and for each variety it is possible to define climates for premium wine production.

The physiological and morphological differences between varieties (genotypes) enable production over a relatively large range of climates and depending upon the suitability to different growing areas the cultivars may be adopted (Rouxinol et al. 2023). An integrated approach with all available options of smart practices will therefore be most appropriate in sustaining the productivity of fruit crops under climate change conditions.

Hydroponics/aeroponics is emerging as a dynamic technology to produce fruit crops (though not so popular, but still need to pick up the popularity) free from diseases, besides improved nutrient- and water-use efficiency (Rubio-Asensio et al. 2020; Rajaseger et al. 2023) driven further by artificial intelligence but demeans the soil as a growth medium for all important biological functions. Hydroponics is put forward as a solution to combat climate change, to reduce the environmental damage and species extinction caused by overexploitation and intensive farming. Nano-fertilizers provide strong opportunities for the development of processes and product, which are impossible to achieve through conventional system for better fruit nutrient load, unless composites of nano-fertilizers are developed depending upon soil fertility constraints matching with delivery system (Srivastava et al. 2022).

1.7.1 Integrated Soil Fertility Management for Climate-Smart Citrus

The long-term experiments conducted around the world indicated that chemical fertilizer alone is not enough to improve or maintain soil fertility at high levels, and the soil acidification problem caused by overapplication of synthetic N fertilizers can be reduced if more N fertilizer is applied as nitrate relative to ammonium- or urea-based N fertilizers (Pathak and Nedwell 2011). Organic fertilizers can improve soil fertility and quality, but long-term application with high rates can also lead to more nitrate leaching, and accumulation of P, if not managed well. Well-managed combination of chemical and organic fertilizers can overcome the disadvantages of applying a single source of fertilizers and substantially achieve higher crop yields, improve soil fertility, alleviate soil acidification problems, and increase nutrient-use efficiency compared with only using chemical fertilizers (Miao et al. 2011). Organic amendments comprising manures have helped in increasing the fruit yield coupled with quality, effectively replacing mineral fertilizers in the nutrient management of commercial fruit tree orchards through associated changes in soil C pool, microbial pool, and available nutrient pool of soil (Baldi et al. 2010; Montanaro et al. 2010). Nutrient (inorganic nutrient source)-microbe as tripartite association on the other hand is better known in fruit crops (Singh and Banik 2011; Srivastava and Ngullie 2009; Khehra and Bal 2014) with respect to both agronomic response and soil health. Such associations have invariably witnessed substantially higher productivity than with use of any single component. However, there is a greater need to expand such plant response advantages using more rhizo-competent microbes preferably in a consortium mode; both plant response and soil health response have to be sustained on a long-term basis (Rogers and Beunett 2004; Srivastava and Bora 2023).

In this background, our long-term studies on microbial consortium (mixture of *Bacillus pseudomycoides*, MF113272; *Acinetobacter radioresistens*, MF113273; *Micrococcus yunnanensis*, MF113274; *Paenibacillus alvei*, MF113275; and *Aspergillus flavus*, MF113270), inorganic fertilizers, and vermicompost-mediated integrated soil fertility management (ISFM) in relation to sustained quality production of Nagpur mandarin (*Citrus reticulata* Blanco) under field conditions during 2007–2016 were initiated with five treatments, viz., module I (T_1), 100% RDF (recommended doses of fertilizers: 600 g N, 200 g P, 300 g K, 200 g $ZnSO_4$, 200 g $FeSO_4$, 200 g $MnSO_4$/tree/year), module II (T_2), 75% RDF + 25% vermicompost; module III (T_3), 75% RDF + 25% vermicompost + microbial consortium; module IV(T_4), 50% RDF + 50% vermicompost; and module V (T_5), 50% RDF + 50% vermicompost + microbial consortium. These modules representing various forms of ISFM were evaluated in terms of vegetative growth, fruit yield, fruit quality, plant available pool of nutrients in soil, leaf nutrient composition, soil carbon stock, and soil microbial load on a black clay soil (Typic Ustocherpt) and presented below through pooled data (Tables 1.3, 1.4, 1.5 and 1.6):

Response on growth, fruit yield, and fruit quality All the ISFM-based treatments displayed significant response on changes in canopy volume (Table 1.3). A maximum cumulative increase in canopy volume was observed with module III (18.67 m^3) followed by module IV (17.16), module V (16.93 m^3), and in part with module II (16.70 m^3) and canopy volume with either module III or module V which were better than either module II or module IV compared to module I which comprised of only inorganic fertilizers. Module I, T_1 (11.50 m^3). Incorporation of microbial consortium either with module III or with module V invariably induced higher

Table 1.3 Growth attributing parameters in response to different vermicompost-based modules of ISFM (pooled data: 2007–2016)

ISFM modules	Plant height (m)	Tree spread (m)		Canopy volume (m^3)	Cumulative increase in canopy volume over 2007–2008 (m^3)
		E-W	N-S		
Module I	3.82 (2.02)	2.52 (1.27)	2.50 (1.20)	13.04 (1.54)	11.50
Module II	5.09 (2.10)	2.98 (1.39)	2.89 (1.25)	18.51 (1.81)	16.70
Module III	3.87 (2.13)	3.00 (1.27)	3.22 (1.23)	20.28 (1.61)	18.67
Module IV	3.76 (2.10)	3.73 (1.30)	2.93 (1.30)	19.78 (1.62)	17.16
Module V	3.86 (1.87)	2.97 (1.18)	2.96 (1.22)	18.33 (1.40)	16.93
C.D. ($P = 0.05$)	0.49	0.38	0.51	1.18	1.32

Source: Srivastava et al. (2019)
Figures in parenthesis indicate the value obtained in 2007–2008

Table 1.4 Fruit yield and quality parameter in response to different vermicompost-based modules of ISFM treatments (pooled data: 2007–2016)

ISFM modules	Yield (kg/tree)	Fruit weight (g/fruit)	Peel thickness (mm)	Fruit quality parameters			
				Juice content (%)	TSS (°Brix)	Acidity (%)	TSS/acid ratio
Module I	67.7	101.4	4.1	40.2	8.4	0.96	8.92
Module II	71.0	104.1	3.0	39.6	8.6	0.93	9.23
Module III	80.5	106.2	3.4	41.5	9.5	0.86	10.05
Module IV	83.2	109.1	3.4	42.7	8.4	0.91	9.56
Module V	88.8	114.6	3.0	44.2	9.3	0.80	11.62
CD ($P = 0.05$)	2.1	1.4	0.50	2.1	0.30	0.10	1.10

Source: Srivastava et al. (2019)

Table 1.5 Changes in soil fertility status in response to different modules of ISFM-based treatments (pooled data: 2007–2016)

ISFM modules	Plant available nutrients (mg/kg)						
	Macronutrients			DTPA micronutrients			
	$KMnO_4$-N	Olsen-P	NH_4OAc-K	Fe	Mn	Cu	Zn
Module I	140.0	9.32	185.2	10.35	10.54	2.2	0.98
Module II	1435	9.27	196.6	12.57	11.54	2.3	1.00
Module III	155.7	9.15	203.6	15.18	12.45	2.7	1.10
Module IV	161.0	9.57	209.2	16.56	11.67	2.8	1.18
Module V	178.5	9.55	212.6	19.50	12.93	2.5	1.32
CD ($P = 0.05$)	6.2	NS	1.3	1.32	1.01	NS	0.46

Source: Srivastava et al. (2019)

RDF stands for recommended doses of fertilizer (600 g N, 200 g P, 300 g K, 200 g $ZnSO_4$, 200 g $FeSO_4$, 200 g $MnSO_4$/tree/year)

Vm stands for vermicompost (nutrient composition: 2.38% N, 0.09% P, 1.42% K, 1072 ppm Fe, 116 ppm Mn, 39 ppm Cu, and 46 ppm Zn)

MC stands for microbial consortium developed by isolating the native microbes (mixture of *Bacillus pseudomycoides* (MF113272), *Acinetobacter radioresistens* (MF113273), *Micrococcus yunnanensis* (MF113274), *Paenibacillus alvei* (MF113275), and *Aspergillus flavus* (MF113270))

Table 1.6 Soil carbon fractions and C/N ratio in response to different modules of ISFM (pooled data: 2007–2016)

ISFM modules	Soil carbon (g/kg)			Soil total *N* (g/kg)	Soil C:N ratio	Soil microbial load ($\times 10^3$ cfu/g)	
	SOC	SiC	TC			Bacterial count	Fungal count
Module I	6.61	1.71	8.32	0.721	11.54	26	12
Module II	6.70	1.71	8.41	0.732	11.48	36	18
Module III	7.02	1.74	8.76	0.741	11.82	54	22
Module IV	7.11	1.75	8.86	0.738	12.00	44	26
Module V	7.43	1.81	9.24	0.748	12.32	86	42
CD ($P = 0.05$)	0.08	NS	0.09	0.006	NS	8.2	4.8

canopy volume suggesting that better response on fruit yield was observed to be significantly affected by different ISFM-based treatments (Table 1.3).

The maximum fruit yield of 88.8 kg/tree was observed with treatment module V which was better than 83.2 kg/tree with T_4 or 80.5 kg/tree with module III than 71.0 kg/tree with treatment module II. However, all these modules were observed far superior in magnitude of response when compared with 100% RDF as module I (67.7 kg/tree). Hence, different ISFM-based modules (71.0–88.8 kg/tree) were much better than exclusive inorganic fertilizer treatment-like module (67.7 kg/tree).

Different fruit quality parameters, except peel thickness (Table 1.4), displayed significant response in relation to different treatments. There was much higher fruit weight with different ISFM-based treatment modules I–V, 101.4–114.6 g, compared to inorganic RDF module I (101.4), showing that incorporation of both organic manure and microbial cultures improved the efficiency of organic fertilizers. The other three fruit quality-related parameters such as juice content, acidity, and TSS expressed a similar pattern of response. However, a significant acidity reduction was observed with those superior treatments, e.g., module III (0.86%) and module V (0.80%) compared to module II (0.93%) and module IV (0.91%) highlighting the favorable changes in different fruit qualities in response to different ISFM-based treatments.

Response on soil health-related parameters: Rational soil use practices allow economically and environmentally sustainable yields, which will only be reached with the maintenance or recovery of the soil health (Lehman et al. 2020). Thus, a healthy soil has been defined as "the continued capacity of soil to function as a vital living system, within ecosystem and land-use boundaries, to sustain biological productivity, promote the quality of air and water environments, and maintain plant, animal and human health" (Doran and Zeiss 2000). Soil health refers to the ecological equilibrium and the functionality of a soil and its capacity to maintain a well--balanced ecosystem with high biodiversity above and below surface and productivity. To understand the use soil health as a tool for production sustainability, physical, chemical, and biological properties must be employed to verify the soil use and management within a desired timescale. Attributes with a rapid response to natural or anthropogenic actions are considered good indicators of soil health (Cardoso et al. 2013). However, most of soil health-related parameters generally have a slow response, when compared to the biological ones, such as microbial biomass C and N, biodiversity, soil enzymes, and soil respiration, in addition to macro- and meso--fauna (Tripathi et al. 2022). Thus, a systemic approach based on different kinds of indicators (physical, chemical, and biological) in assessing soil health would be safer than using only one kind of attribute (Cardoso et al. 2013).

The soil microbial load in terms total bacterial count and fungal count in response to different ISMF-based treatments showed variable higher soil microbial load with module III and module V compared to either module II or module IV including module I. In our study, changes in soil fertility indices with regard to available macro- as well as micronutrients were observed highly significant, but of variable nature in response to different treatments. Among macronutrients, $KMnO_4$-N and Olsen-P showed significant responses, without any significant response on NH_4OAc-K. The treatments such as module III (155.7 and 203.6 mg kg^{-1} $KMnO_4$-N and Olsen-P, respectively) and module V (178.5 and 212.6 mg/kg $KMnO_4$-N and Olsen-P, respectively) were comparatively higher than module II (193.5 and 196.6 mg/ kg $KMnO_4$-N and Olsen-P, respectively) and module IV (161.0 and 209.2 mg/kg $KMnO_4$-N and NH_4OAc-K, respectively). These observations suggested the superiority of ISFM-based treatments than those treatments without microbial consortium. Module I involving exclusive inorganic fertilizers registered lowest test values compared with module V that registers maximum values, validating the supremacy of those treatments that carry all the three components of ISFM (Table 1.5).

These observations showed that changes in soil carbon stock are more governed by organic fraction than inorganic fraction. Maximum SOC and TC of 7.43 g/kg and 9.14 g/kg were observed with module V. Likewise, with module III, much higher SOC and TOC were observed than module II, displaying the significant role of microbial consortium in improving the carbon sink capacity of soil. Different ISFM--based modules were also observed to aid in improving the total soil N stock, being maximum with treatments such as modules III–V (0.741–0.748%) compared to the rest of the treatments such as module I, module II, or module IV (0.721–0.738%). However, the soil C/N ratio in the range of 12.00–12.32 did not display significant changes in response to different treatments (Table 1.6).

These modules provided us the strong success insights about the effectiveness of the developed microbial consortium. We further laid out the field experiment to evaluate in a 10-year-old Nagpur mandarin (*Citrus reticulata* Blanco raised on *Citrus jambhiri* Lush rootstock) during 2007–2021 to study how much nutrient saving can be achieved through microbial consortium, with treatments comprising the following: (i) 50% RDF + 50% RDF equivalent vermicompost + microbial consortium, (ii) 45% RDF + 45% RDF equivalent vermicompost + microbial consortium, (iii) 40% RDF + 40% RDF equivalent vermicompost + microbial consortium, (iv) 35% RDF + 35% RDF equivalent vermicompost + microbial consortium, and (v) 30% RDF + 30% RDF equivalent vermicompost + microbial consortium. Incorporation of microbial consortium in ISFM showed a minimum of 30% of RDF (35% RDF + 35% RDF equivalent vermicompost + microbial consortium) could be saved (Fig. 1.2) without any compromise on either fruit yield, quality, or soil fertility status (Srivastava et al. 2019). These observations were further substantiated by consistent decline in carbon dioxide release from the soil in response to these ISFM treatments (Fig. 1.3) across all seasons, adding an element of strength to the concept that carbon deposited into the passive pool of the soil is a more recalcitrant carbon source, which is a successful guarantee to climate-smart soil fertility management.

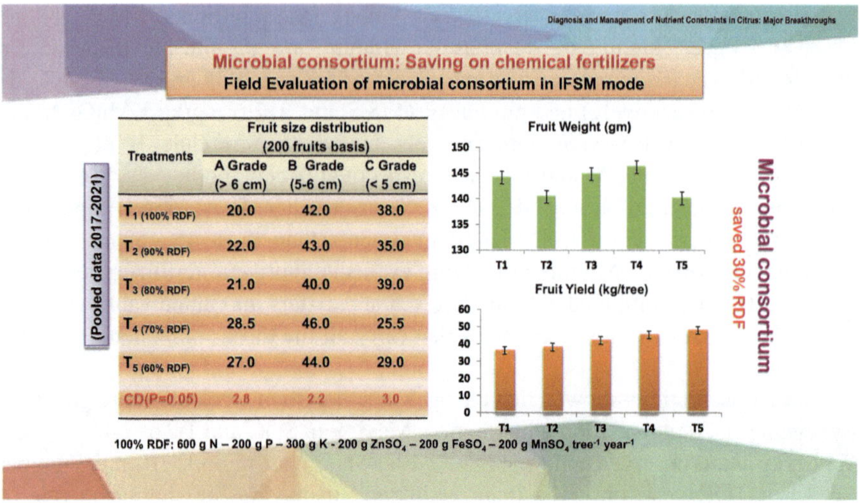

Fig. 1.2 Response of microbial consortium-loaded ISFM treatments versus chemical fertilizers as RDF using increasing substitution of nutrients through RDF. As much as 30% of RDF could be saved with treatment T4 on par with treatment T5

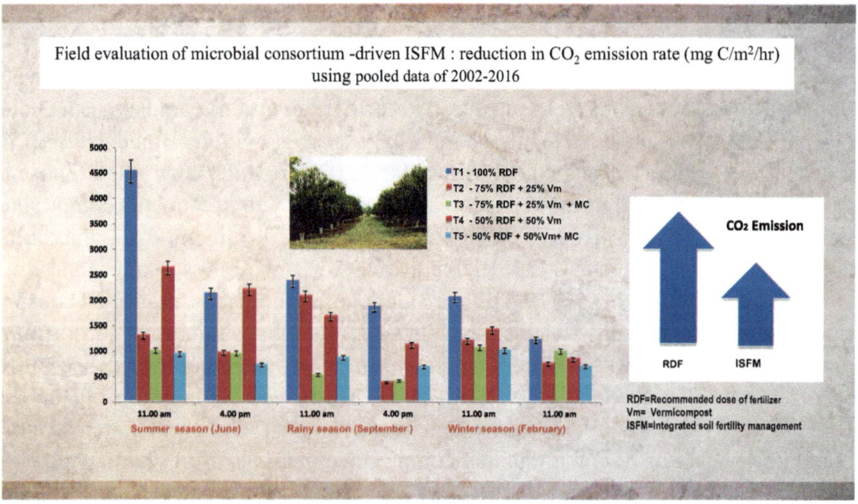

Fig. 1.3 Response of microbial consortium-loaded ISFM treatments versus chemical fertilizers as RDF, interesting to note a sharp decline in C emission from former treatments due to deposition of more recalcitrant carbon into the rhizosphere

1.7.2 Customized Fertilizers or Climate-Smart Fertilizers

Precision management of nutrients in fruit crops is the ultimate objective of all efforts culminating into improved quality production fitting the most appropriate concept of 4R Nutrient Stewardship as a part of climate-smart fertilizer use in order to reduce carbon footprint of crops (Bruulsema 2022a). Such smart fertilizers have (i) the ability to release nutrients on a regulated and slow rate, (ii) increased nutrient--use efficiency, (iii) reduced nutrient loss, (iv) long-lasting effect, (v) improved crop quality, (vi) environment protection, (vii) adopted to soil and climate, (viii) consistent soil adversities, and (ix) economics of production (Shanmugavel et al. 2023), much to the advantage of fruit crops in a tailor-made fashion. Developing customized fertilizers is therefore an integrated part of climate-smart fertilizers to be fitted in fruit nutrition program.

A number of principles are involved in developing customized fertilizers, which include soil fertility-based variogram measuring spatial variability, defining different production management zones along with soil fertility gradient, nutrient budgeting based on nutrient removal and addition, nutrient logging data in relation to phenology of a crop, and blending nutrient-based spatial and temporal crop-specific nutrient removal pattern (Srivastava and Das 2017). It is not necessary to elaborate on the advantages of customized fertilizers compared to straight fertilizers. Unfortunately, not much inroads have been made with respect to developing specialty fertilizers or these fertilizers customized as per crop nutrient demand depending upon soil and climate (Srivastava et al. 2010). We made some sincere efforts to highlight some efforts to develop specialty fertilizers and customize them like citrus, with the help of other fruit crops. These efforts are perhaps more in tune with the ongoing changes in soil fertility regimes vis-à-vis climate change on a real timescale to condition the crop nutrient requirement.

Knowing the nutrients required for all stages of growth and understanding the soil's ability to supply those needed nutrients are critical to profitable fruit crop production. The recommendations on fertilizer application may not, however, produce the same magnitude of results when practiced on an orchard of large area, because of its inability to accommodate large spatial variation in soil fertility (Srivastava et al. 2006). Application of uniform single rate of nutrients, hence, most often result in overapplication of nutrients at some sites and under-application at other sites, cumulatively leading to reduced fertilizer-use efficiency (Jeyabaskaran et al. 2021). Under such circumstances, site-specific nutrient management (SSNM) as a dynamic concept of nutrient management exploits indigenously and spatially available nutrients within bigger orchards requiring differential fertilizer treatments in patches so as to improve the orchard efficiency (average yield of specified trees in relation to average orchard yield) in ultimate terms (Srivastava and Singh 2008). The success of SSNM during the last 15 years has been prominently realized on a number of fruit crops, viz., avocado (Salaza-Garcia and Lazcano-Ferrat 2023) and citrus (Srivastava et al. 2006, 2009), to cite few success stories. The suggested model captured the essence of the law of the minimum in yield and revenue increase only if the limiting nutrients are appropriately adjusted. In the past, various aspects of SSNM have been highlighted in a number reviews (Corwin and Lesch 2005; Elprince 2009; Johnston et al. 2009; Zhang et al. 2010), but there is hardly any comprehensive coverage addressing the perennial fruit crops.

We present below the principles and strategies that could be effectively used in developing customized fertilizer use strategy, leading to development of specialty fertilizers. We developed a specialty micronutrient mixture for Nagpur mandarin (*Citrus reticulata* Blanco as a scion raised *Citrus jambhiri* Lush) grown in Central India and evaluated during 2012–2020 in Vertic Ustochrept soil facing multiple nutrient deficiencies in the form of N, P, Fe, Mn, Zn, and B. Four treatments, viz., T_1, micronutrient mixture (citrus special, ICAR-IIHR Bangalore); T_2, macronutrients (soil application) + micronutrients (foliar application); T_3, micronutrient mixture 1 (soil application) + macronutrients (soil application); and T_4, micronutrient mixture 2 (soil application) + macronutrients (soil application) were tested in the field with six replications during 2007–2019. The responses were evaluated in terms of vegetative growth, changes in soil fertility status, and nutrient composition of index leaves to identify the best treatment during the fifth year of evaluation.

Response on vegetative growth and quality Vegetative growth response using different growth-contributing parameters were evaluated and expressed as canopy volume responded significantly to the different treatments, as in preceding years. Among all the treatments, the most striking increase in canopy volume over 2018–2019 was observed with the treatment T_4 (0.96 m^3) followed by the treatments T_3 (0.92 m^3), T_2 (0.85 m^3), and T_1 (0.46m^3). These responses in canopy volume suggested the superiority of both the micronutrient mixtures (T_3 and T_4) over conventionally used fertilizer (T_1 and T_2) application. Similarly, the yield response with treatments T_4 (58.2 kg tree or 16.2 tons/ha) and T_3 (46.3 kg /tree or 12.9 tons/ha) were of much higher magnitude over T_2 (42.4 kg/tree or 11.8 tons/ha) and T_1 (40.4 kg/tree or 11.1 tons/ha), suggesting that our prepared mixtures have out--yielded the conventionally used fertilizer doses (Fig. 1.4).

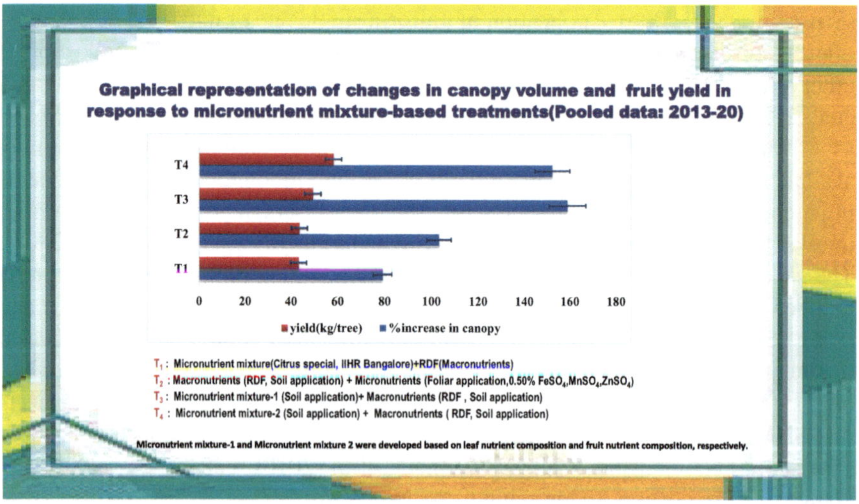

Fig. 1.4 Response of different micronutrient mixtures on increase in tree canopy and fruit yield of Nagpur mandarin (pooled data: 2007–2019)

Changes in soil fertility and plant nutrition The soil fertility (Fig. 1.4) and leaf nutrient composition (Fig. 1.5) were significantly available nutrients, with an exception of DTPA-Cu affected by different treatments. Among plant-available nutrients, with the exception of DTPA-Cu, all the nutrients, viz., $KMnO_4$-N, Olsen-P, NH_4OAc-K, DTPA-Fe, DTPA-Mn, and DTPA-Zn, showed a significant response of different treatments. The treatment T_4 observed significantly higher plant available macronutrient than the treatment T_3 compared to the treatment T_2. These observations suggested the far superior response of micronutrient mixtures (treatments T_3 and T_4) than treatments involving T_1 and T_2. Treatments T_4 and T_3 registered a significantly higher plant-available micronutrient than T_2 and T_1. Similar responses were observed in the leaf nutrient status, where the treatment T_4 shows the best response over T_3, T_2, or T_1. Our observations strongly lend support toward better efficacy of micronutrient mixtures compared to conventionally used fertilizers (Fig. 1.5). Hence, our long-term studies showed that micronutrient mixture developed by simply mixing straight micronutrient fertilizers such as ferrous sulfate, manganese sulfate, zinc sulfate, and borax in the ratio of 1:1:0.50:0.20 stoichiometrically coincided with the similar concentration ratio of micronutrients present in fruits. Interestingly, our developed micronutrient mixture at the rate of 100 g/plant overtaken the response of these micronutrients when applied at the rate of 100 g ferrous sulfate, 100 g manganese sulfate, 100 g zinc sulfate, and 50 g borax/plant/year (total micronutrients at the rate of 350 g/plant proved far inferior over 100 g customized micronutrient mixture).

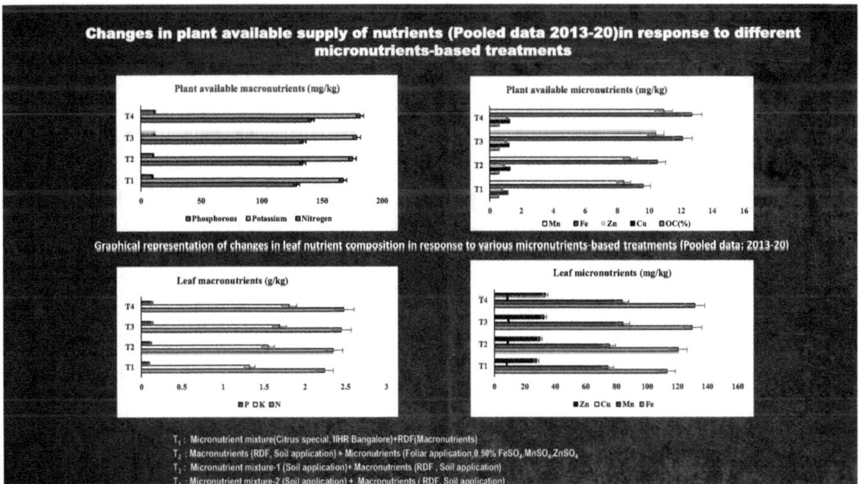

Fig. 1.5 Response of different micronutrient mixtures on changes in plant available nutrients and leaf nutrient composition of Nagpur mandarin (pooled data: 2007–2019) (Source: Srivastava (2012); Srivastava and Pandey (2021))

Changes in fruit quality index A comprehensive fruit quality index was developed based on as many as 14 parameters (disease infestation, yield/tree, uniformity in yield/tree, fruit weight, fruit diameter uniformity in fruit size, peel color, fruit firmness at maturity, shape at maturity, peel/skin smoothness, juice content, TSS, acidity, TSS/acidity ratio, and number of seeds/fruits) for evaluating the fruit quality response of different micronutrient-based treatments (Table 1.7). Evaluating the response of different treatments, the quality index value of 59/100 (T_1) to 69/100

Table 1.7 Fruit quality response with their index values in response to different treatments of customized micronutrient mixtures

Quality parameter	Weightage points	Treatments			
		T_1	T_2	T_3	T_4
Disease infestation	**5**	Moderate (3)	Moderate (3)	Moderate (3)	Moderate (3)
Yield/tree	**3**	Moderate (2)	Moderate (2)	Moderate (2)	Moderate (2)
Uniformity in yield/tree	**5**	Moderately consistent (4)	Moderately consistent (4)	Moderately consistent (4)	Moderately consistent (4)
Fruit weight (gm)	**10**	137.0 (5)	139.2 (5)	144.0 (7)	146.4 (7)
Fruit diameter (mm)	**10**	59.5 (5)	61.5 (7)	62.6 (7)	63 (7)
Uniformity in fruit size	**10**	Moderately consistent (6)	Moderately consistent (6)	Strongly consistent (7)	Strongly consistent (7)
Peel color	**10**	Green (6)	Orange–green (7)	Orange–green (7)	Orange–green (7)
Fruit firmness at maturity	**5**	Tight (5)	Tight (5)	Tight (5)	Tight (5)
Peel/skin smoothness	**5**	Smooth (5)	Smooth (5)	Smooth (5)	Smooth (5)
Juice content (%)	**14**	43 (11)	43.6 (11)	44.75 (11)	45.55 (14)
TSS	**5**	8.45 (1)	8.45 (1)	8.7 (1)	8.71 (1)
Acidity	**5**	0.77 (3)	0.76 (3)	0.72 (3)	0.70 (3)
TSS/acidity ratio	**5**	11.0 (3)	11.1 (3)	12.1 (3)	12.4 (3)
Number of seeds/fruits	**3**	11 (1)	11 (1)	10 (2)	10 (2)
Total (index value)	**100**	**59/100**	**62/100**	**66/100**	**69/100**

Source: Srivastava (2012), Srivastava and Pandey (2021)

T_1, micronutrient mixture (citrus special, IIHR Bangalore); T_2, macronutrients (soil application) + micronutrients (foliar application); T_3, micronutrient mixture 1 (soil application) + macronutrients (soil application); and T_4, micronutrient mixture 2 (soil application) + macronutrients (soil application)

(T_4) was observed (Table 1.7), with the highest fruit quality index of 69/100 with the treatment T_4, suggesting again the development of superior quality fruits with application of customized micronutrient mixture 2 developed on the basis of ionome stoichiometry of fruits, the best indicator of developing customized fertilizers. The same principle could be applicable to other fruit crops as well.

1.8 Important Futuristic Viewpoints

The success of fruit crop, especially under change setup, needs to decipher the mechanistic pathways of both negative and positive crop responses to delve upon an effective mitigation technique. Fruit quality traits involving primary metabolism (flavor and soluble solid/acid ratio defining the external fruit quality) are improved coupled with biochemical pathways (accumulation of antioxidants) relating with plant defense (Bora and Bora 2020; Saikia et al. 2021). Interestingly, the other fruit parameters defining peel/skin color, skin disorder, and nutritional value (nutrient density traits) are invariably negatively affected by climate change to varying proportions. Such kinds of responses to climate change in fact neutralize the increasing intervention of production technologies, unless a holistic strategic framework is developed involving concepts such as biofertigation with subsurface micro--irrigation (Shirgure et al. 2000; Panigrahi et al. 2017), microbial consortia loaded with multi-utility microbes (including mycorrhizas) upscaled to integrated soil fertility options (Srivastava et al. 2015a, b; 2019; Srivastava 2009b, 2011), rhizosphere hybridization (Cheke et al. 2018; Hota et al. 2020; Dzuvichu et al. 2023), fruit crop--based integrated land use (Srivastava and Singh 2005b; Srivastava et al. 2010), biochar-mediated soil fertility regulations (Agegnehu et al. 2017; Mousavi et al. 2023), fruits under protected system coupled with an eye on reduction in carbon and water footprints (Srivastava et al. 2017), plant nutrition-mediated plant defense (Tripathi et al. 2022), and extended postharvest life (Srivastava and Shirgure 2018), further backed up by microbially engineered fruit crops loaded with endophytes for additional soil–plant health resilience (Bora et al. 2013, 2019; Rahman et al. 2023), and dip-digging of climate-smart 4R stewardship (Bruulsema 2022b) besides carbon trade-offs between carbon sequestration and rural development better placed with fruit crops (Palmer and Silber 2012). Some of the major issues needing a fresh relook to better ward off climate change are further highlighted below:

(i) Fruit crops are basically tailor-made for elevated carbon sequestration as a strong plant carbon sink acting as a natural but parallel nutrient sink, but its impact on tolerance against other biotic and abiotic stress is still unexploited, besides its role in extending the tree's productive life. Most of the fruit crops have C3 photosynthetic pathways, and introduction of C4 pathway in C3 fruit crops would be quite an accomplishment in offering a climate-resilient fruit production system. The role of plant breeders and biotechnologists with back-up support of plant physiologists and microbiologists would be quite important in fulfilling this novel objective. Unfortunately, hardly any serious efforts are in progress in this direction.

(ii) Temporal change in crop phenology vis-à-vis climatic gradients and partitioning of fertilizers must be periodically reviewed, instead of considering optimum fertilizer requirement a sometime exercise. A stewardship concept involving 4R × 4W (nutrients × irrigation water) deserves fresh look for fertigation-driven fruit production system. In this regard, the role of growth regulators, anti-transpirants, and orchard floor management is undeniable.

(iii) Crop-based rhizosphere ecology (phytobiome analysis using transcriptomics--metabolomics integrated analysis) needs to be analyzed in relation to different analogues of climate change-related parameters coupled with the role of entophytic lifestyle of microbes to finally identify microbes to be able to impart an additional resilience to fruit crops against any possible adverse effects of climate change. Microbial assembly disruption due to depletion in soil fertility over a period of time makes such exercise quite revealing about the nutrient load of fruit crops.

(iv) Biofortification of fruit crops is often used to define their nutrient supplying capacity, even though they are known to have rich sources of mineral, vitamins, phytonutrients, and antioxidants, but these are highly questionable in nutrient-depleted degraded soils. Moreover, the nutrient density of crops is further reduced due to postharvest losses and processing losses. However, without any well-proven historical data about changes in nutrient density of fruit crops grown on pre-selected sites over a period of time, this issue is interchangeably interpreted using reductionist approaches as industrial biofortification (traits related to consumer acceptance), genetic biofortification (overlooking regional nutritional issues coupled with cultural issues targeting single crop-single nutrient), and agronomic biofortification (looking at frequently occurring deficiencies such as Fe, Zn, and P deficiencies ignoring other nutrients). Nonetheless, a biofortified crop should also look at consumer--related traits and cooking characteristics. Accordingly, a practical meaning of nutrient density of fruit crops should be redefined catering to varied utility purposes.

(v) Looking at the inconsistent response of single or dual microbial inoculation with function-specific microbes, development of microbial consortium using a combination of multi-utility microbes (plant growth promoting ability, antagonists against soil borne pathogens, acaricidal properties, and carbon sequestration ability) has paid far richer dividends in offering a more resilient production system on account of expanded functional corridor, however only with miniscule efforts on fruit crops. The metabolic and other functional pathways as communication signals (defined by secondary metabolites) of different microbial candidates therefore need an in-depth elaboration for developing synthetic microbial community (SymCom) and possibly elevate with green nanocomposites to facilitate the much-needed extended crop response in the field. In this regard, it is most important to ascertain an ideal crop growth stage nurturing the maximum rhizosphere functional traits. Is it possible to develop a schedule of specific function microbial inoculant as per crop growth stages similar to the differential use of nutrients? Is it possible to develop a microbially engineered fruit crop using a microbial gene pool keeping an eye on nutrient accumulation ability of edible plant parts?

(vi) Elevation in nutrient-use efficiency in fruit crops has been the most intensively researched issue in the past, though more for theoretical implications. In this regard, we need some practical interventions to ensure better fertilizer use by fruit crops, e.g., combination of growth regulators and nutrients, as foliar spray or a module involving nutrient sources having dynamic nutrient release kinetics with emphasis on reduced carbon and nutrient footprint.

(vii) Current fruit nutrition program is characterized by fertilizer application strategy addressing straight nutrient interaction overlooking the multi-nutrient interactions (e.g., Fe-Zn-N-P or N-S-Ca-Mg-K, level of interaction to be defined depending upon the nature and properties of soil and crop in question as a part of coordinating nutrient interaction and nutrient stress signaling) that need to be understood at physiological and molecular levels and their cross talks with fruit yield in the field and utilize these in developing alternative strategy of optimizing the fertilization options. The important candidate genes for such nutrient interactions and/or metabolic pathways could be of high utility to plant breeders to improve nutrient-use efficiency and eventually unlock fruit yield barriers.

(viii) Growing fruit crops under protected system is considered a kind of abatement to the ongoing process of climate change with assured quality production, unless supported with suitable crop management technology. Because of the pandemic diseases, such as citrus greening, citrus growers are forced to follow protected cultivation with great ease of success. Let us learn from such success and be prepared for any such disease urgency forcing a complete overhaul of plant nutrition strategy.

(ix) Reduction in plant diseases with improved plant nutrition very often go hand-in-hand, with the result nutrient diagnostics an optimum fertilizer doss coupled with their application schedule earlier developed for disease-free fruit crop, may not be exactly so effective with introduction of diseases. We need to develop an ameliorative strategy to address two contrastingly different production scenarios to cater any adverse addition of climate change.

(x) Various forms of conventional farming, viz., organic farming, natural farming, and regenerative farming, all advocating to the more rational use of agroecology, need a more incisive analysis with emphasis on plant nutrition vis-à-vis production sustainability, keeping an eye on silicon nutrition and postharvest fruit quality.

(xi) An urgent need for improved climate modeling and forecasting through decision support for implementation of adaptation strategies (including traditional knowledge). The support of mass data mining and machine learning could still offer better precision to fruit nutrition to be answerable by multiple objectives of climate change.

(xii) In plant nutrition, silicon, although not considered a nutrient, is a beneficial element that stands out as a mitigating agent for multiple stresses. This element is relatively less studied in fruit trees, but the number of studies on other species has increased significantly in recent years. Considering that climate change amplifies various stresses in crops such as deficiency or excess of nutrients, pests, and salinity, among others, it is common in the field to have simultaneous stresses at the same time that silicon stands out in the mineral nutrition of plants, deserving priority treatment of the research.

These are only a few examples of challenges in fruit crops emphasizing how plant nutrition needs to integrate over different scientific disciplines to meet the societal requests for productive, sustainable, and environmentally sound fruit nutrition program. The other side of the coin is that plant nutritionists face a challenging time with great opportunities for novel developments and exciting discoveries pertaining to a scientifically sound climate-neutral fruit nutrition program.

1.9 Concluding Remarks

Summarizing the entire discussion on a positive note, elevation in carbon dioxide was associated with promoting the fruit yield sugars, total antioxidant capacity, flavonoids, vitamin C, and calcium–phosphorus–potassium levels, but at the expense of protein and minerals such as magnesium, iron, and zinc (which could further exacerbate the increasing menace of global iron and/zinc deficiencies). Such anomalies need to be strictly guarded in terms of carbon dioxide absorbed by the fruit crops and nutrient-use efficiencies.

Marketable fruit quality parameters hold no bar in relation to climate change, even though a major chunk of fruit quality defining parameters remain hugely affected by fluctuating regime in soil environments, though inconsistent response (fruit species and trait specific dependence) could be a major hurdle to frame a clear-cut opinion. Our studies further confirmed the pressing importance of integrated soil fertility concept using nutrient sources with varied nutrient release kinetics. Amidst these changes, we shall shift in our fruit-consuming habits or mitigate them through changes in physiological and genetic responses of fruit crops; the latter is however more challenging and warrants involvement of multidisciplinary integrative research investigations. To cite an example, bioprospecting soil–plant health has to be addressed in tandem than treating them in isolation, and like that, few more examples can be picked to add more visibility to challenges of plant nutrition in fruit crops; the list is simply so long and diversified. Furthermore, additional strategies are also suggested based on the use of beneficial elements, especially silicon, which can contribute in minimizing stress and strengthening the sustainability of crops.

References

Abidi W, Akrimi R, Hajlaui H, Rejeb H, Gorgorcena Y (2023) Foliar fertilization of potassium silicon improved postharvest fruit quality of peach nectarine (*Prunus persica* (L). Batsch) cultivars. Agriculture 13:195

Agegnehu G, Srivastava AK, Bird M (2017) The role of biochar-compost in improving soil quality and crop performance: a review. Appl Soil Ecol 119:156–170

Ahmed N, Srivastava KK, Lal S (2011) Global warming carved niche for apple production. Indian J Hortic 56(1):9–11

Allison SD, Wallenstein MD, Bradford MA (2010) Soil-carbon response to warming dependent on microbial physiology. Nat Geosci 3:336–340

Baldi E, Toselli M, Marcolini G, Quarieri M, Cirillo E, Innocenti A, Marangoni B (2010) Compost can successfully replace mineral fertilizers in the nutrient management of commercial peach orchard. Soil Use Manag 26:346–353

Baranski M, Srednicka-Tober D, Volakakis N, Seal C, Sanderson R, Stewart GB, Benbrook C, Biavati B, Markellou E, Giotis C, Gromadzka-Ostrowska J, Rembialkowska E, Skwarlo-Sonta E, Skwarl Sonta K, Tahyonen R, Janovska D (2014) Higher antioxidant and lower cadmium concentrations and lower incidence of pesticide residues in organically grown crops: a systematic literature review and meta analyses. Braz J Nutr 112(5):794–811

Bharadwaj RL, Parashar A, Parewa HP, Vyas L (2024) An alarming decline in nutritional quality of fruits: the biggest challenge for future generations health. Food 13(6):877. https://doi.org/10.3390/foods13060877

Bhargava R, Singh RS, Pal G, Sharma SK (2011) Physiological disorders in fruits in arid region: a review. Indian J Arid Hortic 6(1–2):1–10

Bhattacharjee P, Warang O, Das S, Das S (2022) Impact of climate change on fruit crops: a review. Cuu World Environ 17(2). https://doi.org/10.12944/CWE.17.2.4

Bindi M, Fibbi L, Miglietta F (2001) Enrichment (FACE II) of grapevine (*Vitis vinifera* L). II. Growth and quality of grape and wine in response to elevated CO_2 concentrations. Eur J Agron 14:145–155

Bora P, Bora LC (2020) Disease management in horticultural crops through microbial interventions: an overview. Indian J Agric Sci 90(8):1389–1396

Bora P, Bora LC, Begum M (2013) Ecofriendly management of soil borne diseases in brinjal through application of antagonists microbial population. J Biol Control 27(1):29–39

Bora P, Saikia K, Hazarika H, Ragesh G (2019) Exploring potential endophytes in disease management of horticultural crops. Curr Hortic 7(2):32–37

Brown PH, Uriu K (1996) Nutrition deficiencies and toxicities: iagnosing imbalances. Almond Production Manual, University of California, Div Agri, Nat Res, Oakland, California, USA. pp 179–188

Brown PH, Zhao F-J, Dobermann A (2022a) What is plant nutrient? Changing definitions of advance science and innovation in plant nutrition. Plant Soil 476:14–23

Brown HB, Zhao F-J, Dobermann A (2022b) What is plant nutrient? Changing definitions to advanced science and innovation in plant nutrition. Plant Soil 476:11–12

Bruulsema T (2022a) Nutrient stewardship: taking 4R further. Crops Soils 55(1):34–40

Bruulsema T (2022b) Climate-smart fertigation in 4R nutrient stewardship. Crops Soils. https://doi.org/10.1002/crso.20191

Cardoso EJBN, Vasconcellos RLF, Bini D, Miyauchi MYH, Alcantara C, Alves PRL, Paula A, Nakatani AS, de Pereira JM, Nogueira M (2013) Soil health; looking for sustainable indicators. What should be considered to assess the effects of use and management on soil health. Sci Agric 70(4):219–303

Cheke AS, Patil VD, Srivastava AK (2018) Studies on rhizoshere hybridization and nutrient dynamics in sweet orange from pot experiment. J Pharmacogn Phytochem 7(15):30077–33082

Chen CC, Paull RE (2001) Fruit temperature and crown removal on the occurrence of pineapple fruits translucency. Sci Hortic 88:85–95

Clonan M, Conchie M, Hall M, Olesen T, Sarkhosh A (2021) Effects of ambient temperatures on floral initiation in Australian mango (*Mangifera indica* L.) selections. Sci Hortic 276:107767. https://doi.org/10.1016/j.scientia.2020.109767

Corato U De, Sanisnenca E (2022) Plant mieral nutrition and disease resistance: a significant linkage for sustaining crop protection. Front Pl Sci 13:133116.

Corwin DL, Lesch SM (2005) Characterizing soil spatial variability with apparent soil electrical conductivity I. Survey protocols. Comput Electron Agric 46:103–133

Darmon N, Darmon N, Maillot M, Drewnowski A (2006) A nutrient density standard for vegetables and fruits: nutrient per calorie or nutrients per unit cost. J Am Diet Assoc 125(12):1871–1887

Datta S (2013) Impact of climate change in Indian horticulture-a review. Int J Sci Environ Technol 2(4):661–671

Davenport T, Campbell C (1977) Stylar-end breakdown in Tahiti lime. Aggravating effect of field heat and fruit maturity. J Am Soc Hort Sci 102(40):484–486

Davis DR (2009) Declining fruits and vegetables nutrient composition: what is the evidence. HortScience 44(1):15–19

De Almeida A-AF, Valle RRW (2007) Physiology of the cacaotre. Braz J Plant Physiol 19:425–448

De Orduna RM (2000) Climate change associated effects on grape wine quality and production. Food Res J 43:1844–1855

Dijkstra FA, Blumenthal D, Morgan JA, Pendall E, Yolima Carrillo Y, Follett RF (2010) Contrasting effects of elevated CO2 and warming on nitrogen cycling in a semiarid grassland. New Phytol 187:426–437

Djomo AN, Knohl A, Gravenhorst G (2011) Estimation of total ecosystem carbon pools distribution and carbon biomass: current annual increment of a moist tropical forest. Forest Ecol Manag 261(8):1448–1459

Dobermann A, Bruulsema T, Cakmak J, Gerard B, Majumdar K, McLaughlin L, Reidsma P, Vanlauwe B, Wollenberg L, Zhang F, Zhang X (2022) Responsible plant nutrition to support food system transformation. Global Food Secur 33:100636. https://doi.org/10.1016/j.gfs-2022.10636

Doddrell NH, Lawson T, Raines CA, Wagstaf C, Sinkin AJ (2023) Feeding the world: impact of elevated (CO_2) on nutrient content of greenhouse grown fruit crops and options for future yield gains. Hortic Res 10(4). https://doi.org/10.1093/hr/uhad026

Dong ZH, Chen MX, Srivastava AK, Mahmood Ul H, Ishfaq M, Shi X, Zhang YQ, Mousaa MG, Li XX, Hu C, Zhang FS (2024) Climate changes altered the citrus fruits quality: a 9-year case study in China. Sci Total Environ 923:17140

Dorans SF, Zeiss MR (2000) Soil health and sustainability: managing the biotic components of soil quality. Appl Soil Ecol 15(1):3–11

Drake B, Gonzalez-Meler M, Long S (1997) More efficient plants: a consequence of rising atmospheric CO_2? Annu Rev Plant Physiol 48:609–639

Dzuvichu M, Alila P, Jamir S, Kanauzia SP, Srivastava AK (2023) Yield response of rhizosphere hybridization in khasi mandarin. Int J Innov Hortic 12(1):102–108

Egea JA, Caro M, García-Brunton J, Gambín J, Egea J, Ruiz D (2022) Agroclimatic metrics for the main stonefruit producing areas in Spain in current and future climate change scenarios: implications from an adaptive point of view. Front Plant Sci 13:842628. https://doi.org/10.3389/fpls.2022.842628

Elbasiouny H, El-Ramady H, Elbehiry F, Rajput VD, Minkina T, Mandzhieva S (2022) Plant nutrition under climate change and soil carbon sequestration. Sustain For 14:914. https://doi.org/10.3390/su14020914

Elprince Adel M (2009) Prediction of soil fertilization maps using logistic modeling and a geographical information system. Soil Sci Soc Am J 73:2032–2042

El-Rhman IE (2010) Physiological studies on cracking phenomenon of pomegranate. J Appl Sci Res 6:696–703

Etesami H, Jeong BR (2020) Importance of silicon in fruit nutrition: agronomic and physiological implications. In: Srivastava AK, Chengxiao H (eds) Diagnosis and management of nutrient constraints in fruit crops. Elsevier, pp 255–277. https://doi.org/10.1016/8978-0-12-818732-6000198

Fan X, Zhou X, Chen H, Tang M, Xie X (2021) Cross-talk between macro- and microutrients uptake and signaling in plants. Front Plant Sci 12. https://doi.org/10.3389/fpls.2021.663477
Fauchereau N, Trzaska M, Rouault M, Richard Y (2003) Rainfall variability and changes in Southern Africa during the 20th century in the global warming context. Nat Hazards 29:139–154
Ferguson I, Snelgar W, Lay-yee M, Watkins C, Brown J (1998) Expression of heat shock protein genes in apple fruit in the field. Funct Plant Biol 25:1955–1963
Fiore A, Lardo E, Montanaro G, Loterza D, Lojudue C, Berloko T, Dichio B, Xiloyannis C (2018) Mitigation of global warming impact of fresh fruit production through climate smart management. J Clean Prod 172:3634–3643
Fischer G, Ramiraz F, Casierra-Posceda F (2016) Ecophysiological aspects of fruit crops in the era of climate change: a review. Agronomia Colombiana 34(2):190–199
Ganeshamurthy AN, Ravindra V, Rupa TR (2019) Carbon sequestration potential of mango orchards in India. Curr Sci 117(2):2006–2012
Gao WQ, Lu LH, Srivastava AK, Wu Q-S, Kuca K (2020) Effects of mycorrhizae on physiological responses and relevant gene expression of peach effected by replant disease. Agronomy 10(2):186
Garcia G, Re B, Orians C, Crone E (2021) By wind or wing: pollination syndrome and alternate bearing in horticulture crops. Philos Trans R Soc Lond Biol Sci 376(1839):202000371
Geetha GA, Shivashankara KS, Reddy N (2016) Varietal variations in temperature response for hermaphrodite flower production and fruit set in mango (*Mangifera indica* L.). South Afr J Bot 106:196–203
Gelaye Y (2024) A systematic review on potential analogy of phytobiomass and soil carbon evaluation methods: Ethiopia insights. Open Agric 9(1):20220313
Gelaye Y, Getahun S (2024) A review of the carbon sequestration potential of fruit trees and their implications for climate change mitigation: the case study of Ethiopia. Cogent Food Agric 10:17–28. https://doi.org/10.1080/23311932.2023.2294544
Gelybo G, Toth E, Farkas C, Horel E, Kasa I, Bakacsi Z (2018) Potential impacts of climate change on soil properties. Agric Soil Sci 67(1):121–141
Gora JS, Verma AK, Singh J, Choudhary DR (2019) Climate change and production of horticultural crops in agricultural impacts of climate change. CRC Press, pp 45–61
Han J-H, Cho JG, Son IC, Kim SH, Lee IB, Choi IM, Kim D (2012) Effects of elevated carbon dioxide and temperature on photosynthesis and fruit characteristics of 'Niitaka' pear (*Pyrus pyrifolia* Nakai). Hortic Environ Biotechnol 53:357–361
Hanumanthaiah MR, Hipparagi K, Renuka DM, Vijendrakumar RC, Santhosa KV, Kumar KK (2015) Effect of soil and foliar application of silicon on fruit quality parameters of banana cv. Neypoovan under hill zone. Plant Arch 15:221–224
Hazarika TK (2013) Climate change and Indian horticulture: opportunities, challenges and mitigation strategies. Int J Environ Eng Manag 4(6):629–630
Heckathorn S, North G, Wang D, Zhu C (2020) Climate change and plant nutrient relations. Front Plant Sci 11. https://doi.org/10.3384/fpls.2020.00869
Higuchi N, Santos J dos, Ribeiro RJ, Minette L, Bioty C (1998) Biomass of the aerial part vegetation of the rotical terra firm forest of the Brazilian Amazon. Acta Amazon 28:153–155
Holmes A, Muller K, Clothier BE, Deurer M (2014) Carbon sequestration of kiwifruit orchard soils at depth to mitigate carbon emission. Commun Soil Sci Plant Anal 46(Supl 1):122–136
Hota D, Srivastava AK, Dahal S, Dubey S (2020) Rhizosphere engineering through microbes enhances agronomic performance of acid lime. Progress Hortic 52(20):120–123
Hu S, Chapin FS, Firestone MK, Field CB, Chiariello NR (2001) Nitrogen limitation of microbial decomposition in a grassland under elevated CO_2. Nature 409:188–191
Huang Y-M, Srivastava AK, Zou Y-N, Ni Q-D, Han Y, Wu Q-S (2014) Mycorrhizal -induced calmodulin mediated changes in antioxidant enzymes and growth response of drought -stressed trifoliate orange. Front Microbiol 5:682–628

Idoso SB, Kimball BA, Shaw PE, Widmer W, Vanderslice JT, Higges DJ, Montanari A, Clark WD (2002) The effect of elevated CO_2 concentration on the vitamin C (sour) orange juice. Agric Agrosyst Environ 90:1–7

Iglesias DJ, Oliver AQ, Font A, Martinez B, Giner MAF, Legaz F, Primo-Millo E (2013) Carbon balance of citrus plantation in eastern Spain. Agric Ecosyst Environ 171:103–111

Jeyabaskaran KJ, Shirgure PS, Srivastava AK, Pandey V, Uma S (2021) Fertigation in horticulture: a guarantee to economized quality production. Indian J Fertilizers 17:364–383

Johnston AM, Khurana HS, Majumdar K, Satyanarayana T (2009) Site specific nutrient management—concept, current research and future challenges in Indian agriculture. J Indian Soc Soil Sci 57:1–10

Jones GV, White MA, Cooper OR, Storchmann K (2005) Climate and global wine quality. Climate Change 73:319–343

Joseph MH, Dhargave TS, Deshpande CP, Srivastava AK (2015) Microbial solubilization of phosphate: Pseudomonas versus Trichoderma. Annals of Plant and Soil Research 17:227–232

Katrodia JS, Rane DA, Salunkhe DK (1985) Biochemical nature of spongy tissue in Alphonso fruits. In: II International Symposium on Mango, vol 23, pp 835–839

Kaur P, Sandhu SP (2020) Climate change: its drivers and their effects on crops. Agric Res J 57(3):297–308

Kavi Kishore PB, Guddimalli R, Kulkarni J, Sigham P, Somanaboina AK, Nandihandalam AT, Patil S, Polavarapu R, Suravajhala P, Sreenivasulu N, Penna S (2023) Impact of climate change on alteration of fruit quality with organoleptic, health benefits, and malnutrition attributes. J Food Agric Chem 71(46):17510–11752

Khehra S, Bal JS (2014) Influence of organic and inorganic nutrient sources on growth of lemon (*Citrus limn* (L.) Burm.) cv Baramasi. J Exp Biol Agric Sci 2:126–129

Kimball BA, Idiso SB, Johnson S, Rillig MC (2007) Seventeen years of carbon dioxide enrichment of sour orange trees: final results. Glob Chang Biol 13:2171–2178

Kirschbaum MUF (1995) The temperature dependence of soil organic matter decomposition and the effect of global warming on soil organic carbon storage. Soil Biol Biochem 27:753–760

Kirschbaum MUF, Fischlin A, Cannell MGR, Cruz RVO, Galinski W (1996) Climate change impacts on forests. In: Watson RT, Zinyowera MC, Moss RH (eds) Climate change 1995: impacts, adaptations and mitigation of climate change: scientific-technical analyses. Contribution of Working Group II to the Second Assessment Report of the Intergovernmental Panel on Climate Change. Cambridge University Press, pp 95–129

Krishna K, Rashid R, Bhat JA, Bhat ZA (2011) Effects of high temperature on fruit crops. Elixir Appl Bot 39:4745–4747

Kumar V, Prasad RK, Suman SN, Tiwari S (2011) Integrated nutrient mannagment for better soil fertility rice rpoductivity. Oryza 48:335–338

Kumar M (2019) Effect of climate change on Indian economy and agriculture. Int J Sci Res Rev 07(01):23–28

Kumar P, Murugan P, Murkute A, Bala Singh S (2010) A carbon sequestration strategy involving temperate fruit crops in the trans-Himalayan region. J Hortic Sci Biotechnol 85(5):405–409

Kumar S, Kumar S, Mohapatra T (2021) Interaction between macro- and micronutrients in plants. Front Plant Sci 12. https://doi.org/10.3389/fpls.20211665583

Laane HM (2017) The effect of the application of foliar sprays with stabilizing silicic acid: an overview of the results from 2003-2024. SILICON 9:803–807

Lal R (2004) Soil carbon sequestration to mitigate climate change. Geoderma 123:1–22

Lalithya KM, Bhagya HP, Bharali K, Choudhary R (2014) Response of silicon and micronutrients on fruit characters and nutrient content in leaf of sapota. Bioscan 2:593–598

Ledesma NA, Nakata M, Sugiyama N (2008) Effect of high temperature stress on reproductive growth strawberry cvs Nyoho and Toyonoka. Sci Hortic 116:186–193

Lehman J, Bassio DN, Kogel-Karbner I, Rillig MC (2020) The concept and future prospects of soil health. Nat Rev Earth Environ 45:544–553

Li QS, Srivastava AK, Zou YN, Wu Q-S (2023) Field inoculation responses of arbuscular mycorrhizal fungi versus endophytic fungi on sugar metabolism associated change in fruit quality of Late navel orange. Sci Hortic 308:11587

Liu C-Y, Srivastava AK, Wu Q-S (2014) Effect of auxin inhibitor and AMF inoculation on growth and rot morphology of trifoliate orange (*Poncirus trifoliata*) seedlings. Indian J Agric Sci 84(11):342–346

Liu C-Y, Srivastava AK, Wu Q-S (2017) Mycorrhizal fungi regulate rot responses and leaf physiological activities in trifoliate orange. Notulae Botanicae Horti Agrobotanici Cluj-Napoca 45(1):17–21

Liu Z, Cheng XF, Zou YN, Srivastava AK, Alqahtani MD, Wu Q-S (2024) Negotiating soil water deficit in mycorrhizal trifoliate orange: a gibberellin pathway. Environ Exp Bot 219:105658

Loladze I (2014) Hidden shift of the ionome of plants exposed to elevated CO2 depletes minerals at the base of human nutrition. elife 3:e02245

Luo Y, Su B, Currie WS, Dukes JS, Finzi A, Hartwig U, Hungate B, McMurtrie RE, Oren F, Parton WJ, Pataki DE, Shaw MR, Zak DR, Field CB (2004) Progressive nitrogen limitation of ecosystem responses to rising atmospheric carbon dioxide. Bioscience 54:731–739

Malhotra SK (2017) Horticultural crops and climate change: a review. Indian J Agric Sci 87(1):12–22

Manjoor MA, Xu Y, Zhengxin L, Au J, Shah IH, Subir IA, Wang Y, Suh W, Lu X, Li W, Liu R, Jiu S, Zhang C (2024) Horticulture crop under pressure: unravelling the impact of climate change on nutrition and fruit cracking. J Environ Manag 357:120759

Marlow GC, Loescher WH (1984) Watercore. Hortic Rev 6:189–251

Mayer A-MB, Trenchard L, Rayns F (2021) Historical changes in the mineral content of fruits and vegetables in the UK from 1940 to 2019: a concern for human nutrition and agriculture. Int J Food Sci Nutr 73(2). https://doi.org/10.1080/09637486.2021.1981831

Medda S, Fadda A, Mulas M (2022) Influence of climate change on metabolism and biological characteristics in perennial woody fruit crops in the Mediterranean environment. Horticulturae 8:273. https://doi.org/10.3390/horticulturae8040273

Melillo J, Steudler P, Aber J, Newkirkm K, Luxm H, Bowles F, Catricalam C, Magill A, Ahrens T, Morrisseau S (2002) Soil warming and carbon-cycle feedbacks to the climate system. Science 298:2173–2176

Miao Y, Bobby A, Stewart H, Zhang F (2011) Long-term experiments for sustainable nutrient management in China. Agron Sustain Develop 31:397–414

Montanaro G, Celano G, Dicho B, Xiloyannis C (2010) Effects of soil protecting agricultural practices on soil organic carbon and productivity of fruit tree orchards. Land Degrad Develop 161:46–54

Montanaro G, Tuzio AC, Xylogiannis E, Kolimenakis A, Dichio B (2016) Carbon budget in a mediterranean peach orchard under different management practices. Agri Ecosyst Environ 238. https://doi.org/10.1016/j.agee.2016.05031

Montgomery DR, Bikile A (2021) Soil health and nutrient density: beyond organic vs conventional farming. Font Sustain Food Syst 5. https://doi.org/10.3389/fusufs2021.699147

More SS, Gokhale MB, Shinde SE, Korake GN (2015) Effect of different sources of silica on nutrient content of leaves of fruit at different growth stages of Alphonso mango (*Mangifera indica* L.) in laterite soil. J Progress Agri 6:27–32

Mousavi SM, Srivastava AK, Cheraghi M (2023) Soil health and crop response of biochar: an updated analysis. Arch Agron Soil Sci 69(7):1085–1110

Naik SK, Sarkar PK, Das B, Singh AK, Bhatt BP (2018) Predictive modes for dry biomass and carbon stock estimation in Litchi chinensis under hot and dry subhumid climate. Arch Agron Soil Sci 64:2100. https://doi.org/10.1080/03650340.2018.1434718

Nardino M, Pernice F, Rossi F, Georgiadis T, Facini O, Motisi A, Drago A (2013) Annual and monthly carbon balance in an intensively mediterannean olive orchard. Photosynthetica 51:63–67

Natali SM, Sanudo-Wilhelmy SA, Manuel T, Lerdau MT (2009) Plant and soil mediation of elevated CO_2 impacts on trace metals. Ecosystems 12:715–727
Nawaz R, Abbasi NA, Hafiz IA, Khalid A, Ahmad T, Aftab M (2019) Impact of climate change on kinnow fruit industry of Pakistan. Agrotechnology 8(186):2
Ngullie E, Singh AK, Sema A, Srivastava AK (2015) Citrus growth and rhizosphere properties. Commun Soil Sci Plant Anal 46(12):1540–1550
Noordwijk MV, Cadisch G (2020) Access and excess problems in plant nutrition. Plant Soil 247(1):25–39
Oh N-H, Richter DD Jr (2004) Soil acidification induced by elevated atmospheric CO_2. Global Change Biol 10:1936–1946
Ortas I (2020) Mycorrhizas in fruit nutrition; important breakthroughs. In: Srivastava AK, Chengxiao H (eds) Diagnosis and management of nutrient constraints in fruit crops. Elsevier, pp 339–351. https://doi.org/10.1016/B978-0-12-818732-6.00024-1
Ortas I, Refique M, Akpinar C, Kacar YA (2017) Growth media and mycorrhizal special effect on acclimatization and nutrient uptake of banana plantlets. Sci Hortic 217:55–60
Palmer C, Silber T (2012) Trade-offs between carbon sequestration and rural incomes in the Nhambita community carbon project, Mozambique. Land Use Policy 29:83–93
Pan H, Shu Z (2007) Temperature affects color and quality characteristics of 'Pink' wax apple fruit discs. Sci Hortic 112:290–296
Panigrahi P, Srivastava AK, Panda DK, Huchche AD (2017) Rainwater, soil, and water conservation for improving productivity of citrus orchards in a drought prone region. Agric Water Manag 115:65–77
Pathak H, Nedwell DB (2011) Nitrous oxide emission from soil with different fertilizer, water levels, and nitrification inhibitors. Water Air Soil Pollut 129:217–228
Patil H, Tank RV, Manoll P (2017) Significance of silicon in fruit crops-a review. Plant Arch 17(2):769–774
Pechan PM, Boule H, Obster F (2023) Reducing vulnerability of fruit orchards to climate change. Agric Syst 210:103713. https://doi.org/10.1016/j.agsy.2023.103713
Peet MM, Wolfe DW (2000) Crop ecosystem response to climate change. Vegetable crops. CABI Publishers, pp 181–296
Pendall E, Bridgham S, Hanson PJ, Hungate B, Kicklighter DW, Johnson DW, Law BE, Luo Megonigal JP, Olsrud M, Ryan MG, Wan S (2004) Below-ground process responses to elevated CO_2 and temperature: a discussion of observations, measurement methods, and models. New Phytol 162:311–322
Peng L, Wang C, He S, Guo C, Yan C (2000) Effects of elevation and climatic factors on the fruit quality of Navel orange. South China Fruits 29(4):3–4
Perez-Piqueres A, Martinez-Alcantara P, Rodriguez-Carretero I, Canet R, Quinones A (2020) Estimating carbon sequestration in fruit crops. In: Srivastava AK, Chengxiao H (eds) Diagnosis and management of nutrient constraints in fruit crops. Elsevier, pp 67–76. https://doi.org/10.1016/B976-0-12-818732-6.00006-X
Quinones A, Martinez-Alcantara B, Font B, Forner-Giner MA, Legaz F, Primo-Millo E, Iglesias DJ (2013) Allelometric models for estimating carbon fixation in citrus trees. Agron J 105(5):1355–1366
Rahman M, Borah SM, Borah PK, Bora P, Sarmah PK, Lal MK, Tiwari RK (2023) Deciphering the antimicrobial activity of multifaceted rhizospheric biocontrol agents of solanaceae crops vi., *Trichoderma harzianum* MC2 and *Trichoderma harzianum* NBG. Front Plant Sci 14:353
Rajaseger G, Chan KL, Tan KY, Ramasamy S, Khin MC, Amaladves A, Haribhai PL (2023) Hydroponics: current trends in sustainable crop production. Bioinformation 19(9):925–938
Ram L, Singh S, Srivastava AK, Huchche AD (2007) Yield prediction in intercropped versus monocropped citrus orchards. Sci Hortic 114:67–70
Ramchandra MK, Sawale DD, Patil AV, Reddy MR (2022) Carbon sequestration assessment in sapota orchards at regional fruit station, Ganeshkhind, Pune, Maharashtra. Pharma Innov J 11(2):4990–4993

Ranasinghe CS, Thimothias KSH (2012) Estimation of carbon sequestration potential in coconut plantation under different agroecological regions and land suitability class. J Nat Sci Found Sri Lanka 40(10):77–93

Rasera JB, de Silva RF, de Filho FAAM, Delbem ACBD, Saraiva AM, Sentelhas PC, Marques PAA (2023) Climate change and citriculture: a bibliometric analysis. Agronomy 13(3):723. https://doi.org/10.3390/agronomy13030723

Reddy AGK, Kumar JS, Maruthi V, Venkatasubbaiah K, Rao CS (2017) Fruit production under climate changing scenario in India: a review. Environ Ecol 35(2B):1010–1017.59

Rietra RPJJ, Heinen M, Dimkpa CO, Bindraban PS (2017) Effects of nutrient antagonism and synergism on yield and fertilizer use efficiency. Commun Soil Sci Plant Anal 48(6):1895–1920

Rogers JR, Beunett PC (2004) Mineral Simulation of subsurface micro-organism: release of limiting nutrients from silicates. Chems Geol 203:91–108

Roussos PA (2024) Climate change challenges in temperate and sub-tropical fruit tree cultivation. Encyclopedia 4:558–582. https://doi.org/10.3390/encyclopedia4010036

Rouxinol MTI, Martins MR, Barroso JM, Rato AE (2023) Wine grapes ripening: a review on climate effect and analytical approach to increase wine quality. Appl Biosci 2(3):347–372

Rubio-Asensio JS, Parra M, Intrigliolo DS (2020) Open field hydroponics in fruit crops: developments and challenges. In: Srivastava AK, Chengxiao H (eds) Diagnosis and management of nutrient constraints in fruit crops. Elsevier, pp 419–430

Sage RF, Zhou X-G (2022) Exploiting the engine of C4 photosynthesis. J Exp Bot 62(9):2989–3000

Saikia S, Bora P, Bora LC (2021) Bioagent mediated management of citrus canker. Indian J Agric Sci 91:198–201

Saini RS, Singh S, Deswal RPS (2004) Effect of micronutrients on plant growth regulators and soil amendments on fruit drop, cracking, yield and quality of bael (*Aegle marmelos* Correa) under rainfed conditions. Indian J Hortic 61(2):175–176

Salaza-Garcia S, Lazcano-Ferrat I (2023) Site specific fertilization increased yield and fruit size in Hass avocado. Better Crops International 17:12–16

Scandellari F, Caruso G, Liguori G, Meggio F, Palese AM, Zanotelli D, Celano G, Gucci R, Inglese P, Pitacco A, Tagliavini M (2016) A survey of carbon sequestration potential vineyard orchards in Italy. Eur J Hortic Sci 81(2):106–114

Schindler Z, Morhart C, Sheppard JP, Frey J, Seifert T (2023) In a nutshell: exploring single tree parameters and above-ground carbon sequestration potential of common walnut (*Juglans regia* L.) in agroforestry systems. Agroforest Syst 97:1007–1024

Schulz H (2000) Climate change and viticulture: an European perspective on climatology, carbon stock, and UV-B effects. Aust J Grape Res 6:2–12

Schultz F, Brock C, Schimdt H, Franz K-P, Feihold G (2013) Development of soil organic matter stocks under different farm types and tillage systems in the organic arable farming. Arch Agron Soil Sci. https://doi.org/10.1080103650340.2013.794935

Schulze ED, Lapshina E, Filippov I, Kuhlmann I, Millicome D (2015) Carbon dynamic sin boreal peatlands of the Yenisey region, Western Siberia. Biogeosciences 12:7057–7070

Shahzad Z, Rouached H (2022) Protecting plant nutrition from the effects of climate change. Curr Biol 32(13):R725–R727

Shanmugavel D, Rusyn I, Solorza-feria O, Kamaraj S-K (2023) Sustainable smart fertilizers in agriculture systems: a review on fundamentals to in-filed applications. Sci Total Environ 904(15):166729

Shao YD, Hu XC, Wu Q-S, Srivastava AK, Zhnag DJ, Gao XB (2021) Mycorrhizas promote P acquisition of tea plants through changes in root morphology and P transfer gene expression. S Afr J Bot 137:455–462

Sharma S, Rana VS, Prasad H, Lakra J, Sharma U (2021) Appraisal of carbon capture, storage, and utilization through fruit crops. Front Environ Sci 9. https://doi.org/10.3389/envs2021700768

Shirgure PS, Srivastava AK, Singh S (2000) Water management in citrus: a review. Agric Rev 21(4):223–230

Singh SR, Banik BC (2011) Response of integrated nutrient management on flowering. Fruits 42:151–154

Singh S, Srivastava AK, Shivankar VJ (1999) Citrus-based cropping system: a review. Agric Rev 20(2):123–128

Soapdmo E (1993) Tropical rain forests as carbon sinks. Chemosphere 27(6):1025–1039

Soares JC, Santos CS, Carvalho SMP, Pintado MM, Vasconcelos MW (2019) Preserving the nutritional quality of crop plants under a changing climate: importance and strategies. Plant Soil 443:1–26

Srivasatva AK (2009a) Integrated nutrient management in citrus: concept and application. In: Tennant P, Beakebhia N (eds) Citrus II. Tree and Forestry Science and Biotechnology, pp 32–58

Srivastava AK (2009b) Integrated nutrient management: concept and application. In: Tennant P, Beakebhia N (eds) Ctrus II (Vol 3). Tree and Forestry Science and Biotechnology, pp 38–58

Srivastva AK (2011) Site specific potassium management for quality for quality production of citrus. Karnataka Journal of Agricultural Sciences 24:60–66

Srivastava AK (2012) Integrated nutrient management in citrus. In: Srivastava AK (ed) Advance sin citrus nutrition. Springer, Heidelberg, pp 368–380

Srivastava AK (2013) Nutrient diagnostics in citrus: are they applicable to current season crop. Agrotechnology 2:104–105. https://doi.org/10.4172/2168P-9881.1000e104

Srivastava AK (2015) Soil fertility dynamics vis-à-vis climate change in citrus. In: Choudhary ML, Patel VB, Siddiqui MW, Verma RB (eds) Climate dynamic in horticultural science. Apple Academic Press, pp 1–34

Srivastava AK (2020) Climate-smart integrated soil fertility management in fruit crops: an overview. In: Srivastava AK, Hu C (eds) Diagnosis and management of nutrient constraints in fruit crops. Elsevier, pp 521–540

Srivastava AK (2023) Integrating natural farming with agroecology for soil health care under fruit production system. Ann Plant Soil Res 25(4):524–533

Srivastava AK, Bora P (2023) Multiple dimensions of agroecology in sustaining agriculture. Indian Farm 73(6):524–533

Srivastava AK, Das SN (2017) Nutrient optima-based productivity zonality and rationale of fertilizer use in citrus. In: Sustainable management of land resources. Apple Academic Press, pp 717–738

Srivastava AK, Hota D (2020) Fruit crop under nutrient capped scenario: a timeless journey. Curr Hortic 8(2):14–17

Srivastava AK, Kohli RR (1999) Agroecological approaches for land use planning of citrus. Agric Rev 20(1):41–47

Srivastava AK, Malhotra SK (2014) Nutrient management in fruit crops -review. Indian J Fertilizers 10(12):72–88

Srivastava AK, Malhotra SK (2017) Nutrient use efficiency in perennial fruit crops -review. JPl Nutr 40(13):1928–1953

Srivastava AK, Ngullie E (2009) Integrated nutrient management: theory and practice. Dyn Soil Dyn Plant 3:1–30

Srivastava AK, Pandey V (2021) Speciality fertilizer centric site specific nutrient management in fruit crops: principles and achievements. Indian J Fertilizers 17(5):458–474

Srivastava AK, Pongener A (2023) Asian citrus: a meta-analysis on soil fertility and plant nutrition constraints. Annals of Plant and Soil Research 25:345–352

Srivastava AK, Shirgure PS (2018) Nutrient diagnosis -based fertilizer prescription in citrus: a perspective analysis. Journal of Eco-Friendly Agriculture 13:1–17

Srivastava AK, Singh S (2002) World citrus: climate and soil analysis. International book Distribution Company, Lucknow, pp 1–290

Srivastava AK, Singh S (2005a) Diagnosis of nutrient constraints in citrus orchards of humid tropical India. J Plant Nutr 29:1061–1076

Srivastava AK, Singh, S (2005b) Diagnosis of nutrient constraints in citrus orchards of humid tropical India. J Plant Nutr 19:5–29

Srivastav AK, Singh S (2006) Citrus nutrition and biochemical markers. J Plant Nutr 29(5):827–855

Srivastava AK, Singh S (2008) Analysis of citrus orchard efficiency in relation to soil properties. J Plant Nutr 30:2077–2090

Srivastava AK, Srivastava OP (1991a) Cation exchange capacity in relation to pH of salt affected soil. J Indian Soc Soil Sci 39:355–357

Srivastava AK, Srivastava OP (1991b) Effect of gypsum and organic matter on the transformation of nitrogen in salt affected soil. Oryza 28:211–213

Srivastava AK, Srivastava OP (1992) Effect of waterlogging on pH changes and nitrate reduction in salt affected soil. J Indian Soc Soil Sci 40:184–186

Srivastava AK, Srivastava OP (1993a) Forms of soil nitrogen in relation to pH. J Indian Soc Soil Sci 41:551–553

Srivastava AK, Srivastava OP (1993b) Cation exchange capacity in relation to soil sodicity in amended saline-alkali soil. J Indian Soc Soil Sci 41:155–157

Srivastava AK, Srivastava OP (1993c) Relation between zinc availability and its uptake as influenced by gypsum and pyrites in Typic Natraqualf. J Indian Soc Soil Sci 41:396–397

Srivastava AK, Srivastava OP (1994) Response of iron and manganese application as affected by their availability in Typic Natraqualf. Arid Soil Res Rehabil 8:301–305

Srivastava AK, Srivastava OP (1995) Transformation and availability of nutrients in Natraqualfs of Uttar Pradesh: a review. Agric Rev 16:203–216

Srivastava AK, Singh S, Marathe RA (2002) Organic citrus: soil fertility and plant nutrition. J Sustain Agric 19:15–29

Srivastava AK, Singh S, Tiwari KN (2006) Site specific nutrient management for Nagpur mandarin (*Citrus reticulata* Blanco). Better Crops 88:22–25

Srivastava AK, Singh S, Diware VS, Singh H (2009) Site-specific nutrient management in 'Mosambi' sweet orange. Better Crops India 3:10–11

Srivastava AK, Singh S, Das SN, Singh H (2010) Delineation of productivity zones in mandarin orchards using DRIS and GIS. Better Crops South Asia 4:13–15

Srivastava AK, Huchche AD, Ram L, Singh S (2015a) Performance of irrigate monocropped versus intercropped Nagpur mandarin orchard on basalt derived tropical soils. Trop Agric 82(3):159–161

Srivastava AK, Malhotra SK, Krishna Kumar NK (2015b) Exploiting nutrient-microbe synergy in unlocking productivity potential of perennial fruits: a review. Indian J Agric Sci 85(4):459–481

Srivastava AK, Shirgure PS, Deshmukh S, Bhoyar P (2017) Soil fertility and soil health care in citrus: a review. Ann Plant Soil Res 19(2):127–136

Srivastava AK, Paithankar DH, Ventataramana KT, Hazarika B, Patil P (2019) INM in fruit crops: sustaining quality production and soi health. Indian J Agric Sci 89(3):379–395

Srivastava AK, Wu Q-S, Mousavi SM, Hota D (2021) Integrated soil fertility management in fruit crops: an overview. Int J Fruit Sci 21(1):413–439

Srivastava AK, Das AK, Jagannadham PTK, Bora P, Ansari FA, Bhate R (2022) Bioprospecting soil and plant health management amidst huanglongbing threat in citrus; a review. Front Plant Sci 13:858842

St Claire SB, Lynch JP (2010) The opening of Pandora box: climate change impacts on soil fertility and crop nutrition in developing countries. Plant Soil 335:101–115

Swierczynski S, Zydlik Z, Kleiber T (2023) The influence of nutrition of apple trees with silicon on growth and yield as well mineral content in leaves and fruits. Agronomy 12(7):1680

Tripathi R, Tewari R, Singh KP, Keswani C, Minkina T, Srivastava AK, De Corato U, Sanisinenea E (2022) Plant mineral nutrition and disease resistance: a significant linkage for sustainable crop production. Front Plant Sci 13:133116

Velazquez-Marti S, Cazco-Logrono C (2018) Structure analysis and biomass models for plum tree (*Prunus domestica* L.) in Ecuador. Exp Agric 54(1):133–141

Wahid A, Gelani S, Ashraf M, Foolad MR (2007) Heat tolerance in plants: an overview. Environ Exp Bot 61:199–223

Wang S, Srivastava AK, Wu Q-S, Fokom R (2014) The effect of mycorrhizal inoculation on the rhizosphere properties of trifoliate orange (*Poncirus trifoliata* L. Raf.). Sci Hortic 170:131–142

Waraich EA, Ahmad R, Halim A, Aziz T (2012) Alleviation of temperature stress by nutrient management in crop plants: a review. J Soil Sci Plant Nutr 12(2):221–244

Warren FJ, Waddington JM, Bourbonniere RA, Day SM (2001) Effect of drought on hydrology and sulphate in temperate swamp. Hydrol Process 15:3133–3150

Wiren NV (2011) Grand challenges in plant nutrition. Front Plant Sci 2:4. https://doi.org/10.3389/fpls.2011.00004

World Health Organization (2023) http://www.who.int

Wu Q-S, Srivastava AK (2012) Rhizosphere microbial communities: isolation, characterization, and value addition for substrate development. In: Srivastava AK (ed) Advances in citrus nutrition. Springer, pp 169–194

Wu T, Wang Y, Yu C, Chiarawipa R, Zhang X, Han Z, Wu L (2012) Carbon sequestration by fruit trees - Chinese apple orchards as an example. PLos One 7(6):e38883. https://doi.org/10.1371/journal.pone0038883

Wu Q-S, Srivastava AK, Zhou Y-N (2013) AMF-induced tolerance to drought stress in citrus: a review. Sci Hortic 164:77–87

Wu Q-S, Srivastava AK, Cao M-Q, Wang J (2014) Mycorrhizal function on soil aggregate stability in root zone and root free hyphae zone of trifoliate orange. Arch Agron Soil Sci. https://doi.org/10.1080/03650340.2014.95226

Wu Q-S, Li Y, Ying- Ning Z, He X-H (2015) Arbuscular mycorrhiza mediates glomalin related soil protein production and soil enzyme actuates in the rhizosphere of trifoliate arrange gowns under different planners. Mycorrhiza 25:121–130

Wu Q-S, Sun P, Srivastava AK (2017) AMF diversity and in citrus rhizoshere. Indian J Agric Sci 87(5):653–656

Yamada K, Xu H-L (2000) Properties and application of an organic fertilizers inoculated with effective microbes. J Crop Prod 3(1):255–268

Yasin G, Nawaz MF, Zubair M, Qadir I, Saleem AR, Ijaz M, Ss G, Bashir MA, Rehim SU, Du Z (2021) Assessing the contribution of citrus orchards in climate change mitigation through carbon sequestration in Sargodha district, Pakistan. Sustainability 13(2):12412. https://doi.org/10.3390/sus/32212412

Zanotelli D, Montagnani L, Manca G, Tagliavini M (2015) Net ecosystem carbon balance of an apple orchard. Eur J Agron 63:97–136

Zhang Q, Yang Z, Li Y, Chen D, Zhang J, Chen M (2010) Spatial variability of soil nutrients and GIS-based nutrient management in Yongji County, China. Int J Geogr Inf Sci 24(7):965–981

Zhang Y, Srivastava AKS, Chi T, Rao G, Wu QS (2014) Evaluation of tolerance to abiotic stress in citrus spp. Curr Hortic 5(1):15–19

Zhang Z-Z, Srivastava AK, Wu Q-S, Li G-H (2015) Growth performance and rhizosphere traits of peach (*Prunus persica*) in response to mycorrhization on replant versus non-replant soil. Indian J Agric Sci 85(10):125–130

Zhao MQ, Li M, Shi YF (2014) Carbon storage and carbon dioxide sequestration of banana plants at different growth stages. Adv Mater Sci 1010(1012):662–665

Zou YN, Srivastava AK, Wu Q-S, Huang YM (2014) Increased tolerance of trifoliate orange (*Poncirus trifoliata*) seedlings to waterlogging after inoculation with arbuscular mycorrhizal fungi. J Anim Plant Sci 24(5):415–420

Zou YN, Chen X, Srivastava AK, Wang P, Xiang L, Wu Q-S (2016) Changes in rhizoshere properties of trifoliate orange in response to mycorrhization and sod culture. Appl Soil Ecol 107:307–312

Chapter 2
Silicon: The Only Element in Plant Nutrition with a Mitigating Effect on Multiple Stresses

Renato de Mello Prado, Deyvielen Maria Ramos Alves, and Anelisa de Aquino Vidal Lacerda Soares

Abstract Silicon (Si) emerges as one of the most relevant elements in plant nutrition for mitigating a wide range of stresses. This review aimed to reinforce the benefits of Si in agriculture, specifically addressing its potential effect on the mitigation of individual and multiple stresses and discussing future research perspectives in this field. Future perspectives suggest that research should deepen studies involving the effects of Si on the soil-microbiota-plant system.

Keywords Silicon · Nano-silica · Mitigator · Deficiency · Toxicity · Nutrient · Multi-stress · Combined stress

2.1 Introduction

Silicon (Si) mitigates biotic or abiotic stresses (Etesami and Jeong 2023). Several reviews address this topic (Thorne et al. 2020; Mir et al. 2022). Thus, Si emerges as one of the most relevant elements in plant nutrition for mitigating a wide range of stresses (Thorne et al. 2020). However, studies have typically evaluated only one stress at a time (Cooke and Leishman 2016), with few studies reviewing its mitigating effect under multiple stresses occurring simultaneously. Therefore, the real benefit of Si in agriculture may be underestimated (Moraes et al. 2022), given the focus on understanding its isolated responses (Cooke and Leishman 2016), sometimes neglecting its capacity to mitigate many simultaneous stresses.

R. de Mello Prado (✉) · D. M. R. Alves
Department of Soil Science, São Paulo State University, Jaboticabal, São Paulo, Brazil
e-mail: rm.prado@unesp.br

A. de Aquino Vidal Lacerda Soares
Agency Paulista Agribusiness Technology-Regional, Bauru, São Paulo, Brazil

R. de Mello Prado et al. (eds.), *Silicon Advances for Sustainable Agriculture and Human Health*, Sustainable Plant Nutrition in a Changing World,
https://doi.org/10.1007/978-3-031-69876-7_2

Plants can employ distinct physiological and biochemical mechanisms to cope with multiple stresses simultaneously (Chapin et al. 1987) compared to individual stressors (Cooke and Leishman 2016). From the initial perception of the stimulus (stress) to the final response in cells, plants use several signalling pathways that vary depending on the challenge (Atkinson and Urwin 2012).

Research involving multiple stresses is notably complex (Nguyen et al. 2016). Moreover, studies conducted in different environments will likely involve interactions with additional stresses beyond the specific one being studied (Maestre et al. 2005). However, it is still unclear whether simultaneous stresses are antagonistic and synergistic or induce susceptibility to a specific type of stress (Pandey et al. 2015). Thus, a new stress may or may not make the plant more resistant to the second stress, inducing or not inducing an increase in the plant's defence system (Roy 2012).

Therefore, studies must include multiple stresses to understand the importance of this element for cropping systems. It is relevant because multiple stresses occur simultaneously, which is common in the field, especially in recent decades. In this context, this review aimed to reinforce the benefits of Si in agriculture, specifically addressing its potential effect on the mitigation of individual and multiple stresses and discussing future research perspectives in this field.

2.2 The Impact of Silicon on Mitigating Isolated Biotic or Abiotic Stressors

Silicon (Si) is widely available in the soil, and typically, plant uptake exceeds the nutrients in several species (Epstein 1994). Despite this, Si has never been considered a nutrient for higher plants, as it does not meet specific criteria of essentiality, such as the plant's inability to complete its life cycle in the element's absence (Kirkby 2011). However, due to its numerous beneficial effects on plants, Si has been recognized as one of the most versatile mineral elements, particularly under stress conditions (Mandlik et al. 2020).

Cultivated plants are exposed to several biotic factors (diseases and pests) and abiotic factors (salinity, temperature, wind, water deficit or excess, mineral deficiency, or heavy metal toxicity), often in combinations. Thus, plants undergo a huge combinatorial complexity with diverse environmental conditions that can cause stress (Farooq and Dietz 2015). However, they activate a series of defence mechanisms (Khan et al. 2021), and Si can modulate stress responses (Campos and Silva 2023; da Silva and de Prado 2023).

In recent years, researchers have been striving to understand the benefits of Si, particularly in plants with a high ability to absorb this element, such as rice and sugarcane (Ma and Takahashi 2002; Matichenkov and Calvert 2002; Teixeira and de Mello Prado 2023). Furthermore, they have been seeking to elucidate the vital role Si plays in alleviating environmental stresses in high-yielding crops, including cereals, legumes, and leafy vegetables. Silicon uptake and accumulation capacity in plants depends on the species and can be an important determinant factor of stress tolerance potential (Ranjan et al. 2021).

Studies demonstrated that Si supplementation reduces heavy metal toxicity in rice and corn (Ranjan et al. 2021; Biswas et al. 2023) and mitigates the negative effects of iron deficiency in common bean (Idoudi et al. 2024) and sugarcane (Teixeira et al. 2023) and ammonia toxicity in plants sensitive to high N concentrations (Campos et al. 2020). Silicon application in wheat increases stress tolerance to high temperatures (Tripathi et al. 2017) and low temperatures in sugarcane (Bocharnikova and Matichenkov 2023).

Several studies focus on the effects of Si in mitigating abiotic stresses in crops in general (Kaur et al. 2015; Patil et al. 2017; Seal et al. 2018; Meena et al. 2022; Čermelj et al. 2021; Mir et al. 2022; Yadav et al. 2023; Campos and Silva 2023) and in specific crops such as fruit trees (Muhammad et al. 2022), sugarcane (Verma et al. 2020; Majumdar and Prakash 2020), and rice (Meharg and Meharg 2015). However, these studies have focused on discussing the importance of Si involving only a single type of stress.

Abiotic stresses, such as drought, salinity, cold, heat, and UV-B radiation, adversely affect growth, resulting in significant yield losses in cultivated plants (Pareek et al. 2010). Moreover, cultivated plants have shown possible stress alleviation mechanisms associated with Si, directly or indirectly, across a wide range of abiotic stress tolerance (Ranjan et al. 2021).

Drought and salinity are the most frequent stress triggers (Ali et al. 2020). Furthermore, drought can threaten global food security (Ullah et al. 2017; Ilyas et al. 2021). Studies report yield losses of around 40% in maize and wheat, primarily due to water stress (Daryanto et al. 2016). One possible Si-mediated tolerance mechanism includes water balance through osmotic adjustment, reducing water loss by transpiration due to the deposition of monosilicic acid in the xylem, cuticle, and stomata of leaves and in bracts of the inflorescence. This Si deposition reduces leaf and panicle transpiration, increases leaf blade area, and promotes better photosynthetic efficiency, favouring plant tolerance to drought, extreme temperatures, and excessive light stress (Li et al. 2018).

Some studies have shown that in roots, Si increases water uptake area and prevents loss to the soil by modulating aquaporins, mainly nodulin 26-like intrinsic proteins (NIPs) and subfamilies of plasma membrane intrinsic proteins (PIPs) (Talakayala et al. 2020; Wang et al. 2021). Other studies suggest that the benefits derived from Si in plants under water stress may be due to symbiosis with arbuscular mycorrhizal fungi. These fungi accumulate Si in their spores and hyphae and transfer it to the roots of host plants, favouring Si and another nutrient uptake (Vega et al. 2023).

In pre-sprouted sugarcane seedlings, Si can be considered a feasible alternative to alleviate the damage caused by the water deficit imposed after transplanting (Teixeira et al. 2020). Regarding canola (*Brassica napus* L.), an important oilseed crop, the study conducted by Moghadam et al. (2022) revealed that Si applied at a dose of 4 g L^{-1} improved plant tolerance to drought in the final period of the growing season.

A study evaluating tebuconazole and nano-silica doses revealed that the applications of these two products increased the relative water content in the leaves, calcium (Ca) content in the roots, and sugar yield. The authors emphasized that, under severe water stress conditions, the significant effects of nano-silica on sugar production are solely due to the improvement of root quality characteristics (Namjoyan et al. 2021).

Another study on the use of fertilizers and water deficit in rice indicated that supplementation with Si alone or combined with the application of nitrogen (N), phosphorus (P), and different ratios of potassium (K) promoted growth, increased grain yield, and improved the plants' physiological characteristics. The authors further emphasized that the inclusion of K and Si with N and P, where the participation of K is at least 50%, could mitigate the harmful impact of water deficit stress (Das et al. 2023).

Research also shows that foliar application of Si on cowpea can mitigate water stress. Thus, applying proline and Si, either together or separately, improved water status, osmoregulation, antioxidant activity, and plant growth maintenance (Santos et al. 2022).

Silicon can mitigate thermal stress, which represents one of the most limiting abiotic risks for plant production, leading to reduced agricultural output (Silpha et al. 2023; Ahmad et al. 2024).

The protective role of Si in thermotolerance may be related to alleviating the impacts caused by this stress by maintaining cellular turgor, membrane stability, and cellular organelle ultrastructure, increasing osmoprotectant accumulation, and improving the efficiency of both enzymatic [ascorbate peroxidase (APX), superoxide dismutase (SOD), and catalase (CAT)] and non-enzymatic (glutathione, tocopherols, carotenoids, and ascorbate) antioxidant defence systems. Silicon also protects the photosynthetic apparatus through its deposition as phytoliths in the leaves, acting as a sunscreen, reducing transpiration, improving gas exchange, and increasing nutrient uptake under stress (Souri et al. 2021; Saha et al. 2021; Parveen et al. 2022; Bishnoi et al. 2023).

In tomato plants, Si applied in the form of sodium metasilicate mitigated the adverse impact of thermal stress by increasing biomass, photosynthetic pigments, and relative water content and stimulating the expression and accumulation of transcripts of heat shock proteins (Khan et al. 2020a, b). In soybeans, the application of the same source maintained osmotic balance under thermal stress by accumulating high levels of proline and glycine betaine (GB), modulating photosynthesis, accumulating additional osmolytes, and regulating the antioxidant system (Sharifi et al. 2022). In wheat, both potassium silicate and application of Si dioxide nanoparticles improved the thermotolerance of the photosynthetic apparatus through positive regulation of genes PsbH, PsbB, and PsbD encoding proteins central to photosystem II (Hassan et al. 2021).

Cold temperatures and frost also negatively impact crop yields. Silicon supply alleviated this stress in sugarcane plants after 6 hours of cold exposure, possibly by reinforcing their immune systems (Bocharnikova and Matichenkov 2023).

In the case of soybean crops grown in cold conditions (8–10 °C), Si decreased oxidative stress, promoting initial growth in the crop (Ahmad et al. 2024). These authors also observed that Si negatively regulated genes in the abscisic acid pathway and osmotic regulation (9-cis-epoxycarotenoid dioxygenase and dehydration-responsive element-binding protein), increasing plant tolerance to low temperatures.

Furthermore, Si also improves the photosynthetic efficiency of plants under UV-B radiation stress by maintaining upright leaf blades (Ma et al. 2001; Perez et al. 2014). The addition of Si also increases chlorophyll content, soluble sugars, anthocyanins, flavonoids, antioxidant compounds, and UV-absorbing substances (Čermelj et al. 2021; Silpha et al. 2023). Biosilicified structures deposited in epidermal leaf cells can reduce heat through a passive cooling mechanism under conditions of high solar irradiation (Wang et al. 2005; Silpha et al. 2023). In soybeans, UV-B radiation severely limited plant growth. However, Si application mitigated stress by increasing leaf area and shoot-to-root ratio and promoting the antioxidant activities of CAT, SOD, and peroxidases (Shen et al. 2010).

Another stress that limits plant growth and productivity is salinity, which causes nutritional imbalance and oxidative, osmotic, and ionic stress, severely impacting fundamental physiological processes in plants. Silicon-mediated tolerance occurs through the regulation of the Na^+/K^+ ratio. In roots, decreased sodium (Na) uptake and transport to the aerial part is attributed to the stimulation of H^+-ATPase activity induced by Si in the plasma membrane and tonoplast, facilitating the export of Na^+ out of the cell and reducing apoplastic transport of Na^+ and Cl^- (Liang et al. 2005; Dhiman et al. 2021). Silicon can also mediate salt stress tolerance by improving the photosynthetic process, maintaining redox homeostasis, and favouring nutrient uptake (Liu et al. 2019).

In fruit trees, the application of Si nanoparticles on leaves and roots inhibited Na^+ and Cl^- uptake by stimulating aquaporins. Silicon also increased root mass and volume under salinity, favouring nutrient uptake and reducing the production of toxic reactive oxygen species (Muhammad et al. 2022).

In wheat cultivation, the combination of Si dioxide nanoparticles and the co-inoculation of mycorrhizae and rhizobacteria were also effective in reducing the negative effects of salt stress from NaCl, improving physiological parameters (Ahmadi-Nouraldinvand et al. 2024). Still, regarding wheat cultivation, Nadeem et al. (2022) observed that silicic acid proved superior to calcium, potassium, and sodium silicates in improving vigour and increasing yield and biochemical processes under salinity stress.

Silicon can also mitigate heavy metal toxicity, with both external and internal effects on plants. External effects include metal inactivation by altering its behaviour upon mixing with silicate materials, co-precipitation of toxic cationic metals with available Si for plants, or complexation in the soil (Bhat et al. 2019; Meena et al. 2022; de Oliveira et al. 2023). For example, the addition of Si in soils with higher concentrations of toxic aluminium (Al^{+3}) results in the formation of aluminium silicate or hydroxyl-aluminium silicate precipitates in solution or at the root apex (Hodson and Evans 2020; Xiao et al. 2021).

Silicon reduced Al toxicity in sugarcane by limiting uptake and transportation to the aerial part, altering root morphoanatomy and promoting greater plant growth (Sousa Junior et al. 2024). Another possibility would be using Si minerals, such as diatomite, which have an alkaline nature and an exchangeable surface that carries more cations such as Na^+ and K^+ than H^+. Thus, when the mineral Si is applied to the soil, the hydrolysis of exchangeable cations becomes stronger than that of H^+ and produces more NaOH in the soil solution, increasing the pH value (Ma et al. 2021) and, consequently, reducing the availability of copper (Cu), cadmium (Cd), chromium (Cr), lead (Pb), iron (Fe), manganese (Mn), and zinc (Zn) (Khan et al. 2021).

The internal effects of Si in plants to mitigate metal toxicity comprise a set of mechanisms, starting with the reduction in metal uptake by the root, translocation to the shoot, the formation of a physical barrier, chelation, and connections between molecular, biochemical, and physiological pathways (Meena et al. 2022; Khan et al. 2021; de Oliveira et al. 2023).

The deposition of silica near the endodermis of the root, for example, acts as a physical barrier and restricts the apoplastic transport of Cd and Na^+. Meanwhile, competition of monosilicic acid with arsenate may decrease the uptake of this element by plants under anaerobic conditions (Li et al. 2018). Silicon also stimulates the production of root exudates that can chelate metals, reduces their uptake by roots, and further reduces the toxicity of metal ions by decreasing the solubility of these metals in the cell wall (Qi et al. 2020; Rajan et al. 2021). Silicon mitigates the excess of macronutrients such as P, which is poorly studied (Alves et al. 2023; Melo et al. 2023).

The regulation of phytohormones mediated by Si stimulates the plant's antioxidant defence system, mitigating stress caused by heavy metals (Zhang et al. 2013; Asgher et al. 2024). When studying several crops, Mir et al. (2022) showed abscisic acid (ABA) levels in Si-treated plants decreased under saline stress. However, they increased when plants were exposed to heavy metals such as Cd and Cu. In rice plants treated with Si and exposed to heavy metals, Lee et al. (2010), Kim et al. (2014), and Liang et al. (2015) observed that the concentration of ABA increased in the early days of development and that after the 14th day, both ABA levels and salicylic acid (SA) and jasmonic acid (JA) decreased.

The application of Si mitigated damages caused by Cd and arsenic (As) in maize due to increased antioxidant activity, resulting in productivity gains (Zhou et al. 2022; Kashif et al. 2021; Asgher et al. 2024). In cotton cultivation, there was a reduction in electrolyte leakage under Pb stress. In rice, Santos et al. (2020) observed that Si addition decreased Fe toxicity in roots and shoots, besides improving photosynthesis and grain yield. Imtiaz et al. (2016) reviewed the effects of Si application in mitigating Cr toxicity, highlighting the increase in plant growth and improvement in photosynthetic parameters.

The additional presence of Si available to plants can inhibit the excessive uptake of P in rice. However, under deficient conditions, Si promotes this nutrient's uptake (Li et al. 2018). The combined supplementation of SA and Si reduces damages caused by boron (B) toxicity by increasing carotenoid and chlorophyll levels and biomass production and reducing oxidative stress in spinach (Eraslan et al. 2008). In beetroot, cauliflower, broccoli, and papaya seedlings, Si mitigated physiological stress caused by excess ammonia (Barreto et al. 2017; Viciedo et al. 2019; Silva et al. 2023) and by Na in sorghum and sunflower (Hurtado et al. 2019).

Among abiotic stresses, nutritional disorders also stand out (Sales et al. 2021), and Si can mitigate them by impacting the uptake, translocation, and availability of many plant nutrients (Prado 2021, 2023) and promoting C/N/P homeostasis (Costa et al. 2023, 2024).

Under nutritional deficiency, Si increases the expression of macronutrient and micronutrient transporters in the root. It promotes translocation to the shoot, further modulating metabolic pathways and improving photosynthetic efficiency (Ali et al. 2020). Wu et al. (2017) demonstrated that Si induced the expression of genes involved in N uptake and translocation under low N supply. However, the expression of the OsAMT1;1 and OsGS1;1 genes was suppressed under high N supply. Silicon also alleviates visual symptoms of N deficiency by increasing the shoot and root dry matter production in quinoa (Sales et al. 2021).

Silicon can increase the availability of P through changes in pH value, reduction of P sorption by soil minerals due to competition between P and Si, or alterations in the dynamics of the microbial community. However, all these effects will depend on the soil type and microbial activity, pH, and the amount of Si supplied (Greger et al. 2018).

In plants, the main mechanism to mitigate P deficiency provided by Si is related to the increased use of this nutrient in plant tissues, as Si can improve the gene expression of P transporters and, consequently, uptake (Pavlovic et al. 2021). The effect of Si under limiting P conditions has been reported in sorghum (Silva and Prado 2021), wheat, rice, maize (Hu et al. 2021), tomato (Zhang et al. 2019), and potato (Soratto et al. 2019). Silicon can also improve the uptake of many other nutrients besides P in oak plants, a European deciduous tree species (Kostic et al. 2023).

Silva et al. (2021) found that Si added to the nutrient solution alleviated Ca deficiency by reducing cell leakage, increasing the content of ascorbic acid and dry matter, and providing firmer leaves due to less water loss after harvest.

The application of Si can mediate K accumulation in the shoot in situations where plants undergo saline stress or when there is K deficiency (Li et al. 2018; Sarah et al. 2021). Silicon mitigated K deficiency in barley crops (Benslima et al. 2022) and arugula (*Eruca sativa*) (Alves et al. 2024). Buchelt et al. (2020) also highlighted the role of Si in mitigating the effects of K, Ca, and magnesium (Mg) deficiency in forage crops due to improved efficiency in using these nutrients.

Silicon can also alleviate Fe deficiency in cucumber, soybean, and sorghum. However, this effect is specific to the plant, the applied dose, and the soil pH (Li et al. 2018; Teixeira et al. 2020; Pavlovic et al. 2021). In oranges, Si nutrition reduced the harmful effects of soil Fe deficiency in environments with higher pH, as observed by the increased Fe content in plants and the reduction of oxidative stress (Mierzaali et al. 2024).

Silicon mitigated the effects of Mn deficiency stress in sorghum plants, and root application was more effective than foliar spraying. The beneficial effect of Si was evidenced by increased SOD activity, reducing oxidative stress, with repercussions on photosynthesis, leaf area, Mn use efficiency, and dry matter production of the plants (de Oliveira et al. 2019).

Silicon mitigated the effects of both B deficiency and toxicity in cotton. The results indicated that foliar-applied Si was crucial for alleviating B deficiency. Meanwhile, Si supplied through the nutrient solution was more effective against B toxicity (Souza Júnior et al. 2019).

Silicon applied to the roots or leaves may promote a better distribution of Zn in the plant, decreasing the effects of nutritional deficiency. Guedes et al. (2022) observed that foliar-applied Si increased Zn accumulation, photosynthetic pigments, and shoot and root dry matter while decreasing cell leakage. In soybeans, Pascual et al. (2016) suggested that adding 0.05 mM of Si to the nutrient solution improved both Zn accumulation in roots and its movement to the shoot, alleviating deficiency symptoms. In cucumber, Lozano-González et al. (2021) highlighted that Si application could reduce the effects of Zn deficiency due to its rapid recovery after the stress period.

Besides mitigating abiotic stresses, Si also plays a crucial role in mitigating biotic stresses, including herbivory, pathogens, and pests (Li et al. 2018). Furthermore, Si has a plant defence-inducing role and is well-known for enhancing crop resistance and reducing damage caused by many pathogenic organisms, thus being efficient in pest and disease management in the field (Khan et al. 2021; Souri et al. 2021; Wang et al. 2020).

Silicon assists plants in improving their performance when subjected to biotic stresses, exerting a physical action, a biochemical defence, and a molecular action (Majundar and Prakash 2020; Souri et al. 2021; Verma et al. 2021a, b; Naidu et al. 2023).

Damage caused by pathogen invasion and insect herbivory can be mitigated by the presence of Si in the plant shoot, comprising morphological barriers of phytoliths in the stem and leaves, promoting resistance mechanisms by increasing tissue hardness in these organs and by developing other physical structures, such as trichomes, which aid in defence (Murali-Baskaran et al. 2021) and weaken the mouthparts of herbivores (Verma et al. 2021a, b; Naidu et al. 2023).

Soluble Si in the cytosol triggers many metabolic pathways by modulating molecular, physiological, and biochemical reactions and signal transduction in which the phytohormones ABA, ethylene (ETH), JA, and SA play an important role (Islam et al. 2023; Song et al. 2021). Silicon also increases the activity of enzymes related to antioxidant defence, such as polyphenol oxidase, glucanase, peroxidase, and phenylalanine ammonia-lyase (PAL). Moreover, it increases the activity of antimicrobial compounds, such as phenolics, flavonoids, phytoalexins, and disease-related proteins in plants (Mvondo-She et al. 2021; Gulzar et al. 2024). The combination of these physical and biochemical processes, mediated by Si, improves plant defence against biotic stresses (Bakhat et al. 2018).

Extensive scientific studies detail the mechanisms involved in resistance to biotic stresses in plants considered accumulators and non-accumulators. Mvondo-She et al. (2021) surveyed the role of Si and potential mechanisms in mitigating citrus diseases. Meanwhile, Verma et al. (2021a, b) published an important review on the influence of Si on biocontrol strategies in several crops. Other research discussed the physiological and molecular bases in improving biotic stress conditions in plants provided by Si (Song et al. 2021) and Si-mediated resistance to herbivorous insects (Alhousari and Greger 2018; Bakhat et al. 2018; Singh et al. 2022).

In sugarcane, known for accumulating high amounts of Si in its tissues, Majumdar and Prakash (2020) elaborated a review providing an overview of the potential of this element in promoting defence against biotic and abiotic stresses for this crop. As a sustainable option, Si nanoparticles (SiNPs) have enormous potential to alter plant growth and development, increase the expression of stress-responsive genes related to defence and enzymatic expression, impact plant morphological and metabolic behaviour, and provide greater resistance against fungi, bacteria, nematodes, and pests (Naidu et al. 2023).

Silicon has even been considered as effective as conventional fungicides in controlling rice diseases, such as leaf scald (*Monographella albescens*), blast (*Magnaporthe grisea*), brown spot (*Cochliobolus miyabeanus*), grain discolouration, and other diseases (Jinger et al. 2017).

Silicon has also increased the detoxification capacity of reactive oxygen species (ROS), modulating antioxidant defence systems and the expression of key genes associated with oxidative stress mitigation and hormonal metabolism (Mostofa et al. 2021), contributing to reducing diseases. In tomato plants, the application of silica nanoparticles for controlling infection by *Ralstonia solanacearum* resulted in increased activities of peroxidase (POD) and CAT enzymes. It positively regulated the expression of genes related to salicylic acid (PR-1, PR-5, and PAL), which provided greater tolerance of plants to bacterial wilt (Wang et al. 2022).

This review clearly demonstrated Si's potential to mitigate a large number of stresses through many action mechanisms. However, these studies predominantly focus on assessments of isolated stresses. Thus, Si stands out as the only element in plant nutrition with extensive literature on the mitigation of dozens of stresses. Consequently, it can be hypothesized that it can mitigate combined stresses. Nevertheless, it needs to be further discussed and understood.

2.3 Impact of Si on the Mitigation of More than One Simultaneous Stress: Whether Abiotic or Biotic

A study conducted in the 1990s comprises one of the earliest reports describing the beneficial effects of Si on plants grown under two combined stresses (water deficit and heat stress) (Agarie et al. 1998). Results indicate that Si alleviated the effects of these stresses due to an increase in leaf polysaccharides and a decrease in the electrolyte leakage index.

Nicolau et al. (2024) evaluated two sugarcane cultivars grown under two stress conditions: leaf scalding and water deficit. They found that Si mitigated both isolated and combined stresses by reducing the malondialdehyde (MDA) content by incrementing different antioxidant enzymes (GPOX, APX, GSH-Px).

Most studies on multiple stresses are limited to dual stresses and have focused on Si-accumulating species under controlled conditions in pots. They mainly focus on abiotic stresses, such as the combination of salinity and water deficit. Some include salinity with potentially toxic heavy metal toxicity and B. Meanwhile, others examine thermal stress and pest or pathogen attacks (Table 2.1).

In situations of combined water and saline stresses in *Glycyrrhiza uralensis*, the application of Si benefited this plant by maintaining the integrity of the anatomical structure and increasing chlorophyll content, photosynthetic rate, transpiration rate, stomatal conductance, and water use efficiency (Zhang et al. 2020).

In another study, Zhang et al. (2019) evaluated the effect of Si on *G. uralensis* in mitigating water deficit and salinity. This element alleviated these dual stresses by reducing Na content and ionic toxicity, improving the plant's water status and maintaining membrane stability by decreasing MDA content. Consequently, it increased plant growth and productivity (Zhang et al. 2019).

In *Glycyrrhiza uralensis* seedlings, the application of Si under combined drought and salinity stress benefited this species by increasing proline content and decreasing hydrogen peroxide (H_2O_2) and MDA, indicating that this element reduced lipid peroxidation through the proline pathway (Cui et al. 2021).

In wheat plants, Si mitigated the combined saline and water deficit stress by enhancing water relations, chlorophyll contents, and the activity of the enzymes SOD, CAT, and POD (Sattar et al. 2018).

Another study was conducted on cucumber plants to evaluate the impact of Si nanoparticles on plants grown under different water regimes and irrigation water salinity. The authors indicated that Si application mitigated this dual stress by reducing Na uptake and, consequently, improving growth, fruit yield, and the efficiency of N and K uptake (Alsaeedi et al. 2019).

Farooq et al. (2015) studied two contrasting rice varieties regarding salinity tolerance, applying Si to mitigate B toxicity and salinity. The authors found that Si application mitigated both stresses by reducing Na uptake, maintaining a balanced K/Na ratio, and increasing photosynthetic efficiency, transpiration rate, leaf water content, and the activity of antioxidant enzymes such as CAT and APX.

Bean crops were cultivated in the field on saline soil contaminated with different heavy metals to investigate the mitigating role of two foliar-supplied Si sources. Biological silica nanoparticles and potassium silicate increased the uptake of N, P, K, and Ca and decreased the uptake of Na, Pb, Cd, and Ni due to the increased activity of the antioxidant enzymes POD, CAT, and APX (El-Saadony et al. 2021). Another similar study aimed to test the mitigating potential of Si and proline under the same salinity condition but with contamination only by Cd. This study observed that the effect of Si with a single application, especially when associated with proline, promoted an increase in pigment content, free proline, glutathione, relative water content, and decreased electrolyte leakage index and MDA (Rady et al. 2019).

Silicon mitigated the combined effect of salinity and Cd toxicity in date palms (*Phoenix dactylifera* L.), as it decreased Cd uptake and increased nutrient uptake. It also reduced hormones associated with stress responses by increasing the expression of genes related to proline transport (proline transporter 2) and energy

Table 2.1 Benefits of Si application in different plant species subjected to two concurrent stressful conditions

Species	Trial	Si content	Application method	Stress	Main improved parameters	Reference
Glycyrrhiza uralensis	Pot	0.1 g	Root	Salinity and water deficit	Leaf anatomy and ultrastructure	Zhang et al. (2020)
Glycyrrhiza uralensis	Petri dishes	2 mM	Seed	Salinity and water deficit	Decreased P5Cs activity and lipid peroxidation	Cui et al. (2021)
Glycyrrhiza uralensis	Pot	0.1 g kg^{-1}	Root	Salinity and water deficit	Increased Na/K ratio, sugars, proline, protein, and antioxidant enzyme activity and decreased oxidative stress	Zhang et al. (2019)
Triticum aestivum	Pot	50 mM	Foliar	Salinity and water deficit	Morphology, water relations, photosynthetic pigments, gas exchange, and antioxidant enzyme activities	Sattar et al. (2018)
Cucumis sativus	Pot	0 to 400 mg kg^{-1}	Root	Salinity and water deficit	Increased height, chlorophyll content, yield, and N and K uptake and decreased Na content	Alsaeedi et al. (2019)
Oryza sativa L.	Pot	0 and 150 mg kg^{-1}	Root	Salinity and boron toxicity	Decreased toxic ions and increased photosynthetic efficiency	Farroq et al. (2015)
Phaseolus vulgaris L.	Field	Bio-Si-NPs (2.5 and 5.0 mmol L^{-1}) and soluble Si (10 mmol L^{-1})	Foliar	Salinity and heavy metal toxicity	Increased productivity and decreased cellular damage and toxic elements	El-Saadony et al. (2021)
Phaseolus vulgaris L.	Pot	6 mM	Foliar	Salinity and cadmium toxicity	Physiological, morphological, and biochemical attributes	Rady et al. (2019)

(continued)

Table 2.1 (continued)

Species	Trial	Si content	Application method	Stress	Main improved parameters	Reference
Chenopodium quinoa Willd.	Pot	Biochar +350 mL of Si	Soil	Arsenic and salinity stress	Antioxidant activity	Alsamadany et al. (2022)
Phoenix dactylifera L.	Pot	1.0 mM	Root	Cadmium and salinity stress	Decreased Cd and increased expression of the proline transporter 2	Khan et al. (2020a, b)
Cucumis sativus L.	Pot	200 mg L^{-1}	Foliar	Salinity and heat stress	Antioxidant capacities, fruit yield, and quality	Shalaby et al. (2021)
Glycine max L. cv. 'Zhonghuang 35'	Pot	1.70 mM	Root	Ultraviolet-B radiation and water deficit	Increased dry matter and distribution of mineral elements in seedlings	Shen et al. (2014)
Oryza sativa L.	Pot	100 ppm	Root	Water deficit and heat stress	Increased desiccation tolerance, thermal stability of membrane lipids, and decreased electrolyte leakage	Agarie et al. (1998)
Triticum aestivum	Pot	500 mg kg^{-1}	Root	Water deficit and heat stress	Increased chlorophyll, carotenoids, antioxidant activity, and nutrients and decreased electrolyte leakage, Na, and Cl	Aouz et al. (2023)
Hordeum vulgare cv. 'Irina'	Pot	0.5 and 1 mM	Root	Osmotic stress and low or high potassium supply	Increased starch, modulation of glycolytic pathways and tricarboxylic acid cycle, ABA, and cytokinin homeostasis	Hosseini et al. (2017)

Saccharum spp.	Pot	124 g; 4000 ppm	Root	Insect attack and water deficit	Reduced number of borers and damage to the stem	Kvedaras et al. (2007)
Triticum aestivum L. 'Borlaug-16'	Pot	Si 0.5 mM and H_2S 0.15 mM + Si 0.3 mM	Seed	Water deficit and infection with *Puccinia triticina*	Increased growth, pigments, and defence enzyme activities and reduced disease and oxidative stress	Naz et al. (2021)
Saccharum spp.	Pot	2 mmol L^{-1}	Root	*Xanthomonas albilineans* stress and water deficiency	Decreased MDA and increased enzyme activity	Nicolau et al. (2024)
Oryza sativa L.	Field	0 to 300 kg ha^{-1} of SiO_2	Root	Pest insects and diseases.	Reduced stem borer damage and leafhopper infestation	Han et al. (2018)

metabolism (GAPDH) and decreasing the expression of the abscisic acid receptor gene (ABA receptor) (Khan et al. 2020a, b).

Biochar doped with Si nanoparticles (SBC) is more effective than simple biochar (without Si) (BC) in alleviating the stress caused by salinity and arsenic (As) toxicity in quinoa plants. The addition of SBC to contaminated soil resulted in increased grain productivity, decreased oxidative stress, and reduced levels of As in the grain, thus reducing risks to human health (Alsamadany et al. 2022).

Cucumbers grown under combined salt and heat stress received the addition of anti-stressors, including Si. Silicon increased chlorophyll content, the activity of antioxidant enzymes CAT and POD, photosynthetic efficiency (Fv/Fm), and fruit production (Shalaby et al. 2021).

Another study with soybean exposed to combined water deficit and ultraviolet radiation (UV-B) found that Si mitigated these stresses by increasing K, Zn, and Mn levels in the roots and K, Ca, Mg, Fe, Zn, and B levels in the leaves and root and plant biomass (Shen et al. 2014).

In wheat, supplementation with Si reduced the effects of combined salt and heat stress. It occurred due to increased nutrient uptake, leaf water content, photosynthetic pigments, osmolyte accumulation, and antioxidant enzyme activity (POD, SOD, CAT, and APX) while decreasing the electrolyte leakage index. Consequently, wheat production increased (Aouz et al. 2023).

Using Si in barley mitigated the dual stress under water deficit and K deficiency by reducing leaf senescence by suppressing the senescence gene HvS40. Si modulates gene expression (Farouk 2022). Silicon also increased chlorophyll and starch content in the shoot, sucrose content in the root, sugar transport to the root, and carbon flux to metabolic pathways (Hosseini et al. 2017). Furthermore, Si decreased ABA levels in the shoot by modulating the transcriptional regulation of ABA biosynthesis and degradation pathways and/or increased active cytokinin levels and modulated the expression of genes related to the conversion of the intermediate product xanthoxin into ABA, such as HvSDR1 and HvSDR2, and the final step of ABA biosynthesis.

An investigation with four sugarcane cultivars studied the effect of Si on the infestation of *Eldana saccharina* Walker associated with water deficit. It was observed that plants treated with Si showed increased resistance to the attack by the *E. saccharina* borer compared to untreated plants, especially in susceptible sugarcane varieties (Kvedaras et al. 2007).

The Si applied to wheat under water deficit and with rust disease mitigated these stresses. It occurred because it reduced the incidence of leaf rust while increasing germination, plant biomass, root length, relative water content, protein content, photosynthetic pigments, and the activity of the antioxidant enzymes SOD, POD, PAL, and polyphenol oxidase (Naz et al. 2021).

In field conditions, Han et al. (2018) assessed the effects of Si on mitigating the incidence of insect pests and diseases in rice. They found that Si fertilization reduced the damage caused by stem borers, the population of leafhoppers, and the incidence of rice blast disease. Overall, these studies describe different mechanisms attributed to Si, ranging from modulation of gene expression associated with osmotic response and carbohydrate metabolism (Bhardwaj and Kapoor 2021; Thorne et al. 2020) to the interaction with hormones such as ABA, which signals stomatal closure and, consequently, reduces water loss (Anjum et al. 2011).

Interestingly, plants respond similarly to induce the defence system against drought and thermal stress (UV-B). Both stresses trigger a series of similar physiological responses due to the overproduction of reactive oxygen species resulting from water imbalance, exacerbating stress, and reducing photosynthetic efficiency (Mavrič Čermelj et al. 2021). This aspect may contribute to the understanding that Si is a potential mitigator of multiple stresses because its role in improving the plant's antioxidant defence system is clear.

In conclusion, future research must adjust to focus on multiple stress factors instead of individual stresses. Multiple stresses are more likely to be present in current agriculture (Gonzalez-Porras et al., 2024). Focusing on multiple stress factors is relevant because they impact global food security (Shalaby et al. 2021).

2.4 Conclusions and Future Perspectives

This review showed that the benefits of Si in stress mitigation are predominantly polarized in studies evaluating isolated stresses, with an extensive range of stresses under study.

Since the beginning of agricultural activity, there have been reports of different stresses in crops, whether abiotic or biotic. However, this situation has caused greater concern due to the increase in the number of multiple stresses in recent years. It has been driven by the intensification of the frequency of damage from climate change worldwide, as there are regions experiencing drought and/or increased temperature and atmospheric CO_2 concentration while others face floods. Other recent issues, such as wars, affect fertilizer-exporting countries, significantly increasing costs and reducing their use in agriculture, potentially causing stresses due to nutritional deficiencies, among other stressors. Furthermore, genetic diversity decreases with the expansion of genetically modified crop cultivation, which could diminish the plant's defence system against stresses, especially biotic ones, leading to extra concern.

In this context, research should advance to expand the use of Si in agriculture, considering the occurrence of multiple stresses. This review discussed that Si can mitigate dual stresses. However, the information is limited. It is polarized towards a few stresses, focusing on salinity, and a minimal number of species evaluated, predominantly addressing Si accumulators. Furthermore, studies do not focus on multidisciplinary approaches. There are standard reports on the benefits of Si in soil, including increasing pH value, reducing the availability of heavy metals, or increasing the availability of nutrients, such as P, and promoting the growth of beneficial microbial communities, thereby enhancing effects on the plant (Khan et al. 2021).

In contrast, current studies involving multiple stresses and chemical element toxicity did not consider these aspects (Table 2.1). Therefore, future perspectives suggest that research should deepen studies involving the effects of Si on the soil-microbiota-plant system. In plants, studies can integrate research from proteomic, physiological, and biochemical levels to improve the understanding of these benefits in processes of nutrient uptake, transport, and utilization by the plant, given the complexity of the effects of multiple stresses on plants. Therefore, research involving the benefits of Si in minimizing simultaneous stresses in agriculture would necessarily have to be at the field level, given the lack of research under these cultivation conditions.

References

Agarie S, Hanaoka N, Ueno O, Miyazaki A, Kubota F, Agata W, Kaufman PB (1998) Effects of silicon on tolerance to water deficit and heat stress in rice plants (Oryza sativa L.), monitored by electrolyte leakage. Plant Prod Sci 1(2):96–103. https://doi.org/10.1626/pps.1.96

Ahmad W, Coffman L, Weerasooriya AD, Crawford K, Khan AL (2024) The silicon regulates microbiome diversity and plant defenses during cold stress in Glycine max L. Front Plant Sci 14:1280251. https://doi.org/10.3389/fpls.2023.1280251

Ahmadi-Nouraldinvand F, Sharifi RS, Siadat SA et al (2024) Fascinating role of silicon dioxide nanoparticles and co-inoculation of mycorrhiza and rhizobacteria to combat NaCl stress: changes in physiological characteristics, uptake of nutrient elements, and enhancing photosystem ii activities in wheat. J Soil Sci Plant Nutr. https://doi.org/10.1007/s42729-024-01640-0

Alhousari F, Greger M (2018) Silicon and mechanisms of plant resistance to insect pests. Plan Theory 7(2):33. https://doi.org/10.3390/plants7020033

Ali N, Réthoré E, Yvin J-C, Hosseini AS (2020) The regulatory role of silicon in mitigating plant nutritional stresses. Plan Theory 9:9121779. https://doi.org/10.3390/plants9121779

Alsaeedi A, El-Ramady H, Alshaal T, El-Garawany M, Elhawat N, Al-Otaibi A (2019) Silica nanoparticles boost growth and productivity of cucumber under water deficit and salinity stresses by balancing nutrients uptake. Plant Physiol Biochem 139:1–10. https://doi.org/10.1016/j.plaphy.2019.03.008

Alsamadany H, Alharby HF, Al-Zahrani HS, Alzahrani YM, Almaghamsi AA, Abbas G, Farooq MA (2022) Silicon-nanoparticles doped biochar is more effective than biochar for mitigation of arsenic and salinity stress in quinoa: insight to human health risk assessment. Front Plant Sci 13:989504. https://doi.org/10.3389/fpls.2022.989504

Alves DMR, de Oliveira JN, de Mello PR, Ferreira PM (2023) Silicon in the form of nanosilica mitigates P toxicity in scarlet eggplant. Sci Rep 13:9190. https://doi.org/10.1038/s41598-023-36412-w

Alves DMR, de Mello Prado R, Barreto RF (2024) Silicon and sodium attenuate potassium deficiency in Eruca sativa Mill. Food Chem 432:137225. https://doi.org/10.1016/j.foodchem.2023.137225

Anjum SA, Xie X, Wang LC, Saleem MF, Man C, Lei W (2011) Morphological, physiological and biochemical responses of plants to drought stress. Afr J Agric Res 6(9):2026–2032. https://doi.org/10.5897/AJAR10

Aouz A, Khan I, Chattha MB, Ahmad S, Ali M, Ali I et al (2023) Silicon induces heat and salinity tolerance in wheat by increasing antioxidant activities, photosynthetic activity, nutrient homeostasis, and osmo-protectant synthesis. Plan Theory 12(14):2606. https://doi.org/10.3390/plants12142606

Asgher M, Rehaman A, Islam ul SN, Khan NA (2024) Multifaceted roles of silicon nano particles in heavy metals-stressed plants. Environ Pollut 341:122886. ISSN 0269-7491, https://doi.org/10.1016/j.envpol.2023.122886

Atkinson NJ, Urwin PE (2012) The interaction of plant biotic and abiotic stresses: from genes to the field. J Exp Bot 63(10):3523–3543. https://doi.org/10.1093/jxb/ers100

Bakhat HF, Bibi N, Zia Z, Abbas S, Hammad HM, Fahad S, Ashraf MR, Shah GM, Rabbani F, Saeed S (2018) Silicon mitigates biotic stresses in crop plants: a review. Crop Prot 104:21–34. https://doi.org/10.1016/j.cropro.2017.10.008

Barreto RF, Júnior AAS, Maggio MA, de Mello Prado R (2017) Silicon alleviates ammonium toxicity in cauliflower and in broccoli. Sci Hortic 225:743–750. https://doi.org/10.1016/j.scienta.2017.08.014

Benslima W, Ellouzi H, Zorrig W et al (2022) Beneficial effects of silicon on growth, nutrient dynamics, and antioxidative response in barley (Hordeum vulgare L.) plants under potassium deficiency. J Soil Sci Plant Nutr 22:2633–2646. https://doi.org/10.1007/s42729-022-00832-w

Bhardwaj S, Kapoor D (2021) Fascinating regulatory mechanism of silicon for alleviating drought stress in plants. Plant Physiol Biochem 166:1044–1053. https://doi.org/10.1016/j.plaphy.2021.07.005

Bhat JA, Shivaraj SM, Singh P, Navadagi DB, Tripathi DK, Dash PK, Solanke AU, Sonah H, Deshmukh R (2019) Role of silicon in mitigation of heavy metal stresses in crop plants. Plan Theory 8(3):71. https://doi.org/10.3390/plants8030071

Bishnoi A, Jangir P, Shekhawat PK, Ram H, Soni P (2023) Silicon supplementation as a promising approach to induce thermotolerance in plants: current understanding and future perspectives. J Soil Sci Plant Nutr 23:34–55. https://doi.org/10.1007/s42729-022-00914-9

Biswas A, Pal S, Paul S (2023) Silicon as a powerful element for mitigation of cadmium stress in rice: a review for global food safety. Plant Stress 100237. https://doi.org/10.1016/j.stress.2023.100237

Bocharnikova E, Matichenkov V (2023) Silicon-induced mitigation of low-temperature stress in sugarcane. In: Verma KK, Song XP, Rajput VD, Solomon S, Li YR, Rao GP (eds) Agro-industrial perspectives on sugarcane production under environmental stress. Springer, Singapore. https://doi.org/10.1007/978-981-19-3955-6_12

Buchelt AC, Teixeira GCM, Oliveira KS, Rocha AMS, de Mello Prado R, Caione G (2020) Silicon contribution via nutrient solution in forage plants to mitigate nitrogen, potassium, calcium, magnesium, and sulfur deficiency. J Soil Sci Plant Nutr 20:1532–1548. https://doi.org/10.1007/s42729-020-00245-7

Campos CNS, Silva Júnior, GBD, Prado RDM., David CHOD, Souza Junior, JPD, Teodoro PE (2020) Silicon mitigates ammonium toxicity in plants. J Agron, 112(2):635–647. https://doi.org/10.1002/agj2.20069

Campos CNS, da Silva BC (2023) Silicon mitigates the effects of nitrogen deficiency in plants. In: Benefits of silicon in the nutrition of plants. Springer International Publishing, Cham, pp 87–100. https://doi.org/10.1007/978-3-031-26673-7_6

Čermelj AM, Golob A, Vogel-Mikuš K, Germ M (2021) Silicon mitigates negative impacts of drought and UV-b radiation in plants. Plants 11(1):91. https://doi.org/10.3390/plants11010091

Chapin FS, Bloom AJ, Field CB, Waring RH (1987) Plant responses to multiple environmental factors. Bioscience 37(1):49–57. https://doi.org/10.2307/1310177

Cooke J, Leishman MR (2016) Consistent alleviation of abiotic stress with silicon addition: a meta-analysis. Funct Ecol 30(8):1340–1357. https://doi.org/10.1111/1365-2435.12713

Costa MG, Prado RM, Sarah MMS, Palaretti LF, Piccolo MC, Júnior JPS (2023) New approaches to the effects of Si on sugarcane ratoon under irrigation in Quartzipsamments, Eutrophic Red Oxisol, and Dystrophic Red Oxisol. BMC Plant Biol 23(51). https://doi.org/10.1186/s12870-023-04077-2

Costa MG, Prado RM, Palaretti LF, de Júnior JPS (2024) The effect of abiotic stresses on plant C:N:P homeostasis and their mitigation by silicon. Crop J. ISSN 2214-5141. https://doi.org/10.1016/j.cj.2023.11.012

Cui G, Xiao X, Zhang W, Lang D, Li Z, Zhang X (2021) Exogenous silicon relieve drought stress and salt stress of Glycyrrhiza uralensis seedlings by regulating proline metabolism and nitrogen assimilation. J Hortic Sci Biotechnol 96(6):728–737. https://doi.org/10.1080/14620316.2021.1921624

da Silva DL, de Prado RM (2023) Silicon mitigates the effects of calcium, magnesium, and sulfur in plants. In: Benefits of silicon in the nutrition of plants. Springer International Publishing, Cham, pp 113–128. https://doi.org/10.1038/s41598-020-80934-6

da Silva GP, de Mello Prado R, Boiago JJ, Chiaia HLJ (2023) Silício como mitigador da toxidez de amônio em mudas de mamoeiro. Revista de Ciências Agrárias Amazonian Journal of Agricultural and Environmental Sciences 66:1–10. https://doi.org/10.5281/zenodo.10261419

da Sousa Junior SSG, Hurtado AC, de Alves RC, Gasparino EC, dos Santos DMM (2024) Silicon attenuates aluminum toxicity in sugarcane plants by modifying growth, roots morphoanatomy, photosynthetic pigments, and gas exchange parameters. Sci Rep 14:4717. https://doi.org/10.1038/s41598-024-53537-8

Daryanto S, Wang L, Jacinthe PA (2016) Global synthesis of drought effects on maize and wheat production. PLoS One 11:e0156362. https://doi.org/10.1371/journal.pone.0156362

Das D, Ullah H, Tisarum R et al (2023) Respostas morfofisiológicas do arroz tropical à fertilização com potássio e silício sob estresse por déficit hídrico. J Soil Sci Plant Nutr 23:220–237. https://doi.org/10.1007/s42729-021-00712-9

de Guedes VHF, de Mello Prado R, Frazão JJ et al (2022) Foliar-applied silicon in sorghum (Sorghum bicolor L.) alleviate zinc deficiency. Silicon 14:281–287. https://doi.org/10.1007/s12633-020-00825-3

de Melo CCF, Silva Amaral D, de Mello Prado R, de Zanine AM, de Ferreira DJ, de Piccolo MC (2023) Nanosilica modulates C: N:P stoichiometry attenuating phosphorus toxicity more than deficiency in Megathyrsus maximus cultivated in an Oxisol and Entisol. Sci Rep 13(10284). https://doi.org/10.1038/s41598-023-37504-3

de Oliveira RLL, de Mello Prado R, Felisberto G, Checchio MV, Gratão PL (2019) Silicon mitigates manganese deficiency stress by regulating the physiology and activity of antioxidant enzymes in sorghum plants. J Soil Sci Plant Nutr 19:524–534. https://doi.org/10.1007/s42729-019-00051-w

de Oliveira LA, Cruz FJR, da Silva DL, Abreu-Junior CH, de Mello Prado R (2023) Silicon mitigates the effects of potentially toxic metals. In: de Mello Prado R (ed) Benefits of silicon in the nutrition of plants. Springer, Cham. https://doi.org/10.1007/978-3-031-26673-7_14

de Souza Junior JP, de Mello Prado R, dos Machado Santos Sarah M, Felisberto G (2019) Silicon mitigates boron deficiency and toxicity in cotton cultivated in nutrient solution. J Plant Nutr Soil Sci 182(5):805–814. https://doi.org/10.1002/jpln.201800398

Dhiman P, Rajora N, Bhardwaj S, Sudhakaran SS, Kumar A, Raturi G, Chakraborty K, Gupta OP, Devanna BN, Tripathi DK, Deshmukh R (2021) Fascinating role of silicon to combat salinity stress in plants: an updated overview. Plant Physiol Biochem 162:110–123. https://doi.org/10.1016/j.plaphy.2021.02.023

dos Santos MS, Sanglard LMVP, Barbosa ML, Namorato FA, de Melo DC, Franco WCG, Pérez-Molina JP, Martins SCV, DaMatta FM (2020) Silicon nutrition mitigates the negative impacts of iron toxicity on rice photosynthesis and grain yield. Ecotoxicol Environ Saf 189:110008. ISSN 0147-6513, https://doi.org/10.1016/j.ecoenv.2019.110008

dos Santos A, Melo Y, de Oliveira L et al (2022) Exogenous silicon and proline modulate osmoprotection and antioxidant activity in cowpea under drought stress. J Soil Sci Plant Nutr 22:1692–1699. https://doi.org/10.1007/s42729-022-00764-5

dos Sarah MMS, de Mello Prado R, de Souza Júnior JP, Teixeira GCM, dos Duarte JCS, de Medeiros RLS (2021) Silicon supplied via foliar application and root to attenuate potassium deficiency in common bean plants. Sci Rep 11:19690. https://doi.org/10.1038/s41598-021-99194-z

El-Saadony MT, Desoky ESM, Saad AM, Eid RS, Selem E, Elrys AS (2021) Biological silicon nanoparticles improve Phaseolus vulgaris L. yield and minimize its contaminant contents on a heavy metals-contaminated saline soil. J Environ Sci 106:1–14. https://doi.org/10.1016/j.jes.2021.01.012

Epstein E (1994) The anomaly of silicon in plant biology. Proc Natl Acad Sci USA 91:11–17. https://doi.org/10.1073/pnas.91.1.11

Eraslan F, Inal A, Pilbeam DJ et al (2008) Interactive effects of salicylic acid and silicon on oxidative damage and antioxidant activity in spinach (Spinacia oleracea L. cv. Matador) grown under boron toxicity and salinity. Plant Growth Regul 55:207–219. https://doi.org/10.1007/s10725-008-9277-4

Etesami H, Jeong BR (2023) How does silicon help alleviate biotic and abiotic stresses in plants? Mechanisms and future prospects. In: Plant stress mitigators. Academic Press, pp 359–402. https://doi.org/10.1016/B978-0-323-89871-3.00031-8

Farooq MA, Dietz KJ (2015) Silicon as versatile player in plant and human biology: overlooked and poorly understood. Front Plant Sci 6:994. https://doi.org/10.3389/fpls.2015.00994

Farouk S (2022) Silicon-mediated modulations of genes and secondary metabolites in plants. In: Silicon and nano-silicon in environmental stress management and crop quality improvement. Academic Press, pp 77–90. https://doi.org/10.1016/B978-0-323-91225-9.00014-5

Gonzalez-Porras CV, Teixeira GCM, Prado RDM, Ferreira PM, Palaretti LF, Oliveira KS (2024) Silicon via fertigation with and without potassium application, improve physiological aspects of common beans cultivated under three water regimes in field. Sci Rep 14(1):2051. https://doi.org/10.1038/s41598-024-52503-8

Greger M, Landberg T, Vaculík M (2018) Silicon influences soil availability and accumulation of mineral nutrients in various plant species. Plan Theory 7(2):41. https://doi.org/10.3390/plants7020041

Gulzar N, Lone R, Al-Sadi AM, Azeez A (2024) Interactive role of silicon and phenolics in biotic stress regulation in plants and expression of phenylpropanoid pathway genes. In: Lone R, Khan S, Mohammed Al-Sadi A (eds) Plant phenolics in biotic stress management. Springer, Singapore. https://doi.org/10.1007/978-981-99-3334-1_12

Han YQ, Wen JH, Peng ZP, Zhang DY, Hou ML (2018) Effects of silicon amendment on the occurrence of rice insect pests and diseases in a field test. J Integr Agric 17(10):2172–2181. https://doi.org/10.1016/S2095-3119(18)62035-0

Hassan H, Alatawi A, Abdulmajeed A, Emam M, Khattab H (2021) Roles of Si and SiNPs in improving thermotolerance of wheat photosynthetic machinery via upregulation of PsbH, PsbB and PsbD genes encoding PSII core proteins. Horticulturae 7(2):16. https://doi.org/10.3390/horticulturae7020016

Hodson MJ, Evans DE (2020) Aluminium–silicon interactions in higher plants: an update. J Exp Bot 71:6719–6729. https://doi.org/10.1093/jxb/eraa024

Hosseini SA, Maillard A, Hajirezaei MR, Ali N, Schwarzenberg A, Jamois F, Yvin JC (2017) Induction of barley silicon transporter HvLsi1 and HvLsi2, increased silicon concentration in the shoot and regulated starch and ABA homeostasis under osmotic stress and concomitant potassium deficiency. Front Plant Sci 8:1359. https://doi.org/10.3389/fpls.2017.01359

Hu AY, Xu SN, Qin DN, Li W, Zhao XQ (2021) Role of silicon in mediating phosphorus imbalance in plants. Plan Theory 10(1):51. https://doi.org/10.3390/plants10010051

Hurtado AC, Chiconato DA, de Mello PR, Junior GDSS, Felisberto G (2019) Silicon attenuates sodium toxicity by improving nutritional efficiency in sorghum and sunflower plants. Plant Physiol Biochem 142:224–233. https://doi.org/10.1016/j.plaphy.2019.07.010

Idoudi M, Slatni T, Laifa I, Rhimi N, Rabhi M, Hernández-Apaolaza L, Zorrig W, Abdelly C (2024) Silicon (Si) mitigates the negative effects of iron deficiency in common bean (Phaseolus vulgaris L.) by improving photosystem activities and nutritional status. Plant Physiol Biochem 206:108236. https://doi.org/10.1016/j.plaphy.2023.108236

Ilyas M, Nisar M, Khan N, Hazrat A, Khan AH, Hayat K, Fahad S, Khan A, Ullah A (2021) Drought tolerance strategies in plants: a mechanistic approach. Plant Growth Regul 40:926–994. https://doi.org/10.1007/s00344-020-10174-5

Imtiaz M, Rizwan MS, Mushtaq MA, Ashraf M, Shahzad SM, Yousaf B, Saees DA, Rizwan M, Nawaz MA, Mehmood S, Tu S (2016) Silicon occurrence, uptake, transport and mechanisms of heavy metals, minerals and salinity enhanced tolerance in plants with future prospects: a review. J Environ Manag 183:521–529. https://doi.org/10.1016/j.jenvman.2016.09.009

Islam W, Tauqeer A, Waheed A, Ali H, Zeng F (2023) Silicon in plants mitigates damage against pathogens and insect pests. In: Benefits of silicon in the nutrition of plants. Springer International Publishing, Cham, pp 347–378. https://doi.org/10.1007/978-3-031-26673-7_19

Jinger D, Devi MT, Dhar S, Dass A, Rajanna GA, Upadhaya P, Raj R (2017) Silicon in mitigating biotic stresses in rice (Oryza sativa L.) a review. Ann Agric Res 38(1):1–14

Kashif M, Sattar A, Ul-Allah S, Sher A, Ijaz M, Butt M, Qayyum A (2021) Silicon alleviates arsenic toxicity in maize seedlings by regulating physiological and antioxidant defense mechanisms. J Soil Sci Plant Nutr 21(3):2032–2040. https://doi.org/10.1007/s42729-021-00499-9

Kaur S, Kaur N, Siddique KHM, Nayyar H (2015) Beneficial elements for agricultural crops and their functional relevance in defence against stresses. Arch Agron Soil Sci. https://doi.org/10.1080/03650340.2015.1101070

Khan A, Khan AL, Imran M, Asaf S, Kim YH, Bilal S, Numan M, Al-Harrasi A, Al-Rawahi A, Lee IJ (2020a) Silicon-induced thermotolerance in Solanum lycopersicum L. via activation of antioxidant system, heat shock proteins, and endogenous phytohormones. BMC Plant Biol 20(1):1–18. https://doi.org/10.1186/s12870-020-02456-7

Khan A, Bilal S, Khan AL, Imran M, Al-Harrasi A, Al-Rawahi A, Lee IJ (2020b) Silicon-mediated alleviation of combined salinity and cadmium stress in date palm (Phoenix dactylifera L.) by regulating physio-hormonal alteration. Ecotoxicol Environ Saf 188:109885. https://doi.org/10.1016/j.ecoenv.2019.109885

Khan I, Awan SA, Rizwan M, Ali S, Hassan MJ, Brestic M, Zhang X, Huang L (2021) Effects of silicon on heavy metal uptake at the soil-plant interphase: a review. Ecotoxicol Environ Saf 222:112510. ISSN 0147-6513, https://doi.org/10.1016/j.ecoenv.2021.112510

Kim Y-H, Khan AL, Kim D-H, Lee S-Y, Kim K-M, Waqas M et al (2014) Silicon mitigates heavy metal stress by regulating P-type heavy metal ATPases, Oryza sativa low silicon genes, and endogenous phytohormones. BMC Plant Biol 14:1–13. https://doi.org/10.1186/1471-2229-14-13

Kirkby E (2011) Introduction, definition and classification of nutrients. In: Marschner P (ed) Marschner's mineral nutrition of higher plants. Academic Press, London, pp 3–5

Kostic I, Nikolic N, Milanovic S, Milenkovic I, Pavlovic J, Paravinja A, Nikolic M (2023) Silicon modifies leaf nutriome and improves growth of oak seedlings exposed to phosphorus deficiency and Phytophthora plurivora infection. Front Plant Sci 14:1265782. https://doi.org/10.3389/fpls.2023.1265782

Kvedaras OL, Keeping MG, Goebel FR, Byrne MJ (2007) Water stress augments silicon-mediated resistance of susceptible sugarcane cultivars to the stalk borer Eldana saccharina (Lepidoptera: Pyralidae). Bull Entomol Res 97(2):175–183

Lee SK, Sohn EY, Hamayun M, Yoon JY, Lee IJ (2010) Effect of silicon on growth and salinity stress of soybean plant grown under hydroponic system. Agrofor Syst 80:333–340. https://doi.org/10.1007/s10457-010-9299-6

Li Z, Song Z, Yan Z, Hao Q, Song A, Liu L, Yang X, Xia S, Liang Y (2018) Silicon enhancement of estimated plant biomass carbon accumulation under abiotic and biotic stresses. A meta-analysis. Agron Sustain Dev 38:26. https://doi.org/10.1007/s13593-018-0496-4

Liang Y, Zhang W, Chen, Q., & Ding, R. (2005). Effects of silicon on H+-ATPase and H+-PPase activity, fatty acid composition and fluidity of tonoplast vesicles from roots of salt-stressed barley (Hordeum vulgare L.). Environ Experiment Botany 53(1), 29–37. https://doi.org/10.1016/j.envexpbot.2004.02.010

Liang Y, Nikolic M, Bélanger R, Gong H, Song A (2015) Effect of silicon on crop growth, yield and quality. Silicon Agric 2015:209–223. https://doi.org/10.1007/978-94-017-9978-2_11

Liu B, Soundararajan P, Manivannan A (2019) Mechanisms of silicon-mediated amelioration of salt stress in plants. Plan Theory 8(9):307. https://doi.org/10.3390/plants8090307

Lozano-González JM, Valverde C, Hernández CD, Martin-Esquinas A, Hernández-Apaolaza L (2021) Beneficial effect of root or foliar silicon applied to cucumber plants under different zinc nutritional statuses. Plan Theory 10(12):2602. https://doi.org/10.3390/plants10122602

Ma, J. F., Miyake, Y., & Takahashi, E. (2001). Silicon as a beneficial element for crop plants. Plant Sci Res 8:17–39. https://doi.org/10.1016/S0928-3420(01)80006-9

Ma JF, Takahashi E (2002) Silicon uptake and accumulation in plants. In: Soil, fertilizer and plant silicon research in Japan, 1st edn. Elsevier Science, Amsterdam, pp 73–106. ISBN 0 444 51166 0

Ma C, Ci K, Zhu J, Sun Z, Liu Z, Li X, Zhu Y, Cheng T, Wang P, Liu Z (2021) Impacts of exogenous mineral silicon on cadmium migration and transformation in the soil-rice system and on soil health. Sci Total Environ 759:143501. https://doi.org/10.1016/j.scitotenv.2020.143501

Maestre FT, Valladares F, Reynolds JF (2005) Is the change of plant–plant interactions with abiotic stress predictable? A meta-analysis of field results in arid environments. J Ecol 93(4):748–757. https://doi.org/10.1111/j.1365-2745.2005.01017.x

Majumdar S, Prakash NB (2020) An overview on the potential of silicon in promoting defence against biotic and abiotic stresses in sugarcane. J Soil Sci Plant Nutr 20:1969–1998. https://doi.org/10.1007/s42729-020-00269-z

Mandlik R, Thakral V, Raturi G, Shinde S, Nikolić M, Tripathi DK, Sonah H, Deshmukh R (2020) Significance of silicon uptake, transport, and deposition in plants. J Exp Bot 71(21):6703–6718. https://doi.org/10.1093/jxb/eraa301

Matichenkov VV, Calvert DV (2002) Silicon as a beneficial element for sugarcane. J Am Soc Sugarcane Technol 22(2):21–30

Mavrič Čermelj A, Golob A, Vogel-Mikuš K, Germ M (2021) Silicon mitigates negative impacts of drought and UV-b radiation in plants. Plan Theory 11(1):91. https://doi.org/10.3390/plants11010091

Meena V, Dotaniya ML, Saha JK et al (2022) Potencial do silício para mitigar o estresse por metais pesados nas plantas para a agricultura sustentável: uma revisão. Silício 14:4447–4462. https://doi.org/10.1007/s12633-021-01200-6

Meharg C, Meharg AA (2015) Silicon, the silver bullet for mitigating biotic and abiotic stress, and improving grain quality, in rice? Environ Exp Bot 120:8–17. https://doi.org/10.1016/j.envexpbot.2015.07.001

Mierzaali T, Abdolzadeh A, Sadeghipour HR (2024) Silicon ameliorates iron deficiency in sour orange seedlings grown under different pHs. Sci Hortic 323:112489. https://doi.org/10.1016/j.scienta.2023.112489

Mir RA, Bhat BA, Yousuf H, Islam ST, Raza A, Rizvi MA et al (2022) Multidimensional role of silicon to activate resilient plant growth and to mitigate abiotic stress. Front Plant Sci 13:819658. https://doi.org/10.3389/fpls.2022.819658

Moghadam MSK, Rad AHS, Khodabin G et al (2022) Application of silicon for improving some physiological characteristics, seed yield, and oil quality of rapeseed genotypes under late-season drought stress. J Soil Sci Plant Nutr 22:2872–2890. https://doi.org/10.1007/s42729-022-00852-6

Moraes SPDO, Rodrigues JFM, Escobar MEO, Araújo FS, Oliveira TSD (2022) Beneficial effects of silicon in plants under abiotic stress conditions: a new approach. Rev Ciênc Agron 53:e20218213. https://doi.org/10.5935/1806-6690.20220052

Mostofa MG, Rahman MM, Ansary MMU, Keya SS, Abdelrahman M, Miah MG, Phan Tran LS (2021) Silicon in mitigation of abiotic stress induced oxidative damage in plants. Crit Rev Biotechnol 41(6):918–934. https://doi.org/10.1080/07388551.2021.1892582

Muhammad HMD, Abbas A, Ahmad R (2022) Fascinating role of silicon nanoparticles to mitigate adverse effects of salinity in fruit trees: a mechanistic approach. SILICON 14:8319–8326. https://doi.org/10.1007/s12633-021-01604-4

Murali-Baskaran RK, Senthil-Nathan S, Hunter WB (2021) Anti-herbivore activity of soluble silicon for crop protection in agriculture: a review. Environ Sci Pollut Res 28:2626–2637. https://doi.org/10.1007/s11356-020-11453-0

Mvondo-She MA, Gatabazi A, Laing MD, Ndhlala AR (2021) A review on the role of silicon treatment in biotic stress mitigation and citrus production. Agronomy 11(11):2198. https://doi.org/10.3390/agronomy11112198

Nadeem M, Anwar-ul-Haq M, Saqib M et al (2022) Ameliorative effect of silicic acid and silicates on oxidative, osmotic stress, and specific ion toxicity in spring wheat (Triticum aestivum L.) genotypes. J Soil Sci Plant Nutr 22:2334–2345. https://doi.org/10.1007/s42729-022-00812-0

Naidu S, Pandey J, Mishra LC, Chakraborty A, Roy A, Singh IK, Singh A (2023) Silicon nanoparticles: synthesis, uptake and their role in mitigation of biotic stress. Ecotoxicol Environ Saf 255:114783. https://doi.org/10.1016/j.ecoenv.2023.114783

Namjoyan S, Sorooshzadeh A, Rajabi A et al (2021) Melhorando a qualidade da raiz e o rendimento da beterraba sacarina por meio de nano-silício e tebuconazol sob irrigação limitada. J Soil Sci Plant Nutr 21:3378–3386. https://doi.org/10.1007/s42729-021-00613-x

Naz R, Batool S, Shahid M, Keyani R, Yasmin H, Nosheen A, Muhammad NHA, Saqib M, Siddiqui MH (2021) Exogenous silicon and hydrogen sulfide alleviates the simultaneously occurring drought stress and leaf rust infection in wheat. Plant Physiol Biochem 166:558–571. https://doi.org/10.1016/j.plaphy.2021.06.034

Nguyen D, Rieu I, Mariani C, van Dam NM (2016) How plants handle multiple stresses: hormonal interactions underlying responses to abiotic stress and insect herbivory. Plant Mol Biol 91:727–740. https://doi.org/10.1007/s11103-016-0481-8

Nicolau MCM, Bini AP, Prado RM, Piccolo MC, Gratão PL (2024) Silicon induces resistance to leaf scald in sugarcane under water deficit. Research Square Platform LLC. https://doi.org/10.21203/rs.3.rs-3875720/v1

Pandey P, Ramegowda V, Senthil-Kumar M (2015) Shared and unique responses of plants to multiple individual stresses and stress combinations: physiological and molecular mechanisms. Front Plant Sci 6:153997. https://doi.org/10.3389/fpls.2015.00723

Parveen A, Mumtaz S, Saleem MH, Hussain I, Perveen S, Thind S (2022) Silicon and nanosilicon mediated heat stress tolerance in plants. In: Silicon and nano-silicon in environmental stress management and crop quality improvement. Academic Press, pp 153–159. https://doi.org/10.1016/B978-0-323-91225-9.00001-7

Pareek SK. Sopory HJ, Bohnert (Eds.) (2010) Abiotic stress adaptation in plants, abiotic stress adaptation in plants. Springer Netherlands, Dordrecht. https://doi.org/10.1007/978-90-481-3112-9

Pascual MB, Echevarria V, Gonzalo MJ, Hernández-Apaolaza L (2016) Silicon addition to soybean (Glycine max L.) plants alleviate zinc deficiency. Plant Physiol Biochem 108:132–138. https://doi.org/10.1016/j.plaphy.2016.07.008

Patil H, Tank RV, Manoli P (2017) Significance of silicon in fruit crops—a review. Plant Arch 17:769–774

Pavlovic J, Kostic L, Bosnic P et al (2021) Interactions of silicon with essential and beneficial elements in plants. Front Plant Sci 12:697592. https://doi.org/10.3389/fpls.2021.697592

Perez CE, Rodrigues FA, Moreira WR, DaMatta FM (2014) Leaf gas exchange and chlorophyll a fluorescence in wheat plants supplied with silicon and infected with Pyricularia oryzae. Phytopathology. https://doi.org/10.1094/PHYTO-06-13-0163-R

Prado RM (2021) Mineral nutrition of tropical plants, vol 1. Springer, Cham

Prado, RM (Ed.). (2023). Benefits of silicon in the nutrition of plants. Springer. https://doi.org/10.1007/978-3-031-26673-7

Qi X, Tam NF-Y, Li WC, Ye Z (2020) The role of root apoplastic barriers in cadmium translocation and accumulation in cultivars of rice (Oryza sativa L.) with different Cd-accumulating characteristics. Environ Pollut 264:114736. https://doi.org/10.1016/j.envpol.2020.114736

Rady MM, Elrys AS, El-Maati MFA, Desoky ESM (2019) Interplaying roles of silicon and proline effectively improve salt and cadmium stress tolerance in Phaseolus vulgaris plant. Plant Physiol Biochem 139:558–568. https://doi.org/10.1016/j.plaphy.2019.04.025

Ranjan A, Sinha R, Bala M, Pareek A, Singla-Pareek SL, Singh AK (2021) Silicon-mediated abiotic and biotic stress mitigation in plants: underlying mechanisms and potential for stress resilient agriculture. Plant Physiol Biochem 163:15–25. https://doi.org/10.1016/j.plaphy.2021.03.044

Roy J (2012) Response of plants to multiple stresses. Academic Press

Sales AC, Campos CNS, de Souza Junior JP, da Silva DL, Oliveira KS, de Mello Prado R et al (2021) Silicon mitigates nutritional stress in quinoa (Chenopodium quinoa Willd.). Sci Rep 11(1):14665. https://doi.org/10.1038/s41598-021-94287-1

Sattar A, Cheema MA, Sher A, Abbas T, Irfan M, Ijaz M et al (2018) Foliage applied silicon alleviates the combined effects of salinity and drought stress on wheat seedlings. Int J Agric Biol 20:2537–2543. https://doi.org/10.17957/IJAB/15.0800

Saha G, Mostofa MG, Rahman MM, Tran LSP (2021) Silicon-mediated heat tolerance in higher plants: A mechanistic outlook. Plant Physiol Biochem 166:341–347. https://doi.org/10.1016/j.plaphy.2021.05.051

Seal P, Das P, Biswas AK (2018) Versatile potentiality of silicon in mitigation of biotic and abiotic stresses in plants: a review. Am J Plant Sci 9(7):1433–1454. https://doi.org/10.4236/ajps.2018.97105

Shalaby TA, Abd-Alkarim E, El-Aidy F, Hamed ES, Sharaf-Eldin M, Taha N, El-Ramady H, Bayoumi Y, Dos Reis AR (2021) Nano-selenium, silicon and H2O2 boost growth and productivity of cucumber under combined salinity and heat stress. Ecotoxicol Environ Saf 212:111962. https://doi.org/10.1016/j.ecoenv.2021.111962

Sharifi P, Amirnia R, Shirani Bidabadi S (2022) Role of silicon in mediating heat shock tolerance in soybean. Gesunde Pflanzen 74:397–411. https://doi.org/10.1007/s10343-021-00617-8

Shen X, Li X, Li Z, Li J, Duan L, Eneji AE (2010) Growth, physiological attributes and antioxidant enzyme activities in soybean seedlings treated with or without silicon under UV-B radiation stress. J Agron Crop Sci 196(6):431–439. https://doi.org/10.1111/j.1439-037X.2010.00428.x

Shen X, Li Z, Duan L, Eneji AE, Li J (2014) Silicon effects on the partitioning of mineral elements in soybean seedlings under drought and ultraviolet-B radiation. J Plant Nutr 37(6):828–836. https://doi.org/10.1080/01904167.2013.873458

Shilpha J, Manivannan A, Soundararajan P, Jeong BR (2023) Heat stress mitigation by silicon nutrition in plants: a comprehensive overview. In: de Mello Prado R (ed) Benefits of silicon in the nutrition of plants. Springer, Cham. https://doi.org/10.1007/978-3-031-26673-7

Silva JLFD, Prado RDM (2021) Elucidating the action mechanisms of silicon in the mitigation of phosphorus deficiency and enhancement of its response in sorghum plants. J Plant Nutr 44(17):2572–2582. https://doi.org/10.1080/01904167.2021.1918155

Silva DL da, de Mello Prado R, Tenesaca LFL, da Silva JLF, & Mattiuz BH (2021) Silicon attenuates calcium deficiency by increasing ascorbic acid content, growth and quality of cabbage leaves. Sci Rep 11(1):1770. https://doi.org/10.1038/s41598-020-80934-6

Singh S, Sahoo MR, Acharya GC, Junger D, Nayak P (2022) Silicon: a potent nutrient in plant defense mechanisms against arthropods. SILICON 14:6493–6505. https://doi.org/10.1007/s12633-021-01427-3

Song XP, Verma KK, Tian DD, Zhang XQ, Liang YJ, Huang X, Li CN, Li YR (2021) Exploration of silicon functions to integrate with biotic stress tolerance and crop improvement. Biol Res 54:19. https://doi.org/10.1186/s40659-021-00344-4

Soratto RP, Fernandes AM, Pilon C, Souza MR (2019) Phosphorus and silicon effects on growth, yield, and phosphorus forms in potato plants. J Plant Nutr 42(3):218–233. https://doi.org/10.1080/01904167.2018.1554072

Souri Z, Khanna K, Karimi N et al (2021) Silicon and plants: current knowledge and future prospects. J Plant Growth Regul 40:906–925. https://doi.org/10.1007/s00344-020-10172-7

Talakayala A, Ankanagari S, Garladinne M (2020) Role of silicone transportation through aquaporin genes for abiotic stress tolerance in plants. In: Protective chemical agents in the amelioration of plant abiotic stress: biochemical and molecular perspectives, pp 622–631. https://doi.org/10.1002/9781119552154.ch31

Teixeira GCM, de Mello Prado R (2023) Silicon mitigates the effects of water deficit in tropical plants. In: Benefits of silicon in the nutrition of plants. Springer International Publishing, Cham, pp 275–298. https://doi.org/10.1007/978-3-031-26673-7_16

Teixeira GCM, de Mello PR, Oliveira KS, Damião VD, da Junior GSS (2020) Silicon increases leaf chlorophyll content and iron nutritional efficiency and reduces iron deficiency in sorghum plants. J Soil Sci Plant Nutr 20:1311–1320. https://doi.org/10.1007/s42729-020-00214-0

Teixeira GCM, Prado RM, Rocha AMS, Princi MB, Andrade CSD (2023) Silicon mitigates iron deficiency in two energy cane cultivars by modulating physiological and nutritional mechanisms. Front Plant Sci 14:1204836. https://doi.org/10.3389/fpls.2023.1204836

Thorne SJ, Hartley SE, Maathuis FJ (2020) Is silicon a panacea for alleviating drought and salt stress in crops? Front Plant Sci 11:543584. https://doi.org/10.3389/fpls.2020.01221

Tripathi DK, Singh S, Singh VP, Prasad SM, Dubey NK, Chauhan DK (2017) Silicon nanoparticles more effectively alleviated UV-B stress than silicon in wheat (Triticum aestivum) seedlings. Plant Physiol Biochem 110:70–81. https://doi.org/10.1016/j.plaphy.2016.06.026

Ullah H, Datta A, Shrestha S, Ud Din S (2017) The effects of cultivation methods and water regimes on root systems of drought-tolerant (RD6) and drought-sensitive (RD10) rice varieties of Thailand. Arch Agron Soil Sci 63:1198–1209. https://doi.org/10.1080/03650340.2016.1266077

Vega I, Pontigo S, Nunes-Nesi A et al (2023) Interaction between silicon and arbuscular mycorrhizal symbiosis: an ecologically sustainable tool to improve crop fitness under a drought scenario? J Soil Sci Plant Nutr 23:125–138. https://doi.org/10.1007/s42729-021-00701-y

Verma KK, Singh P, Song XP et al (2020) Mitigating climate change for sugarcane improvement: role of silicon in alleviating abiotic stresses. Sugar Tech 22:741–749. https://doi.org/10.1007/s12355-020-00831-0

Verma KK, Song X-P, Tian D-D, Guo D-J, Chen Z-L, Zhong C-S, Nikpay A, Singh M, Rajput VD, Singh RK, Minkina T, Li Y (2021a) Influence of silicon on biocontrol strategies to manage biotic stress for crop protection, performance, and improvement. Plan Theory 10(10):2163. https://doi.org/10.3390/plants10102163

Verma KK, Song X-P, Tian D-D, Guo D-J, Chen Z-L, Zhong C-S, Nikpay A, Singh M, Rajput VD, Singh RK et al (2021b) Influence of silicon on biocontrol strategies to manage biotic stress for crop protection, performance, and improvement. Plants 10(10):2163. https://doi.org/10.3390/plants10102163

Viciedo DO, de Mello PR, Lizcano Toledo R, dos Santos LCN, Hurtado AC, Nedd LLT, Gonzalez LC (2019) Silicon supplementation alleviates ammonium toxicity in sugar beet (Beta vulgaris L.). J Soil Sci Plant Nutr 19:413–419. https://doi.org/10.1007/s42729-019-00043-w

Wang L, Nie Q, Li M, Zhang F, Zhuang J, Yang W, Li T, Wang Y (2005) Biosilicified structures for cooling plant leaves: a mechanism of highly efficient midinfrared thermal emission. Appl Phys Lett 87(19). https://doi.org/10.1063/1.2126115

Wang J, Xue R, Ju X, Yan H, Gao Z, Esmail Abdalla Elzaki M, Hu L, Zeng R, Song Y (2020) Silicon-mediated multiple interactions: simultaneous induction of rice defense and inhibition of larval performance and insecticide tolerance of chilo suppressalis by sodium silicate. Ecol Evol 10(11):4816–4827. https://doi.org/10.1002/ece3.6235

Wang D, Hou L, Zhang L, Liu P (2021) The mechanisms of silicon on maintaining water balance under water deficit stress. Physiol Plant 173(3):1253–1262. https://doi.org/10.1111/ppl.13520

Wang L, Pan T, Gao X, An J, Ning C, Li S, Cai K (2022) Silica nanoparticles activate defense responses by reducing reactive oxygen species under Ralstonia solanacearum infection in tomato plants. NanoImpact 28:100418. https://doi.org/10.1016/j.impact.2022.100418

Wu XY, Yu YG, Baerson SR, Song YY, Liang GH, Ding CH, Niu JB, Pan ZQ & Zeng RS (2017). Interactions between nitrogen and silicon in rice and their effects on resistance toward the brown planthopper Nilaparvata lugens. Front Plant Sci 8:28. https://doi.org/10.3389/fpls.2017.00028

Xiao Z, Yan G, Ye M, Liang Y (2021) Silicon relieves aluminum-induced inhibition of cell elongation in rice root apex by reducing the deposition of aluminum in the cell wall. Plant Soil 462:189–205. https://doi.org/10.1007/S11104-021-04850-Y

Yadav M, George N, Dwibedi V (2023) Emergence of toxic trace elements in plant environments: insights into potential of silica nanoparticles for mitigation of metal toxicity in plants. Environ Pollut 122112. https://doi.org/10.1016/j.envpol.2023.122112

Zhang Q, Yan C, Liu J, Lu H, Wang W, Du J, Duan H (2013) Silicon alleviates cadmium toxicity in Avicennia marina (Forsk.) Vierh. Seedlings in relation to root anatomy and radial oxygen loss. Mar Pollut Bull 76:187–193. https://doi.org/10.1016/j.marpolbul.2013.09.005

Zhang Y, Liang Y, Zhao X, Jin X, Hou L, Shi Y, Ahammed GJ (2019) Silicon compensates phosphorus deficit-induced growth inhibition by improving photosynthetic capacity, antioxidant potential, and nutrient homeostasis in tomato. Agronomy 9(11):733. https://doi.org/10.3390/agronomy9110733

Zhang WJ, Zhang XJ, Lang DY, Li M, Liu H, Zhang XH (2020) Silicon alleviates salt and drought stress of Glycyrrhiza uralensis plants by improving photosynthesis and water status. Biol Plant 64:302–313. https://doi.org/10.32615/bp.2019.136

Zhou Q, Cai Z, Xian P, Yang Y, Cheng Y, Lian T, Ma Q, Nian H (2022) Silicon-enhanced tolerance to cadmium toxicity in soybean by enhancing antioxidant defense capacity and changing cadmium distribution and transport. Ecotoxicol Environ Saf 241:113766. https://doi.org/10.1016/j.ecoenv.2022.113766

Chapter 3
Silicon Dynamics and Crop Responses in Tropical Soils

Clístenes Williams Araújo do Nascimento and Karina Patrícia Vieira da Cunha

Abstract Silicon (Si) is an important yet often overlooked element in agriculture. It impacts soil-plant dynamics and crop productivity, particularly in Si-depleted tropical soils. This deficiency, given the role Si plays in enhancing tolerance to biotic and abiotic stresses, poses challenges for plant growth and yield. Silicon applications have shown promise in mitigating various stressors, including diseases, pests, water deficits, and heavy metals. However, the integration of Si into fertilization routines remains limited, partly due to growers' hesitance. Understanding the complex interplay of Si in soil-plant systems is essential for optimizing its benefits. This involves recognizing various sources and sinks of Si in soils, as well as its uptake mechanisms and deposition within plant tissues. To harness Si full potential, strategies for Si management must be tailored to specific soil conditions and crop requirements. This necessitates a comprehensive knowledge of Si chemistry, extraction methods, and fertilizer applications. By embracing Si as a crucial component of fertilization practices, agricultural systems can enhance resilience and productivity in Si-limited environments such as those found in the tropics.

Keywords Tropical soils · Silicon · Nanosilica · Si-depleted extraction methods

3.1 Introduction

Silicon (Si), after oxygen, is the most abundant element in the Earth's crust, occurring in more than 370 rock-forming minerals. Silicon content can vary from <1 to 45% of dry weight in soil, meaning that 1 kilogram of soil can contain up to 450

C. W. A. do Nascimento (✉)
Federal Rural University of Pernambuco, Recife, Brazil
e-mail: clistenes.nascimento@ufrpe.br

K. P. V. da Cunha
Federal University of Rio Grande do Norte, Natal, Brazil

R. de Mello Prado et al. (eds.), *Silicon Advances for Sustainable Agriculture and Human Health*, Sustainable Plant Nutrition in a Changing World,
https://doi.org/10.1007/978-3-031-69876-7_3

grams of silicon. The element is found in soil solution as monosilicic acid (H_4SiO_4), most of it in undissociated form, promptly available to plants. Due to desilicatization caused by intense weathering and leaching of tropical soils, the forms of Si most found in these soils are quartz, opal ($SiO_2.nH_2O$), and other Si forms unavailable to plants. Therefore, tropical soils are generally Si-depleted, which can negatively impact the yield and quality of several important crops. In addition, the continual removal of Si at harvest promotes soil desilicatization by crops of 210–224 million tons of Si worldwide each year (Matichenkov and Bocharnikova 2001).

The main transformations and processes influencing the Si concentration in soil solution are summarized in Fig. 3.1. Monosilicic acid is at the center of these interactions and transformations, as a product of the dissolution of Si-rich minerals (Lindsay 1979). The sources of monosilicic acid for the soil solution are the decomposition of crop residues, dissociation of polysilicic acids, desorption of Si from Fe and Al oxides and hydroxides, dissolution of crystalline and non-crystalline minerals, addition of Si fertilizers, and irrigation water. The main sinks include the Si precipitation, the polymerization of monosilicic acids forming polysilicic acid, leaching, adsorption on Fe and Al oxides and hydroxides, and plant uptake.

Silicon is not regarded as an essential element, yet it plays a relevant role in plant development and yield, notably when plants are under biotic or abiotic stress. The research on Si in agriculture has increased in recent decades due to the beneficial effects of the Si application on crops, mainly caused by tolerance toward plant diseases and pests, water deficit, and heavy metal stress (Guntzer et al. 2012; Ferreira et al. 2015; Silva et al. 2016; Crusciol et al. 2018;

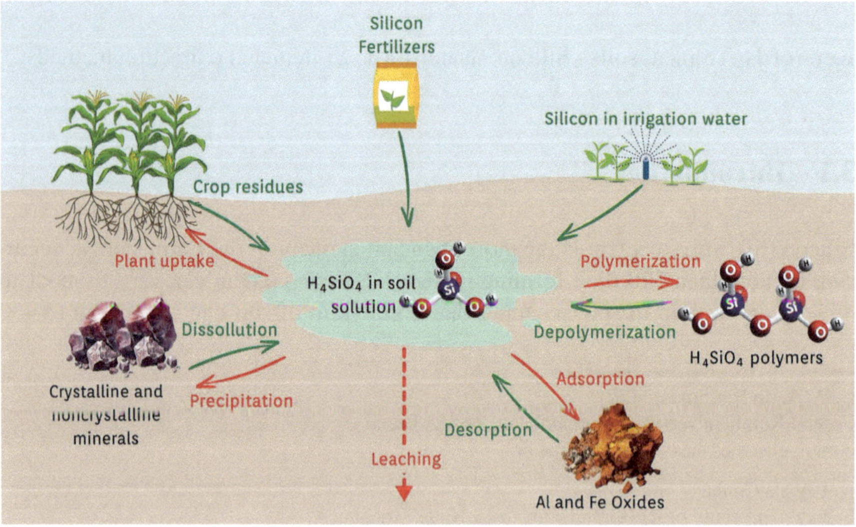

Fig. 3.1 Transformations and processes affecting the silicon concentration in the soil solution

Zarebanadkouki et al. 2024), although growers have been reticent in integrating Si into their fertilization routine.

Plants uptake Si as monosilicic acid [$Si(OH)_4$], an undissociated form present in soil solution in concentrations of 0.1 to 0.6 mmol L^{-1}, and, to a lesser extent, as $Si(OH)_3O^-$, predominant ionic form at pH > 9 (Epstein 1994). According to the species, Si accumulation in plant tissues varies considerably (between 1% and 10% of dry weight) (Currie and Perry 2007). The wide variation in Si accumulation between plant species is likely due to the differential regulation of Si absorption and distribution throughout the plant tissues (Epstein 1994; Zellner et al. 2021). Most species absorb Si by passive diffusion so that Si reaches the xylem and the aerial part following the transpiration flow. However, species from the Poaceae, Equisetaceae, and Cyperaceae families, which have high Si accumulation (>4% Si in dry weight), actively absorb Si (Currie and Perry 2007). In this case, Si is absorbed via specific membrane proteins, which guarantees the accumulation of Si by the plant, regardless of the concentration gradient (Ma et al. 2007; Yamaji et al. 2008).

Silicon absorbed by the roots is transported to the shoot and deposited extracellularly in plant tissues as hydrated amorphous silica (Fig. 3.2). In grasses, such as maize, rice, and sorghum, silica is deposited in the form of siliceous bodies, mainly in epidermal, siliceous, and bulliform cells and stomata and leaf trichomes (Currie and Perry 2007). In many species, a dense layer formed by the deposition of silica can be found below the cuticle. The formation of this layer has been of great importance in conditions of biotic and abiotic stress, contributing to reducing water loss through transpiration and increasing water use efficiency, serving as a mechanical barrier to the penetration of pathogens and herbivore chewing (Savant et al. 1999; Oliva et al. 2021).

Although the best known Si effect in plants is the high accumulation of the element by gramineous species such as sugarcane and rice, increased yields due to Si application have also been reported in tropical horticultural crops, including melon (Nascimento et al. 2020), onions (Nascimento et al. 2022a), and table grapes (Nascimento et al. 2022b). Silicon is also known to influence soil nutrient availability and the efficiency of nutrient use by plants (Neu et al. 2017; Nascimento et al. 2022a). Indeed, Jana and Jeong (2014) have highlighted Si as the most underappreciated element in horticulture.

Many studies suggest that Si may positively impact other plant groups than grasses, such as legumes (Fabaceae). In the legume-rhizobia symbiosis, for example, Si supply increases nodule number per plant, specific nodule flavonoid concentrations, and contents of foliar nitrogenous compounds (Putra et al. 2020; Shamshiripour et al. 2022; Lamsaadi et al. 2023). Another symbiosis enhanced using Si is the symbiosis with arbuscular mycorrhizal fungi, known as two of the most effective and economical ways to increase plant productivity, especially when subjected to stress (Etesami et al. 2022).

Through these countless interactions, Si supply is an opportunity for sustainable agriculture for a series of crops and cropping systems in years to come, which is essential to safeguarding the population's food security. Furthermore, Si is a

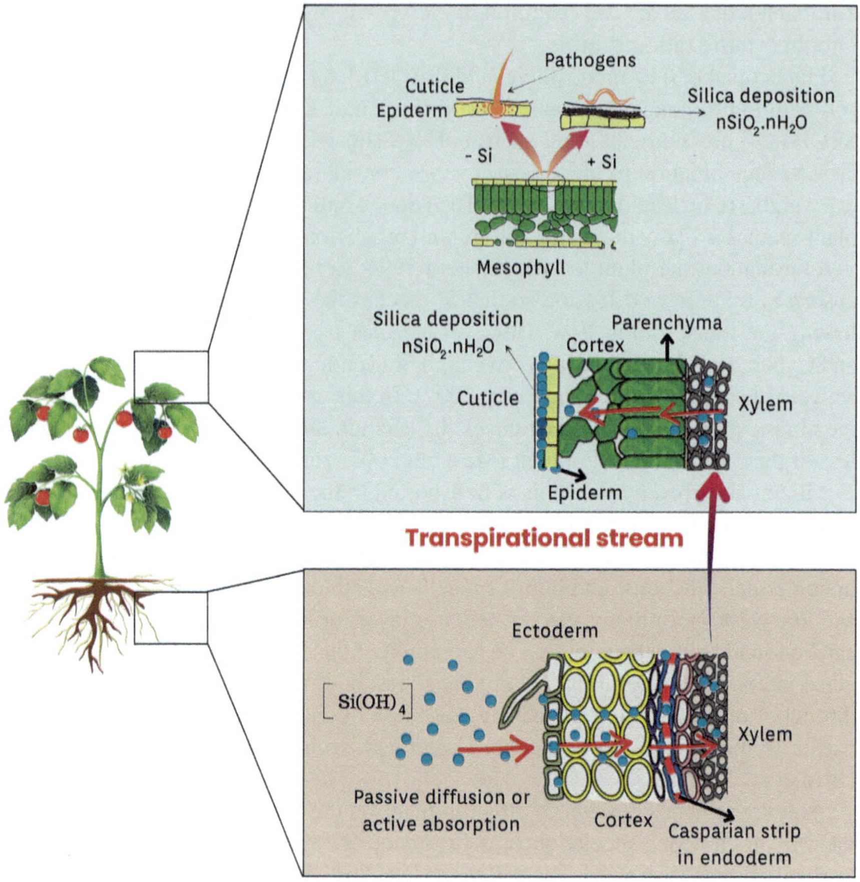

Fig. 3.2 Silicon ([$Si(OH)_4$]) uptake from the soil, transport to shoots, and deposition of silica ($SiO_2.nH_2O$) in the epidermis

beneficial element for human health, as it strengthens bone and connective tissue, enhances immune and inflammatory responses, and regulates the absorption of mineral elements (Pritchard and Nielsen 2024).

The positive impact of Si on crop yields hinges on its availability in soils. Therefore, a comprehensive understanding of the chemistry of Si in the soil-plant system is crucial to incorporating this still underestimated element into fertilization routines alongside macro- and micronutrients. In this chapter, we delve into the dynamics of Si in soils, focusing on tropical soils that are Si-depleted and responsive to the element. We also provide information on the Si extraction methods and the primary Si sources used as fertilizers and their reactions in soils and response by crops, underscoring the need for a nuanced approach to Si management.

3.2 Silicate Minerals in the Soil Environment

Silicate minerals, the most common of Earth's minerals, are the main source of Si in soils, which contain the element in several primary minerals such as quartz, mica, feldspar, olivine, biotite, muscovite, and albite (Haynes 2017). Chemical weathering of these minerals takes place with Si losses and forms the secondary silicate minerals illite, montmorillonite, smectite, and kaolinite (Li et al. 2020; Tombeur et al. 2021). Lastly, the high degree of chemical weathering in the tropics drives the formation of Al and Fe (oxy)hydroxides, leaving the tropical soils impoverished in Si (Gilkes and Prakongkep 2016).

During weathering, H_4SiO_4 dissolved from minerals can precipitate as secondary quartz or react with metals in the soil solution, forming metal silicate minerals. For example, Al ions react with Si ions to form poorly crystalline aluminosilicates, such as imogolite ($Al_2SiO_3(OH)_4$) and allophane ($Al_2O_3{\cdot}(SiO_2)_{1.3–2}{\cdot}(2.5–3)H_2O$), which are typically found in volcanic materials or podzolized soils (Gustafsson et al. 2001; Harsh et al. 2002; Saentho et al. 2023). Amorphous silica, in turn, can have mineralogical (e.g., bound to Fe oxides) or biogenic (diatoms; phytoliths) origin (Sauer and Burghardt 2006; Katz et al. 2021).

All solid Si forms in soils can dissolve at varying rates and provide Si into the soil solution. Such dissolution varies with soil conditions (e.g., pH and organic matter content) and the type of mineral, but all lead to the release of H_4SiO_4, which is available for plant uptake and has a maximum solubility of 1 mmol L^{-1}, with a maximum soluble portion from amorphous, biogenic silica (Monger and Kelly 2002). Chemically active forms of Si in soil are represented by soluble and weakly adsorbed monosilicic acid, polysilicic acid, and organosilicon compounds (Matichenkov and Calvert 2002; Schaller et al. 2021).

Soil minerals can also adsorb Si and decrease its solution concentration. For example, Fe and Al oxides are the primary pools for adsorbing monosilicic acid in acidic soils (Haynes 2017), with poorly crystalline or amorphous minerals retaining more Si than crystalline ones (Delstanche et al. 2009). Such Si-adsorbed species can be exchanged with such oxyanions as phosphates and become available for plant uptake (Ma and Takahashi 1991; Neu et al. 2017).

Adsorption of Si onto mineral surfaces is a reversible process and depends on the form of Si adsorbed. For instance, polysilicic acid's adsorption to mineral surfaces takes just some minutes, while monosilicic acid adsorption occurs within weeks because polysilicic acid has a higher binding affinity than monosilicic acid. For this reason, monosilicic acid is more abundant in natural systems than polysilicic acid. Depolymerization also occurs as polysilicic acid is decomposed to monosilicic acid and released into soil solution, which can increase Si mobility and plant availability. Silicon adsorption also depends on the mineral considered and increases, for example, from hematite, goethite, ferrihydrite, and amorphous iron hydroxide to gibbsite (Dietzel 2002).

Silicon K-edge X-ray absorption near-edge structure (XANES) spectra of clay samples have been used to improve the knowledge of Si speciation and solubility in soil. For instance, Li et al. (1995) identified several silicate minerals using Si K-edge. They demonstrated that the Si-edge energy could explain the degree of Si polymerization, with higher energy showing greater Si polymerization. On the other hand, lower-edge energy indicated a high degree of Al replacement for Si in the tetrahedral structure. The data of Saentho et al. (2023) showed that structural Si in the minerals beidellite, montmorillonite, and kaolinite are likely significant pools for Si in tropical soils. Data on the solid phase speciation of Si adsorbed by soil constituents using modern analytical techniques can represent an essential step to understanding the availability of the Si fractions retained on mineral surfaces.

3.3 Soluble Silica Species in Soils

The presence and content of different Si species in soil solutions are critical to understanding Si availability to plants. Numerous Si forms are found in the solution, varying from monosilicic acid (monomer) to polymers with two or more silicic acid units (dimers, trimers, tetramers, and oligomers), the so-called polysilicic acids (Dietzel 2002), with a release of an H_2O molecule for each condensation stage (Fig. 3.3). Polymerization of monosilicic acids can significantly influence soil texture, adsorption capacity, and stability of soil organic matter (Savant et al. 1999; Matichenkov and Calvert 2002). The polymerization minimizes the abundance of uncondensed Si-OH bonds in favor of more condensed O-Si-O bonds (Iler 1979). The more silicic acid units forming polymers, the higher the tendency toward gel formation and precipitation (Schaller et al. 2021).

The Si polymerization process is influenced by several factors, including the pH, temperature, ionic strength, presence of other elements in the solution, and degree of saturation regarding amorphous silica (Icopini et al. 2005). The polymerization rate increases from acidic to near-neutral pH and in higher ionic strength solutions (Icopini et al. 2005; Schaller et al. 2021). This is consistent with the results by Wonisch et al. (2008), who found increased polymerization in a B1 horizon of a Dystric Cambisol, where pH was lower and Al content was higher than in upper soil horizons. Stein et al. (2020) showed that such metals as lead (Pb^{+2}), copper (Cu^{+2}), and cadmium (Cd^{+2}) can interact with monosilicic acid and increase polymerization and precipitation. In this situation, polymerizing silica contributes to reduced risks of heavy metal mobility in the environment.

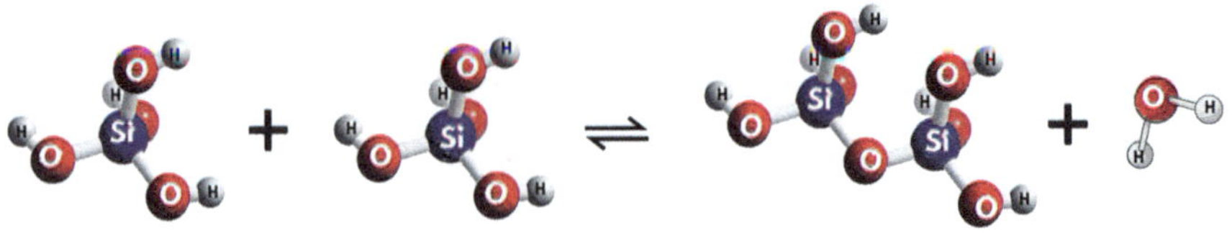

Fig. 3.3 Formation of a polysilicic acid (dimer) from two units of monosilicic acid, with the release of a water molecule

3.4 Single and Sequential Extractions of Si to Estimate Plant Availability

Soil analysis calibration for plant-available Si is a basis for making Si fertilizer recommendations, ensuring optimum soil Si levels, enhancing crop yield and quality, and avoiding excessive Si fertilization. Several extractants with varying success have been used to assess the Si availability in soils; the most studied are shown in Table 3.1. The amount of Si extracted varies with each extractant used because of several factors: dissolution of soluble, exchangeable, and specifically adsorbed Si; dissolution of unavailable forms of Si such as polymerized and precipitated Si; type and nature of the extractant used; soil-to-solution ratio; extractant pH; and shaking period (Narayanaswamy and Prakash 2009; Wang et al. 2004; Babu et al. 2018).

The most common methods to estimate plant-available Si in tropical soils use $CaCl_2$ 10 mmol L^{-1} and acetic acid 0.5 mol L^{-1}. $CaCl_2$ is the weakest extractant after water and is considered to access only the easily soluble Si fraction in soils. Indeed, some authors suggested that water could bring into solution the same amount of Si as $CaCl_2$ (Wang et al. 2004; Babu et al. 2018). Acetic acid is supposed to remove soluble Si and some exchangeable Si from soils, consistently extracting higher amounts of Si than $CaCl_2$. Acid extraction results in the dissolution of Al and Fe oxy-hydroxide surfaces with the release of previously adsorbed silicate plus some release of Si from amorphous aluminosilicates (Haynes 2014). However, the relatively low pH of the extractant can overestimate Si through the dissolution of unavailable Si pools (Nonaka and Takahashi 1990; Haynes 2014). This is particularly true for soils receiving alkaline slags because acid extractants solubilize residual undissolved Si, which is not plant-available (Sumida 2002).

Results with these two extractants in tropical soils are sometimes contradictory. For example, Korndörfer et al. (1999) found that acetic acid was more efficient than $CaCl_2$ in predicting the Si concentration in rice leaves, while Crusciol et al. (2018) indicated that neutral extractants such as KCl and $CaCl_2$ more adequately predicted the Si availability for sugarcane. As highlighted by Babu et al. (2018), it is unlikely that there is a single universal extraction procedure for Si for all soil types, and more research on Si extraction methods is warranted. For example, biogenic silica, an important fraction in many soils (Saccone et al. 2007), is not extracted by available Si extractants (neutral salts and acidic reagents), making estimating Si availability more difficult. Haynes and Zhou (2020) suggested the use of alkaline extractants

Table 3.1 Most used single extractants and procedures to estimate available silicon in soils

Extractant	Procedure	Reference
Water (H_2O)	10 g: 100 mL/4 h/room temperature	Fox et al. (1967)
10 mmol L^{-1} $CaCl_2$	10 g: 100 mL/1 h/room temperature	Korndörfer et al. (1999)
0.5 M NH_4-acetate; pH 4.8	1 g: 10 mL/1 h/room temperature	Fox et al. (1967)
0.5 mol L^{-1} Acetic acid	1 g: 10 mL/1 h/room temperature	Snyder (1991)
NH_4 citrate	10 g: 25 mL/80 h/room temperature	Sauer and Burghardt (2006)

such as Na_2CO_3 or Tiron to overcome this drawback. However, such extractants have not been routinely used to estimate Si available pools in soils and deserve more studies.

Studying different Si forms in soils using sequential extraction methods can help us understand the processes that govern Si dynamics in soils and Si availability to plants. Georgiadis et al. (2013) developed a sequential extraction method that split soil Si into six fractions: (1) readily plant-available, mobile Si; (2) adsorbed Si; (3) Si bound to organic matter; (4) Si occluded in pedogenic (hydr)oxides; (5) Si in biogenic and minerogenic amorphous silica; and (6) crystalline Si (the difference between total Si and the sum of the Si fractions obtained in extraction steps 1 to 5). Although this sequential extraction method needs to be proven by comprehensive tests using different soil types and regions, it is promising. We could, for example, use it to understand the Si fractions accessed by the single extractants used to assess the plant-available Si, which helps recommend an extractant for different types or groups of soils.

The determination of Si in chemical extracts can be performed by colorimetry and spectroscopy. Most evaluations on plant availability of soil Si in the literature were carried out with the molybdenum blue colorimetry (Snyder 2001; Korndorfer et al. 2004; Nakamura et al. 2020), which requires the preparation of specifically optimized reagents and is highly time-sensitive. Therefore, the use of other methods, such as atomic absorption (AA), inductively coupled plasma atomic emission spectrometry (ICP-AES), and near-infrared spectroscopy (NIRS), has been proposed as time-efficient and accurate approaches to estimate Si in soil extracts (Wang et al. 2004; Quigley et al. 2016).

The determination of Si in the supernatant of six plant-available extractants (HCl, citric acid, acetic acid, acetate buffer, and NH_4OAc) measured by ICP-AES was highly correlated with the colorimetric analysis (Wang et al. 2004). Other works also demonstrated the efficiency of ICP-AES in measuring Si in soil extracts (Quigley et al. 2016; Delvigne et al. 2021; Puppe et al. 2024). Thus, Si determination by ICP-AES can replace the more laborious and time-consuming colorimetric method for Si dosage in soil extracts. However, despite ICP-AES offering a significant advantage in terms of speed, and delivering rapid analysis compared to other methods, the high cost of the equipment and its limited availability means that many laboratories conducting routine Si analyses do not have access to it.

3.5 Silicon Sources for Soil Amendments

Despite the abundance of Si in the Earth's crust, the Si plant availability in the soil is low, especially in highly weathered tropical soils, due to the desilication process and the depletion of Si readily available through crop removal (Camargo et al. 2013; Keeping 2017; Liang et al. 2015). For this reason, important crops such as maize, sugarcane, rice, wheat, and others, have benefited from Si fertilization, with significant increases in their productivity. For the Si materials to be effective as fertilizers, they need to fulfill several requirements. These include having a significant amount of soluble Si, being in a physical form that allows for easy storage and application and being free from soil pollutants such as heavy metals (Gascho 2001). Because the chapter focuses on soil chemistry related to Si, only soil Si fertilizers will be discussed here.

Silicon-deficient soils can be corrected by several sources that vary widely in total and plant-available silicon (Keeping 2017). These include slag-based silicate fertilizer; fused magnesium phosphate silicate fertilizer; fused potassium magnesium silicate; porous hydrated calcium silicate; micron SiO_2; soluble silicates such as silicic acid, sodium silicate, and potassium silicate; and other natural sources such as wollastonite, diatoms, olivine, rice husk, straw of wheat, and grains and plant byproducts (Vasanthi et al. 2014; Liang et al. 2015; Nascimento et al. 2022b).

Silicate fertilizers ($CaSiO_3$ and $MgSiO_3$) are certainly one of the most used Si sources. Tropical soils usually have low pH, high Al^{3+} content, low base saturation, and high phosphorus (P) fixation capacity. Particularly in these soils, silicate fertilizers can influence crops in two ways: (1) by improving chemical properties and soil fertility and (2) by having a direct effect on plant growth and development (Matichenkov and Calvert 2002). Calcium/magnesium silicates behave in the soil similarly to Ca/Mg carbonates. They can raise the pH, neutralize exchangeable Al, and are associated with an increase in the availability of soluble Si and exchangeable Ca and Mg contents, promoting increased base saturation and lower Al saturation (Savant et al. 1999). Silicate fertilizers are usually neutral to slightly alkaline (Lindsay 1979). The corrective effect of soil acidity promoted by silicates occurs through the reactions of SiO_3^{-2} anions with H^+ protons in the soil solution (Fig. 3.4).

Another critical aspect of the effect of silicates on soil properties concerns their interaction with P (Matichenkov and Calvert 2002) and NPK fertilization (Lima-Filho et al. 1999). Zhang et al. (2017) state that using silicate fertilizers increases NPK fertilization efficiency. Silicates have good adsorption properties and promote less leaching of potassium and other mobile nutrients in the surface horizon. According to Matichenkov and Calvert (2002), silicate fertilizers adsorb P, reducing its leaching by around 40–90% without reducing its availability to plants since P is maintained in exchangeable forms on the surface of the silicates. With the addition of silicates to the soil, chemical exchange reactions occur between silicate and phosphates of Ca, Al, and Fe with consequent formation of their respective silicates and concomitant release of the phosphate ion into the soil solution.

In addition to the exchange reactions with phosphates, the silicate anion can displace the phosphate anion from the adsorption sites on the oxides or

Fig. 3.4 Dissolution of $CaSiO_3$ and reactions of SiO_3^{2-} anions with protons (H^+) in the soil solution

$$CaSiO_3 + H_2O \leftrightarrow Ca^{2+} + SiO_3^{2-}$$

$$SiO_3^{2-} + 2H^+ \leftrightarrow H_2SiO_3$$

$$H_2SiO_3 + H_2O \leftrightarrow H_4SiO_4$$

Fig. 3.5 Exchange reactions involving calcium silicate and phosphates in soils

$$CaHPO_4 + Si(OH)_4 \leftrightarrow CaSiO_3 + H_2O + H_3PO_4$$

$$2Al(H_2PO_4)_3 + 2Si(OH)_4 + 5H^+ \leftrightarrow Al_2Si_2O_5 + 5H_3PO_4 + 5H_2O$$

$$2FePO_4 + Si(OH)_4 + 2H^+ \leftrightarrow Fe_2SiO_4 + 2H_3PO_4$$

preferentially occupy them. The exchange reactions between silicate and calcium, aluminum, and iron phosphates are as follows (Fig. 3.5).

As many silicate fertilizers are derived from slags (or slags are directly applied to soils), concerns about heavy metals contamination arise, and prudent use of these sources is advised. Using fertilizers based on natural amorphous silica (diatomaceous Earth or diatomite) may improve fertilization efficiency while reducing the risks of heavy metals input into the agroecosystem. Biogenic silica originates from the accumulation of fossil shells of diatom algae and is predominantly composed of amorphous silica (Sandhya and Prakash 2019). Amorphous silica (AS) is suggested to have a higher solubility in soil solution than other commonly used silicon sources; also, it can contribute to the retention of water and cations in the soil (Pati et al. 2016; Sandhya and Prakash 2019; Gokavi et al. 2021).

Studies have demonstrated the efficiency of AS in supplying Si and improving the efficiency of nutrient use for different crops. However, data on amorphous silica-based fertilizers are scarce in tropical soils. In Brazil, sandy soils with neutral to alkaline pH and Si content ($CaCl_2$ 10 mmol L^{-1}) below 10 mg dm^{-3} are more responsive to the application of amorphous silica sources. For example, Nascimento et al. (2022a) applying an AS to soil improved the nutritional status and ameliorated abiotic stresses in two table grape cultivars and significantly affected the yield and the quality of berries in a semiarid setting, climate where abiotic stresses are intense and limit the yield and quality of table grapes. The same amorphous silica-based fertilizer was also tested in melon (Nascimento et al. 2020), sugarcane (Oliva et al. 2021), and onion (Nascimento et al. 2022b), with the three crops showing increases in yield and reduction in abiotic and biotic stresses. In addition, AS amendments strongly increased the soil water holding capacity, potentially decreasing the adverse effects of drought on crop performance (Schaller et al. 2021; Zarebanadkouki et al. 2024).

Rice husk biochar (RHB) is another Si source investigated to supply the element to crops. Approximately 80 million tons of rice husk/hulls are produced yearly, comprising over three million tons of Si. When converted into biochar, RHB can be an effective soil ameliorant (Chaudhary et al. 2002; Ebe et al. 2019). For example, Lima et al. (2022) showed that applying 30 Mg ha^{-1} of RHB to tropical soil increased the soil mean values of organic carbon, cationic exchangeable capacity (CEC), and P by 27%, 30%, and 107%, respectively. Bakhat et al. (2021) found that RHB decreased insect infestation and increased Si concentration in the leaves of eggplant.

In recent studies, silica nanoparticles have demonstrated the potential to boost plant growth and bolster resistance to diseases (Suriyaprabha et al. 2012; El-Shetehy et al. 2021; Goswami et al. 2022). These nanoparticles could emerge as an additional Si source for agricultural purposes in the future. However, comprehensive research is necessary to determine optimal concentrations, assess particle mobility within plant tissue, and ascertain the longevity of particle integrity (Elmer and White 2018).

Because Si is strongly associated with the carbon cycle, Si fertilizers have been associated with an increased carbon pool in the terrestrial ecosystem at the global scale (Chen et al. 2024). Incorporating Si in plants fosters carbon accumulation through enhanced mesophyll conductance, chloroplast CO_2 concentration, and the maximum carboxylation rate (Detmann et al. 2012). As a result, the net photosynthetic rate of plants experiences a marked upsurge upon applying Si fertilizer (Ouzounidou et al. 2016). Moreover, the decomposition rate of plant litter, such as straw, significantly accelerates with Si fertilizer application due to the increased amount of plant-available silicon, which stimulates microbial growth and augments the surface area for microbial infiltration into plant litter (Marxen et al. 2016; Chen et al. 2024). However, the mechanisms by which Si fertilizers influence gains and losses in the terrestrial system need to be studied more, and they may play a role in global warming mitigation.

3.6 Final Considerations

Tropical soils often have low natural Si content, which limits the impact of the yield of several crops in the region. Such low availability is due to high weathering rates, intense Si leaching, acidic pH, and high levels of Al in soils. On the other hand, the Si impoverishment of the soils in the tropics makes them prone to respond to Si fertilization since improved soil management practices to enhance Si availability and strategies to mitigate the effects of soil acidity and aluminum toxicity on Si uptake by plants are considered. Additionally, it is crucial to identify the specific conditions under which Si fertilization will be most beneficial for increased crop yields and disease and pest management. Addressing these challenges will be essential to unlock the full potential of Si as a sustainable solution in the tropics and worldwide.

The low adoption of Si as a fertilizer in the tropics suffers from the same limitation as in other regions: lack of awareness of most growers on the Si benefits for crops and soils; limited investigation on Si in agriculture, with a handful number of research groups in a few tropical countries (e.g., Australia, Brazil, and India); lack of soil laboratories routinely testing Si in soils, plants, and fertilizers; costs involved in applying an extra fertilizer other than NPK and micronutrients (blending Si with NPK has certainly overcome this drawback); regulatory issues, as Si is not included in fertilizer regulations in many tropical countries; and, more extension service to prove the economic impact of applying Si for crops. Addressing these limitations requires targeted research and management strategies tailored to the specific characteristics of tropical soils, which will undoubtedly advance the practice of using Si in routine fertilization programs.

References

Babu TB, Tubana L, Datnoff P et al (2018) Survey of the plant-available silicon status of agricultural soils in Louisiana. J Plant Nutr 41(3):273–287

Bakhat HF, Bibi N, Fahad S et al (2021) Rice husk biochar improves brinjal growth decreases insect infestation by enhancing silicon uptake. SILICON 13:3351–3360

Camargo MS, Rocha G, Korndörfer GH (2013) Silicate fertilization of tropical soils: silicon availability and recovery index of sugarcane. Rev Bras Cien Solo 37:1267–1275

Chaudhary DS, Jollands MC, Cser F (2002) Understanding rice hull ash as fillers in polymers: a review. Silicon Chem 1:281–289

Chen Y, He D, Wu H et al (2024) Silicon fertiliser application increases the terrestrial ecosystem carbon pool at the global scale. Geoderma 442:116806

Crusciol CA, de Arruda DP, Fernandes AM et al (2018) Methods and extractants to evaluate silicon availability for sugarcane. Sci Rep 8:916

Currie HA, Perry CC (2007) Silica in plants: biological, biochemical and chemical studies. Ann Bot 100(7):1383–1389

Delstanche S, Opfergelt S, Cardinal D et al (2009) Silicon isotopic fractionation during adsorption of aqueous monosilicic acid onto iron oxide. Geochim Cosmochim Acta 73(4):923–934

Delvigne C, Guihou A, Schuessler JA et al (2021) Silicon isotope analyses of soil and plant reference materials: an inter-comparison of seven laboratories. Geostand Geoanal Res 45(3):525–538

Detmann KC, Araújo WL, Martins SCV (2012) Silicon nutrition increases grain yield, which, in turn, exerts a feed-forward stimulation of photosynthetic rates via enhanced mesophyll conductance and alters primary metabolism in rice. New Phytol 196:752–762

Dietzel M (2002) Interaction of polysilicic and monosilicic acid with mineral surfaces. In: Stober I, Bucher K (eds) Water-rock interaction. Water Science and Technology Library, vol 40. Springer, Dordrecht. https://doi.org/10.1007/978-94-010-0438-1_9

Ebe S, Ohike T, Okanami M et al (2019) Components of rice husk biochar in promoting the growth, sporulation and iturin A production of Bacillus sp. strain IA. Z Naturforsch C J Biosci 74:211–217

Elmer W, White JC (2018) The future of nanotechnology in plant pathology. Annu Rev Phytopathol 56:111–133

El-Shetehy M, Moradi A, Maceroni M et al (2021) Silica nanoparticles enhance disease resistance in Arabidopsis plants. Nat Nanotechnol 16:344–353

Epstein E (1994) The anomaly of silicon in plant biology. Proc Natl Acad Sci 91(1):11–17

Etesami H, Li Z, Maathuis FJ et al (2022) The combined use of silicon and arbuscular mycorrhizas to mitigate salinity and drought stress in rice. Environ Exp Bot 201:104955

Ferreira HA, Nascimento CWA, Datnoff LE et al (2015) Effects of silicon on resistance to bacterial fruit blotch and growth of melon. Crop Prot 78:277–283

Fox RL, Silva J, Younge O et al (1967) Soil and plant silicon and silicate response by sugar cane. Soil Sci Soc Am J 31:775–779

Gascho GJ (2001) Silicon sources for agriculture. In: Datnoff LE, Snyder GH, Korndorfer GH (eds) Silicon in agriculture. Elsevier, Amsterdam

Georgiadis A, Sauer D, Herrmann L et al (2013) Development of a method for sequential Si extraction from soils. Geoderma 209:251–261

Gilke RJ, Prakongkep N (2016) How the unique properties of soil kaolin affect the the fertility of tropical soils. Appl Clay Sci 131:100–106

Gokavi N, Jayakumar M, Mote K et al (2021) Diatomaceous earth as a source of silicon and its impact on soil physical and chemical properties, yield and quality, pests and disease incidence of Arabica Coffee cv Chandragiri. SILICON 13:4583–4600

Goswami P, Mathur J, Srivastava N (2022) Silica nanoparticles as novel sustainable approach for plant growth and crop protection. Heliyon 8(7):e09908

Guntzer F, Keller C, Meunier JD (2012) Benefits of plant silicon for crops: a review. Agron Sustain Dev 32:201–213

Gustafsson JP, Berggren D, Simonsson M, Zysset M, Mulder J (2001) Aluminium solubility mechanisms in moderately acid Bs horizons of podzolized soils. Eur J Soil Sci 52:655–665
Harsh J, Chorover J, Nizeyimana E (2002) Allophane and imogolite. Soil Mineral Environ Appl 7:291–322
Haynes RJ (2014) A contemporary overview of silicon availability in agricultural soils. J Plant Nutr Soil Sci 177(6):831–844
Haynes RJ (2017) The nature of biogenic Si and its potential role in Si supply in agricultural soils. Agric Ecosyst Environ 245:100. https://doi.org/10.1016/j.agee.2017.04.021
Haynes RJ, Zhou Y (2020) Silicate sorption and desorption by a Si-deficient soil—Effects of pH and period of contact. Geoderma 365:114204
Icopini GA, Brantley SL, Heaney PJ (2005) Kinetics of silica oligomerization and nanocolloid formation as a function of pH and ionic strength at 25 C. Geochim Cosmochim Acta 69:293–303
Iler RK (1979) The chemistry of silica. Wiley, New York
Jana S, Jeong BR (2014) Silicon: the most under-appreciated element in horticultural crops. Trends Hortic Res 4:1–19
Katz O, Puppe D, Kaczorek D et al (2021) Silicon in the soil–plant continuum: intricate feedback mechanisms within ecosystems. Plan Theory 10:652
Keeping MG (2017) Uptake of silicon by sugarcane from applied sources may not reflect plant-available soil silicon and total silicon content of sources. Front Plant Sci 8:760
Korndörfer GH, Coelho N, Snyder GH et al (1999) Avaliação de métodos de extração de silício em solos cultivados com arroz de sequeiro. Rev Bras Ciênc Solo 23:101–106
Korndörfer GH, Pereira HS, Nolla A (2004) Análise de Silício: Solo, Planta e Fertilizante. Instituto de Ciências Agrárias, Universidade Federal de Uberlândia, Uberlândia
Lamsaadi N, Hidri R, Zorrig W et al (2023) Exogenous silicon alleviates salinity stress in fenugreek (Trigonella foenum-graecum L.) by enhancing photosystem activities, biological nitrogen fixation and antioxidant defence system. S Afr J Bot 159:344–355
Li D, Bancroft GM, Fleet ME et al (1995) Silicon K-edge XANES spectra of silicate minerals. Phys Chem Miner 22:115–122
Li Z, Guo F, Cornelis J-T, Song Z, Wang X, Delvaux B (2020) Combined silicon-phosphorus fertilization affects the biomass and Phytolith stock of Rice plants. Front Plant Sci 11:67. https://doi.org/10.3389/fpls.2020.00067
Liang Y, Nikolic M, Bélanger R et al (2015) Silicon—mediated tolerance to other abiotic stresses. In: Datnoff LE, Snyder GH, Korndörfer GH (eds) Silicon in agriculture. Springer, Dordrecht, pp 161–179
Lima VL, da Silva FBV, dos Santos NM et al (2022) Phytoattenuation of Cd, Pb, and Zn in a slag contaminated soil amended with rice straw biochar and grown with energy maize. Environ Manag 69:196–212
Lima-Filho OF, Lima MTG, Tsai SM (1999) O silício na agricultura. Informações Agronômicas 87:1–7
Lindsay WL (1979) Chemical equilibria in soils. Wiley-Interscience, New York
Ma JF, Takahash E (1991) Effect of silicate on phosphate availability for rice in a P-deficient soil. Plant Soil 133:151–155
Ma JF, Yamaji N, Mitani N et al (2007) An efflux transporter of silicon in rice. Nature 448(7150):209–212
Marxen A, Klotzbücher T, Jahn R et al (2016) Interaction between silicon cycling and straw decomposition in a silicon deficient rice production system. Plant Soil 398:153–163
Matichenkov VV, Bocharnikova EA (2001) The relationship between silicon and soil physical and chemical properties. In: Datnoff LE, Snyder GH, Korndorfer GH (eds) Silicon in agriculture. Studies in plant science, vol 8. Elsevier, Amsterdam, pp 209–219
Matichenkov VV, Calvert DV (2002) Silicon as a beneficial element for sugarcane. J Am Soc Sugarcane Technol 22:21–30
Monger HC, Kelly EF (2002) Silica minerals. In: Dixon JB, Schulze DG (eds) Soil mineralogy with environmental applications. Soil Science Society of America, Madison, pp 611–636

Nakamura R, Cornelis JT, de Tombeur F et al (2020) Comparative analysis of borate fusion versus sodium carbonate extraction for quantification of silicon contents in plants. J Plant Res 133(2):271–277

Narayanaswamy C, Prakash NB (2009) Calibration and categorization of plant available silicon in rice soils of south India. J Plant Nutr 32:1237–1254

Nascimento CWA, Souza NGH, Preston HAF et al (2020) Influence of silicon fertilization on nutrient accumulation, yield and fruit quality of melon grown in Northeastern Brazil. SILICON 12:937–942

Nascimento CWA, da Silva FBV, Lima LHV et al (2022a) Silicon application to soil increases the yield and quality of table grapes (Vitis vinifera L.) grown in a semiarid climate of Brazil. SILICON 13:2129

Nascimento CWA, de Barros MKD, da Silva FBV et al (2022b) NDVI, nutrient accumulation and yield of onion fertilized with amorphous silica in a semiarid setting in Brazil. SILICON 14:2222

Neu S, Schaller J, Dudel EG (2017) Silicon availability modifies nutrient use efficiency and content, C:N:P stoichiometry, and productivity of winter wheat (Triticum aestivum L.). Sci Rep 7:40829

Nonaka K, Takahashi K (1990) A method of assessing the need of silicate fertilizers in paddy soils. In: International Congress Soil Science, Proceedings, pp 513–514

Oliva KME, da Silva FBV, Araújo PRM et al (2021) Amorphous silica-based fertilizer increases stalks and sugar yield and resistance to stalk borer in sugarcane grown under field conditions. J Soil Sci Plant Nutr 21:543

Ouzounidou G, Giannakoula A, Ilias I et al (2016) Alleviation of drought and salinity stresses on growth, physiology, biochemistry and quality of two Cucumis sativus L. cultivars by Si application. Braz J Bot 39:531–539

Pati S, Pal B, Badole S et al (2016) Effect of silicon fertilization on growth, yield, and nutrient uptake of rice. Commun Soil Sci Plant Anal 47:284–290

Pritchard A, Nielsen BD (2024) Silicon supplementation for bone health: an umbrella review attempting to translate from animals to humans. Nutrients 16(3):339. https://doi.org/10.3390/nu16030339. PMID: 38337624; PMCID: PMC10857027

Puppe D, Buhtz C, Kaczorek D, Schaller J, Stein M (2024) Microwave plasma atomic emission spectroscopy (MP-AES) - a useful tool for the determination of silicon contents in plant samples? Front Environ Sci 12:1378922

Putra R, Powell JR, Hartley SE et al (2020) Is it time to include legumes in plant silicon research? Funct Ecol 34(6):1142–1157

Quigley KM, Althoff AG, Donati GL (2016) Inductively coupled plasma optical emission spectrometry as a reference method for silicon estimation by near infrared spectroscopy and potential application to global-scale studies of plant chemistry. Microchem J 129:231–235

Saccone L, Conley DJ, Koning BE et al (2007) Assessing the extraction and quantification of amorphous silica in soils of forest and grassland ecosystems. Eur J Soil Sci 58:1446–1459

Saentho A, Sjöstedt C, Prakongkep N et al (2023) Fate of silicon in tropical agricultural soil clays using XANES spectroscopy. Appl Clay Sci 245:107145

Sandhya K, Prakash NB (2019) Bioavailability of silicon from different sources and its effect on the yield of rice in acidic, neutral, and alkaline soils of Karnataka, South India. Commun Soil Sci Plant Anal 50:295–306

Sauer D, Burghardt W (2006) The occurrence and distribution of various forms of silica and zeolites in soils developed from wastes of iron production. Catena 65:247–257

Savant NK, Korndorter GH, Datnoff LE et al (1999) Silicon nutrition and sugarcane production: a review. J Plant Nutr 22:1853–1903

Schaller J, Puppe D, Kaczorek D et al (2021) Silicon cycling in soils revisited. Plants 10(2):295

Shamshiripour M, Motesharezadeh B, Rahmani HA et al (2022) Optimal concentrations of silicon enhance the growth of soybean (Glycine max L.) cultivars by improving nodulation, root system architecture, and soil biological properties. SILICON 14:5333–5345

Silva AJ, Nascimento CWN, Gouveia-Neto AS (2016) Assessment of cadmium phytotoxicity alleviation by silicon using chlorophyll a fluorescence. Photosynthetica (Praha) 55:18–23
Snyder G (1991) Development of a silicon soil test for Histosol-grown rice. Belle Glade EREC Res Rep EV-Fla Univ Agric Res Educ Cent (USA) 1:29–39
Snyder GH (2001) Methods for silicon analysis in plants, soils and fertilizers. In: Datnoff LE, Snyder GH, Korndorfer GH (eds) Silicon in agriculture. Elsevier, Amsterdam
Stein M, Georgiadis A, Gudat D, Rennert T (2020) Formation and properties of inorganic Si-contaminant compounds. Environ Pollut 265:115032
Sumida H (2002) Plant available silicon in paddy soils. In: Proc 2nd silicon in agriculture conf, Tsuruoka, Japan, Japanese Society of Soil Sci
Suriyaprabha R, Karunakaran G, Yuvakkumar R et al (2012) Growth and physiological responses of maize (Zea mays L.) to porous silica nanoparticles in soil. J Nanopart Res 14:1294
Tombeur F, Roux P, Cornelis JT (2021) Silicon dynamics through the lens of soil plant-animal interactions: perspectives for agricultural practices. Plant Soil 467:1–28
Vasanthi N, Saleena LM, Raj SA (2014) Silicon in crop production and crop protection—a review. Agric Rev 35:14–23
Wang JJ, Dodla SK, Henderson RE (2004) Soil silicon extractability with seven selected extractants in relation to colorimetric and ICP determination. Soil Sci 169:861–870
Wonisch H, Gérard F, Dietzel M et al (2008) Occurrence of polymerized silicic acid and aluminum species in two forest soil solutions with different acidity. Geoderma 144:435–445
Yamaji N, Mitatni N, Ma JF (2008) A transporter regulating silicon distribution in rice shoots. Plant Cell 20:1381–1389
Zarebanadkouki M, Al Hamwi W, Abdalla M et al (2024) The effect of amorphous silica on soil–plant–water relations in soils with contrasting textures. Sci Rep 14:10277
Zellner W, Tubana B, Rodrigues FA et al (2021) Silicon's role in plant stress reduction and why this element is not used routinely for managing plant health. Plant Dis 105:2033–2049
Zhang M, Liang Y, Chu G (2017) Applying silicate fertilizer increases both yield and quality of table grape (Vitis vinifera L.) grown on calcareous grey desert soil. Sci Hortic (Amsterdam) 225:757–763

Chapter 4
Innovations in Studies on the Quantification of the Bioavailability of Silicon in the Soil

Angélica Cristina Fernandes Deus, Ana Paula Rodrigues da Silva, Lucas Jónatan Rodrigues da Silva, Rosemary Marques de Almeida Bertani, Anelisa de Aquino Vidal Lacerda Soares, Cristiaini Kano, Dirceu Maximino Fernandes, and Leonardo Theodoro Büll

Abstract Silicon (Si) use in agriculture has grown in recent years. However, one of the challenges lies in the difficulty of defining a method for extracting Si from the soil that is both efficient and allows for advances in this element's interpretation. This chapter reviews the main methodologies used over the years, contributing to a deeper understanding of Si availability assessment. Here, we present a basis for a current understanding of existing methodologies, consolidating important information for future improvements in studies on the bioavailability of Si in soil. Advanced techniques using synchrotron radiation, near-infrared spectroscopy, and chemical fractionation provide possibilities for innovation in the quantification of available Si in the soil.

Keywords Silicon measurement · Extraction methods · Spectroscopy techniques · Silicon dynamics

---

A. C. F. Deus (✉)
Department of Plant Protection, Rural Engineering and Soils, College of Engineering, São Paulo State University, Ilha Solteira, São Paulo, Brazil
e-mail: angelica.deus@unesp.br

A. P. R. da Silva · L. J. R. da Silva · D. M. Fernandes · L. T. Büll
Department of Forest Science, Soils and Environment, São Paulo State University, Botucatu, São Paulo, Brazil

R. M. de Almeida Bertani · A. de Aquino Vidal Lacerda Soares
Agency Paulista Agribusiness Technology-Regional, Bauru, São Paulo, Brazil

C. Kano
Embrapa Territorial, Campinas, São Paulo, Brazil

R. de Mello Prado et al. (eds.), *Silicon Advances for Sustainable Agriculture and Human Health*, Sustainable Plant Nutrition in a Changing World,
https://doi.org/10.1007/978-3-031-69876-7_4

4.1 Introduction

Silicon (Si) is a beneficial element for plants. It reduces biotic and abiotic stresses, and its use in worldwide agriculture has been the subject of many studies in recent decades (Epstein 1999; Wang et al. 2017; Debona et al. 2017; Li et al. 2018). The literature reports that applying Si reduces the incidence of pests and diseases (Panda et al. 2024; Qi et al. 2024), increases nutrient absorption (da Silva et al. 2023), mitigates the stress caused by nutritional deficiency (Prado 2021; Pavlovic et al. 2021; Mierzaali et al. 2024), and acts in water deficit situations (Verma et al. 2021; Olivera-Viciedo et al. 2024).

In some situations, especially in soils with low Si availability and/or when growing plants that absorb more than 1% Si, fertilization may be required (Epstein 1994; Tubana et al. 2016; Debona et al. 2017). Some countries, such as Japan, have already adopted this practice. Japanese rice growers have been using Si for a long time (Ma and Takahashi 2002). Meanwhile, in the United States, farmers have applied it to crops such as rice and sugarcane (Tubaña et al. 2016). Other countries, such as Australia, Brazil, and South Africa, have also used it in sugarcane crops (Berthelsen et al. 2003; Camargo and Keeping 2021; Crusciol et al. 2018a, b; Meyer and Keeping 2001).

In order to indicate a recommendation for silicate fertilizer, it is essential to know the critical limits of Si in the soil that can affect plant growth. These limits are obtained from a methodology that quantifies the Si content in the soil and has a high correlation with the content in the plant, providing accuracy and identifying the degree of deficiency of Si in the soil and the amount to be applied as fertilizer (Buck et al. 2010; Tubaña and Heckman 2015; Tubaña 2023). The difficulty in choosing a method to assess the availability of Si in the soil has been a limiting factor for using Si in agriculture and in developing research related to this element (Crusciol et al. 2018a, b).

Even though the importance of Si is well-known with its increasingly frequent use in agriculture, there is no standardization of extraction methodologies to assess its availability in the soil. Thus, this chapter aims to present the factors that can affect the assessment of Si bioavailability in the soil, the main methodologies that have been developed over time, and the innovations that have been studied to improve the techniques for assessing available Si. Therefore, we reviewed the methodologies used to extract different fractions of Si from soils developed over the last 50 years. The first topic in this chapter addresses concepts regarding the dynamics of Si in the soil, followed by a presentation of what is considered when assessing the bioavailability of Si in the soil and the main methodologies worldwide. Finally, we present the latest innovations in assessing the bioavailability of Si in the soil and the gaps that must be filled to create a standardized and efficient assessment method.

4.2 Silicon Chemistry in the Soil

Silicon is the second most abundant element in the earth's crust. In soils, its composition varies between <1 and 45% of Si (Epstein 1999; Sommer et al. 2006). Si that is naturally present in soils comes from the chemical weathering of silicate-containing minerals, such as quartz, opal, and clay minerals of the 2:1 and 1:1 types. This process makes soluble Si available in the form of monosilicic acid (H_4SiO_4), polysilicic acid, or Si complexed with other inorganic components (Tubana et al. 2016; Schaller et al. 2021; Khalifa et al. 2020; Menegale et al. 2015). Silicon can also be incorporated into the soil system through fertilization, a practice still considered recently but which is becoming common in agriculture (Barão 2023; Rea et al. 2022).

Silicon is found in three phases in the soil: liquid, adsorbed, and solid. In the solid phase, Si can be found in crystalline or microcrystalline forms (primary/secondary silicates and amorphous silica). When Si is in the form of amorphous silica, it may have originated through the decomposition of biogenic materials, such as plant waste or microorganisms, or litho-pedogenic, such as Si bound to Fe oxides, Al hydroxides, or heavy metals (Tubana et al. 2016). Studies show that biogenic silica contributes to the concentration of Si in the soil solution, being up to 17 times more soluble than quartz (Fraysse et al. 2006). The liquid and adsorbed phases contain Si in the form of monosilicic acid, polysilicic acid, or complexed with organic and inorganic compounds (Tubana et al. 2016) (Fig. 4.1).

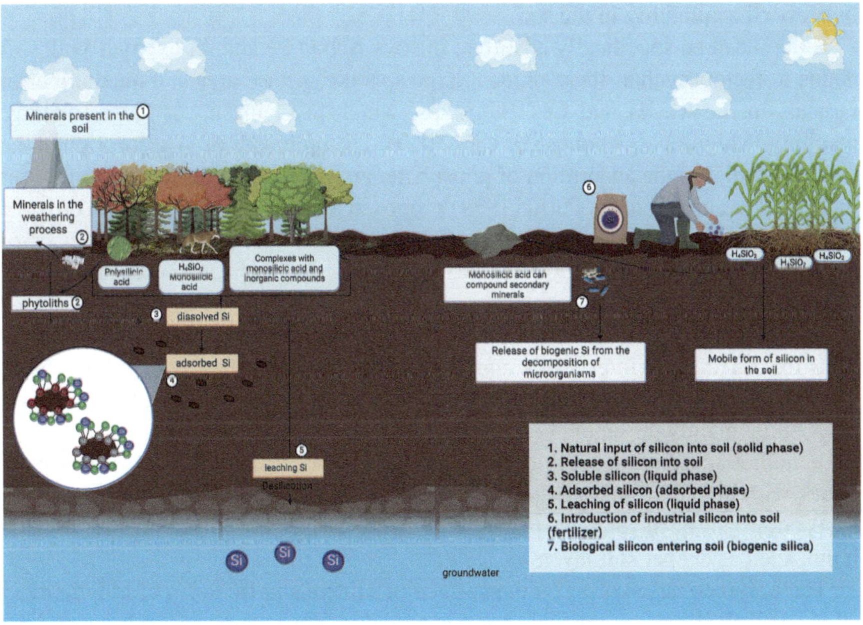

Fig. 4.1 A representative diagram of Si in the soil and its respective phases

Silicon from weathering processes can result in (i) silicates; (ii) precipitate in the form of amorphous silica on the surface of minerals; (iii) phytogenic Si from plant absorption, returning to the soil through the decomposition of plant residues; (iv) translocation between soil profiles; and (v) loss of Si through leaching (Sommer et al. 2006). The Si leaching process is called *desilication*. The literature reports that this phenomenon is more common in tropical soil regions (Lucas and Chauvel 1992; Schaller et al. 2021).

Even though Si is solubilized by weathering, this is a slow process and can involve its adsorption and desorption in the soil solid phase. Furthermore, it can be absorbed or assimilated by plants and microorganisms. Combining these factors can reduce the levels of available Si in the soil solution (Cornelis et al. 2011).

Several factors can influence the content of Si available in the soil. Texture is among the main factors (Makabe et al. 2009). Clay soils have higher levels of available Si than sandy soils. This behavior is mainly related to the source material since sandy soils have a significant quartz predominance, making a small amount of soluble Si available (Demattê et al. 2011).

The soil's pH also influences the Si levels in the soil solution. Under low pH conditions, Si availability tends to increase (Tavakkoli et al. 2011). Moreover, soils cultivated without a Si supply reduce the levels in the soil since the plants extract the Si (Korndörfer and Darnoff 1995). Haynes and Zhou (2020) showed that Si adsorption in redoxic hydrosol was at its maximum under high pH values (9–10), and the greatest Si availability in the soil occurred in the pH range between 5.0 and 6.5. Agronomically, this information is interesting in tropical soils since this is the right pH range for most crops, indicating that agricultural practices such as liming can increase Si availability in the soil.

Silicon can be specifically adsorbed in soils rich in Fe and Al oxyhydroxides. It is due to factors such as these oxides' large specific surface area and the number of radicals in the Al.OH and Fe.OH groups, which favor the exchange of ligands, resulting in high-energy bonds (Haynes 2019). The processes of Si adsorption in the soil are similar to the adsorption of phosphate, arsenate, and other elements present in the soil in anionic form. Parfit (1979) proposed a model that explains the order of adsorption according to affinity to the adsorption sites present on the surfaces of oxyhydroxides: phosphate > arsenate > selenite = molybdate > silicate > sulfate = fluoride > chloride > nitrate.

Klotzbucher et al. (2020) studied Si adsorption on synthetic goethite and solutions extracted from forest soils and its interactions with other components (e.g., phosphate and dissolved organic matter). These authors observed that Si adsorption on goethite mainly presented Fe-O-Si groups without polymerization. Furthermore, these authors describe that Si mobility can be favored in superficial soil compartments with high deposition of organic or inorganic material. On the other hand, Si adsorption can be favored in places with more adsorption sites and low residue input, such as deeper soil layers.

The polymerization of Si on the surface of minerals in the form of polymerized silicic acid (PS), i.e., monosilicic species bound together, can directly influence the availability of Si in the soil. Wonisch et al. (2008) have shown that PS in forest soils can account for up to 10% of total Si and that its origin is related to the weathering processes in these soils. Moreover, the high correlation between the occurrence of aluminosilicates may indicate that PS participates directly in the formation of polymerized Si on the minerals' surface. The polymerization of Si in the soil is pH-dependent, with soluble Si in the form of monosilicic acid at acidic pH (below 4) having a slow polymerization process, which can take months. Meanwhile, this phenomenon occurs rapidly at high pH conditions, altering the binding affinity of Si with other minerals in the soil (Christl et al. 2012; Schaller et al. 2021).

The polymerization process is reversible. It is called depolymerization. Dietzel (2000) evaluated the depolymerization of Si in solution and found that at pH 3–5.5, depolymerization took 3 days to a year. Meanwhile, saline solution at pH between 6.1 and 8.2 took around an hour. Depolymerization occurred in less than a minute in alkaline solution (pH between 9.3 and 12.9). Thus, this process can take years under tropical soil conditions, where the pH is more acidic. However, in temperate climate soils, it occurs quickly.

Understanding the dynamics of Si in the soil is fundamental to understanding its biogeochemical cycle since the bioavailability of Si is directly related to factors such as weathering of the source material, and absorption by plants and microorganisms, followed by release in (bio)mineral forms of Si, which are subdivided into many Si reservoirs in the soil (Shepard et al. 2023). However, the lack of an established standard method for assessing available Si and its adsorbed forms in the soil hinders a better understanding of Si chemistry in the soil.

4.3 Assessment of Silicon Bioavailability in Soils

Three compartments in the soil are important in assessing the bioavailability of Si: (1) the intensity, represented by the concentration of Si in the soil solution; (2) the quantity, formed by the exchangeable solid phase that supplies Si to the soil solution; and (3) the capacity factor indicating the soil's potential to retain the element (Haynes 2014). In order to quantify an element's bioavailability, we must assess the combination of the intensity and capacity compartments. However, for Si, the relationship between the different compartments is still not fully understood (Haynes 2014).

The quantification of available Si in the soil has been carried out using different methodologies that vary regarding the extractant used (salts, buffer solution, acetate-based solutions, and acids), concentration of the extractant solution, soil/extractant solution ratio, laboratory procedures (shaking time and speed, resting time, and filtering or not), and the element's quantification (colorimetry, inductively coupled plasma (ICP) optical emission spectroscopy, near-infrared spectroscopy (NIRS)) (Haynes 2014; Rodrigues et al. 2022).

The extractants used to assess Si availability in the soil include water (Korndörfer et al. 1999; Takahashi and Nonaka 1986), calcium chloride (Hayson and Chapman 1975; Kondorfer et al. 2004; Berthelsen et al. 2002), sodium acetate plus acetic acid at pH 4.0 (Imaizumi and Yoshidai 1958), acetic acid (Korndörfer et al. 1999), ammonium acetate (Fox et al. 1967), sulfuric acid (Hurney 1973), and citric acid (Patil et al. 2018), among others.

Currently, different methodologies are in use in countries such as Brazil, the United States, Japan, Korea, China, and Australia (Table 4.1). For example, Brazil uses acetic acid 0.5 mol L^{-1} and $CaCl_2$ 0.01 mol L^{-1} (Korndörfer et al. 2004), the United States uses acetic acid 0.5 mol L^{-1} (Snyder 2001), Japan and Korea use sodium acetate (Rodrigues et al. 2003), and Australia uses calcium chloride ($CaCl_2$) 0.01 mol L^{-1} (Berthelsen and Korndörfer 2003). In China, studies show the use of citric acid, calcium chloride, and sodium acetate extractants (pH 4.0) (Xu et al. 2001; Yang et al. 2021). Thus, there is no universal standard method.

The methodologies extract different forms of Si from the soil. As a result, they differ in the quantified content. Studies have found that acid extractants such as acetic acid, citric acid, and sulfuric acid can extract greater amounts of Si from the soil than water and $CaCl_2$ extractants (Fox et al. 1967; Korndörfer et al. 1999; Babu et al. 2018; Paye et al. 2018).

Extraction with water evaluates the most easily soluble and readily available Si for the plant (Fox et al. 1967; Elawad et al. 1982). However, some studies show that this extraction significantly dilutes the ionic strength of the soil solution. Thus, the amounts of Si extracted differ from those in the soil solution (Babu et al. 2018; Haynes, 2014). Furthermore, the water extractor causes high dispersion of the clay particles, requiring a longer decanting time before filtering (Korndorfer et al. 1999).

Calcium chloride extractant ($CaCl_2$) was used as an alternative to overcome the limitations observed in extractions with water. Calcium chloride has an ionic strength similar to that of the soil solution and helps to prevent the dispersion of clay, making it easier to filter the extract (Haynes 2014). The $CaCl_2$ extractant quantifies the readily available Si present in the soil solution (Haysom and Chapman 1975).

Another extraction method involves acetic acid (CH_3COOH). It extracts soluble, exchangeable, and adsorbed Si from Al and Fe oxyhydroxides (Pereira et al. 2004; Tubaña and Heckman 2015; Sauer et al. 2006; Höhn et al. 2008). However, it may overestimate the availability of Si in the soil since it also has the potential to extract the adsorbed form, which is unavailable to the plant (Paye et al. 2018).

The citric acid and the sodium acetate buffer solutions quantify soluble, exchangeable, and adsorbed Si in the soil (Sauer et al. 2006). Citric acid extracts high amounts of Si because the extractant is acidic in nature, and at the same time, it can solubilize aluminosilicates and hydrated Al and Fe oxides. Thus, it displaces the adsorbed silicate, preventing the formation of Al and Fe complexes with Si during extraction (Sauer et al. 2006; Babu et al. 2018).

Sulfuric acid (H_2SO_4) is a strong acid that can extract many forms of elements present in the soil, establishing the potential content of some plant nutrients (Teixeira et al. 2017). Hurney (1973) evaluated the extraction of Si using H_2SO_4 0.005 M and continuous shaking for 16 h. The author found that sulfuric acid can access the available, exchangeable, and specifically adsorbed Si in clay minerals. Therefore, the Si available to plants can be overestimated (Hurney 1973). Sauer et al. (2006) reported a higher extraction potential for H_2SO_4 and citric acid and a lower potential for distilled water.

Table 4.1 Results of research evaluating Si extractors

Si extractor	Country	Shaker time	Silicon content on soil	Si fertilizer application	Correlation with Si plant content	Reference
Water	Brazil	1 h	0.7–8.4 mg kg^{-1}	With application	0.97	Fox et al. (1967)
		2 h	9 mg L^{-1}	With application	0.23	Barbosa-Filho et al. (2001)
			8 mg L^{-1}	No application	0.49	
		1 h	7.3 mg dm^{-3}	With application	0.84	Kornörfer et al. (1999)
		4 h	1.56 mg kg^{-1}	No application	–	
			14.25 mg kg^{-1}	With application	–	
	United States	1 h	21.6 mg kg^{-1}	With application		Paye et al. (2018)
				No application	–	
0.01 M $CaCl_2$	Australia	16 h	17.83 mg kg^{-1}	No application	–	Tavakkoli et al. (2011)
			22.91 mg kg^{-1}	With application	–	
		–	8.87 mg kg^{-1}	No application	–	Haynes and Zhou (2018)
		–	27.48 mg kg^{-1}	With application	–	
	Brazil	1 h	7.3 mg dm^{-3}	With application	0.70	Kornörfer et al. (1999)
	India	1 h	22.4–55.9 mg kg^{-1}	With application	–	Narayanaswamy and Prakash (2009)
0.5 M acetic acid	Australia	16 h	47 mg kg^{-1}	With application	–	Tavakkoli et al. (2011)
			51.91 mg kg^{-1}			
	Brazil	1 h	27 mg L^{-1}	No application	0.94	Barbosa-Filho et al. (2001)
			34 mg L^{-1}	With application	0.85	
	India	72 min	11.3–73.8 mg kg^{-1}	With application	–	Narayanaswamy and Prakash (2009)
		1 h	13.2–60.2 mg kg^{-1}		–	
		2 h	22.1–102.5 mg kg^{-1}		–	
		1 h	70.31 mg kg^{-1}	No application	–	Patil et al. (2015)
			114.55 mg kg^{-1}	With application	–	

(continued)

Table 4.1 (continued)

Si extractor	Country	Shaker time	Silicon content on soil	Si fertilizer application	Correlation with Si plant content	Reference
0.1 M citric acid	Brazil	2 h	62 mg L^{-1}	No application	0.75	Barbosa-Filho et al. (2001)
			80 mg L^{-1}	With application	0.76	
	China	5 h	160 mg kg^{-1}	No application	0.85	Xu et al. (2001)
	India	Shaking for 2 h, resting overnight, and then shaking for 1 h	[a]854.05 Kg ha^{-1}	With application	0.79	Patil et al. (2018)
0.005 M H_2SO_4	India	16 h	118–263.2 mg kg^{-1}	With application	–	Narayanaswamy and Prakash (2009)
0.5 M ammonium acetate buffer (pH 4.8)	India	1 h	264.18 mg kg^{-1}	No application	–	Patil et al. (2015)
			326.12 mg kg^{-1}	With application	–	
0.1 mol L^{-1} sodium acetate buffer solution (pH 4.0)	China	5 h	7.6 mg kg^{-1}	–	–	Xu et al. (2001)
				With application	–	Yang et al. (2021)
	Japan		79.7 mg kg^{-1}	–	–	Yanai et al. (2016)
	Korea		[a]149 mg kg^{-1}	–	–	Park et al. (2019)
0.19 M Na_2CO_3	China	30 min	72.8 mg kg^{-1}	With application	–	Xu et al. (2001)
Modified 0.5 M $NaHCO_3$ (pH 8.5)	China	5 h	2.3 mg kg^{-1}	With application	–	Xu et al. (2001)
0.02 mol L^{-1} NaH_2PO_4 (pH 6.9)	Japan	5 h	48.8 mg kg^{-1}	–	–	Yanai et al. (2016)

– = Information not found
[a]Mean of value

Other methods, such as acetate and potassium chloride (KCl), have also been studied to assess the Si bioavailability in the soil. Ammonium acetate (NH_4OAc) is an inorganic salt that can extract the available and exchangeable Si in the soil (Ayres 1966; Fox et al. 1967; Haynes 2014). However, it also promotes the dissolution of amorphous oxide-hydroxides of Al and Fe and amorphous and highly soluble crystalline aluminosilicates, releasing Si, leading to the desorption of the adsorbed silicate anion from the acetate anion. Thus, the assessed Si availability may be overestimated (Crusciol et al. 2018a, b; Haynes 2014; Höhn et al. 2008; Sauer et al. 2006).

The quantification of available Si in the soil with the potassium chloride extractant at 1 M L^{-1} (KCl) follows the same methodology used for the simultaneous extraction of the exchangeable cations calcium (Ca), magnesium (Mg), and aluminum (Al) (Cantarella et al. 2001; Embrapa 2017). Crusciol et al. (2018a, b) concluded that extraction with KCl was no more effective than other methods, such as $CaCl_2$, deionized water, Na acetate buffer (pH 4.0), and acetic acid in assessing Si availability.

One of the criteria for choosing a method to assess the bioavailability of a given element involves the significant correlation between the estimate of its availability in the soil and the amount of the element absorbed by the plants. If there is a significant correlation, the extracted element is considered to be the fraction "available" in the soil for plant absorption. The best extraction method will be the one that provides the highest correlation (Wadt 2011).

One of the pioneering studies evaluating the availability of Si in the soil used ammonium acetate, water, and phosphate acetate extractants in different soils. The authors found high correlation values with Si absorption by the sugarcane plant (Fox et al. 1967).

Several studies correlating methodologies for extracting Si from the soil and its absorption by the plant have been carried out over more than 50 years. However, the results still differ. Furthermore, silicate fertilization can alter the effectiveness of some extractors, and the same extractor can show variable correlations in conditions with or without fertilization (Table 4.1). The shaking time is modified in some methodologies, showing that there is still no standard set out (Table 4.1).

For the $CaCl_2$ extractant solution at 0.01 mol L^{-1}, results indicate both significant correlation (Pereira et al. 2007; Meirelles et al. 2023) and low or nonsignificant correlations (Babu et al. 2018; Crusciol et al. 2018a; Deus and Bull 2013; Rodrigues et al. 2003). The low correlation can be attributed to the fact that $CaCl_2$ extracts only the fraction that is in the soil solution (Fox et al. 1967; Narayanaswamy and Prakash 2009).

In Brazil, Korndöfer et al. (1999) evaluated the availability of Si in four soils with different textural classes using the following extractants: acetic acid at 0.5 ML^{-1}, buffer solution (pH 4.0), calcium chloride at 0.0025 M L^{-1}, and water. All the methods showed a high correlation with Si accumulation in rice plants. These authors observed that $CaCl_2$ and water showed the lowest levels of available Si. Meanwhile, acetic acid at 0.5 M L^{-1} showed the highest levels. These results corroborate those of Vidal (2005), who evaluated these extractants in different soil types by applying silicon, calcium carbonate, and phosphorus.

The acetic acid at 0. 5 M, sodium acetate, and $CaCl_2$ at 0.01M extractants were tested on 137 soils collected in 31 countries. The results indicated that acetate showed the highest extraction of Si, from 0 to 509 mg kg^{-1}. Acetic acid was extracted from 1 to 239 mg kg^{-1}. Finally, calcium chloride showed low extraction compared to the other methods, with Si contents ranging from 3 to 109 mg kg^{-1} (Rodrigues et al. 2003).

Narayanaswamy and Prakash (2009) found that the 0.005 M H_2SO_4 extractant solution was not suitable for assessing the availability of Si in the soil due to the high levels obtained when compared to other methods. The extraction proceeded in the following descending order: 0.005 M H_2SO_4 > 0.1 M citric acid > N NaOAc > $CaCl_2$ 0.01 M > NH_4Oac 0.5 M > distilled H_2O. Babu et al. (2016) also evaluated different Si extraction methodologies in soils that received slag and wollastonite. The results indicated that depending on the extractant used, the content obtained is not proportional to the dose of the element applied to the soil. The authors also observed the following descending order of extraction: citric acid 0.1 M > acetic acid 0.5 M > NaOAc 1 M > NH_4OAc 0.5 M > deionized water > $CaCl_2$ 0.01 M.

Other studies using 0.5 M acetic acid, 0.01 M calcium chloride ($CaCl_2$), 1 M sodium acetate (NaOAc), 0.5 M ammonium acetate (NH_4OAc), citric acid, and water extractants evaluated the bioavailability of Si in 212 soils from Louisiana, USA (Babu et al. 2018). The authors observed a wide range of Si contents, varying from 10 to 3323 μg g^{-1}. They classified the extractants from the strongest to the weakest: 0.1 M citric acid >0.5 M acetic acid >1 M NaOAc >0.5 M NH_4Oac > deionized water >0.01 M $CaCl_2$. The deionized water and $CaCl_2$ extractants also showed low correlations. The work suggests that extraction with $CaCl_2$ may reflect only a transient state of soluble Si in the soil, similar to the deionized water extraction procedure. Moreover, it suggests that extraction with acetic acid may reflect the effects of sorption/desorption reactions, extracting what is readily available and also what may be released slowly, providing a true measure of availability.

The studies above show a large variation between the Si contents extracted from the soil with the different evaluation methodologies, as reported by Babu et al. (2013) and Pereira et al. (2007). The variations observed in Si extraction are attributed to the quantification of different forms of Si, both soluble and the extraction of some forms of Si present in the soil and not available to the plants. In addition to the pH and composition of the extracting solution, laboratory procedures such as soil/solution ratio, shaking time, and speed influence the element's extraction. The longer the soil/extractor contact time and shaking speed, the greater the extraction of Si since silicate clay minerals are affected chemically and mechanically, and Si availability can be overestimated, especially by strong extractants such as citric acid and sulfuric acid (Narayanaswamy and Prakash 2009; Sauer et al. 2006). The soil characteristics, especially the sesquioxide and clay content, pH, and the presence of other anions, can interfere with silicon adsorption and consequently influence the extraction process of the available silicon (Berthelsen and Korndörfer 2003).

Studies show that all acid extractants extract higher levels of Si when compared to $CaCl_2$ and water (Babu et al. 2018; Narayanaswamy and Prakash 2009; Fox et al. 1967) because acidic solutions can extract the Si that is adsorbed to Al and Fe oxy-hydroxides more intensely than neutral solutions such as water and $CaCl_2$ (Sauer et al. 2006; Höhn et al. 2008). Overall, the more acidic the extracting solution, the greater the power to extract Si from the soil (Narayanaswamy and Prakash 2009). The sulfuric acid and citric acid extractants can desorb specifically adsorbed Si fractions and quantify some Si fractions that are not easily available for plant absorption, such as amorphous, microcrystalline, and crystalline forms (Narayanaswamy and Prakash 2009; Sauer et al. 2006). Meanwhile, the acetic acid and acetate extractants quantify exchangeable and adsorbed forms of Al and Fe oxyhydroxides (Narayanaswamy and Prakash 2009; Haynes 2014; Höhn et al. 2008; Sauer et al. 2006). In contrast, the water and $CaCl_2$ extractors only quantify the easily soluble or exchangeable forms of Si (Fox et al. 1967; Haynes 2014).

Thus, in some situations, acid extractors result in an overestimation of soluble Si content (Sumida 2002), which can be even greater in soils that have received silicate fertilization and/or liming because the element can react with amorphous surfaces (Berthelsen and Korndörfer 2003). Moreover, acid extractors remove forms of Si dissolved in calcium silicates, which are soluble in acids but not easily absorbed by plants (Xu et al. 2001). As a result, using acid extractors is more suitable for soils that have not received sources that alter the soil pH (Xu et al. 2001; Gutierrez et al. 2011).

Over the years, researchers from many countries have tried to find an effective universal method for assessing the availability of Si in soils. However, despite many advances, there is still no method effective for all soil conditions. The following topic will present the most recent innovations aimed at assessing the bioavailability of Si in the soil.

4.4 Innovations in Soil Science to Assess Silicon Bioavailability

Even though the methodologies presented throughout the chapter have the same objective (to quantify the available Si), they show different capacities for extracting the element since they extract different forms of Si. Thus, they generate different interpretations of the element's availability, which makes it difficult to interpret and recommend Si fertilization. Given these difficulties, several studies have sought to understand and overcome the obstacles to assessing the bioavailability of Si in the soil.

In order to find the best Si extraction method for soils in tropical conditions, Meirelles et al. (2023) evaluated the bioavailability of Si in 18 soils from the state of São Paulo in Brazil. They studied adaptations in shaking speed and time regarding the standard methodologies of 0.01 mol L^{-1} $CaCl^{-1}$ and acetic acid 0.05 mol L^{-1}.

They also used different concentrations for the acetic acid extractant. The authors also studied the possibility of using ion exchange resin as a Si extractor, with promising results, showing a high correlation in the methodology of 12 h of shaking at 200 rpm. Ion exchange resin is commonly used in Brazil to extract P from soils in the state of São Paulo.

There were high correlations with Si accumulation in rice plants for acetic acid and $CaCl_2$, regardless of the adaptations to the methodologies studied. The same study also evaluated the reading method, colorimetry, and ICP and found that colorimetry was more efficient.

The authors concluded that all the extractants used are efficient in assessing the bioavailability of Si in soils without Si application. For $CaCl_2$, the authors suggest the methodology of 1 h of shaking at 220 rpm due to the short period of shaking and adequate Si extraction capacity. For acetic acid, a shaking period of 1 h at 220 rpm and a concentration of 0.5 mol L^{-1} were indicated because they showed high correlation and high power to extract Si from the soil. The resin method with 12 h of shaking shows potential for extracting Si from the soil, as evidenced by the high significant correlation. These results show that the resin could be a promising Si extractant, especially in regions that use the method to quantify other elements, as it could easily be included in routine laboratory analyses. However, more studies must be conducted to validate this alternative.

Crusciol et al. (2018a, b) studied the possibility of using potassium chloride solution (KCl) to quantify the available Si. Combined with KCl, the authors compared the $CaCl_2$ 0.01 mol L^{-1}, acetic acid 0.5 mol L^{-1}, sodium acetate buffer, and deionized water extractants in soils of different textures cultivated with sugarcane. In addition to the different extractors, the authors studied two extraction methods (horizontal table shaking and microwave oven). The results indicated that conventional extraction on a shaking table with KCl was no more effective than other methods in assessing Si availability. However, there was less variation in the estimate of soluble Si between soils with different textural classes (Crusciol et al. 2018a, b).

When evaluating the efficiency of Si extractants in soil, Oliva (2019) used acetic acid 0.5 mol L^{-1}, $CaCl_2$ 0.01 mol L^{-1}, KCl 1 mol L^{-1}, and distilled water in 56 soil samples with varying clay contents cultivated with sugarcane in the northeast region of Brazil. The author found that the lowest levels of available Si were obtained when using the KCl 1mol L^{-1} extractant, and the least variation occurred when using water. Furthermore, the correlations between the available Si and Si absorbed by sugarcane indicated a positive correlation for the four extractants evaluated. However, the best correlations were obtained by KCl (R^2 = 0.76) and $CaCl_2$ (R^2 = 0.72), indicating the efficiency of these extractants in the types of soils evaluated.

As mentioned in the Sect. 4.2 of this chapter, one of the obstacles to defining the most suitable method for assessing the available Si is the lack of knowledge regarding its dynamics in the soil. Sequencing the forms of Si present in the soil can help to deepen this knowledge and, consequently, to choose a suitable method for assessing its availability.

Georgiadis et al. (2013) developed a method for the sequential extraction of Si from the soil. They propose the extraction of Si fractions considered to be mobile, adsorbed, bound to organic matter, bound to Fe and Al oxides, and amorphous silica of biogenic and mineralogical origin. The mobile fraction is extracted by 0.01 M $CaCl_2$, the adsorbed Si is obtained with 0. 01 M acetic acid through anion exchange, Si bound to organic matter is evaluated by oxidizing the organic matter with H_2O_2, Si occluded in Fe, and Al oxides is released with ammonium oxalate and oxalic acid solution under UV light. In a further sequence, the samples are split in two, and the bio-opal is separated using a 0.2 M sodium hydroxide solution for 168 h to determine the amorphous silica of biogenic and minerogenic origin. In the last stage, the total Si is determined by X-ray fluorescence analysis.

Through fractionation, the authors observed that Si adsorption varies according to the types of clay minerals present in the soil. The adsorption of monosilicic acid occurs through a superficial reaction with the hydroxyls on the mineral surfaces. Smectites have the highest adsorption capacity, followed by goethite, kaolinite, and gibbsite. The high Si adsorption capacity of smectite can be partly explained by the mineral's alkaline pH (10.5) since, under alkaline conditions, monosilicic acid can polymerize and adsorb to mineral surfaces as polysilicic acid or precipitate. These data show that the difficulty in extracting Si may be related to Si adsorption and polymerization phenomena. Thus, acid extractors can access adsorbed or polymerized Si and may overestimate the available Si content.

The sequencing of the different forms of Si proposed by Georgiadis et al. (2013) was studied in six humid temperate soils, and the lowest amounts of Si in the soils were found in the mobile and adsorbed Si fractions. The largest contribution to total Si in the soil comes from primary and secondary silicate minerals, around 86–99%. Meanwhile, the second largest contribution is related to Si in the amorphous minerogenic phase, which represents 0.1–1.5% of the total Si (Georgiadis et al. 2014). The results showed a close relationship between the mobile and adsorbed Si fractions extracted by $CaCl_2$ and acetic acid, respectively. Furthermore, the results showed a significant correlation between these fractions and the clay content and the Si content present in the amorphous minerogenic silica. These results suggest that the clay fraction of the soil acts as the main adsorbent of silica, and amorphous silica may also be an important source of easily mobilized Si, possibly due to the dissolution of silica and the desorption of silicic acid, which is reversibly adsorbed on the surface of amorphous silica. The results also show a dynamic balance and possible overlap between the mobile and adsorbed fractions since the weakly adsorbed Si can be desorbed in the extraction step with $CaCl_2$. According to Georgiadis et al. (2013), the balance between the solid phase and the mobile Si varies depending on the properties of the soil, such as bio-opal and clay content.

Georgiadis et al. (2017) also used the method developed by Georgiadis et al. (2013) to evaluate the Si fractions present in floodplain soils (Mollic Fluvisols) in Germany. The authors reported some limitations in the method for evaluating silicon bound to organic matter and biogenic silicon. Moreover, they found that the dynamics of Si in soils flooded by water for a long period in the floodplain differ from non-flooded soils. These results raise the hypothesis that the methods used to assess the bioavailability of Si in non-flooded soils may be different from those used in floodplain soils. However, studies should be conducted comparing extraction methods and the factors influencing the dynamics of available silicon in these soils.

Pallavi and Prakash (2019) also studied the sequential extraction of Si in rice cultivation in three soils in India. Five fractions of Si were determined: mobile Si, adsorbed Si, Si bound to organic matter, Si occluded in pedogenic oxides/hydroxides, and amorphous Si. The results indicated that not only the fractions of mobile and adsorbed Si are related to the Si available to plants. Occluded and amorphous Si are also important compartments for the supply of Si to rice plants. It was observed for all the soils that the amorphous Si reservoir is the largest, while the mobile and adsorbed Si reservoirs are smaller.

Shepard et al. (2023) performed fractionation of fragipan and non-fragipan soil in the central region of the United States. This study sequentially extracted Si using water (4 h and 21 days), sodium pyrophosphate, and dithionite citrate. This fractionation determined labile Si (water), Si associated with amorphous Fe and Al oxides (Na-pyrophosphate), and Si adsorbed on secondary crystalline Fe and Al oxides (citrate-dithionite). The methods used differentiate the Si fractions of the soils studied. However, these fractions may overlap, considering that one fraction may overestimate the other or vice versa. Thus, a more rigorous fractionation scheme, such as the one developed by Georgiadis et al. (2013), may be necessary to differentiate these Si reservoirs better.

Even though the fractionation of Si in the soil has contributed to advances in the knowledge regarding this element's dynamics, the methodology still needs further study. The methodology has shown some limitations, such as the variable efficiency of the second extraction phase, which is influenced by the composition of the soil, and the overestimation of Si in the soil's organic matter, which can occur due to the partial dissolution of clay minerals, pedogenic oxides, and amorphous silica by H_2O_2. In some situations, Si is underestimated in pedogenic oxides and hydroxides due to the incomplete destruction of highly crystalline pedogenic oxides or nodules (Georgiadis et al. 2013).

Despite these limitations, sequential extraction for Si in soils is an important prerequisite for studies on the interactions of these fractions and understanding the bioavailability of Si in soils. It is worth noting that most of the studies involving Si fractionation have been conducted in soils under temperate climates. Thus, there is a need for more studies with soils under different conditions.

The use of visible diffuse reflectance spectroscopy in the near-infrared (vis-NIR) and analyses based on synchrotron radiation (SR) represent innovative possibilities for advancing techniques for quantifying Si in the soil (Guo et al. 2022). These techniques have already been used in soil science to quantify other elements. These techniques are known for their high precision and resolution.

Research using visible diffuse reflectance spectroscopy in the near-infrared (vis-NIR) has been used to predict soil properties due to the correlation between spectral responses and soil attributes. Calibrations for total and organic carbon are probably the most frequent, followed by clay content (Rossel et al. 2011). These techniques allow for a faster and more economical evaluation process without using chemical reagents (Rossel et al. 2006; Stenberg et al. 2010). Some of the frequently used proximal detection techniques include visible (Vis, 350–700 nm), near-infrared (NIR, 700–1100 nm), short-wave infrared (SWIR, 1100–2500 nm), and mid-infrared (MIR, 2500–50,000 nm), which have shown good results in determining the soil's physical, chemical, and mineralogical properties (Zeng et al. 2022; Stenberg et al. 2010; Rossel et al. 2006). Predictions of Si in soils have been successful, with results obtained using MIR ranging from moderately successful to excellent ($R^2 = 0.82$ and 0.97 for oxalate-extractable Si and total Si, respectively). However, few determinations have been reported, and further confirmation is still required (Soriano-Disla et al. 2014). Rodrigues et al. (2022) confirm the potential of Vis-NIR-SWIR spectroscopy to predict soil fertility attributes that actively interact with the soil matrix. However, they conclude that additional research is required to develop more robust models for attributes such as pH, P, S-SO_4^{2-}, Si, Mn^{2+}, and Zn^{2+}.

Synchrotron radiation (SR)-based analyses comprise micro−/nanoscale tools for investigating the interaction between elements of interest and the soil matrix (Hettiarachchi et al. 2017). According to Saentho et al. (2023), using X-ray absorption near-edge structure (XANES) to assess the bioavailability of silicon can help to elucidate the interactions of this element (e.g., adsorption, polymerization, and formation of aluminosilicates) and provide information on the primary forms of Si, present in clay minerals, adsorbed and in Fe and Al oxyhydroxides. These authors carried out a study to create a Si speciation library to identify the forms of Si, mainly in the clay fraction of agricultural soils, using the Si K-edge XANES technique. Speciation has shown that Si in tropical agricultural soils can be adsorbed by clay minerals such as beidellite, montmorillonite, or kaolinite. Studies using synchrotron radiation techniques such as XANES open up a new way of understanding the dynamics of Si in agricultural soils in tropical regions. In the future, in line with other synchrotron radiation techniques, it could be a tool for further research into methods of quantifying available Si by identifying chemical species that can reduce or overestimate the Si contents quantified by colorimetry or ICP, for example.

4.5 Final Considerations

The studies published over the last 50 years indicate that the investigations carried out to date reflect the need for an in-depth understanding of the chemistry of Si between the different soil compartments, which influences the amount of Si available to plants.

It is indisputable that we now need to establish standardized methods for assessing the availability of Si in soils, both in natural and agricultural systems. In this context, methodological innovations such as fractionation, near-infrared spectroscopy, and synchrotron radiation represent alternatives to be studied to help assess the bioavailability of Si in the soil.

This chapter presents a synthesis of the main existing methodologies, showing their limitations and highlighting innovations. Thus, it seeks to guide new research and contribute to advances in the assessment of the bioavailability of Si. Identifying the best extraction method will allow us to define a more appropriate recommendation of Si for use in agriculture, given its abundance in the earth's crust and its benefits for plants.

References

Ayres AS (1966) Calcium silicate slag as a growth stimulant for sugarcane on low-silicon soils. Soil Sci 101:216–227

Babu T et al (2013) Silicon status of Louisiana soils grown to different field crops. In: Abstracts of the ASA, CSSA & SSSA International Annual Meetings, Tampa.

Babu T, Tubana B, Paye W et al (2016) Establishing soil silicon test procedure and critical silicon level for rice in Louisiana soils. Commun Soil Sci Plant Anal 47(12):1578–1597. https://doi.org/10.1080/00103624.2016.1194996

Babu T, Tubana B, Datnof L et al (2018) Survey of the plant-available silicon status of agricultural soils in Louisiana. J Plant Nutr 41:273–287. https://doi.org/10.1080/01904167.2017.1346668

Barão L (2023) The use of Si-based fertilization to improve agricultural performance. J Soil Sci Plant Nutr 23:1096–1108. https://doi.org/10.1007/s42729-022-01106-1

Barbosa-Filho M, Snyder GH, Elliott CL, Datnoff LE (2001) Evaluation of soil test procedures for determining rice-available silicon. Commun Soil Sci Plant Anal 32(11–12):1779–1792. https://doi.org/10.1081/CSS-120000249

Berthelsen S, Korndorfer GH (2003) Methods for silicon analysis in soil, plant and fertilizers. In: Paper presented at the Proceedings of the third International Conference on Silicon in Agriculture, pp. 85–91

Berthelsen S et al (2002) Effect of Ca-silicate amendments on soil chemical properties under a sugarcane cropping system. In: Abstracts of the Silicon in agriculture conference, Japanese Society of Soil Science and Plant Nutrition, Tsuruoka, p. 57

Berthelsen S et al. (2003) Improving yield and ccs in sugarcane through the application of silicon based amendments. Final Report, Sugar Research and Development Corporation Project CLW009. https://elibrary.sugarresearch.com.au/handle/11079/12957. Accessed 20 Feb 2024

Buck GB, Korndörfer GH, Datnoff LE (2010) Extractors for estimating plant available silicon from potential silicon fertilizer sources. J Plant Nutr 34(2):272–282. https://doi.org/10.1080/01904167.2011.533327

Camargo MS, Keeping MG (2021) Silicon in sugarcane: availability in soil, fertilization, and uptake. SILICON 13:3691–3701. https://doi.org/10.1007/s12633-020-00935-y

Cantarella H et al (2001) Determination of exchangeable aluminum, calcium, and magnesium in the potassium chloride extract. In: Raij BV et al (eds) Chemical analysis for fertility evaluation of tropical soils. Instituto Agronômico, Campinas, pp 213–224

Christl I, Brechbühl Y, Graf M, Kretzschmar R (2012) Polymerization of silicate on hematite surfaces and its influence on arsenic sorption. Environ Sci Technol 46(24):13235–13243. https://doi.org/10.1021/es303297m

Cornelis JT, Delvaux B, Georg RB et al (2011) Tracing the origin of dissolved silicon transferred from various soil-plant systems towards rivers: a review. Biogeosciences 8(1):89–112. https://doi.org/10.5194/bg-8-89-2011

Crusciol CAC, de Arruda DP, Fernandes AM et al (2018a) Methods and extractants to evaluate silicon availability for sugarcane. Sci Rep 8:916. https://doi.org/10.1038/s41598-018-19240-1

Crusciol CAC, Arruda DP, Fernandes AM et al (2018b) Evaluation of soil extractants for silicon availability for sugarcane. J Plant Nutr 41:2241–2255. https://doi.org/10.1080/01904167.2018.1500590

da Silva APR, da Silva LJR, Deus ACF et al (2023) Silicon application methods influence the nutrient uptake of maize plants in tropical soil. SILICON 15:7327–7733. https://doi.org/10.1007/s12633-023-02592-3

Debona D, Rodrigues FA, Datnoff LE (2017) Silicon's role in abiotic and biotic plant stresses. Annu Rev Phytopathol 55:85–107. https://doi.org/10.1146/annurev-phyto-080516-035312

Demattê JLI, Paggiaro CM, Beltrame JA, Ribeiro SS (2011) Uso de silicatos em cana-de-açúcar. Inf Agron 133:7–12. http://www.ipni.net/PUBLICATION/IA-BRASIL.NSF/0/CF832453F89BC9D483257A8F005F59AE/$FILE/Page7-12-133.pdf

Deus ACF, Büll LT (2013) Eficiência de escórias de siderurgia na cultura do feijoeiro em sistema de semeadura direta. Cienc Rural 43:1783–1789. https://doi.org/10.1590/S0103-84782013005000120

Dietzel M (2000) Dissolution of silicates and the stability of polysilicic acid. Cosmochim Acta 64(19):3275–3281. https://doi.org/10.1016/S0016-7037(00)00426-9

Elawad SH, Gascho GJ, Street JJ (1982) Response of sugarcane to silicate source and rate. I Growth yield. Agric J 74:481–484

Epstein E (1994) The anomaly of silicon in plant biology. PNAS 91(1):11–17. https://doi.org/10.1073/pnas.91.1.11

Epstein E (1999) Silicon. Annu Rev Plant Biol 50(1):641–664

Fox RL, Silva JA, Younge OR et al (1967) Soil and plant silicon and silicate response by sugarcane. Soil Sci Soc Am J 31(6):775–779. https://doi.org/10.2136/sssaj1967.03615995003100060021x

Fraysse F, Cantais F, Pokrovsky OS et al (2006) Aqueous reactivity of phytoliths and plant litter: physico-chemical constraints on terrestrial biogeochemical cycle of silicon. J Geochem Explor 88(1–3):202–205. https://doi.org/10.1016/j.gexplo.2005.08.039

Georgiadis A, Sauer D, Herrmann L et al (2013) Development of a method for sequential Si extraction from soils. Geoderma 209:251–261. https://doi.org/10.1016/j.geoderma.2013.06.023

Georgiadis A, Sauer D, Herrmann L et al (2014) Testing a new method for sequential silicon extraction on soils of a temperate–humid climate. Soil Res 52(7):645–657. https://doi.org/10.1071/SR14016

Georgiadis A, Rinklebe J, Straubinger M, Rennert T (2017) Silicon fractionation in mollic fluvisols along the central elbe river, Germany. Catena 153:100–105. https://doi.org/10.1016/j.catena.2017.01.027

Guo L, Chen A, Li C et al (2022) Solution chemistry mechanisms of exogenous silicon influencing the speciation and bioavailability of cadmium in alkaline paddy soil. J Hazard Mater 438:129526. https://doi.org/10.1016/j.jhazmat.2022.129526

Gutierrez RS, Inocêncio MF, Novelino JO (2011) Extração de silício em solos fertilizados com fosfato e silicato. Pesqui Agropecu Trop 41:01–07

Haynes RJ (2014) A contemporary overview of silicon availability in agricultural soils. J Plant Nutr Soil Sci 177(6):831–844. https://doi.org/10.1002/jpln.201400202

Haynes RJ (2019) What effect does liming have on silicon availability in agricultural soils? Geoderma 337:375–383. https://doi.org/10.1016/j.geoderma.2018.09.026

Haynes RJ, Zhou YF (2018) Effect of pH and added slag on the extractability of Si in two Si-deficient sugarcane soils. Chemosphere 193:431–437. https://doi.org/10.1016/j.geoderma.2018.09.026

Haynes RJ, Zhou YF (2020) Silicate sorption and desorption by a Si-deficient soil–effects of pH and period of contact. Geoderma 365:114204. https://doi.org/10.1016/j.geoderma.2020.114204

Haysom MBC, Chapman LS (1975) Some aspects of the calcium silicate trials at Mackay. Proc Queens Soc Sugar Cane Technol 42:117–122

Hettiarachchi GM, Donner E, Doelsch E (2017) Application of synchrotron radiation-based methods for environmental biogeochemistry: introduction to the special section. J Environ Qual 46(6):1139–1145. https://doi.org/10.2134/jeq2017.09.0349

Hohn A, Sommer M, Kaczorek D et al (2008) Silicon fractions in histosols and gleysols of a temperate grassland site. J Plant Nutr Soil Sci 171:409–418

Hurney AP (1973) A progress report on the calcium silicate investigations. Proc Aust Sugar Cane Technol 40:109–113

Imaizumi K, Yoshidai S (1958) Edaphological studies on the silicon supplying power of paddy soils. Bull Natl Inst Agric Sci B 8:261–304

Khalifa AZ, Cizer Ö, Pontikes Y et al (2020) Advances in alkali-activation of clay minerals. Cem Concr Res 132:106050. https://doi.org/10.1016/j.cemconres.2020.106050

Klotzbücher T, Treptow C, Kaiser K et al (2020) Sorption competition with natural organic matter as mechanism controlling silicon mobility in soil. Sci Rep 10:11225. https://doi.org/10.1038/s41598-020-68042-x

Korndörfer GH, Datnoff LE (1995) Adubacao com silicio: uma alternativa no controle de doencas canade-açúcar e do arroz. Inf Agron 70:70

Korndörfer GH, Coelho NM, Snyder GH, Mizutani CT (1999) Evaluation of soil extractants for silicon availability in upland rice. Rev Bras Ciênc Solo 23:101–106. https://doi.org/10.1590/S0100-06831999000100013

Korndörfer GH, Pereira HS, Nolla A (2004) Análise de silício no solo, planta e fertilizante. GPSi, Uberlândia, p 50

Li Z, Song Z, Yan Z et al (2018) Silicon enhancement of estimated plant biomass carbon accumulation under abiotic and biotic stresses. A meta-analysis. Agron Sustain Dev 38(26):1–19. https://doi.org/10.1007/s13593-018-0496-4

Lucas Y, Chauvel A (1992) Soil formation in tropically weathered terrains. In: CRM B, Zeegers H (eds) Handbook of exploration geochemistry–in tropical and subtropical terrains. Elsevier, pp 57–77

Ma JF, Takahashi E (2002) Brief history of silicon research in Japan—Birth of silicate fertilizer. In: Ma JF, Takahashi E (eds) Soil, fertilizer, and plant silicon research in Japan, pp 1–3. https://doi.org/10.1016/B978-044451166-9/50001-4

Makabe S, Kakuda KI, Sasaki Y et al (2009) Relationship between mineral composition or soil texture and available silicon in alluvial paddy soils on the Shounai Plain, Japan. Soil Sci Plant Nutr 55(2):300–308. https://doi.org/10.1111/j.1747-0765.2008.00352.x

Meirelles GC, Deus ACF, Fernandes DM, Bull LT (2023) Evaluation of silicon bioavailability in soil with different chemical extractants. SILICON 15(2):925–935

Menegale MDC, Castro GSA, Mancuso MAC (2015) Silício: interação com o sistema solo-planta. J Agron Sci 4:435–454. http://www.alice.cnptia.embrapa.br/alice/handle/doc/1025188

Meyer JH, Keeping MG (2001) Past, present and future research of the role of silicon for sugarcane in southern Africa. In: Datnoff LE, Snyder GH, Korndörfer GH (eds) Silicon in agriculture. Elsevier, Amsterdam, pp 257–275

Mierzaali T, Abdolzadeh A, Sadeghipour HR (2024) Silicon ameliorates iron deficiency in sour orange seedlings grown under different pHs. Sci Hortic 323:112489. https://doi.org/10.1016/j.scienta.2023.112489

Narayanaswamy C, Prakash NB (2009) Evaluation of selected extractants for plant-available silicon in rice soils of southern India. Commun Soil Sci Plant Anal 41:977–989. https://doi.org/10.1080/00103621003646063

Oliva KME (2019) Extração do Si e efeitos da fertilização silicatada na cana-de-açúcar, cultivada em solo arenoso. Dissertation, Universidade Federal Rural de Pernambuco

Olivera-Viciedo D, Oliveira KS, Prado RM et al (2024) Silicon uptake and utilization on Panicum maximum grass modifies C: N: P stoichiometry under warming and soil water deficit. Soil Tillage Res 235:105884. https://doi.org/10.1016/j.still.2023.105884

Pallavi T, Prakash NB (2019) Pools of silicon in soils and their contribution to rice. J Indian Soc Soil Sci 67(2):211–220. https://doi.org/10.5958/0974-0228.2019.00023.9

Panda S, Das A, Das AB et al (2024) Silicon amendment escalates cellular si deposition and antioxidant enzyme defense against scirpophaga incertulas (Walker) attack in Rice. SILICON. https://doi.org/10.1007/s12633-023-02838-0

Parfitt RL (1979) Anion adsorption by soils and soil materials. Adv Agron 30:1–50. https://doi.org/10.1016/S0065-2113(08)60702-6

Park WP, Song KC, Koo BJ, Hyun HN (2019) Distribution of available silicon of volcanic ash soils in Jeju Island. Appl Environ Soil Sci 2019:1–10. https://doi.org/10.1155/2019/2729694

Patil AA, Durgude AG, Pharande AL (2015) Evaluation of suitable extractant for soil available silicon in Inceptisols and Vertisols. Ecoscan 7:215–220

Patil AA, Durgude AG, Pharande AL et al (2018) Suitability of extractant for soil available silicon and silicon response to upland paddy grown on alkaline soils of central India. J Plant Nutr 41(18):2298–2308. https://doi.org/10.1080/01904167.2018.1500593

Pavlovic J, Kostic L, Bosnic P et al (2021) Interactions of silicon with essential and beneficial elements in plants. Front Plant Sci 12:697592. https://doi.org/10.3389/fpls.2021.697592

Paye W, Tubana B, Harrell D et al (2018) Determination of critical soil silicon levels for rice production in Louisiana using different extraction procedures. Commun Soil Sci Plant Anal 49(17):2091–2102. https://doi.org/10.1080/00103624.2018.1495731

Pereira HS, Korndorfer GH, Vidal AA, Camargo MS (2004) Fontes de silício para a cultura do arroz. Sci Agric 16:522–528. https://doi.org/10.1590/S0103-90162004000500010

Pereira HS, Barbosa NC, Carneiro MAC, Korndorfer GH (2007) Avaliação de fontes e de extratores de silício no solo. Pesq Agrop Brasileira 42:239–247. https://doi.org/10.1590/S0100-204X2007000200013

Prado RM (2021) Mineral nutrition of tropical plants. Springer, Cham. https://doi.org/10.1007/978-3-030-71262-4_12

Qi X, Xue X, Wang Z et al (2024) Silicon application enhances wheat defence against Sitobion avenae F. by regulating plant physiological-biochemical responses. Basic Appl Ecol 74:13–23. https://doi.org/10.1016/j.baac.2023.11.003

Rea RS, Islam MR, Rahman MM et al (2022) Growth, nutrient accumulation, and drought tolerance in crop plants with silicon application: a review. Sustain For 14(8):4525. https://doi.org/10.3390/su14084525

Rodrigues LL, Daroub SH, Rice RW, Snyder GH (2003) Comparison of three soil test methods for estimating plant-available silicon. Commun Soil Sci Plant Anal 34:2059–2071. https://doi.org/10.1081/CSS-120024048

Rodrigues M, Argenta JC, Cezar E et al (2022) The use of Vis-NIR-SWIR spectroscopy in the prediction of soil available ions after application of rock powder. Inf Process Agric 11. https://doi.org/10.1016/j.inpa.2022.07.001

Rossel RV, Walvoort DJJ, McBratney AB et al (2006) Visible, near infrared, mid infrared or combined diffuse reflectance spectroscopy for simultaneous assessment of various soil properties. Geoderma 131(1–2):59–75. https://doi.org/10.1016/j.geoderma.2005.03.007

Rossel RV, Adamchuk VI, Sudduth KA et al (2011) Proximal soil sensing: an effective approach for soil measurements in space and time. Adv Agron 113:243–291. https://doi.org/10.1016/B978-0-12-386473-4.00005-1

Saentho A, Sjöstedt C, Prakongkep N et al (2023) Fate of silicon in tropical agricultural soil clays using XANES spectroscopy. Appl Clay Sci 245:107145. https://doi.org/10.1016/j.clay.2023.107145

Sauer D, Saccone L, Conley DJ, Herrmann L, Sommer M (2006) Review of methodologies for extracting plant-available and amorphous Si from soils and aquatic sediments. Biogeochemistry 80:89–108. https://doi.org/10.1007/s10533-005-5879-3

Schaller J, Puppe D, Kaczorek D et al (2021) Silicon cycling in soils revisited. Plants 10(2):295. https://doi.org/10.3390/plants10020295

Shepard C, Sapp B, Ramsey RC (2023) Silicon fractionation of fragipan and non-fragipan horizons in the Central United States. Biogeochemistry 162:409–426. https://doi.org/10.1007/s10533-023-01013-2

Snyder GH (2001) Methods for silicon analysis in plants, soils and fertilizers. In: Datnoff LE, Snyder GH, Korndorfer GH (eds) Silicon in agriculture. Elsevier, Amsterdam, pp 185–196

Sommer M, Kaczorek D, Kuzyakov Y, Breuer J (2006) Silicon pools and fluxes in soils and landscapes—a review. Plant Nutr Soil Sci 69(3):310–329. https://doi.org/10.1002/jpln.200521981

Soriano-Disla JM, Janik LJ, Viscarra Rossel RA et al (2014) The performance of visible, near-, and mid-infrared reflectance spectroscopy for prediction of soil physical, chemical, and biological properties. Appl Spectrosc Rev 49(2):139–186. https://doi.org/10.1080/05704928.2013.811081

Stenberg B, Rossel RAV, Mouazen AM, Wetterlind J (2010) Visible and near infrared spectroscopy in soil science. Adv Agron 107:163–215. https://doi.org/10.1016/S0065-2113(10)07005-7

Takahashi K, Nonaka K (1986) Available silicates in paddy soils. Part 2. Development of method of measuring available silicates and its application to soil silicate analysis. Bull Shikoku Agric Exp Stn 47:16–39 (In Japanese with English summary)

Sumida H (2002) Plant available silicon in paddy soils. In: Abstracts of the silicon in agriculture conference. Japanese Society of Soil Science and Plant Nutrition, Tsuruoka, pp 43–49

Tavakkoli E, Lyons G, English P, Guppy CN (2011) Silicon nutrition of rice is affected by soil pH, weathering and silicon fertilization. J Plant Nutr Soil Sci 174(3):437–446. https://doi.org/10.1002/jpln.201000023

Teixeira PC, Donagemma GK, Fontana A, Teixeira WG (2017) Manual de Métodos de Análise de Solo, Revisada e Ampliada, 3rd edn. Embrapa, Brasília. http://www.infoteca.cnptia.embrapa.br/infoteca/handle/doc/1085209

Tubana BS (2023) Dynamics of silicon in soil and plant to establish silicate fertilization. In: Prado RM (ed) Benefits of silicon in the nutrition of plants. Springer, pp 57–73. https://doi.org/10.1007/978-3-031-26673-7_4

Tubana BS, Heckman JR (2015) Silicon in soils and plants. In: Rodrigue FA, Datnoff LE (eds) Silicon and plant diseases. Springer International Publishing, Cham, pp 7–51. https://doi.org/10.1007/978-3-319-22930-042

Tubana BS, Babu T, Datnoff LE (2016) A review of silicon in soils and plants and its role in US agriculture: history and future perspective. Soil Sci 181:393–411. https://doi.org/10.1097/SS.0000000000000179

Verma KK, Song XP, Lin B et al (2021) Silicon induced drought tolerance in crop plants: physiological adaptation strategies. SILICON 2021:1–15. https://doi.org/10.1007/S12633-021-01071-X

Vidal A (2005) Efeito do pH na solubilidade de silício em solos da região do Triângulo Mineiro. Dissertation, Universidade Federal de Uberlândia

Wadt PGS (2011) Diagnose foliare recomendação de adubação para lavouras comerciais. Embrapa Acre, Rio Branco

Wang M, Gao L, Dong S et al (2017) Role of silicon on plant–pathogen interactions. Front Plant Sci 8:701. https://doi.org/10.3389/fpls.2017.00701

Wonisch H, Gérard F, Dietzel M et al (2008) Occurrence of polymerized silicic acid and aluminum species in two forest soil solutions with: different acidity. Geoderma 144(3–4):435–445. https://doi.org/10.1016/j.geoderma.2007.11.022

Xu G, Zhan X, Li C et al (2001) Assessing methods of available silicon in calcareous soils. Commun Soil Sci Plant Anal 32(5–6):787–801. https://doi.org/10.1081/CSS-100103909

Yanai J, Taniguchi H, Nakao A (2016) Evaluation of available silicon content and its determining factors of agricultural soils in Japan. Soil Sci Plant Nutr 62(5–6):511–518. https://doi.org/10.1080/00380768.2016.1232601

Yang X, Song Z, Van Zwieten L et al (2021) Spatial distribution of plant-available silicon and its controlling factors in paddy fields of China. Geoderma 401:115215. https://doi.org/10.1016/j.geoderma.2021.115215

Zeng R, Rossiter DG, Zhang J et al (2022) How well can reflectance spectroscopy allocate samples to soil fertility classes? Agronomy 12:1964. https://doi.org/10.3390/agronomy12081964

Chapter 5
Biological Cycle of Silicon and Its Role in Agricultural Systems

Alin Song, Xian Zeng, Fenliang Fan, Xiaomin Yang, Hassan Etesami, and Zimin Li

Abstract Silicon (Si) is the second most abundant element on the Earth's surface. Its biological cycle holds a significant importance in maintaining the functionality and stability of agricultural ecosystems. Due to weathering and dissolution of silica minerals in soils, the roots absorb dissolved Si (DSi) from the soil solution and then transfer it to plant tissues. During this process, Si may participate in biomineralization and deposition to form phytoliths. Silicon uptake and phytolith formation affect crop growth and development by improving resistance to abiotic stresses, such as drought, salinity, and heavy metals, as well as biotic stresses, such as microbial or insect diseases. They can also promote the uptake of other essential nutrients. Noteworthy issue is the biological cycle of Si, which relates to Si uptake, phytolith formation and dissolution, and DSi release, which can reduce the use of pesticides in agricultural production to a certain extent, improve soil quality, and protect the environment. This chapter review provides a theoretical foundation for obtaining high agricultural yields.

A. Song · X. Zeng · F. Fan
State Key Laboratory of Efficient Utilization of Arid and Semi arid Arable Land in Northern China, Key Laboratory of Plant Nutrition and Fertilizer, Ministry of Agriculture, Institute of Agricultural Resources and Regional Planning, Chinese Academy of Agricultural Sciences, Beijing, China

X. Yang
Key Laboratory of Karst Georesources and Environment (Guizhou University), Ministry of Education, Guiyang, China

College of Resources and Environmental Engineering, Guizhou University, Guiyang, China

H. Etesami
Department of Soil Science, University of Tehran, Tehran, Iran

Z. Li (✉)
State Key Laboratory of Loess and Quaternary Geology, Institute of Earth Environment, Chinese Academy of Sciences, Xi'an, Shaanxi, China

National Observation and Research Station of Earth Critical Zone on the Loess Plateau, Xi'an, Shaanxi, China
e-mail: lizimin@ieecas.cn

R. de Mello Prado et al. (eds.), *Silicon Advances for Sustainable Agriculture and Human Health*, Sustainable Plant Nutrition in a Changing World,
https://doi.org/10.1007/978-3-031-69876-7_5

Keywords Silicon · Agriculture · Phytolith · Crop production · Biological silicon cycle

5.1 Introduction

Silicon (Si) is a vital element for plant growth and development. It plays a crucial role in various physiological and biochemical processes. In agricultural systems, the biological cycle of Si is a key component in improving plant productivity, resilience to biotic and abiotic stresses, and overall soil health. This cycle involves the uptake, translocation, deposition, and recycling of Si within plant tissues and soil environments. Understanding the dynamics of the biological cycle of Si provides valuable insights into optimizing agricultural practices and promoting sustainable farming methods. This chapter explores the importance of the biological cycle of Si in agricultural systems and its impact on crop production.

5.2 Silicon Pool and Soil Mobility

Silicon is the second most abundant element on the Earth's surface. Its concentration varies from ~5% to ~47% (McKeague and Cline 1963; Swain and Rout 2017). Silicon is divided into two categories: crystalline silica and amorphous silica (Cornelis and Delvaux 2016; Schaller et al. 2021). Crystalline silica includes silicate minerals (e.g., olivine, mica, zeolite, etc.), which are formed by the combination of Si with other elements, as well as simple silica minerals (e.g., quartz, cristobalite quartz, lepidolite quartz). Silica minerals can also be divided into reticulated silicates, page silicates, chain silicates, and island silicates. Amorphous silica includes mainly soil-forming amorphous aluminosilicates, bio-amorphous silica, and adsorbed silica, more than 95% of which is in the form of crystalline silica in most soils (Georgiadis et al. 2013; Georgiadis et al. 2017; Yang et al. 2020a, b). Although crystalline silica can be dissolved due to weathering and release dissolved Si (DSi) for plant uptake, the dissolution rate of crystalline silica is slower than that of amorphous silica (Fraysse et al. 2006; Fraysse et al. 2009; Li et al. 2022). In addition, leaching of DSi is also strong in areas with a strong weathering of crystalline silica (Cornelis and Delvaux 2016; Schaller et al. 2021). Therefore, crystalline silica has little significance as the source of DSi for crop uptake on short timeframes (Cornelis and Delvaux 2016; Schaller et al. 2021) compared to amorphous silica, such as biogenic silica or phytoliths (Derry et al. 2005; Sommer et al. 2006; Puppe et al. 2017; Li et al. 2022, 2023). Despite of the great potential of amorphous silica to replenish the DSi pool, multiple interconversion processes between them may affect the mobility of DSi in soils, such as polymerization/depolymerization and adsorption/desorption (Sommer et al. 2006; Schaller et al. 2021). DSi, as monosilicic acid (H_4SiO_4), can be partly polymerized into polysilicic acid when saturated in certain soil environments (Schaller et al. 2021). Aluminum hydroxide and iron hydroxide can also interact with DSi by adsorption, leading to the formation of amorphous aluminosilicates (Li et al. 2022). This process may affect the overall DSi pool (Georgiadis et al. 2017; Yang et al. 2020a, b). Soil properties and processes also drive DSi mobility (Cornelis and Delvaux 2016; Meunier et al. 2018; Li et al. 2022, 2023). Markedly, intensive cultivation in agricultural systems significantly causes depletion of the DSi source (Struyf et al. 2010; Guntzer et al. 2012). Thus, Si feralization may be a significant challenge in the supplement of DSi pool for crop uptake in agricultural systems.

5.3 Silicon Uptake in Crops

In soils, roots absorb Si as monosilicic acid (i.e., H_4SiO_4, DSi) and transport it upward to various plant parts via xylem flow and transpiration (Ma et al. 2006). In terrestrial ecosystems, the variation of Si content in different plants usually ranges from 1 to 150 g SiO_2 kg^{-1} (Carnelli et al. 2001; Epstein 1994; Hodson et al. 2005). This contrasting variation is mainly attributed to plant species, which have different capacities to absorb Si (Hodson et al. 2005; Yang et al. 2015) (Cornelis et al. 2011). The uptake of Si by plants includes active, repulsive, and passive absorption (Mitani and Ma 2005; Takahashi et al. 1990). According to the Si uptake capacity, terrestrial plant can be divided into Si-loving plants, Si-averse plants, and Si-neutral plants (Ma and Takahashi 2002). Among these plants, crop plants are the most Si-loving plants. For example, sugarcane (Parr et al. 2009), wheat (Parr and Sullivan 2011), and rice (Li et al. 2013) are typical Si-loving plants, which can take up more than 50 g SiO_2 kg^{-1} or even 150 g SiO_2 kg^{-1}. In addition to crop species, climatic and environmental conditions may also affect the uptake and contents of Si in plant tissues (Liang et al. 2015). In addition, the ability of plants to take up Si varies according to stages of the growth (Ru et al. 2018; Sun et al. 2008). The highest absorption capacity is usually strongest during plant maturity.

5.4 Phytolith Formation and Biological Cycle of Si in Agricultural Systems

Silicon-loving plants absorb H_4SiO_4 from the soil solution through active absorption. This substance is mainly accumulated in plants as phytoliths (amorphous silica, SiO_2-H_2O) due to transpiration (Guntzer et al. 2012; Haynes 2017). Phytoliths in plants present a variety of shapes, mainly square, oval, and round (Liang et al. 2015). The amount of phytoliths varies greatly according to crop (Song et al. 2014; Li and Delvaux 2019) and their amounts in each part of a same crop can also vary. Although the amounts of phytoliths vary greatly in different parts of plants, phytoliths are mainly stored in the plant's straw at the end of crop growth (Yang et al. 2024). After returning crop straws into the soils, phytoliths can be released from plant tissues as they degrade. These phytoliths provide a source of DSi in the soil, which can be taken up by plants (Matichenkov et al. 2008; Wickramasinghe and Rowell 2006). This is due to the high dissolution rate of phytoliths (Fraysse et al. 2009; Puppe et al. 2017; Li et al. 2022). Thus, the biological cycle of Si relates to Si uptake, phytolith formation and dissolution, and DSi release (Cornelis and Delvaux 2016; Li and Delvaux 2019).

The biological cycle of Si induced by phytoliths is an important process in regulating the biogeochemical cycle of Si in agricultural systems. The biogeochemical cycle of Si is a long process in which the Si produced from the weathering of rock minerals replenishes the DSi in the soil to a certain extent, affecting crop growth (Alexandre et al. 1997; Bartoli 1985). During this process, most DSi absorbed by plants from the soil in agricultural ecosystems comes from phytolith dissolution (Struyf et al. 2010; Guntzer et al. 2012; Marxen et al. 2016; Haynes 2017; Li et al. 2020). Yet, this contribution depends largely upon soil properties and processes (Cornelis and Delvaux 2016), as well as land use (Vander Linden and Delvaux 2019). With an increasing soil weathering degree, the contribution of supplied phytoliths to plant Si uptake markedly increases from 26 to 80%, of which the rock mineral contribution surpasses that of plant Si uptake (Li et al. 2020). In this regard, Li and Delvaux (2019) have reported a global potential of phytolith-rich biochar produced using major crop residues and manure as a potential Si fertilizer, positively responding to the biological cycle of Si in agricultural systems. This clearly suggests that adding Si fertilizer or any other approaches altering soil Si bioavailability may be a significance challenge in improving the biological cycle of Si in agricultural systems (Li et al. 2019; de Tombeur et al. 2021). Considering that improving this Si cycle promotes a sustainable development of agriculture by reducing abiotic and biotic stresses (Liang et al. 2015; Song et al. 2017a, b), its contribution to responding to climate change and improving agricultural production efficiency needs to be further studied.

5.5 The Role of Biological Cycle of Silicon in Sustainable Agriculture

Silicon is not usually considered as a nutritional element for plant growth. Yet, in recent years, Si has been regarded as an essential and beneficial element for plant growth, especially for crops, as the biological cycle of Si has a great potential as a friendly and environmentally conscious biogeochemical process to promote the sustainable development of agriculture, such as reducing the use of pesticides and improving crop resistance to diseases, thus improving soil quality, protecting the environment, responding to climate change, and improving agricultural production efficiency (Liang et al. 2015; Coskun et al. 2019).

5.5.1 Reducing Pesticide Use and Improving Crop Resistance to Diseases

Recently, unreasonable agricultural practices, such as continuous cropping of single plants and improper fertilization, have sharply weakened the soil's ecological function. As a result, the complexity of the microecological composition around crops has lowered, which consequently has decreased the soil's resistance to soilborne diseases. These issues affect greatly the growth of crops. The application of Si can effectively mitigate plant diseases caused by bacteria or fungi, as previous studies have reported (Kaiser et al. 2005, 2015). Si fertilization inhibits the growth of pathogenic fungi, including *Phytophthora camphora*, *Sclerotinia sclerotinia*, and *Fusarium* in various crops by altering soil properties (e.g., pH and Si bioavailability). There is resistance of tomatoes grown with Si to bacterial wilt caused by *R. solanacearum*, as Si controls the number of soil bacteria and actinomycetes, reduces the ratio of fungi to bacteria, and controls bacterial wilt (Wang et al. 2013). Rice blast is one of the most relevant rice diseases; it can cause a significant reduction in rice yield. The effect of Si on resistance to rice blast is attributed to breaking the invasion of pathogens by siliconizing leaf epidermal cells (Abed-Ashtiani et al. 2012).

The application of Si also has a positive effect on the control of insects and other pests and diseases, reducing the toxicity of pests and diseases to plants, increasing crop yields. There are some reasons for this. Firstly, Si application improves plant resistance to insect pests. The plants develop a stronger attraction to the natural enemies of pests after Si application. This helps to control adverse effects of pests on plants (Kvedaras et al. 2010). Secondly, Si fertilization promotes a large accumulation of silica in leaves that enhances leaf hardness, resisting pests' damage to leaves and other tissues (Gomes et al. 2005; Griffin et al. 2015). Thirdly, Si induces a large increase in non-protein nitrogen that prevents the normal synthesis of insect protein, thus changing the metabolism of insect pests, which in turn damages the growth and development of pests (Ferreira and Moraes 2011).

5.5.2 Improving Soil Quality and Protecting the Environment

Silicon has a beneficial effect on improving soil quality and protecting the environment mainly via alleviation of Al toxicity and metal toxicity. The mechanisms of Si-mediated alleviation of Al toxicity in higher plants involve Al adsorption on Si-O− sites, formation of Si-O-Al-O-Si bonds, and further co-precipitation as hydroxyaluminosilicate (HAS) complexes (Hodson and Evans 1995; Cooke and Leishman 2016; Liang et al. 2015). Consequently, fertilization with Si efficiently alleviates Al toxicity through rapid HAS co-precipitation (Liang et al. 2015). In addition, the liming effect of Si also alleviates Al phytotoxicity due to Al release from soil minerals in acidic conditions (Hodson and Evans 1995; Barhebwa et al.

2024). Si application has indeed a significant depressive effect on exchangeable Al^{3+} content in the soil (Liang et al. 2007). In particular, due to the liming effect of Si fertilization, the hydrolysis of Al^{3+} converts it into $Al(OH)^{2+}$ and $Al(OH)^{2+}$ monomers, which can be strongly adsorbed by the amended soil organic substances through surface complexation (Mimmo et al. 2009).

Silicon fertilizer such as Si-rich biochar has the function as low-cost sorbents for soil contaminants, organic or metallic in nature (Babel and Kurniawan 2003; Liang et al. 2007). Alleviating the toxicity of organic substances happens in colloidal and amorphous silica minerals, but only after neoformation or co-precipitation processes (Chorover et al. 2008). Because of biochar high surface area, high internal porosity, and ion exchange capacity, these neo-formed or precipitated silicas with heavy metals significantly alleviate heavy metal toxicity in contaminated soils (Adrees et al. 2015; Fuller et al. 2016). In addition to its surface properties, Si-rich biochar fertilization can exert a liming effect, which increases soil pH. This increased pH promotes the deprotonation of organic ligands, thus enhancing their binding with heavy metals, especially lead (Pb) and copper (Cu) (Ma and Takahashi 2002; Liang et al. 2007, 2015; Cooke and Leishman 2016; Huang et al. 2024).

5.5.3 Fighting Climate Change and Improving Agricultural Carbon Sequestration

In agroecosystems, Si-loving crops such as sugarcane, wheat, maize, rice, millet, and sorghum have a great potential in driving the biogeochemical cycle of Si. In this cycle, the formation of phytoliths promotes sustainable C sequestration, which positively responds to climate change (Zuo and Lu 2011; Parr and Sullivan 2011; Song et al. 2014; Yang et al. 2024). During the process of phytolith formation, some organic C can be encapsulated within the phytoliths of crop tissues (Bartoli and Wilding 1980; Carter 2009; Santos et al. 2012; Li et al. 2013; Alexandre et al. 2015). These encapsulated C is known as phytolith-occluded C (PhytOC) (Parr and Sullivan 2005). They generally lead to a great stability under the protection of phytoliths; they are conserved in soils on a long-term scale (Wilding 1967; Kelly et al. 1991; Parr and Sullivan 2005; Hodson 2019). Therefore, PhytOC positively improves soil C sequestration in terrestrial ecosystems (Parr and Sullivan 2005; Song et al. 2014, 2018). Among terrestrial ecosystems, agricultural systems show the greatest potential in sustainable CO_2 sequestration to positively respond to climate change at a global scale because of high yields of phytolith and PhytOC (Song et al. 2017a, b). However, this potential depends largely upon the type of crop system. In sugarcane system, the C flux of PhytOC production was estimated as 123~364 kg CO_2 ha^{-1} $year^{-1}$ (Parr et al. 2009), which is the greatest among other systems such as wheat (12.9~129 g kg^{-1}, Parr and Sullivan 2011), rice (30~130 kg CO_2 ha^{-1} $year^{-1}$, Li et al. 2013), and millet (20~30 kg CO_2 ha^{-1} $year^{-1}$, Zuo and Lu 2011). This difference among these ecosystems is attributed to their contrasting variation of phytolith and PhytOC. For example, in China's farmland ecosystems, phytolith C sequestration of rice, wheat, and maize accounted for 40%, 18%, and 30% of the total farmland phytolith sequestration, respectively (Song et al. 2014). Based on the extensive and available literatures, Song et al. (2013, 2017a, b) further estimated that 26.35×10^6 t of CO_2 could be fixed annually by phytoliths in global cropland ecosystems. The three major food crops are rice, wheat, and maize, contributing with 25%, 19%, and 23%, respectively; the three major food producing regions are Asia, the Americas, and Europe, contributing with 49%, 17%, and 16%, respectively. Among the world's three major food producers, the C sequestration flux of phytoliths in China's croplands (36 kg CO_2 ha^{-1} $year^{-1}$) was comparable to that of the United States (34 kg CO_2 ha^{-1} $year^{-1}$) and higher than that of India (19 kg CO_2 ha^{-1} $year^{-1}$). Agricultural ecosystems are closely related to human production activities, survival, and development, significantly affecting climate change and the C cycle (Hu et al. 2022; Zhang et al. 2022). This means that human activities are key in regulating the yield of phytolith C sequestration (Song et al. 2013, 2014). This positively contributes to agricultural C sinks, ultimately enhancing long-term biogeochemical C sequestration in terrestrial ecosystems. Despite the great potential of Si fertilization in this matter (Song et al. 2015), previous studies have limited the analyses to the assessment of experimental results from controlled experiments (Li et al. 2020; Yang et al. 2024), but not from field experiments. This has led to a limited understanding of the intrinsic mechanisms driving the potential of phytolith C sequestration.

5.6 Conclusions

The biological cycle of Si plays a crucial role in promoting sustainable agriculture by reducing pesticide use, improving resistance to diseases, enhancing soil quality, protecting the environment, and fighting climate change through C sequestration. The intricate processes of Si uptake and phytolith formation and dissolution contribute to the overall health and productivity of agricultural systems. The biological cycle of Si and its role in agricultural systems have achieved certain results. Yet, there are still many challenges that need solution. Due to the increasing awareness of the importance of Si in agriculture and the continuous transformation of agricultural production methods, Si cycle should receive more attention and research. First, the mechanism and potential of phytoliths replacing Si fertilizers need to be further explored to improve the use efficiency of Si. Secondly, the specific mechanism of Si on crop growth and phytolith C sequestration under various stresses need to be further studied to better assess Si resources and trace Si sources. Thus, producers should be expected to explore new approaches to achieve a sustainable development of agriculture by conducting an in-depth study of the mechanism of biological Si cycle in agriculture, optimization of Si fertilizer technology, and promotion of a wide application of phytolith C sequestration under different environmental and climatic conditions.

References

Abed-Ashtiani F, Kadir JB, Selamat A-B, Hanif AHB-M, Nasehi A (2012) Effect of foliar and root application of silicon against rice blast fungus in MR219 rice variety. Plant Pathol J 28:164–171

Adrees M, Ali S, Rizwan M, Zia-ur-Rehman M, Ibrahim M, Abbas F, Farid M, Qayyum MF, Irshad MK (2015) Mechanisms of silicon-mediated alleviation of heavy metal toxicity in plants: a review. Ecotoxicol Environ Saf 119:186–197

Alexandre A, Meunier JD, Colin F, Koud JM (1997) Plant impact on the biogeochemical cycle of silicon and related weathering processes. Geochim Cosmochim Acta 61:677–682

Alexandre A, Basile-Doelsch I, Delhaye T, Borshneck D, Mazur JC, Reyerson P, Santos GM (2015) New highlights of phytolith structure and occluded carbon location: 3-D X-ray microscopy and NanoSIMS results. Biogeosciences 12(3):863–873

Babel S, Kurniawan TA (2003) Low-cost adsorbents for heavy metals uptake from contaminated water: a review. J Hazard Mater 97(1–3):219–243

Barhebwa F, Li Z, Bragard C, Vanlauwe B, Delvaux B (2024) Is silicon beneficial for cassava (Manihot esculenta Crantz)? Plant Soil 500:1–21

Bartoli F (1985) Crystallochemistry and surface properties of biogenic opal. J Soil Sci 36:335–350

Bartoli F, Wilding LP (1980) Dissolution of biogenic opal as a function of its physical and chemical properties. Soil Sci Soc Am J 44:873–878

Carnelli AL, Madella M, Theurillat JP (2001) Biogenic silica production in selected alpine plant species and plant communities. Ann Bot 87:425–434

Carter JA (2009) Atmospheric carbon isotope signatures in phytolith-occluded carbon. Quat Int 193(1–2):20–29

Chorover J, Choi S, Rotenberg P, Serne RJ, Rivera N, Strepka C, Thompson A, Mueller KT, O'Day PA (2008) Silicon control of strontium and cesium partitioning in hydroxide-weathered sediments. Geochim Cosmochim Acta 72(8):2024–2047

Cooke J, Leishman MR (2016) Consistent alleviation of abiotic stress with silicon addition: a meta-analysis. Funct Ecol 30(8):1340–1357

Cornelis JT, Delvaux B (2016) Soil processes drive the biological silicon feedback loop. Funct Ecol 30(8):1298–1310

Cornelis J-T, Delvaux B, Georg RB, Lucas Y, Ranger J, Opfergelt S (2011) Tracing the origin of dissolved silicon transferred from various soil-plant systems towards rivers: a review. Biogeosciences 8:89–112. https://doi.org/10.5194/bg-8-89-2011

Coskun D, Deshmukh R, Sonah H, Menzies JG, Reynolds O, Ma JF, Kronzucker HJ, Bélanger RR (2019) The controversies of silicon's role in plant biology. New Phytol 221(1):67–85

de Tombeur F, Roux P, Cornelis JT (2021) Silicon dynamics through the lens of soil-plant-animal interactions: perspectives for agricultural practices. Plant Soil 467(1):1–28

Derry LA, Kurtz AC, Ziegler K, Chadwick OA (2005) Biological control of terrestrial silica cycling and export fluxes to watersheds. Nature 433(7027):728–731

Epstein E (1994) The anomaly of silicon in plant biology. Proc Natl Acad Sci USA 91:11–17

Ferreira RS, Moraes JC (2011) Silicon influence on resistance induction against Bemisia tabaci biotype B (Genn.)(Hemiptera: Aleyrodidae) and on vegetative development in two soybean cultivars. Neotrop Entomol 40:495–500

Fraysse F, Pokrovsky OS, Schott J, Meunier JD (2006) Surface properties, solubility and dissolution kinetics of bamboo phytoliths. Geochim Cosmochim Acta 70:1939–1951

Fraysse F, Pokrovsky OS, Schott J, Meunier JD (2009) Surface chemistry and reactivity of plant phytoliths in aqueous solutions. Chem Geol 258:197–206

Fuller AJ, Shaw S, Peacock CL, Trivedi D, Burke IT (2016) EXAFS study of Sr sorption to illite, goethite, chlorite, and mixed sediment under hyperalkaline conditions. Langmuir 32(12):2937–2946

Georgiadis A, Sauer D, Herrmann L, Breuer J, Zarei M, Stahr K (2013) Development of a method for sequential Si extraction from soils. Geoderma 209:251–261

Georgiadis A, Rinklebe J, Straubinger M, Rennert T (2017) Silicon fractionation in Mollic Fluvisols along the Central Elbe River, Germany. *Catena* 153:100–105

Gomes FB, de Moraes JC, dos Santos CD, Goussain MM (2005) Resistance induction in wheat plants by silicon and aphids. Sci Agric 62:547–551

Griffin M, Hogan B, Schmidt O (2015) Silicon reduces slug feeding on wheat seedlings. J Pest Sci 88:17–24

Guntzer F, Keller C, Poulton PR, McGrath SP, Meunier JD (2012) Long-term removal of wheat straw decreases soil amorphous silica at Broadbalk, Rothamsted. Plant Soil 352:173–184

Haynes RJ (2017) The nature of biogenic Si and its potential role in Si supply in agricultural soils. Agric Ecosyst Environ 245:100–111

Hodson MJ (2019) The relative importance of cell wall and lumen phytoliths in carbon sequestration in soil: a hypothesis. Front Earth Sci 7:167

Hodson MJ, Evans DE (1995) Aluminium/silicon interactions in higher plants. J Exp Bot 46(2):161–171

Hodson MJ, White PJ, Mead A, Broadley MR (2005) Phylogenetic variation in the silicon composition of plants. Ann Bot 96:1027–1046

Hu Y, Li D, Deng J, Yue Y, Zhou J, Chai Y, Li Y (2022) Mechanisms controlling water-level variations in the middle Yangtze River following the operation of the three gorges dam. Water Resour Res 58(10):e2022WR032338

Huang F, Li Z, Yang X, Liu H, Chen L, Chang N, He H, Zeng Y, Qiu T, Fang L (2024) Silicon reduces toxicity and accumulation of arsenic and cadmium in cereal crops: a meta-analysis, mechanism, and perspective study. Sci Total Environ 918:170663

Kaiser C, Van der Merwe R, Bekker TF, Labuschagne N (2005) In-vitro inhibition of mycelial growth of several phytopathogenic fungi, including Phytophthora cinnamomi by soluble silicon. S Afr Avocado Growers' Assoc Yearb 28(1):70–74

Kaiser C, Kilburn MR, Clode PL, Fuchslueger L, Koranda M, Cliff JB, Solaiman ZM, Murphy DV (2015) Exploring the transfer of recent plant photosynthates to soil microbes: mycorrhizal pathway vs direct root exudation. New Phytol 205:1537–1551. https://doi.org/10.1111/nph.13138

Kelly EF, Amundson RG, Marino BD, Deniro MJ (1991) Stable isotope ratios of carbon in phytoliths as a quantitative method of monitoring vegetation and climate change. Quat Res 35(2):222–233

Kvedaras OL, An M, Choi YS, Gurr GM (2010) Silicon enhances natural enemy attraction and biological control through induced plant defences. Bull Entomol Res 100:367–371

Li Z, Delvaux B (2019) Phytolith-rich biochar: a potential Si fertilizer in desilicated soils. GCB Bioenergy 11(11):1264–1282

Li Z, Song Z, Parr JF, Wang H (2013) Occluded C in rice phytoliths: implications to biogeochemical carbon sequestration. Plant Soil 370:615–623

Li Z, Unzué-Belmonte D, Cornelis JT et al (2019) Effects of phytolithic rice-straw biochar, soil buffering capacity and pH on silicon bioavailability. Plant Soil 438:187–203. https://doi.org/10.1007/s11104-019-04013-0

Li Z, Cornelis JT, Vander Linden C, Van Ranst E, Delvaux B (2020) Neoformed aluminosilicate and phytogenic silica are competitive sinks in the silicon soil–plant cycle. Geoderma 368:114308

Li Z, Meunier JD, Delvaux B (2022) Aggregation reduces the release of bioavailable silicon from allophane and phytolith. Geochim Cosmochim Acta 325:87–105

Li Z, Meunier JD, Delvaux B (2023) Goethite affects phytolith dissolution through clay particle aggregation and pH regulation. Geochim Cosmochim Acta 349:11–22

Liang Y, Sun W, Zhu YG, Christie P (2007) Mechanisms of silicon-mediated alleviation of abiotic stresses in higher plants: a review. Environ Pollut 147(2):422–428

Liang Y, Nikolic M, Belanger R, Gong H, Song A (2015) Silicon in agriculture: from theory to practice. Springer

Ma JF, Takahashi E (2002) Soil, fertilizer, and plant silicon research in Japan. Elsevier

Ma JF, Tamai K, Yamaji N, Mitani N, Konishi S, Katsuhara M, Ishiguro M, Murata Y, Yano M (2006) A silicon transporter in rice. Nature 440:688–691

Marxen A, Klotzbücher T, Jahn R, Kaiser K, Nguyen VS, Schmidt A, Schädler M, Vetterlein D (2016) Interaction between silicon cycling and straw decomposition in a silicon deficient rice production system. Plant Soil 398:153–163

Matichenkov VV, Bocharnikova EA, Kosobryukhov AA, Biel KY (2008) Mobile forms of silicon in plants: doklady biological sciences: proceedings of the Academy of Sciences of the USSR. Biol Sci Sect 418:39–40

McKeague JA, Cline MG (1963) Silica in soils. Adv Agron 15:339–396

Meunier JD, Sandhya K, Prakash NB, Borschneck D, Dussouillez P (2018) pH as a proxy for estimating plant-available Si? A case study in rice fields in Karnataka (South India). Plant Soil 432:143–155

Mimmo T, Sciortino M, Ghizzi M, Gianquinto G, Gessa CE (2009) The influence of aluminium availability on phosphate uptake in Phaseolus vulgaris L. and Phaseolus lunatus L. Plant Physiol Biochem 47(1):68–72

Mitani N, Ma JF (2005) Uptake system of silicon in different plant species. J Exp Bot 56(414):1255–1261

Molotoks A, Stehfest E, Doelman J, Albanito F, Fitton N, Dawson TP, Smith P (2018) Global projections of future cropland expansion to 2050 and direct impacts on biodiversity and carbon storage. Glob Chang Biol 24(12):5895–5908

Parr JF, Sullivan LA (2005) Soil carbon sequestration in phytoliths. Soil Biol Biochem 37(1):117–124

Parr JF, Sullivan LA (2011) Phytolith occluded carbon and silica variability in wheat cultivars. Plant Soil 342:165–171

Parr J, Sullivan L, Quirk R (2009) Sugarcane phytoliths: encapsulation and sequestration of a long-lived carbon fraction. Sugar Tech 11:17–21

Piperno DR (1988) Phytolith analysis: an archaeological and geological perspective. Academic, San Diego

Puppe D, Höhn A, Kaczorek D, Wanner M, Wehrhan M, Sommer M (2017) How big is the influence of biogenic silicon pools on short-term changes in water-soluble silicon in soils? Implications from a study of a 10-year-old soil–plant system. Biogeosciences 14(22):5239–5252

Ru N, Yang X, Song Z, Liu H, Hao Q, Liu X, Wu X (2018) Phytoliths and phytolith carbon occlusion in aboveground vegetation of sandy grasslands in eastern Inner Mongolia, China. Sci Total Environ 625:1283–1289

Santos GM, Alexandre A, Southon JR, Treseder KK, Corbineau R, Reyerson PE (2012) Possible source of ancient carbon in phytolith concentrates from harvested grasses. Biogeosciences 9(5):1873–1884

Schaller J, Puppe D, Kaczorek D, Ellerbrock R, Sommer M (2021) Silicon cycling in soils revisited. Plan Theory 10(2):295

Sommer M, Kaczorek D, Kuzyakov Y, Breuer J (2006) Silicon pools and fluxes in soils and landscapes—a review. J Plant Nutr Soil Sci 169(3):310–329

Song Z, Parr JF, Guo F (2013) Potential of global cropland Phytolith carbon sink from optimization of cropping system and fertilization. PLoS One 8:e73747

Song Z, Wang H, Strong PJ, Guo F (2014) Phytolith carbon sequestration in China's croplands. Eur J Agron 53:10–15

Song A, Ning D, Fan F et al (2015) The potential for carbon bio-sequestration in China's paddy rice (Oryza sativa L.) as impacted by slag-based silicate fertilizer. Sci Rep 5:17354. https://doi.org/10.1038/srep17354

Song A, Fan F, Yin C, Wen S, Zhang Y, Fan X, Liang Y (2017a) The effects of silicon fertilizer on denitrification potential and associated genes abundance in paddy soil. Biol Fertil Soils 53:627–638

Song Z, Liu H, Strömberg CA, Yang X, Zhang X (2017b) Phytolith carbon sequestration in global terrestrial biomes. Sci Total Environ 603:502–509

Song Z, Liu C, Müller K, Yang X, Wu Y, Wang H (2018) Silicon regulation of soil organic carbon stabilization and its potential to mitigate climate change. Earth Sci Rev 185:463–475. https://doi.org/10.1016/j.earscirev.2018.06.020

Struyf E, Smis A, Van Damme S, Garnier J, Govers G, Van Wesemael B, Conley DJ, Batelaan O, Frot E, Clymans W, Vandevenne F (2010) Historical land use change has lowered terrestrial silica mobilization. Nat Commun 1(1):129

Sun L, Wu LH, Ding TP, Tian SH (2008) Silicon isotope fractionation in rice plants, an experimental study on rice growth under hydroponic conditions. Plant Soil 304:291–300

Swain R, Rout GR (2017) Silicon in agriculture. In: Lichtfouse E (ed) Sustainable agriculture reviews, vol 25, pp 233–260

Takahashi E, Ma JF, Miyake Y (1990) The possibility of silicon as an essential element for higher plants. Comments Agric Food Chem 2(2):99–102

Vander Linden C, Delvaux B (2019) The weathering stage of tropical soils affects the soil-plant cycle of silicon, but depending on land use. Geoderma 351:209–220

Wang L, Cai K, Chen Y, Wang G (2013) Silicon-mediated tomato resistance against Ralstonia solanacearum is associated with modification of soil microbial community structure and activity. Biol Trace Elem Res 152:275–283

Wickramasinghe DB, Rowell DL (2006) The release of silicon from amorphous silica and rice straw in Sri Lankan soils. Biol Fertil Soils 42:231–240

Wilding LP (1967) Radiocarbon dating of biogenetic opal. Science 156(3771):66–67

Yang X, Song Z, Liu H et al (2015) Plant silicon content in forests of north China and its implications for phytolith carbon sequestration. Ecol Res 30:347–355. https://doi.org/10.1007/s11284-014-1228-0

Yang X, Song Z, Qin Z, Wu L, Yin L, Van Zwieten L, Song A, Ran X, Yu C, Wang H (2020a) Phytolith-rich straw application and groundwater table management over 36 years affect the soil-plant silicon cycle of a paddy field. Plant Soil 454:343–358

Yang X, Song Z, Yu C, Ding F (2020b) Quantification of different silicon fractions in broadleaf and conifer forests of northern China and consequent implications for biogeochemical Si cycling. Geoderma 361:114036

Yang X, Ni Y, Li Z, Yue K, Wang J, Li Z, Yang X, Song Z (2024) Silicon in paddy fields: benefits for rice production and the potential of rice phytoliths for biogeochemical carbon sequestration. Sci Total Environ 929:172497

Zhang W, Qiu X, Wang C, Zhong L, Fu F, Zhu J, Zhang Z, Qin Y, Yang D, Xu CC (2022) Lignin derived carbon materials: current status and future trends. Carbon Res 1(1):14

Zuo X, Lu H (2011) Carbon sequestration within millet phytoliths from dry-farming of crops in China. Chin Sci Bull 56:3451–3456

Chapter 6
The Use of Silicon-Based Fertilization in Agricultural Fields for Effective Management of Carbon Sequestration

Mahboobeh Jalali, Keyvan Valizadeh Rad, and Hassan Etesami

Abstract The increase in greenhouse gases in the atmosphere causes climate change and poses a major threat to the stability of the Earth's ecosystem. Therefore, striving for a safe and efficient carbon sequestration mechanism in the long term is very important. In agricultural lands, crop harvesting and removal of agrarian land residues often lead to the imbalance of carbon and nutrient biogeochemical cycles. Silicon (Si) serves various ecological roles that aid plants in adapting to challenging environments. While Si application has been shown to mitigate abiotic and biotic stresses and boost biomass accumulation, its impact on carbon fixation and nutrient uptake in agriculture remains uncertain. Therefore, carbon retention in the soil plays a vital role in maintaining soil fertility and preventing climate change. Silicon facilitates plant growth and substrate decomposition, indicating that carbon cycling in croplands is vital for the biogeochemical cycles of carbon and nutrients. Therefore, managing Si in croplands scientifically can help maintain sustainable agricultural development.

Keywords Silicon · Carbon sequestration · Soil · Deforestation · Agricultural land

6.1 Introduction

The escalation in atmospheric greenhouse gas concentrations stands as the primary driver of climate alteration and poses a substantial peril to the terrestrial ecosystem's sustainability. A notable surge in global carbon dioxide (CO_2) concentration, reaching up to 31.5 Gt, has been witnessed. The uptick in atmospheric temperature

M. Jalali (✉) · K. V. Rad
Faculty of Agriculture, Lorestan University, Khorramabad, Iran
e-mail: jalali.mah@lu.ac.ir

H. Etesami
Department of Soil Science, University of Tehran, Tehran, Iran

R. de Mello Prado et al. (eds.), *Silicon Advances for Sustainable Agriculture and Human Health*, Sustainable Plant Nutrition in a Changing World,
https://doi.org/10.1007/978-3-031-69876-7_6

directly impacts soil respiration, thereby expediting the liberation of a significant volume of carbon from soil (AminiTabrizi et al. 2020). Consequently, strategies for mitigating climate change are imperative, either through curtailing emissions from sources or by diminishing atmospheric CO_2 concentration via carbon sinks. Despite the substantial carbon reservoir in soil, owing to altered land usage, intricate carbon storage processes, and perpetually evolving environmental conditions, much of the organic carbon in soil is unable to endure over protracted periods. Thus, the quest for a secure and efficient long-term carbon sequestration mechanism is paramount (Polidoro et al. 2021).

The trapping of carbon within phytoliths, known as biotic carbon sequestration, is seen as a practical method to combat climate change. Phytolith-occluded carbon (PhytOC) has been shown to have a significant impact on capturing and storing carbon over the long term. Phytoliths, which are silicon-coated carbon structures also known as amorphous silica (Si), are formed within plant cells through a process called biosilicification. Silicon is absorbed by the roots as soluble silicic acid [$Si(OH)_4$] and then transported throughout the plant via the vascular system, where it accumulates in various plant parts such as leaves, stems, and roots. While silicon is not essential for plant growth, it has many beneficial effects on plant development, yield, nutrition, and overall health, offering protection against various stresses. Phytoliths also play a role in influencing element cycling during the decomposition of plant litter. The aggregation of phytoliths within plant parts is aided by processes such as evaporation and water transport (Davamani et al. 2022).

Although there is little information about the mechanisms preventing the combination and precipitation of silicic acid in the xylem, according to the characteristics of silicic acid, which begins to form a polymer at a high concentration of 2 mM and a temperature of 25 °C, if silicic acid loses water during movement in aerial organs along with transpiration flow (due to its concentration), the flow of silicic acid polymerization occurs. The polymerization process includes converting silicic acid to colloidal silicic acid and finally to silica gel, which leads to the storage of polymerized silicon in the cuticle and special cells of the plant (Ma and Takahashi 2002). When the concentration of silicon in colloidal form together with silicic acid is about 140–230 mg/L of silicon and in other words, more than 90% of the total silicon in the aerial parts of the accumulating plant, silicon is stored in different types of phytoliths or silicic structures. Cellulose has been transformed into amorphous silicon (Yoshida et al. 1959) and appears as silica gel in the structure. Phytoliths, which are deposited in the cell wall, intracellular space and capillaries, have a very diverse structure (Ma and Yamaji 2006; Mazumdar 2011; Cooke and Leishman 2011) so that a plant can synthesize several forms of it (Mica 1986).

Meanwhile, silicon-rich structures in nanometer (molecular), microscopic (superstructural), and macroscopic (mass) dimensions are synthesized in plants (Mann and Perry 1986). Biogenic silicon is in the form of unique cellular structures or is produced intracellularly at the nanometer level (Mann and Ozin 1996). Silica cells are present in vascular and dumbbell-shaped organs, while silica organs are observed in bulliform cells of leaves. When the silicon concentration in the shoot reaches less than 5% of the total silicon, we only see the formation of silica cells.

Still, as the concentration of silicon increases, the number of silica cells increases, indicating the silicification process in epidermal cells. It is by changing the shape of silica cells into silica organs (Ma and Takahashi 2002).

Phytolith-occluded carbon (PhytOC) is a stable organic carbon fraction trapped within phytoliths, making up 0.1–5.8% of the total. Phytoliths are formed when silicon is deposited on plant cell walls, playing a vital role in cell structure. The composition of phytoliths, including PhytOC, is influenced by factors such as plant species, soil type, climate, pH, temperature, and where in the plant they are located. PhytOC has the potential to sequester carbon over long periods, with some studies showing it can make up to 82% of carbon in well-drained soils after 2000 years. The amount of PhytOC in different plants varies significantly due to their ability to accumulate phytoliths (Nawaz et al. 2019). The aim of this book chapter is to explore the use of silicon-based fertilization for effective carbon sequestration in agricultural fields. It discusses how silicon enhances plant growth, stress tolerance, and photosynthesis, ultimately leading to increased carbon sequestration. Factors influencing plant-available silicon and PhytOC estimation, such as soil type, pH levels, climate, and soil microorganisms, are also considered. Understanding these factors is crucial for maximizing the benefits of silicon-based fertilization in carbon sequestration efforts.

6.2 Concept of Carbon Dioxide Deposition in Nature

Terrestrial ecosystems are vital in the global carbon cycle by absorbing a significant portion of human-made CO_2 emissions since the 1960s. The increase in atmospheric CO_2 levels has diverse effects on terrestrial ecosystems, including warming the climate, enhancing photosynthesis, reducing plant transpiration, and altering the composition of carbon, nitrogen, and phosphorus in ecosystem carbon pools. These responses underscore the importance of terrestrial ecosystems in the carbon cycle and their role as crucial carbon sinks (Peñuelas et al. 2013). Carbon dioxide deposition in nature refers to the natural processes through which carbon dioxide (CO_2) is removed from the atmosphere and stored in various natural reservoirs for an extended period. These reservoirs include the oceans, forests, soils, and geological formations (Arora et al. 2019). Carbon dioxide deposition occurs in different natural systems. Plants absorb CO_2 from the atmosphere during photosynthesis, converting sunlight, water, and CO_2 into glucose and oxygen. The carbon captured from CO_2 is used to build plant tissues and roots. Forests, grasslands, and other vegetation act as carbon sinks sequestering significant amounts of carbon through photosynthesis. The world's oceans also absorb CO_2 from the atmosphere through dissolution. CO_2 dissolves in seawater, reacting with water molecules to form carbonic acid, bicarbonate, and carbonate ions. Marine organisms such as phytoplankton also play a role by using CO_2 for photosynthesis. The oceans store vast amounts of carbon in the form of dissolved inorganic carbon, organic carbon, and sediments, making them crucial carbon sinks (Basu and Mackey 2018). In addition, carbon is stored in soils through various processes, including decomposing plant and animal matter, root exudates, and microbial activity. When organic matter decomposes in the soil, some carbon is converted into stable forms such as humus, which can persist in soils for long periods. Healthy soils with high levels of organic matter can sequester significant amounts of carbon, making sustainable land management practices essential for maximizing soil carbon storage (Gougoulias et al. 2014).

Carbon dioxide (CO_2) is stored in soils primarily through two main processes: carbon mineralization and physical adsorption. Carbon mineralization, or soil organic matter decomposition, is how organic carbon compounds in plant residues, root exudates, and microbial biomass are broken down by soil microorganisms, releasing CO_2 into the soil atmosphere. This CO_2 can then be stored in soil in several ways. Soil microorganisms decompose organic matter through respiration, releasing CO_2 as a byproduct. This CO_2 can accumulate in soil pore spaces or diffuse into the soil atmosphere, contributing to soil CO_2 storage (Wiesmeier et al. 2019). Some of the CO_2 released during microbial respiration can dissolve in soil water, forming dissolved inorganic carbon (DIC). DIC can be transported through soil pore water and stored in deeper soil layers, contributing to long-term carbon storage (Liu et al. 2020a). Despite microbial decomposition, a fraction of the organic carbon compounds may resist degradation and become stabilized in soil organic matter. This stabilized organic matter can persist in soils for extended periods, effectively storing CO_2 over long timescales (Lützow et al. 2006).

Soil minerals, such as clays and oxides, can adsorb CO_2 molecules onto their surfaces through physical interactions. This process, known as physical adsorption or physisorption, allows soils to store CO_2 in mineral-associated forms. The extent of CO_2 adsorption depends on soil texture, mineralogy, and environmental conditions (Chouikhi et al. 2019). Clay minerals such as montmorillonite and saponite have high surface areas and negatively charged surfaces that can attract and retain CO_2 molecules. CO_2 adsorption onto clay surfaces contributes to soil carbon storage and can enhance the stabilization of soil organic matter (Gómez-Pozuelo et al. 2019). Iron and aluminum oxides in soils also have reactive surfaces that can adsorb CO_2. These mineral oxides contribute to CO_2 storage in soils, particularly in soils with high iron and aluminum oxide concentrations. In some soils, CO_2 can react with calcium or magnesium ions in solution to form carbonate minerals, such as calcite or dolomite. These carbonate minerals act as long-term sinks for atmospheric CO_2, storing carbon in solid mineral forms. Physical adsorption of CO_2 onto soil minerals provides another mechanism for storing CO_2 in soils, complementing the biological processes of carbon mineralization (Keykha et al. 2023). Over geological timescales, carbon dioxide can be removed from the atmosphere through weathering and mineralization processes. Carbon dioxide reacts with minerals in rocks, forming carbonate minerals such as calcite and dolomite. These carbonates are then deposited in sedimentary rocks, effectively locking away carbon for millions of years. Geological carbon sequestration is a natural process but occurs over much longer timescales than other carbon sinks.

6.3 Importance of Carbon Dioxide Deposition for Mitigating Climate Change

Carbon dioxide deposition in nature is crucial for mitigating climate change because it removes carbon dioxide (CO_2) from the atmosphere and holds it in various natural reservoirs (Prajapati et al. 2023). Carbon dioxide deposition for mitigating climate change is essential for (i) reducing atmospheric CO_2 concentrations. Carbon dioxide contributes to global warming by entrapping heat in the Earth's atmosphere. By removing CO_2 from the atmosphere and keeping it in natural reservoirs such as forests, oceans, and soils, carbon dioxide deposition helps reduce the concentration of greenhouse gases:

(i) Slowing down the rate of global warming (McLaughlin et al. 2023).
(ii) Balancing the carbon cycle. Carbon dioxide deposition plays a vital role in balancing the Earth's carbon cycle. The carbon cycle involves the exchange of carbon between the atmosphere, oceans, land, and living organisms. Natural processes such as photosynthesis, oceanic uptake, and soil carbon storage help maintain equilibrium in the carbon cycle, preventing excessive buildup of CO_2 in the atmosphere (McLaughlin et al. 2023).
(iii) Preserving ecosystem services. Forests, wetlands, and grasslands as natural carbon sinks provide essential ecosystem services that support biodiversity, water regulation, and soil fertility. By promoting carbon dioxide deposition in these ecosystems, we mitigate climate change, protect valuable habitats, and maintain the services they provide to human societies (Prajapati et al. 2023).
(iv) Enhancing resilience to climate change. Healthy ecosystems with high levels of carbon storage are more resilient to the impacts of climate change. Forests, for example, can help regulate local climate patterns, prevent soil erosion, and mitigate the effects of extreme weather events such as floods and droughts. By preserving and restoring natural carbon sinks, we can enhance the resilience of ecosystems and communities to climate change impacts (Prajapati et al. 2023).
(v) Contributing to carbon neutrality. Achieving carbon neutrality, where the net emissions of CO_2 are balanced by removal or offsetting measures, is essential for meeting climate goals and limiting the rise in global temperature. Carbon dioxide deposition in nature provides a natural and cost-effective way to sequester carbon and offset emissions from human activities such as burning fossil fuels, deforestation, and industrial processes.

6.4 Silicon in Soil

Silicon is the second most abundant element in the Earth's crust and is widely distributed in various mineral forms (Liu et al. 2020a). Within the soil, silicon exists in two distinct phases: liquid and solid. In its liquid phase, silicon dissolves in soils as monosilicic acid and polysilicic acid, as well as in complexes involving inorganic and organic compounds (Sauer et al. 2006). In the solid phase, silicon takes on diverse forms, including amorphous silicon (ASi), biogenic ASi (encompassing diatom shells, phytoliths, testate amoeba shells, radiolarian shells, and silicified sponge spicules), and minerogenic forms (such as iron oxide silica, pedogenic oxide silica, and silica nodules) (Sauer et al. 2006). The processes involving polymerization, depolymerization, and complexation of silicic acid with organic and inorganic ligands, along with adsorption and desorption onto or from mineral or organic surfaces, govern the concentration of silicic acid in the soil solution (Sjöberg 1996). Agricultural soils generally exhibit significantly lower silicon concentrations than nonagricultural soils due to the continual removal of silicon through crop harvesting (Schaller et al. 2021). The quantity of organic carbon potentially associated with phytoliths relies on the availability of plant-available silicon (PAS) (Li et al. 2013), highlighting the importance of regulating PAS in soils. In silicon-deficient soils, silicon fertilization can boost crop yield and enhance both phytolith content and estimated PhytOC fluxes (Sun et al. 2022).

6.4.1 Role of Silicon in Soil and Geological Formations

Silicon plays a crucial role in soil and geological formations due to its abundance and various chemical properties. In soil, silicon contributes to the formation of soil particles, particularly in the form of silicate minerals such as quartz, feldspar, and mica. These minerals provide structural stability to soil particles, influencing soil texture, porosity, and water retention capacity. Silicon is also incorporated into soil aggregates through interactions with organic matter, clay minerals, and metal oxides. Silicon can form complexes with organic molecules, facilitating the binding of soil particles and organic matter into stable aggregates. These aggregates improve soil structure by enhancing stability, porosity, and water infiltration rates. Silicon also promotes the formation of microaggregates, which are clusters of soil particles held together by organic compounds and mineral coatings. These microaggregates contribute to soil stability and prevent soil erosion by wind and water (Matichenkov and Bocharnikova 2001). Silicon contributes to the stability of soil aggregates by promoting the formation of bridges between soil particles and organic matter. Silicate minerals, such as clay minerals and silica gels, can adsorb onto soil particle surfaces and act as cementing agents, binding soil particles together. This binding effect increases the cohesion and mechanical strength of soil aggregates, reducing their susceptibility to disruption by physical forces such as water erosion and compaction. Additionally, silicon enhances the stability of soil organic matter by forming complexes with organic molecules, thereby reducing the rate of organic matter decomposition and preserving soil structure over time (Matichenkov and Bocharnikova 2001). Silicon influences soil porosity by promoting the formation of soil aggregates with well-defined pore spaces. These pore spaces provide habitats for soil microorganisms, facilitate the movement of water and air through the soil, and promote root growth and nutrient uptake by plants. Silicon enhances soil porosity by increasing the stability of soil aggregates and preventing compaction, which can impede water infiltration and root penetration. Improved soil porosity enhances soil aeration, water retention, and drainage, contributing to overall soil health and productivity (Schaller et al. 2020). Silicon plays a critical role in reducing soil erosion by enhancing soil stability and aggregation. Stable soil aggregates formed with the help of silicon are less susceptible to detachment and transport by wind and water erosion processes. Silicon-treated soils exhibit greater resistance to erosion and maintain their structural integrity under erosive forces. By reducing soil erosion, silicon contributes to the preservation of soil fertility, the prevention of sedimentation in water bodies, and the protection of terrestrial ecosystems (Matichenkov and Bocharnikova 2001). Silicon can also act as a buffer against changes in soil pH, helping to maintain soil stability and structure. Silicate minerals such as clay minerals and silica gels can neutralize acidic or alkaline soil conditions by releasing or absorbing hydrogen ions (H^+) or hydroxide ions (OH^-). This buffering effect stabilizes soil pH within a suitable range for soil microbial activity, nutrient availability, and plant growth, thereby promoting soil structure and fertility (Sirisuntornlak et al. 2021). Silicon containing minerals undergo physical and chemical weathering processes, contributing to soil formation. Physical weathering involves breaking rocks into smaller particles, while chemical weathering involves the alteration of minerals through processes such as hydrolysis and dissolution. Silicon is released during these weathering processes, contributing to the soil's chemical composition.

The increase in the absorption of elements such as potassium, phosphorus, and nitrogen by adding silicon to the food solution has been stated in the research of some researchers (Wang et al. 2001). Based on the results of some research, silicon boosts the mobility of phosphorus in the soil, as the beneficial effect of silicon in conditions of phosphorus deficiency has been reported in many plants, such as rice and barley (Marschner 1995; Valizadeh-rad et al. 2022). Due to the competition of usual anions in surface absorption, silicon can cause the release of phosphate, thus increasing the mobility and usability of phosphorus for plants. Also, silicon can positively affect plants' photosynthesis by mobilizing soil phosphorus (Valizadeh-Rad et al. 2023). Mechanisms are known that silicon (through up-regulating the expression of transporter genes to absorb inorganic phosphorus and increase the secretion of malate and citrate) (Kostic et al. 2017) can increase the availability of phosphorus to the plant and, ultimately, its absorption by the plant. Research studying the effect of silicon on corn cultivated in a hydroponic environment reported an increase in iron as a result of the use of silicon. It was determined that iron absorption in plants treated with silicon is related to the increase in the activity of iron chelate reductase FRO (Fe(III)-chelate reductase) is in the root (Gottardi et al. 2012). It has been reported that the application of silicon reduces chlorosis caused by iron deficiency in plants treated with silicon by increasing the transfer of iron from root to stem, balancing other nutritional elements such as the ratio of iron to manganese, and the accumulation of mobile iron compounds such as citrate (in xylem vessels and root and stem tissues). Moreover, it decreased the increase in iron chelate reductase activity (Etesami and Jeong 2018). Pavlovic et al. (2013) stated that adding silicon to the cucumber nutrient solution regulates the protein gene expression level and, as a result, increases the mobility of iron and absorption in the cucumber plant. Also, adding silicon to the nutrient solution reduced the symptoms of iron deficiency chlorosis in the plant (Gonzalo et al. 2013; Bityutskii et al. 2014). Also, when plants suffer from zinc deficiency in the growth environment, silicon can solve zinc deficiency by increasing citrate in the plant and changing zinc distribution (Pascual et al. 2016). When silicon and zinc are in the vicinity of the root endoderm (Gu et al. 2011), they precipitate in the form of zinc silicate, which prevents the transport of zinc to the xylem (da Cunha and do Nascimento 2009). Finally, it accumulates inside the vacuole (Neumann and Zur Nieden 2001) and prevents excessive transfer of some metals to sensitive plant organs (da Cunha and do Nascimento 2009). Researchers have found that low molecular weight chelating agents (such as citrate) increase zinc solubility by forming complexes with zinc (Pavlovic et al. 2013; Bityutskii et al. 2014).

Silicon is a critical component of silica minerals, such as quartz and amorphous silica. These minerals are abundant in various geological formations, including sedimentary rocks such as sandstone and conglomerate and igneous rocks such as granite and basalt. Silica minerals contribute to the strength and durability of rocks and are essential in producing various construction materials, including concrete and ceramics. Most of the silicon in the soil consists of silicon in the solid phase, which is found in the structure of clay minerals and amorphous silicates. It is the most important primary source of silicon needed by plants, which is obtained through ligand exchange or permeation (Valizadeh-rad et al. 2022). The anionic form is absorbed on the surface and is considered the most immediate source of soil-soluble silicon (Ranganathan et al. 2006). The crystalline form includes primary silicates (quartz, mica, feldspar, olivine, and pyroxene) in the sand and silt component and secondary silicates (clay minerals including phyllosilicates) in the soil clay component. The weak crystalline and microcrystalline forms include allophane, opal, imogolite, and secondary quartz. The amorphous form includes the biogenic form (vegetable waste and the remains of microorganisms and phytoliths) or the lithopoedogenic form (complexes of silica with iron and aluminum and heavy metals and soil organic matter substances) (Sauer et al. 2006).

6.5 How Silicon Interacts with Carbon Dioxide in Soil

In soil, silicon interacts with carbon dioxide (CO_2) primarily through chemical weathering processes. Silicon is commonly found in soil in the form of silicate minerals, such as quartz, feldspar, and clay minerals. When these silicate minerals are exposed to CO_2 dissolved in water, they undergo weathering reactions, releasing silicon ions into the soil solution and consuming CO_2 in the process (Marschner 1995). The chemical weathering reactions involving silicon and CO_2 in soil can be described as follows:

Dissolution of silicate minerals Silicate minerals in soil, such as feldspar and olivine, contain silicon-oxygen tetrahedra linked together in various structures. When these minerals are exposed to CO_2 dissolved in water, carbonic acid (H_2CO_3) forms, which can react with the mineral surfaces. This leads to the dissolution of silicate minerals and the release of silicon ions into the soil solution. For example, the dissolution of feldspar ($KAlSi_3O_8$) can be represented as:

$$\mathrm{KAlSi_3O_8 + 4CO_2 + 4H_2O \rightarrow Al^{3+} + 4HCO_3^- + K^+ + 3Si(OH)_4}$$

This reaction consumes CO_2 from the soil solution, effectively sequestering carbon.

Formation of secondary minerals The released silicon ions can further react with other dissolved ions in the soil solution to form secondary minerals such as clay minerals. These secondary minerals may also contain carbonate ions (CO_3^{2-}) that have reacted with CO_2, further contributing to carbon storage in the soil. For instance, the formation of clay minerals such as kaolinite ($Al_2Si_2O_5(OH)_4$) can involve the reaction between dissolved silicon ($Si(OH)_4$) and aluminum (Al^{3+}) ions in the soil solution.

Additionally, carbonate minerals such as calcite ($CaCO_3$) may precipitate in soil and release CO_2 as a result of reactions between calcium ions (Ca^{2+}) and carbonate ions ($2HCO_3^-$).

6.6 Different Silica Mechanisms for Carbon Sequestration

Plants primarily take up Si in the form of H_4SiO_4 and subsequently polymerize it into a form called phytoliths, which are composed of SiO_2 (Epstein and Bloom 2005; Piperno 2006; Wani et al. 2022). During the formation of phytoliths, about 0.2–5.8% of the C becomes entrapped within them, known as phytolith-occluded carbon (PhytOC) (Parr and Sullivan 2011; Song et al. 2016). Because of the high resistance of phytoliths, the PhytOC can be stored in soils for thousands of years (Rashid et al. 2019; Li et al. 2020, 2022), with up to 82% of the total C remaining even after 2000 years of litter decomposition (Parr and Sullivan 2005). The accumulation rate of PhytOC in soils is determined by various factors, including the geochemical stability of soils, turnover rates in different ecosystems, land-use type, climate, anthropogenic activities, lithology, hydrology, and soil physicochemical properties (Pan et al. 2017; Song et al. 2016; Nguyen et al. 2019). However, the composition of the overlying vegetation is the primary factor that determines the PhytOC pool in the soil. Considering this, many studies have estimated or modeled the potential function of different crops (Song et al. 2017), grasses (Liu et al. 2020a), and forests (Yang et al. 2019) in PhytOC production. Globally, the total phytolith

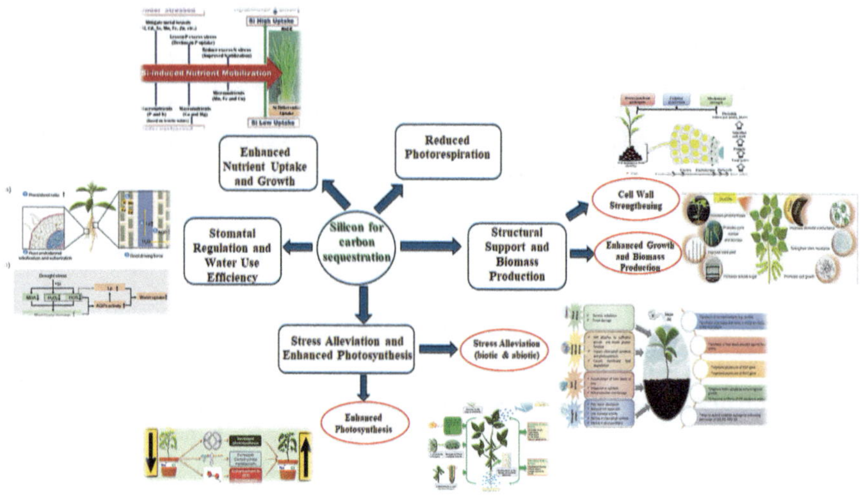

Fig. 6.1 Some physiological mechanisms by which silicon indirectly contributes to carbon sequestration

carbon sequestration in terrestrial biomes has been estimated to be 156.7 ± 91.6 Tg CO_2 $yr.^{-1}$ (Song et al. 2017).

Climate change and environmental stresses can change vegetation types and species abundance, thereby restricting carbon fixation and influencing the distribution of silicon in crops. These changes affect both primary (carbohydrate, N, and polyol) and secondary (flavonol glycosides, phenolic, and phytohormone) crop metabolisms (Kuokkanen et al. 2001; Garamvölgyi and Hufnagel 2013; Ode et al. 2014; Johnson and Hartley 2018). The physiology of silicon (Si) in the context of carbon sequestration in plants involves several intricate processes that influence plant growth, stress responses, and carbon assimilation. While silicon is not directly involved in carbon fixation, its presence in plant tissues can affect various physiological mechanisms that indirectly contribute to carbon sequestration (Fig. 6.1).

6.6.1 Structural Support and Biomass Production

Silicon plays a crucial role in enhancing the structural integrity of plants by accumulating in cell walls and forming silicon-based structures, such as silica bodies or phytoliths. This deposition of silicon strengthens cell walls, making them more rigid and resistant to physical stress (Singh et al. 2023). As a result, plants are better equipped to support their aboveground structures, which can lead to increased biomass production and carbon assimilation through photosynthesis. Silicon plays a crucial role in providing structural support to plants, which indirectly influences biomass production and carbon sequestration. Silicon is deposited within plants' cell walls, forming a layer of silica (SiO_2). This process, called silicification, enhances the rigidity and strength of cell walls. As a result, plant tissues become more resistant to physical stressors such as wind, rain, and mechanical damage. This structural support is significant for upright-growing plants, such as grasses and cereals, as it helps them maintain their posture and withstand environmental challenges. Within plant cells, silicon is deposited in the cell walls, primarily in amorphous silica ($SiO_2 \cdot nH_2O$) (Currie and Perry 2007). This deposition occurs in both primary and secondary cell walls. Silicon deposition is facilitated by specialized transporters that facilitate the movement of silicic acid to the cell walls. The presence of silica in the cell walls enhances their mechanical strength and rigidity. This reinforcement makes the cell walls more resistant to physical stresses, such as bending, compression, and abrasion. As a result, plant cells can better maintain their structural integrity under various environmental conditions. Silicon deposition in the cell wall can affect the activity and expression of enzymes and proteins involved in cell wall metabolism and remodeling. Silicon has been shown to influence the synthesis and deposition of cell wall components such as hemicellulose, cellulose, pectin, lignin, and structural proteins such as extensions and hydroxyproline-rich glycoproteins (HRGPs) (Sheng and Chen 2020). These changes in cell wall composition and architecture can impact various physiological processes, including growth, development, and stress responses.

Recent studies have reported a trade-off strategy between silicon (Si) and C components in silicon-accumulating crop species under sub-optimal environmental conditions. Silica deposition in plant tissues plays a crucial function in strengthening crop erectness, particularly in wetland species, providing a functional impact similar to that of carbon-based structural components, such as phenolics and lignin (Frew et al. 2016; Johnson and Hartley 2018). In response to this, the silicon components in crop tissues tend to partially substitute for carbon-based structural components to provide mechanical support. This trade-off phenomenon provides valuable insights into crop acclimation strategies and is evidence for the coupled cycling of silicon and carbon in plant systems.

Silicon indirectly promotes overall growth and biomass production by providing structural support to plants. Plants with stronger cell walls are better equipped to allocate resources toward vegetative growth, such as stem elongation, leaf expansion, and root development. Moreover, the enhanced stress resistance conferred by silicon allows plants to allocate more energy toward growth rather than defense mechanisms against biotic and abiotic stresses. As a result, silicon-treated plants often exhibit increased biomass accumulation compared to plants grown without silicon supplementation (Sah et al. 2022).

The increased biomass production facilitated by silicon contributes to carbon sequestration in plants. During photosynthesis, plants assimilate atmospheric carbon dioxide (CO_2) and convert it into organic carbon compounds, primarily in the form of carbohydrates. The additional biomass produced by silicon-treated plants represents an additional pool of stored carbon, which is sequestered from the atmosphere. This process helps mitigate the impacts of climate change by reducing the concentration of CO_2 in the atmosphere, a major greenhouse gas responsible for global warming. Furthermore, the deposition of silicon in the form of phytoliths within plant tissues contributes to long-term carbon storage. Phytoliths are mineral structures composed of amorphous silica, which are highly resistant to decomposition (Wani et al. 2023). When plants containing phytoliths die and decompose, the silica remains in the soil for extended periods, effectively sequestering carbon in the form of organic matter and associated phytoliths. This process helps to maintain soil carbon stocks and mitigate carbon losses from land-use changes and ecosystem disturbances (Rehman et al. 2022).

6.6.2 Stress Alleviation and Enhanced Photosynthesis

Silicon uptake by crops has been shown to improve stress tolerance, particularly under abiotic stress conditions such as drought (Gong et al. 2003), salinity, and metal toxicity. By enhancing stress tolerance, silicon helps plants maintain optimal photosynthetic activity even under unfavorable environmental conditions. This sustained photosynthetic performance contributes to higher carbon assimilation rates and increased carbon sequestration in plant tissues (Liang et. al. 2015; Pu et al. 2023). Silicon is primarily absorbed and translocated within plants as monomeric silicic acid (H_4SiO_4), which then undergoes polymerization to form silica gel nanoparticles. These silica structures are subsequently deposited in plant tissues as phytoliths, serving as a mechanical barrier against various environmental stresses (Ma et al. 2007; Klančnik et al. 2014; Amin et al. 2016). The deposition of silica in crop leaves contributes to improved light capture by strengthening cell membrane integrity and leaf blades. This enhanced structural integrity leads to increased photosynthetic assimilation of CO_2 and accumulation of biomass carbon under diverse abiotic and biotic stress conditions (Li et al. 2018; Wang et al. 2019). Silicon improves plant tolerance to various abiotic stresses, such as drought, salinity, heat, and heavy metal toxicity. Under water-limited conditions, silicon-treated plants exhibit enhanced water uptake efficiency, reduced transportational water loss, and improved stomatal regulation. These adaptations help plants maintain adequate leaf turgor pressure and avoid dehydration-induced reductions in photosynthetic activity. Similarly, silicon alleviates salt stress by promoting ion homeostasis, osmotic adjustment, and antioxidative defense mechanisms, thereby sustaining photosynthetic performance under saline conditions (Saja-Garbarz et al. 2024). Silicon mitigates oxidative stress in plants by enhancing antioxidant defense systems and scavenging reactive oxygen species (ROS) (Zargar et al. 2019). Environmental stresses such as high light intensity, drought, and pollutants can generate ROS, which damage cellular components and impair photosynthetic efficiency. Silicon-treated plants exhibit higher activities of antioxidant enzymes (e.g., superoxide dismutase, catalase, and peroxidase) (Gong et al. 2005) and increased levels of nonenzymatic antioxidants (e.g., glutathione and ascorbate), resulting in enhanced ROS detoxification and protection against oxidative damage. Silicon preserves chloroplast function, photosynthetic electron transport, and carbon dioxide assimilation by minimizing oxidative stress. Silicon modulates hormonal and signaling pathways involved in stress responses and plant growth regulation. Silicon application activates defense-related signaling cascades, including salicylic acid, jasmone acid, and ethylene signaling pathways, which conform crop responses to biotic and abiotic stresses (Li et al. 2019). Additionally, silicon influences the biosynthesis and signaling of phytohormones such as abscisic acid (ABA) and gibberellins (GA), which regulate stomatal conductance, water relations, and stress tolerance mechanisms. By orchestrating hormonal and signaling networks, silicon enhances stress resilience and maintains carbon dioxide fixation in plants exposed to adverse environmental conditions (Li et al. 2019). Silicon also helps plants defend against biotic stresses such as pest attacks and diseases. It strengthens cell walls, making it more difficult for pathogens to penetrate plant tissues. Additionally, silicon can induce systemic resistance in plants, triggering defense mechanisms that protect against pathogens (Ranjan et al. 2021).

Silicon supplementation has been linked to increased photosynthetic efficiency in many plant species. It enhances chlorophyll synthesis, leading to higher chlorophyll content in leaves. This, in turn, promotes greater light capture and utilization during photosynthesis. Silicon also influences stomatal conductance and transpiration rates. Studies have shown that silicon-treated plants often have more open stomata, improving CO_2 uptake from the atmosphere. Additionally, silicon can regulate transpiration, reducing water loss through stomatal closure under drought conditions while maintaining photosynthetic rates (Rastogi et al. 2021). Silicon has been found to boost the activity and efficiency of enzymes involved in photosynthesis, particularly ribulose-1,5-bisphosphate carboxylase/oxygenase (Rubisco) (Tripathi et al. 2017). Rubisco is responsible for capturing atmospheric carbon dioxide (CO_2) and incorporating it into organic molecules during the initial step of the Calvin cycle. By increasing the activity of Rubisco and other photosynthetic enzymes, silicon-treated plants can fix more CO_2 from the atmosphere, leading to increased carbon assimilation and subsequent sequestration in plant biomass. Stress alleviation and enhanced photosynthesis facilitated by silicon can increase plant biomass production. As plants grow, they assimilate atmospheric carbon dioxide (CO_2) through photosynthesis, converting it into organic carbon compounds such as carbohydrates, proteins, and lipids (Avila et al. 2020; Hassani et al. 2021).

6.6.3 Stomatal Regulation and Water Use Efficiency (WUE)

Silicon has been reported to influence stomatal behavior in plants, leading to improved water use efficiency and reduced transpiration rates. Stomata are microscopic pores on the leaf surface that regulate gas exchange, including carbon dioxide (CO_2) uptake for photosynthesis. By modulating stomatal conductance, silicon helps plants optimize CO_2 uptake while minimizing water loss through transpiration (Malik et al. 2021). This efficient use of water resources enhances carbon assimilation and biomass production, contributing to plant carbon sequestration. Stomata are tiny pores on the surface of leaves that regulate gas exchange between the crop and its environment, including the uptake of carbon dioxide (CO_2) for photosynthesis and the release of water vapor during transpiration (Sohag et al. 2020). Silicon has been found to influence stomatal behavior in plants. Research suggests that silicon-treated plants often exhibit more open stomata, allowing for increased uptake of CO_2 from the atmosphere. This can enhance photosynthetic rates and carbon assimilation in plants (Zhang et al. 2020; Verma et al. 2020). Additionally, silicon may regulate stomatal aperture in response to environmental stresses such as drought and temperatures. By maintaining more open stomata under stress conditions, silicon-treated plants can continue to photosynthesize and assimilate carbon more efficiently than non-treated plants.

Water use efficiency (WUE) refers to the amount of biomass produced per unit of water transpired by a plant. It reflects the plant's ability to optimize water use for growth and biomass production. Silicon has been shown to improve WUE in plants by several mechanisms. One way is through its influence on stomatal regulation. By maintaining more open stomata, silicon-treated plants can continue to photosynthesize while reducing water loss through transpiration. This allows them to achieve higher rates of carbon assimilation per unit of water transpired, thus increasing WUE. Additionally, silicon can enhance the efficiency of water uptake by plants, leading to improved water use efficiency. It may alter root morphology and physiology, increasing root surface area or enhancing water transport efficiency within the plant (Alayafi et al. 2022). Stomatal regulation and improved water use efficiency facilitated by silicon can lead to boosted biomass production in crops. As plants grow, they assimilate atmospheric CO_2 through photosynthesis, converting it into organic carbon compounds. The additional biomass produced by silicon-treated plants represents a significant pool of stored carbon, effectively sequestering carbon from the atmosphere. Silicon-treated plants often exhibit reduced transpiration rates compared to untreated plants. Transpiration is the process by which plants lose water vapor through stomata on the surfaces of leaves. Silicon application can lead to the closure of stomata or the regulation of stomatal conductance, thereby reducing the rate of water loss through transpiration. By conserving water, silicon-treated plants can maintain adequate hydration levels and sustain photosynthesis even under conditions of limited water availability, such as drought. This reduced transportational loss improves water use efficiency, allowing plants to maximize the utilization of available water resources for photosynthesis and biomass production. Silicon supplementation can modulate plant responses to osmotic stress, including accumulating osmolytes and compatible solutes. These osmolytes help plants maintain cellular hydration and osmotic balance under water stress conditions, preventing dehydration and metabolic imbalances. By regulating osmotic stress responses, silicon-treated plants can maintain photosynthetic activity and carbon assimilation, ultimately contributing to increased carbon sequestration in plant biomass (Dhiman et al. 2021; Yassen et al. 2017).

6.6.4 Induction of Defense Mechanisms

Silicon supplementation can stimulate the plant's innate defense mechanisms against biotic stresses, such as insect herbivory and fungal pathogens. Silicon-treated plants often exhibit increased production of defense-related compounds, such as phenolics, lignin, and phytoalexins. These compounds not only protect plants from pests and diseases but also influence carbon partitioning within the plant, potentially leading to increased carbon allocation toward defensive tissues and away from respiratory losses. This defense induction can indirectly contribute to carbon sequestration through several mechanisms. Silicon deposition in plant tissues strengthens cell walls, making them more resistant to mechanical damage caused by pests and pathogens. By forming a layer of silica (SiO_2) in the cell walls, silicon creates physical barriers that inhibit the penetration of pathogens and deter feeding by herbivores. The structural complexity of crop cell wall components can provide a rich source of oligomeric and polymeric substrates for the development of various bio-based materials, such as thermosets, composites, thermoplastics, cellulose nanocrystals, and nanofibers. These plant-derived components can be effectively utilized in the production of thermoset materials, including epoxy, silicone, and polyurethane. Furthermore, the incorporation of lignin- and carbohydrate-derived monomers into polymer structures can create new bio-based materials with enhanced performance characteristics compared to their fossil fuel-derived counterparts (Zhao and Abu-Omar 2015; Jiang et al. 2018; Chen et al. 2020). Silicon treatment can induce systemic resistance in plants, triggering defense mechanisms that protect against biotic stresses. This induction involves the activation of various signaling pathways and the upregulation of defense-related genes. For example, silicon-treated plants may produce higher defense-related proteins, antimicrobial compounds, and secondary metabolites that help deter pests and pathogens (Sun et al. 2022). Studies have shown that silicon supplementation can reduce the severity of pest infestations in crops. By enhancing plant defenses, silicon-treated plants are less susceptible to damage by herbivores such as insects and mites. This reduction in pest pressure can lead to higher crop yields and biomass production, contributing to carbon sequestration through increased plant growth (Lin et al. 2020; Elshahawy et al. 2021; Xue et al. 2021). Silicon has also been shown to suppress the development of various plant diseases caused by fungal, bacterial, and viral pathogens. By bolstering plant defenses, silicon-treated plants are better equipped to resist pathogen attacks and reduce disease severity. This protection against diseases can promote healthier plant growth and improve overall biomass production, which aids in carbon sequestration (Lin et al. 2020; Elshahawy et al. 2021; Xue et al. 2021). The enhanced defense mechanisms induced by silicon can promote the growth and vigor of plants, leading to increased biomass accumulation over time. As plants grow, they assimilate atmospheric carbon dioxide (CO_2) through photosynthesis, converting it into organic carbon compounds. The additional biomass produced by silicon-treated plants represents a significant pool of stored carbon, contributing to long-term carbon sequestration in plant tissues and soils.

6.6.5 Enhanced Nutrient Uptake and Growth

Silicon has been linked to improved nutrient uptake and plant utilization, particularly for essential nutrients such as nitrogen, phosphorus, and potassium. Optimal nutrient availability is essential for sustaining the photosynthetic machinery and maximizing carbon assimilation rates. Silicon-mediated nutrient uptake and growth improvements can lead to increased carbon sequestration in plant tissues, as more carbon is allocated toward growth and biomass production. This is how this mechanism aids in carbon sequestration. Silicon supplementation can enhance the availability and uptake of essential nutrients such as potassium (K), nitrogen (N), phosphorus (P), calcium (Ca), magnesium (Mg), and others. Silicon-treated plants often exhibit increased root biomass and surface area, facilitating greater exploration and absorption of nutrients from the soil. Silicon may improve nutrient availability by enhancing soil fertility and nutrient cycling processes. It can alter soil properties such as pH, cation exchange capacity, and microbial activity, which influence the availability and solubility of nutrients in the rhizosphere (Pavlovic et al. 2021). By facilitating nutrient uptake, silicon promotes overall plant growth and development. Nutrients are essential for various physiological processes such as photosynthesis, respiration, protein synthesis, and enzyme activity. Silicon-treated plants often exhibit greater vegetative growth, including boosted shoot and root biomass, leaf area, and tillering. This enhanced growth results in greater carbon assimilation through photosynthesis, as more biomass is produced for carbon storage (Ranjan et al. 2021). Silicon supplementation has been shown to improve photosynthetic efficiency in plants. By promoting nutrient uptake and optimizing metabolic processes, silicon-treated plants exhibit higher rates of CO_2 assimilation and carbon fixation during photosynthesis. Increased photosynthetic efficiency leads to greater carbon assimilation and biomass production, contributing to carbon sequestration in plant tissues (Hao et al. 2023).

6.6.6 Reduced Photorespiration

Photorespiration is a process in plants where oxygen competes with CO_2 for binding to Rubisco, leading to the release of CO_2 and decreased photosynthetic efficiency. Silicon application has been reported to reduce the rate of photorespiration in plants, potentially by improving Rubisco specificity for CO_2 over oxygen. By minimizing photorespiration, silicon-treated plants can maintain higher rates of CO_2 fixation and increase carbon sequestration in plant tissues (Vieira-Filho and Monteiro 2022).

6.6.7 *Increased Root Biomass and Root Exudates*

Silicon fertilization has been shown to stimulate root growth and increase root biomass. As roots grow and expand, they release organic compounds known as root exudates into the soil. These exudates contain carbon compounds derived from photosynthesis. Some of these carbon compounds are not readily decomposed by soil microbes, leading to their accumulation in the soil as organic carbon, a process known as rhizodeposition (Mišúthová et al. 2021; Ashfaq et al. 2023). Silicon-induced enhancement of root growth and exudation can thus contribute to increased carbon inputs to the soil. Silicon fertilization can also influence soil microbial communities and activity. Certain microorganisms in the soil, such as mycorrhizal fungi, can form symbiotic relationships with plants and boost nutrient uptake, including silicon (Wang et al. 2020). Additionally, silicon can influence the composition and activity of soil microbial communities involved in organic matter decomposition. Some studies suggest that silicon may promote the growth of microbial communities that are more efficient at decomposing organic matter, leading to increased turnover rates of soil organic carbon. However, other research indicates that silicon may also inhibit microbial activity in certain contexts, potentially leading to carbon stabilization in the soil (Deng et al. 2021).

6.6.8 *Improvement of Soil Properties*

Silicon can enhance soil aggregation and stability, improving soil structure and increasing soil organic carbon sequestration. Improved soil structure facilitates the formation and stabilization of soil aggregates, which protect organic matter from decomposition and promote its long-term storage in the soil. Additionally, silicon can enhance soil water retention capacity, potentially promoting plant growth and organic carbon inputs to the soil (Barbosa et al. 2024; Schaller et al. 2019, 2022). Silicon fertilization has been shown to strengthen plant cell walls and enhance resistance to biotic and abiotic stresses, including herbivory and disease. Stronger plant structures can improve plant stand and canopy cover, reducing soil erosion caused by wind and water. Reduced soil erosion helps retain organic carbon within the soil profile, contributing to long-term carbon sequestration (Barbosa et al. 2024; Schaller et al. 2019).

6.7 Function of Si in Phytolith-Occluded Carbon (PhytOC) Sequestration

The uptake of silicon by crop roots occurs in the form of monosilicic acid (H_4SiO_4) from the soil solution, which is then deposited throughout cellular structures, leading to the formation of amorphous silicon particles referred to as "phytoliths"

(Wani et al. 2023). During the process of silicon polymerization, a portion of organic carbon, about 0.2–5.8%, can become occluded within phytoliths, known as PhytOC (Song et al. 2016). PhytOC is proposed as an effective method of carbon sequestration, capable of persisting in soils for millennia as dead plant material decomposes and releases phytoliths into the soil (Parr and Sullivan 2005). The addition of silicon fertilizers boosts the uptake of soluble silicon by crop roots, thereby increasing phytolith production in crop tissues and ultimately augmenting the influx of phytolith-associated organic carbon into the soil through litterfall (Liu et al. 2020b; Rehman et al. 2022).

Silicon, an essential element for the formation of phytoliths, plays a crucial role in the sequestration of PhytOC (Nawaz et al. 2019). The accumulation of silicon in plants is closely associated with the development of phytoliths and the concentration of PhytOC (Meunier et al. 2017). The external addition of silicon boosts the availability of silicon to plants and subsequently promotes the formation of phytoliths and PhytOC (Rehman et al. 2022). Silica deposition in crop leaves enhances light capture by reinforcing cell membrane integrity and leaf blades, thereby improving the photosynthetic assimilation of carbon (Wang et al. 2019). The silicon content of plants and their transpiration rate may directly influence the compaction and specific surface area of phytoliths, which in turn affects their ability to occlude carbon (Li et al. 2013). Silica also plays a vital function in strengthening plant structure, similar to carbon-based structural components such as lignin and phenolics (Johnson and Hartley 2018). This strengthening increases leaf transpiration, facilitating the transport and deposition of silicon to various plant parts in the form of phytoliths (Liu et al. 2020b). Studies have shown that the development of phytoliths in the leaves of silicon-accumulating crops boosts with silicon uptake and transpiration (Li et al. 2017). It is speculated that the increase in PhytOC due to silicon fertilization is likely attributed to enhanced cell growth, partially increased cell volume, and reduced specific surface area of phytoliths (Li et al. 2020). Silicon fertilization indirectly enhances the storage of phytolith-occluded carbon (PhytOC) through increased plant biomass production and litter deposition (Huang et al. 2014). The improved PhytOC sequestration with silicon application is primarily driven by the increased accumulation of phytoliths in plant tissues, rather than higher carbon content within individual phytoliths (Liu et al. 2020b). This suggests that the elevated phytolith formation, even with a relatively low carbon concentration per phytolith, leads to a higher overall PhytOC pool in the plant biomass (Anjum and Nagabovanalli 2021). Additionally, it has been observed that silicon fertilization decreases phytolith dissolution and enhances the accumulation of PhytOC (Huang et al. 2020).

Applying silicon fertilizer boosted the sequestration of PhytOC within the crop-soil system by promoting phytolith formation and altering its distribution among physical fractions in a subtropical plantation (Huang et al. 2020; Liu et al. 2020b). Liu et al. (2020a) illustrate Si′s capacity to facilitate plant carbon absorption through PhytOC formation and biomass carbon accumulation in Si-dominant grasslands. Plant species are classified into accumulator, non-accumulator, and intermediate categories based on differences in silicon uptake and accumulation rates. The choice of cultivar plays a crucial role in PhytOC sequestration. Therefore, in silicon-deficient soils, the use of silicon fertilizer not only provides silicon for crop uptake and deposition to meet growth requirements and enhances carbon sequestration through increased PhytOC production.

6.8 Factors Influencing Crop-Available Si and PhytOC Estimation

Factors affecting crop-available silicon and the estimation of PhytOC (phytolith-occluded carbon) involve a combination of environmental, soil-related, and plant-specific factors (Matichenkov and Bocharnikova 2001; Huang and Hseu 2021; Anggria et al. 2020).

6.8.1 Soil Type and Composition

Different soil types vary in their silicon content and availability. Sandy soils generally have lower silicon availability compared to clay soils. The composition of soil minerals also influences silicon availability, with soils rich in silicate minerals generally providing more plant-available silicon. Different soil types have varying amounts of silicon. Soils derived from silicate-rich parent materials, such as granite or basalt, generally contain higher concentrations of silicon compared to soils derived from other materials such as limestone or sandstone. Therefore, the inherent silicon content of the soil influences the potential availability of silicon for plant uptake. The solubility of silicon in soil solution is influenced by soil type. Sandy soils, characterized by larger particle sizes and lower surface area, typically have lower silicon solubility compared to clay soils, which have smaller particle sizes and higher surface area. Clay soils can adsorb and retain more silicic acid, making silicon more available to plants. Soil texture, which refers to the relative proportions of sand, clay, and silt particles, influences the soil's ability to retain moisture and nutrients, including silicon. Additionally, soil structure, which refers to the arrangement of soil particles into aggregates, affects water infiltration and root penetration, consequently impacting plant access to silicon.

6.8.2 pH Levels

Soil pH affects the solubility and availability of silicon (Yanai et al. 2016). Silicon is present in soils primarily as silicic acid ($Si(OH)_4$) in soluble form and silicate minerals such as quartz and feldspar. In acidic soils (pH < 6), silicon availability tends to be lower due to increased aluminum and iron toxicity, which can inhibit silicon uptake by plants. Additionally, in acidic soils, silicate minerals tend to be less soluble. In neutral to alkaline soils (pH > 6), silicon availability is generally higher as aluminum and iron toxicity decreases, and silicate minerals become more soluble, releasing more silicon for plant uptake. Thus, soil pH influences the availability of silicon for plants, affecting their growth and development, as silicon is beneficial for plant structure, disease resistance, and stress tolerance (Makabe et al. 2013). Estimating PhytOC in soils is crucial for understanding carbon cycling, soil fertility, and ecosystem functioning. Soil pH influences the decomposition rate of organic matter. In acidic soils, decomposition tends to be slower due to microbial activity being suppressed by the low pH. Conversely, in neutral to alkaline soils, decomposition rates are typically higher. Higher pH can enhance microbial activity, accelerating the breakdown of organic matter and increasing the release of PhytOC into the soil. Additionally, soil pH affects the chemical properties of organic matter, such as its charge and ability to complex with minerals, which can further influence the dynamics of PhytOC in soils. Therefore, when estimating PhytOC in soils, accounting for soil pH is essential as it can significantly impact the decomposition and stabilization of organic carbon.

6.8.3 Temperature

Soil temperature is a critical factor that affects various soil processes and properties, including the availability of nutrients and organic carbon. Soil temperature strongly influences microbial activity. As temperature increases, microbial activity generally increases as well, up to a certain point. Microbes play a crucial function in nutrient cycling and organic matter decomposition, which indirectly affect the availability of silicon and the estimation of PhytOC. Soil temperature affects the rates of nutrient mineralization and immobilization. Higher temperatures generally accelerate these processes, potentially affecting the availability of silicon and other nutrients for plant uptake. Temperature-mediated changes in nutrient availability can impact plant growth and subsequently influence the demand for silicon. Soil temperature influences the solubility of various compounds in soil, including silicon. Generally, higher temperatures can enhance the dissolution of silicate minerals, potentially increasing the availability of silicon for plant uptake. However, this relationship can be complex and dependent on other soil properties such as pH and moisture content (Schaller and Puppe 2021). Soil temperature directly influences root growth and activity. Warmer temperatures typically promote root growth, which can impact the uptake of silicon by plants. Increased root activity can lead to higher demand for nutrients, including silicon, thereby affecting its availability in the soil solution (Schaller and Puppe 2021). Soil temperature affects the decomposition rates of soil organic matter. Higher temperatures generally accelerate decomposition, leading to increased mineralization of organic carbon and other nutrients (Barros et al. 2021). This can indirectly influence the estimation of PhytOC, as higher temperatures may result in a more rapid turnover of organic carbon pools within the soil.

6.8.4 Soil Organic Matter

Organic matter in soil can influence silicon availability through its interactions with silicic acid and other forms of silicon. Higher levels of organic matter can enhance silicon availability by forming complexes that prevent silicon from leaching or binding to minerals (Klotzbücher et al. 2018). Soil organic matter serves as a reservoir of nutrients, including silicon. It undergoes decomposition by microbial activity, releasing nutrients into the soil solution. As SOM decomposes, it releases silicon in forms that are available for plant uptake. Therefore, the quantity and quality of SOM can directly affect the availability of silicon to plants. Organic matter can complex with silicon ions in the soil solution, affecting their mobility and availability to plants. Some organic compounds can form complexes with silicon, potentially making it more or less available for plant uptake depending on the specific chemical interactions involved (McCray et al. 2011; McCray and Shangning 2012). Soil organic matter provides a substrate for microbial communities. Microbes are crucial in mineralizing organic matter and releasing nutrients, including silicon, into forms accessible to plants. SOM quantity and quality changes can influence microbial activity, impacting silicon cycling and availability. SOM contributes to soil aggregation and structure, influencing water retention and drainage. Proper soil structure facilitates root growth and nutrient uptake, including silicon. Additionally, SOM improves soil water-holding capacity, indirectly affecting plant growth and nutrient availability, thus influencing silicon uptake. Soil organic matter is the primary source of PhytOC in soil ecosystems. Accurate estimation of PhytOC involves considering the quantity and quality of SOM in the soil. Changes in SOM content or composition can affect the estimation of PhytOC, as they impact the quantity of organic carbon stored in the soil and its turnover rates (Zhang et al. 2008).

6.8.5 Climate and Weathering

Climate influences the rate of weathering of parent rocks and minerals, which in turn affects the release of silicon into the soil solution. Warm and humid climates typically promote faster weathering, leading to higher silicon availability. Climate and weathering are crucial factors that influence the availability of plant-available silicon (Si) and the estimation of plant organic carbon (PhytOC) in soil ecosystems. Climate influences the rates of physical and chemical weathering processes, which in turn affect the release of silicon from mineral sources. Higher temperatures generally accelerate weathering rates, leading to increased dissolution of silicate minerals and subsequent release of silicon into the soil solution. Precipitation patterns also play a role, as water facilitates the chemical weathering of minerals. Regions with high rainfall may experience more intense weathering, resulting in higher silicon availability in the soil (Landré et al. 2020). Climate interacts with parent material to influence soil formation processes. Different climates favor the weathering of specific rock types, leading to the formation of soils with varying silicon content. For example, in humid tropical climates, intense weathering of basaltic rocks may lead to the formation of soils rich in clay minerals and plant-available silicon. Climate affects the rate of organic matter decomposition, which in turn influences the cycling of silicon and organic carbon in soil ecosystems. Warmer climates generally accelerate microbial activity and decomposition rates, leading to faster turnover of organic matter and potentially affecting the estimation of PhytOC. Additionally, climate influences the types of vegetation present, which can further impact organic matter inputs to the soil. Climate determines the availability of soil moisture, which affects the rates of weathering processes and silicon dissolution. Adequate soil moisture is essential for promoting chemical weathering and the release of silicon from mineral sources. Additionally, soil moisture influences plant growth and root activity, which can indirectly affect silicon uptake by plants. Climate influences ecosystem processes such as plant productivity, litterfall rates, and nutrient cycling, all of which affect the availability of plant-available silicon and the estimation of PhytOC. Changes in climate can alter these ecosystem processes, leading to shifts in silicon cycling dynamics and organic carbon storage in soils.

6.8.6 *Plant Species and Genotypes*

Different plant species vary in their ability to uptake and accumulate silicon (Mitani and Ma 2005). Some plants, known as silicon accumulators, have specialized mechanisms for efficient silicon uptake and transport. Additionally, within a species, different genotypes may exhibit variations in silicon uptake capacity. Different plant species and genotypes exhibit varying degrees of efficiency in taking up silicon from the soil. Some plant species, known as silicon accumulators, have evolved mechanisms to actively take up and accumulate silicon in their tissues, while others may have lower silicon uptake rates. The presence of silicon accumulators in an ecosystem can influence the overall availability of plant-available silicon by depleting it from the soil or enhancing its cycling through biomass. Plant species and genotypes differ in their root morphology and exudation patterns, which can impact the rhizosphere environment and influence silicon availability. Certain plant species and genotypes may release organic acids or other compounds from their roots, which can enhance the weathering of silicate minerals and increase the availability of silicon in the soil solution. Different plant species and genotypes can alter soil properties through their root exudates, litter composition, and decomposition rates, leading to plant-soil feedback effects. These feedbacks can influence soil organic carbon dynamics, microbial activity, and nutrient cycling, ultimately affecting the availability of plant-available silicon and the estimation of PhytOC. Plant species and genotypes vary in their ability to incorporate silicon into their tissues and to translocate it within the plant. Some species may deposit silicon in specific tissues such as epidermal cells or vascular tissues, while others may distribute it more evenly throughout their biomass. The efficiency of silicon incorporation and translocation can impact the amount of silicon sequestered in plant biomass, which affects silicon cycling and availability in the soil. Plant species and genotypes differ in their rates of organic carbon production and decomposition, leading to variations in the amount and quality of organic matter input to the soil. Differences in litter composition and decomposition rates can influence the estimation of PhytOC and its turnover dynamics in soil ecosystems (Ma and Takahashi 2002; Hodson et al. 2005).

6.8.7 Management Practices

Agricultural practices such as crop rotation, tillage, and the use of silicon fertilizers can influence plant-available silicon (Sierra et al. 2021). Silicon fertilization, in particular, can directly increase the silicon content of soils and enhance silicon uptake by plants. Management practices employed in agricultural and soil management systems significantly influence the availability of plant-available silicon and the estimation of plant organic carbon (PhytOC) in soil ecosystems. Various management practices impact Si availability and PhytOC estimation. For example, using fertilizers can affect soil pH, nutrient availability, and microbial activity, influencing silicon cycling and availability. Certain fertilizers may alter soil pH, affecting the solubility of silicon in the soil solution. Additionally, fertilizers containing silicon can directly increase plant-available silicon levels in soils (Fujii et al. 2023). Tillage practices can also affect soil structure, microbial communities, and organic matter decomposition rates. Intensive tillage can disrupt soil aggregates and accelerate the decomposition of organic matter, potentially leading to decreased silicon availability and altered PhytOC dynamics. Conservation tillage practices, on the other hand, may help preserve soil organic carbon and promote silicon retention in the soil Crop rotation and diversification practices can influence soil organic matter inputs, nutrient cycling, and microbial activity. Different crops have varying demands for silicon and organic carbon inputs to the soil. Rotating silicon-accumulating crops with other plant species can help maintain soil silicon levels and influence PhytOC dynamics by varying organic matter inputs. Irrigation management affects soil moisture levels, which in turn influence microbial activity, organic matter decomposition rates, and nutrient availability, including silicon. Proper irrigation practices can help maintain optimal soil moisture conditions for silicon cycling and organic carbon preservation. Applying organic amendments such as compost, manure, or crop residues can impact soil organic carbon levels and microbial activity. Organic amendments can also contribute to soil silicon inputs and influence plant uptake availability (Kang et al. 2022). Additionally, organic amendments may contain silica, further enhancing plant-available silicon levels. Practices for soil conservation, such as conserving soil through cover crops, mulching, and erosion control measures, can affect soil carbon and silicon levels. These practices help preserve soil structure, reduce erosion, and promote organic matter retention, influencing silicon cycling and PhytOC estimation.

6.8.8 Soil Microorganism

Soil microorganisms are fundamental components of soil ecosystems and play critical roles in influencing the availability of plant-available silicon (Si) and the estimation of plant organic carbon (PhytOC). Soil microorganisms contribute to silicon cycling through various processes such as weathering, mineral dissolution, and organic matter decomposition. Certain microbial species produce organic acids and enzymes that facilitate the weathering of silicate minerals, releasing silicon into the soil solution and making it available for plant uptake. Some soil microorganisms, particularly silicon-solubilizing bacteria and fungi, possess the ability to solubilize insoluble forms of silicon, such as silicate minerals. These microorganisms secrete organic acids and chelating compounds that dissolve silicates, thereby increasing the availability of plant-available silicon in the soil. Soil microorganisms play key roles in nutrient cycling processes, including the decomposition of organic matter. As microorganisms decompose organic materials, they release nutrients, including silicon, into the soil. Microbial decomposition of plant residues and organic matter contributes to the release of silicon that becomes available for plant uptake. Soil microorganisms interact with plant roots in the rhizosphere, the soil zone influenced by root exudates. These interactions can influence silicon availability by altering soil pH, organic matter decomposition rates, and mineral weathering processes. Root exudates may stimulate microbial activity, affecting silicon cycling dynamics in the rhizosphere. Soil microorganisms are intimately involved in the decomposition of organic matter and the formation of plant organic carbon (Thepbandit and Athinuwat 2024). Microbial decomposition of plant residues leads to organic carbon accumulation in soil, contributing to the estimation of PhytOC. Additionally, microorganisms play roles in stabilizing organic carbon in soil aggregates, affecting its turnover rates and sequestration potential. Soil microorganisms can influence plant growth and silicon uptake through symbiotic or pathogenic plant interactions. Certain beneficial microorganisms can enhance plant growth and silicon uptake, while pathogenic microorganisms may suppress plant growth and reduce silicon uptake efficiency. These interactions can ultimately affect soil silicon availability and plant-derived organic carbon inputs.

6.8.9 Analytical Techniques

The estimation of PhytOC involves various analytical techniques, including chemical extractions, microscopy, and spectroscopy (Sinha et al. 2022). Different methods may yield different estimates of PhytOC, and the choice of method depends on factors such as sample type, the purpose of analysis, and available resources. Analytical techniques are essential for accurately assessing plant-available silicon (Si) and estimating plant organic carbon (PhytOC) in soil ecosystems. Different analytical methods offer various levels of precision, sensitivity, and applicability, influencing the estimation of these parameters. Colorimetric methods involve the reaction of silicon with specific reagents to produce a color change, which is then quantified spectrophotometrically. Colorimetric methods are relatively simple and cost-effective but may need more sensitivity and specificity, leading to potential inaccuracies in Si estimation. Atomic absorption spectroscopy (AAS) and inductively coupled plasma optical emission spectroscopy (ICP-OES) are highly sensitive techniques that accurately quantify silicon concentrations in soil extracts. They offer high precision and specificity, making them suitable for precise Si estimation in soil samples. X-ray fluorescence (XRF) spectroscopy allows for nondestructive analysis of silicon and other elemental concentrations in soil samples. It offers rapid analysis capabilities but may have sensitivity and sample depth limitations.

The Walkley-Black method involves oxidizing soil organic carbon using potassium dichromate and subsequent titration with ferrous sulfate. It measures total organic carbon content in soil samples but may overestimate organic carbon due to interference from other soil components. Combustion techniques (e.g., CHN Analyzer) involve the complete combustion of soil samples to release carbon dioxide, which is then quantified using infrared detection methods. These techniques offer high precision and accuracy in estimating organic carbon content but may be costly and require specialized equipment. Soil organic matter fractionation involves separating soil organic matter into different fractions based on their solubility or density. These methods provide insights into the composition and stability of organic carbon pools in soil but may be labor-intensive and require complex analytical procedures. Particle size fractionation techniques involve separating soil samples into different size fractions to assess the distribution of silicon and organic carbon across soil particle sizes. This helps understand the associations between soil minerals, organic matter, and plant-available silicon. Chemical extraction methods involve extracting silicon and organic carbon from soil samples using specific reagents. Different extraction methods may target different forms of silicon and organic carbon, leading to variations in their estimated concentrations.

6.9 Conclusions and Future Research Recommendations

In conclusion, the use of silicon-based fertilization in agricultural fields shows great potential for enhancing carbon sequestration through various mechanisms such as promoting plant growth, stress tolerance, and photosynthesis. Factors influencing plant-available silicon and PhytOC estimation play a crucial role in determining the effectiveness of silicon fertilization in carbon sequestration. Future research in this area should focus on optimizing silicon application rates and methods for different soil types and crop species, exploring the interactions between silicon and other nutrients, and investigating the long-term effects of silicon fertilization on soil organic carbon dynamics. Additionally, studying the role of soil microorganisms in silicon cycling and PhytOC sequestration can provide valuable insights for further improving carbon sequestration strategies in agricultural systems.

References

Alayafi AH, Al-Solaimani SGM, Abd El-Wahed MH, Alghabari FM, Sabagh AE (2022) Silicon supplementation enhances productivity, water use efficiency and salinity tolerance in maize. Front Plant Sci 13:953451. https://doi.org/10.3389/fpls.2022.953451

Amin M, Ahmad R, Ali A, Hussain I, Mahmood R, Aslam M et al (2016) Influence of silicon fertilization on maize performance under limited water supply. Silicon 10:177–183. https://doi.org/10.1007/s12633-015-9372-x

AminiTabrizi R, Wilson RM, Fudyma JD, Hodgkins SB, Heyman HM, Rich VI, Saleska SR, Chanton JP, Tfaily MM (2020) Controls on soil organic matter degradation and subsequent greenhouse gas emissions across a permafrost thaw gradient in Northern Sweden. Front Earth Sci 8:557961

Anggria L, Husnain H, Masunaga T (2020) The controlling factors of silicon solubility in soil solution. Agric 32(2):83–94. https://doi.org/10.24246/agric.2020.v32.i2.p83-94

Anjum M, Nagabovanalli PB (2021) Assessing production of phytolith and phytolith occluded carbon in aboveground biomass of intensively cultivated rice ecosystems in India. Carbon Manag 12:509–519. https://doi.org/10.1080/17583004.2021.1978552

Arora V, Saran RK, Kumar R, Yadav S (2019) Separation and sequestration of CO2 in geological formations. Mater Sci Energy Technol 2(3):647–656. https://doi.org/10.1016/j.mset.2019.08.006

Ashfaq W, Brodie G, Fuentes S et al (2023) Silicon improves root system and canopy physiology in wheat under drought stress. Plant Soil. https://doi.org/10.1007/s11104-023-06202-4

Avila RG, Magalhães PC, da Silva EM. et al. (2020) Silicon supplementation improves tolerance to water deficiency in sorghum plants by increasing root system growth and improving photosynthesis. Silicon 12:2545–2554. https://doi.org/10.1007/s12633-019-00349-5

Barbosa LP, Stein M, Gerke HH et al (2024) Synergistic effects of organic carbon and silica in preserving structural stability of drying soils. Sci Rep 14:8330. https://doi.org/10.1038/s41598-024-58916-9

Barros N, Rodríguez-Añon JA, Proupín J, Pérez-Cruzado C (2021) The effect of extreme temperatures on soil organic matter decomposition from Atlantic oak forest ecosystems. iScience 24(12):103527. https://doi.org/10.1016/j.isci.2021.103527

Basu S, Mackey KRM (2018) Phytoplankton as key mediators of the biological carbon pump: their responses to a changing climate. Sustain For 10(3):869. https://doi.org/10.3390/su10030869

Bityutskii N, Pavlovic J, Yakkonen K, Maksimović V, Nikolic M (2014) Contrasting effect of silicon on iron, zinc and manganese status and accumulation of metal-mobilizing compounds in micronutrient-deficient cucumber. Plant Physiol Biochem 74:205–211

Chen C, Wang Y, Wu Q, Wan Z, Li D, Jin Y (2020) Highly strong and flexible composite hydrogel reinforced by aligned wood cellulose skeleton via alkali treatment for muscle-like sensors. Chem Eng J 400:125876. https://doi.org/10.1016/j.cej.2020.125876

Chouikhi N, Cecilia JA, Vilarrasa-García E, Besghaier S, Chlendi M, Franco Duro FI, Rodriguez Castellon E, Bagane M (2019) CO2 adsorption of materials synthesized from clay minerals: a review. Fortschr Mineral 9(9):514. https://doi.org/10.3390/min9090514

Cooke J, Leishman MR (2011) Silicon concentration and leaf longevity: is silicon a player in the leaf dry mass spectrum? Funct Ecol 25(6):1181–1188

Currie HA, Perry CC (2007) Silica in plants: biological, biochemical and chemical studies. Ann Bot 100(7):1383–1389. https://doi.org/10.1093/aob/mcm247

da Cunha KPV, do Nascimento CWA (2009) Silicon effects on metal tolerance and structural changes in maize (Zea mays L.) grown on a cadmium and zinc enriched soil. Water Air Soil Pollut 19(1):323–330

Davamani V, Sangeetha Piriya R, Rakesh SS, Parameswari E, Paul Sebastian S, Kalaiselvi P, Maheswari M, Santhi R (2022) Phytolith-occluded carbon sequestration potential of oil palm plantation in Tamil Nadu. ACS Omega 7(3):2809–2820. https://doi.org/10.1021/acsomega.1c05592

Deng Q, Yu T, Zeng Z, Ashraf U, Shi Q, Huang S, Lian T, Chen J, Muzaffar W, Shen W (2021) Silicon application modulates the growth, rhizosphere soil characteristics, and bacterial community structure in sugarcane. Front Plant Sci 12:710139. https://doi.org/10.3389/fpls.2021.710139

Dhiman P, Rajora N, Bhardwaj S, Sudhakaran SS, Kumar A, Raturi G, Chakraborty K, Gupta OP, Devanna BN, Tripathi DK et al (2021) Fascinating role of silicon to combat salinity stress in plants: an updated overview. Plant Physiol Biochem 162:110–123. [Google Scholar] [CrossRef]

Elshahawy IE, Osman SA, Abd-El-Kareem F (2021) Protective effects of silicon and silicate salts against white rot disease of onion and garlic, caused by *Stromatinia cepivora*. J Plant Pathol 103:27–43. https://doi.org/10.1007/s42161-020-00685-1

Epstein E, Bloom AJ (2005) Mineral nutrition of plants: principles and perspectives. Sinauer Associates Inc, Shuderland, Massachusetts. https://doi.org/10.1016/j.jplph.2005.06.002

Etesami H, Jeong BR (2018) Silicon (Si): review and future prospects on the action mechanisms in alleviating biotic and abiotic stresses in plants. Ecotoxicol Environ Saf 147:881–896

Frew A, Powell JR, Sallam N, Allsopp PG, Johnson SN (2016) Trade-offs between silicon and phenolic defenses may explain enhanced performance of root herbivores on phenolic-rich plants. J Chem Ecol 42:768–771. https://doi.org/10.1007/s10886-016-0734-7

Fujii K, Ichinose Y, Arai K, Komatsu K, Hayakawa C, de Guzman Alvindia D et al (2023) Effects of soil types and fertility management practices on soil silicon availability and banana silicon uptake. Soil Sci Plant Nutr 69(3):183–189. https://doi.org/10.1080/00380768.2023.2175177

Garamvölgyi Á, Hufnagel L (2013) Impacts of climate change on vegetation distribution. No. 1: climate change induced vegetation shifts in the palearctic region. Appl Ecol Env Res 11:79–122. https://doi.org/10.2166/wst.2013.207. https://doi.org/10.15666/aeer/1101_079122

Gómez-Pozuelo G, Sanz-Pérez ES, Arencibia A, Pizarro P, Sanz R, Serrano DP (2019) CO2 adsorption on amine-functionalized clays. Microporous Mesoporous Mater 282:38–47

Gong H, Chen K, Chen G, Wang S-M, Zhang C (2003) Effects of silicon on growth of wheat under drought. J Plant Nutr 26:1055–1063. https://doi.org/10.1081/PLN-120020075

Gong HJ, Zhu XY, Chen KM, Wang SM, Zhang CL (2005) Silicon alleviates oxidative damage of wheat plants in pots under drought. Plant Sci 169:313–321. https://doi.org/10.1016/j.plantsci.2005.02.023

Gonzalo MJ, Lucena JJ, Hernández-Apaolaza L (2013) Effect of silicon addition on soybean (Glycine max) and cucumber (Cucumis sativus) plants grown under iron deficiency. Plant Physiol Biochem 70:455–461

Gottardi S, Iacuzzo F, Tomasi N, Cortella G, Manzocco L, Pinton R, Römheld V, Mimmo T, Scampicchio M, Dalla Costa L (2012) Beneficial effects of silicon on hydroponically grown corn salad (Valerianella locusta (L.) Laterr) plants. Plant Physiol Biochem 56:14–23

Gougoulias C, Clark JM, Shaw LJ (2014) The role of soil microbes in the global carbon cycle: tracking the below-ground microbial processing of plant-derived carbon for manipulating carbon dynamics in agricultural systems. J Sci Food Agric 94(12):2362–2371. https://doi.org/10.1002/jsfa.6577

Gu H, Qiu H, Tian T, Zhan S, Deng THB, Chaney R, Wang S (2011) Mitigation effects of silicon rich amendments on heavy metal accumulation in rice (Oryza sativa L.) planted on multi-metal contaminated acidic soil. Chemosphere 83:1234–1240. https://doi.org/10.1016/j.chemosphere.2011.03.014

Hao Y, Yu Y, Sun G, Gong X, Jiang Y, Lv G, Zhang Y, Li L, Zhao Y, Sun D, Gu W, Qian C (2023) Effects of multi-walled carbon nanotubes and nano-silica on root development, leaf photosynthesis, active oxygen and nitrogen metabolism in maize. Plants (Basel) 12(8):1604. https://doi.org/10.3390/plants12081604

Hassani A, Azapagic A, Shokri N (2021) Global predictions of primary soil salinization under changing climate in the 21st century. Nat Commun 12:6663. https://doi.org/10.1038/s41467-021-26907-3

Hodson MJ, White PJ, Mead A, Broadley MR (2005) Phylogenetic variation in the silicon composition of plants. Ann Bot 96:1027–1046

Huang Y-C, Hseu Z-Y (2021) Silicon availability in relation to soil properties in Inceptisols on uncultivated lands and paddy fields in Taiwan. Geoderma Reg 26:e00406. https://doi.org/10.1016/j.geodrs.2021.e00406

Huang ZT, Li YF, Jiang PK, Chang SX, Song ZL, Liu J (2014) Long-term intensive management increased carbon occluded in phytolith (PhytOC) in bamboo forest soils. Scientific reports 4:3602. https://doi.org/10.1038/srep03602

Huang C, Wang L, Gong X, Huang Z, Zhou M, Li J (2020) Silicon fertilizer and biochar effects on plant and soil PhytOC concentration and soil PhytOC stability and fractionation in subtropical bamboo plantations. Sci Total Environ 715:136846. https://doi.org/10.1016/j.scitotenv.2020.136846

Jiang Y, Ding D, Zhao S, Zhu H, Kenttämaa HI, Abu-Omar MM (2018) Renewable thermoset polymers based on lignin and carbohydrate derived monomers. Green Chem 20(5):1131–1138. https://doi.org/10.1039/C7GC03552G

Johnson SN, Hartley SE (2018) Elevated carbon dioxide and warming impact silicon and phenolic-based defences differently in native and exotic grasses. Glob Change Biol 24:3886–3896. https://doi.org/10.1111/gcb.13971

Kang Y, Ma Y, Wenli W, Zeng S, Jiang S, Yang H, Li Y, Wang Z, Dong C, Yangchun X, Shen Q (2022) Bioorganic and silicon amendments alleviate early defoliation of pear trees by improving the soil nutrient bioavailability, microbial activity, and reshaping the soil microbiome network. Appl Soil Ecol 173:104383. https://doi.org/10.1016/j.apsoil.2021.104383

Keykha HA, Zangani A, Romiani HM, Asadi A, Kawasaki S, Radmanesh N (2023) Characterizing microbial and CO2-induced carbonate minerals: implications for soil stabilization in Sandy environments. Fortschr Mineral 13(7):976. https://doi.org/10.3390/min13070976

Klančnik K, Vogel-Mikuš K, Gaberščik A (2014) Silicified structures affect leaf optical properties in grasses and sedges. J Photoch Photobio B 130:1–10. https://doi.org/10.1016/j.jphotobiol.2013.10.011

Klotzbücher T, Klotzbücher A, Kaiser K, Merbach I, Mikutta R (2018) Impact of agricultural practices on plant-available silicon. Geoderma 331:15–17. https://doi.org/10.1016/j.geoderma.2018.06.011

Kostic L, Nikolic N, Bosnic D, Samardzic J, Nikolic M (2017) Silicon increases phosphorus (P) uptake by wheat under low P acid soil conditions. Plant Soil 419(1):447–455

Kuokkanen K, Julkunen-Tiitto R, Keinänen M, Niemelä P, Tahvanainen J (2001) The effect of elevated CO2 and temperature on the secondary chemistry of Betula pendula seedlings. Trees Struct Funct 15:378–384. https://doi.org/10.1007/s004680100108

Landré A, Cornu S, Meunier J-D, Guerin A, Arrouays D, Caubet M, Ratié C, Saby NPA (2020) Do climate and land use affect the pool of total silicon concentration? A digital soil mapping approach of French topsoils. Geoderma 364:114175. https://doi.org/10.1016/j.geoderma.2020.114175

Li Z, Meunier JD, Delvaux B (2022) Aggregation reduces the releaseof bioavailable silicon from allophane and phytolith. GCA325:87–105. https://doi.org/10.1016/j.gca.2022.03.025

Li ZM, Song ZL, Li BL (2013) The production and accumulation of phytolith-occluded carbon in Baiyangdian reed wetland of China. Appl Geochem 37:117–124. https://doi.org/10.1016/j.apgeochem.2013.07.012

Li RC, Fan J, Carter J, Jiang N, Gu YS (2017) Monthly variations of phytoliths in the leaves of the bamboo Dendrocalamus ronganensis (Poaceae: Bambusoideae). Rev Palaeobot Palyno 246:62–69. https://doi.org/10.1016/j.revpalbo.2017.06.006

Li ZC, Song ZL, Yan ZF, Hao Q, Song AL, Liu LN et al (2018) Silicon enhancement of estimated plant biomass carbon accumulation under abiotic and biotic stresses. A meta-analysis Agron Sustain Dev 38:26. https://doi.org/10.1007/s13593-018-0496-4

Li N, Han X, Feng D, Yuan D, Huang LJ (2019) Signaling crosstalk between salicylic acid and ethylene/Jasmonate in plant defense: do we understand what they are whispering? Int J Mol Sci 20(3):671. https://doi.org/10.3390/ijms20030671

Li Z, de Tombeur F, Vander Linden C, Cornelis JT, Delvaux (2020) Soil microaggregates store phytoliths in a sandy loam. Geoderma 360:114037. https://doi.org/10.1016/j.geoderma.2019.114037

Liang Y, Nikolic M, Bélanger R, Gong H, Song A (2015) Silicon in agriculture: from theory to practice. Springer, Berlin

Lin WP, Jiang NH, Peng L, Fan XY, Gao Y, Wang GP et al (2020) Silicon impacts on soil microflora under *Ralstonia solanacearum* inoculation. J Integr Agr 19:251–264. https://doi.org/10.1016/S2095-3119(18)62122-7

Liu J, Zhong J, Ding H, Yue FJ, Li C, Xu S, Li S (2020a) Hydrological regulation of chemical weathering and dissolved inorganic carbon biogeochemical processes in a monsoonal river. Hydrol Process 34:2780–2792

Liu L, Song Z, Yu C, Yu G, Ellam RM, Liu H, Singh BP, Wang H (2020b) Silicon effects on biomass carbon and phytolith-occluded carbon in grasslands under high-salinity conditions. Front Plant Sci 11:657. https://doi.org/10.3389/fpls.2020.00657

Lützow M, Kögel-Knabner I, Ekschmitt K, Matzner E, Guggenberger G, Marschner B, Flessa H (2006) Stabilization of organic matter in temperate soils: mechanisms and their relevance under different soil conditions—a review. Eur J Soil Sci 57:426–445. https://doi.org/10.1111/j.1365-2389.2006.00809.x

Ma JF, Takahashi E (2002) Soil, fertilizer, and plant silicon research in Japan. Elsevier

Ma JF, Yamaji N (2006) Silicon uptake and accumulation in higher plants. Trends Plant Sci 11(8):392–397

Ma JF, Yamaji N, Mitani N, Tamai K, Konishi S, Fujiwara T et al (2007) An efflux transporter of silicon in rice. Nature 448:209–212. https://doi.org/10.1038/nature05964

Makabe SS, Kakuda KI, Sasaki Y, Ando H (2013) Effect of slag silicate fertilizer on dissolved silicon in soil solution based on the chemical properties of Gleysols. Soil Sci Plant Nutr 59:271–277

Malik MA, Wani AH, Mir SH, Rehman IU, Tahir I, Ahmad P, Rashid I (2021) Elucidating the role of silicon in drought stress tolerance in plants. Plant Physiol Biochem 165:187–195. https://doi.org/10.1016/j.plaphy.2021.04.021

Mann S, Ozin GA (1996) Synthesis of inorganic materials with complex form. Nature 382(6589):313–318

Mann S, Perry CC (1986) Structural aspects of biogenic silica. Ciba Foundation Symposium 121:40–58. https://doi.org/10.1002/9780470513323.ch4

Marschner H (1995) Mineral nutrition of higher plants (2nd edn, pp 99–101). Academic Press, London

Matichenkov VV, Bocharnikova EA (2001) Chapter 13: The relationship between silicon and soil physical and chemical properties. In: Datnoff LE, Snyder GH, Korndörfer GH (eds) Studies in plant science, vol 8. Elsevier, pp 209–219, ISSN 0928-3420. https://doi.org/10.1016/S0928-3420(01)80017-3

Mazumdar J (2011) Phytoliths of pteridophytes. S Afr J Bot 77(1):10–19

McCray JM, Shangning J (2012) Calibration of sugarcane response to calcium silicate on Florida Histosols. J Plant Nutr 35:1192–1209. https://doi.org/10.1080/01904167.2012.676131

McCray JM, Rice RW, Baucum LE (2011) Calcium silicon recommendations for sugarcane on Florida organic soils. Available at https://edis.ifas.ufl.edu/sc092. Accessed 8 Jan 2020

McLaughlin H, Littlefield AA, Menefee M, Kinzer A, Hull T, Sovacool BK, Griffiths S (2023) Carbon capture utilization and storage in review: sociotechnical implications for a carbon reliant world. Renew Sust Energ Rev 177:113215

Meunier JD, Barboni D, Anwar-ul-Haq M, Levard C et al. (2017) Effect of phytoliths for mitigating water stress in durum wheat. New Phytologist 215(1):229–239. https://doi.org/10.1111/nph.14554

Mica V (1986) Rezidua sodiku a zevera ve frakcioxidu kremiciteho pri rozforech rostein. Agrochemia 26(9):270–272

Mišúthová A, Slováková Ľ, Kollárová K, Vaculík M (2021) Effect of silicon on root growth, ionomics and antioxidant performance of maize roots exposed to As toxicity. Plant Physiol Biochem 168:155–166. https://doi.org/10.1016/j.plaphy.2021.10.012

Mitani N, Ma JF (2005) Uptake system of silicon in different plant species. J Exp Bot 56(414):1255–1261. https://doi.org/10.1093/jxb/eri121

Nawaz MA, Zakharenko AM, Zemchenko IV, Haider MS, Ali MA, Imtiaz M, Chung G, Tsatsakis A, Sun S, Golokhvast KS (2019) Phytolith formation in plants: from soil to cell. Plants (Basel) 8(8):249. https://doi.org/10.3390/plants8080249

Neumann D, Zur Nieden U (2001) Silicon and heavy metal tolerance of higher plants. Phytochemistry 56(7):685–686

Nguyen MN, Dultz S, Meharg A, Pham QV, Hoang AN, Dam TT, Nguyen VT, Nguyen KM, Nguyen HX, Nguyen NT (2019) Phytolith content in Vietnamese paddy soils in relation to soil properties. Geoderma 333:200–213

Ode PJ, Johnson SN, Moore BD (2014) Atmospheric change and induced plant secondary metabolites–are we reshaping the building blocks of multi-trophic interactions? Curr Opin Insect Sci 5:57–65. https://doi.org/10.1016/j.cois.2014.09.006

Pan W, Song Z, Liu H, Müeller K, Yang X, Zhang X, Li Z, Liu X, Qiu S, Hao Q, Wang H (2017) Impact of grassland degradation on soil phytolith carbon sequestration in Inner Mongolian steppe of China. Geoderma 308(4):86–92. https://doi.org/10.1016/j.geoderma.2017.08.037

Parr JF, Sullivan LA (2005) Soil carbon sequestration in phytoliths. Soil Biol Biochem 37:117–124. https://doi.org/10.1016/j.soilbio.2004.06.013

Parr JF, Sullivan LA (2011) Phytolith occluded carbon and silica variability in wheat cultivars. Plant Soil 342(1–2):165–171. https://doi.org/10.1007/s11104-010-0680-z

Pascual MB, Echevarria V, Gonzalo MJ, Hernández-Apaolaza L (2016) Silicon addition to soybean (Glycine max L.) plants alleviate zinc deficiency. Plant Physiol Biochem 108:132–138

Pavlovic J, Samardzic J, Maksimović V, Timotijevic G, Stevic N, Laursen KH, Hansen TH, Husted S, Schjoerring JK, Liang Y (2013) Silicon alleviates iron deficiency in cucumber by promoting mobilization of iron in the root apoplast. New Phytol 198(4):1096–1107

Pavlovic J, Kostic L, Bosnic P, Kirkby EA, Nikolic M (2021) Interactions of silicon with essential and beneficial elements in plants. Front Plant Sci 12:697592. https://doi.org/10.3389/fpls.2021.697592

Peñuelas J, Poulter B, Sardans J et al. (2013) Human-induced nitrogen–phosphorus imbalances alter natural and managed ecosystems across the globe. Nat Commun 4:2934. https://doi.org/10.1038/ncomms3934

Piperno DR (2006) Phytoliths: a comprehensive guide for archaeologists and paleoecologists. Rowman Altamira

Polidoro JC, de Freitas PL, Hernani LC, Anjos LHCD, Rodrigues RDAR, Cesário FV, Andrade AGD, Ribeiro JL (2021) Potential impact of plans and policies based on the principles of conservation agriculture on the control of soil erosion in Brazil. Land Degrad Dev 32:3457–3468

Prajapati S, Choudhary S, Kumar V, Dayal P, Srivastava R, Gairola A, Borate R (2023) Carbon sequestration: A key strategy for climate change mitigation towards a sustainable. Future 2:1–14. https://doi.org/10.18782/2583-4770.128

Pu J, Ma J, Li J, Wang S, Zhang W (2023) Organosilicon and inorganic silica inhibit polystyrene nanoparticles uptake in rice. J Hazard Mater 442:130012. https://doi.org/10.1016/J.JHAZMAT.2022.130012

Ranganathan S, Suvarchala V, Rajesh Y, Prasad MS, Padmakumari A, Voleti S (2006) Effects of silicon sources on its deposition, chlorophyll content, and disease and pest resistance in rice. Biol Plant 50(4):713–716

Ranjan A, Sinha R, Bala M, Pareek A, Singla-Pareek SL, Singh AK (2021) Silicon-mediated abiotic and biotic stress mitigation in plants: underlying mechanisms and potential for stress resilient agriculture. Plant Physiol Biochem 163:15–25, ISSN 0981-9428. https://doi.org/10.1016/j.plaphy.2021.03.044

Rashid I, Mir SH, Zurro D, Dar RA, Reshi ZA (2019) Phytoliths asproxies of the past. Earth Sci Rev 194:234–250. https://doi.org/10.1016/j.earscirev.2019.05.005

Rastogi A, Yadav S, Hussain S, Kataria S, Hajihashemi S, Kumari P, Yang X, Brestic M (2021) Does silicon really matter for the photosynthetic machinery in plants…? Plant Physiol Biochem 169:40–48. https://doi.org/10.1016/j.plaphy.2021.11.004

Rehman IU, Malik MA, Rashid I, Sheergojri IA, Dar RA (2022) Silicon fertilization increases carbon sequestration by augmenting phytoc production in wheat. J Soil Sci Plant Nutr 29:1–7

Sah SK, Reddy KR, Li J (2022) Silicon enhances plant vegetative growth and soil water retention of soybean (Glycine max) plants under water-limiting conditions. Plants (Basel) 11(13):1687. https://doi.org/10.3390/plants11131687

Saja-Garbarz D, Libik-Konieczny M, Janowiak F (2024) Silicon improves root functioning and water management as well as alleviates oxidative stress in oilseed rape under drought conditions. Front Plant Sci 15:1359747. https://doi.org/10.3389/fpls.2024.1359747

Sauer D, Saccone L, Conley DJ, Herrmann L, Sommer M (2006) Review of methodologies for extracting plant-available and amorphous Si from soils and aquatic sediments. Biogeochemistry 80(1):89–108

Schaller J, Puppe D (2021) Heat improves silicon availability in mineral soils. Geoderma 386:114909. https://doi.org/10.1016/j.geoderma.2020.114909

Schaller J et al (2019) Silicon increases the phosphorus availability of arctic soils. Sci Rep 9:1–11

Schaller J, Frei S, Rohn L, Gilfedder BS (2020) Amorphous silica controls water storage capacity and phosphorus mobility in soils. Front Environ Sci 8:94. https://doi.org/10.3389/fenvs.2020.00094

Schaller J, Puppe D, Kaczorek D, Ellerbrock R, Sommer M (2021) Silicon cycling in soils revisited. Plants 10(2):295. https://doi.org/10.3390/plants10020295

Schaller J et al (2022) Silicon as a potential limiting factor for phosphorus availability in paddy soils. Sci Rep 12:1–8

Sheng H, Chen S (2020) Plant silicon cell wall complexes: identification, model of covalent bond formation and biofunction. Plant Physiol Biochem 155:13–19. https://doi.org/10.1016/j.plaphy.2020.07.020

Sierra M, Schmid T, Guirado M, Escolano O, Millán R (2021) How management practices affect silicon uptake by Hordeum vulgare grown in a highly calcareous soil. Soil Use Manag 38. https://doi.org/10.1111/sum.12723

Singh P, Kumar V, Sharma A (2023) Interaction of silicon with cell wall components in plants: a review. J Appl Nat Sci 15:480–497. https://doi.org/10.31018/jans.v15i2.4352

Sinha D, Mukherjee S, Chowdhury S (2022) Methods of extraction of phytochemicals. In: Isolation, characterization, and therapeutic applications of natural bioactive compounds. IGI Global. https://doi.org/10.4018/978-1-6684-7337-5.ch010

Sirisuntornlak N, Ullah H, Sonjaroon W et al (2021) Interactive effects of silicon and soil pH on growth, yield and nutrient uptake of maize. Silicon 13:289–299. https://doi.org/10.1007/s12633-020-00427-z

Sjöberg S (1996) Silica in aqueous environments. J Non-Cryst Solids 196:51–57. https://doi.org/10.1016/0022-3093(95)00562-5

Sohag AAM, Tahjib-Ul-Arif M, Brestic M, Afrin S, Sakil MA, Hossain MT, Hossain MA, Hossain MA (2020) Exogenous salicylic acid and hydrogen peroxide attenuate drought stress in rice. Plant Soil Environ 66(1):7–13

Song Z, McGrouther K, Wang H (2016) Occurrence, turnover and carbon sequestration potential of phytoliths in terrestrial ecosystems. Earth Sci Rev 158:19–30. https://doi.org/10.1016/j.earscirev.2016.04.007

Song Z, Liu H, Strömberg CA, Yang X, Zhang X (2017) Phytolithcarbon sequestration in global terrestrial biomes. Sci Total Environ 603:502–509. https://doi.org/10.1016/j.scitotenv.2017.06.107

Sun S, Yang Z, Song Z, Wang N, Guo N, Niu J, Liu A, Bai B, Ahammed GJ, Chen S (2022) Silicon enhances plant resistance to Fusarium wilt by promoting antioxidant potential and photosynthetic capacity in cucumber (Cucumis sativus L.). Front Plant Sci 13:1011859. https://doi.org/10.3389/fpls.2022.1011859. PMID: 36311065; PMCID: PMC9608603

Thepbandit W, Athinuwat D (2024) Rhizosphere microorganisms supply availability of soil nutrients and induce plant defense. Microorganisms 12(3):558. https://doi.org/10.3390/microorganisms12030558

Tripathi DK, Singh S, Singh VP, Prasad SM, Dubey NK, Chauhan DK (2017) Silicon nanoparticles more effectively alleviated UV-B stress than silicon in wheat (Triticum aestivum) seedlings. Plant Physiol Biochem 110:70–81

Valizadeh-rad K, Motesharezadeh B, Alikhani HA, Jalali M (2022) Direct and residual effects of water deficit stress, different sources of silicon and plant-growth promoting bacteria on silicon fractions in the soil. SILICON 14(7):3403–3415

Valizadeh-Rad K, Motesharezadeh B, Alikhani HA, Jalali M, Etesami H, Javadzarin I (2023) Morphophysiological and nutritional responses of canola and wheat to water deficit stress by the application of plant growth-promoting bacteria, nano-silicon, and silicon. J Plant Growth Regul 42(6):3615–3631

Verma KK, Anas M, Chen Z, Rajput VD, Malviya MK, Verma CL, Singh RK, Singh P, Song XP, Li YR (2020) Silicon supply improves leaf gas exchange, antioxidant defense system and growth in saccharum officinarum responsive to water limitation. Plan Theory 9(8):1032

Vieira-Filho LO, Monteiro FA (2022) Silicon improves photosynthetic activity and induces antioxidant enzyme activity in Tanzania Guinea grass under copper toxicity. Plant Stress 3:100045. https://doi.org/10.1016/j.stress.2021.100045

Wang G, Dobermann A, Witt C, Sun Q, Fu R (2001) Performance of site-specific nutrient management for irrigated rice in Southeast China. Agron J 93(4):869–878

Wang YW, Zhang BB, Jiang DX, Chen GX (2019) Silicon improves photosynthetic performance by optimizing thylakoid membrane protein components in rice under drought stress. Environ Exp Bot 158:117–124. https://doi.org/10.1016/j.envexpbot.2018.11.022

Wang B, Chu C, Wei H, Zhang L, Ahmad Z, Wu S, Xie B (2020) Ameliorative effects of silicon fertilizer on soil bacterial community and pakchoi (Brassica chinensis L.) grown on soil contaminated with multiple heavy metals. Environ Pollut 267:115411. https://doi.org/10.1016/j.envpol.2020.115411

Wani AH, Mir SH, Kumar S, Malik MA, Tyub S, Rashid I (2022) Silicon en route - from loam to leaf. Plant Growth Regul 99. https://doi.org/10.1007/s10725-022-00931-9

Wani AH, Mir SH, Kumar S, Malik MA, Tyub S, Rashid I (2023) Silicon en route-from loam to leaf. Plant Growth Regul 99:465–476

Wiesmeier M, Urbanski L, Hobley E, Lang B, von Lützow M, Marin-Spiotta E, van Wesemael B, Rabot E, Ließ M, Garcia-Franco N et al (2019) Soil organic carbon storage as a key function of soils—a review of drivers and indicators at various scales. Geoderma 333:149–162

Xue XJ, Geng TT, Liu HF, Yang W, Zhong WR, Zhang ZL et al (2021) Foliar application of silicon enhances resistance against *Phytophthora infestans* through the ET/JA- and NPR1-dependent signaling pathways in potato. Front Plant Sci 12:609870. https://doi.org/10.3389/fpls.2021.609870

Yanai J, Taniguchi H, Nakao A (2016) Evaluation of available silicon content and its determining factors of agricultural soils in Japan. Soil Sci Plant Nutr 62:1–8. https://doi.org/10.1080/00380768.2016.1232601

Yang Y, Wang F, Jin J, Hu S, Gao Y (2019) Carbon storage potential and seasonal dynamics of phytolith from different vegetation types in a subtropical region, China. Environ Sci Pollut Res 26:29834–29844. https://doi.org/10.1007/s11356-019-06143-5

Yassen A, Abdallah E, Gaballah M, Zaghloul S (2017) Role of silicon dioxide Nano fertilizer in mitigating salt stress on growth, yield and chemical composition of cucumber (Cucumis sativus L.). Int J Agric Res 12:130–135

Yoshida S, Onishi Y, Kitagishi K (1959) The chemical nature of silicon in rice plant. Soil Sci Plant Nutr 5(1):23–27

Zargar SM, Mahajan R, Bhat JA, Nazir M, Deshmukh R (2019) Role of silicon in plant stress tolerance: opportunities to achieve a sustainable cropping system. 3 Biotech 9(3):73. https://doi.org/10.1007/s13205-019-1613-z

Zhang Y, Shi X-z, Zhao Y-c, Yu D, Wang H-j, Sun W-x (2008) Estimates and affecting factors of soil organic carbon storages in Yunnan-Guizhou-Guangxi region of China. Huan jing ke xue 29:2314–2319

Zhang WJ, Zhang XJ, Lang DY, Li M, Liu H, Zhang XH (2020) Silicon alleviates salt and drought stress of Glycyrrhiza uralensis plants by improving photosynthesis and water status. Biol Plant 64:302–313

Zhao S, Abu-Omar M (2015) Biobased Epoxy Nanocomposites Derived from Lignin-Based Monomers. Biomacromolecules 16. https://doi.org/10.1021/acs.biomac.5b00670

Chapter 7
The Multifaceted Role of Silicon and Silicon-Solubilizing Bacteria in Sustainable Agriculture

Amir Hosein Yadegari and Hassan Etesami

Abstract Silicate-solubilizing bacteria (SSB) play a crucial role in weathering and converting insoluble Si minerals into plant-absorbable forms, thereby improving Si nutrition and plant performance. This chapter provides an overview of the multifaceted role of Si and SSB in sustainable agriculture. It begins by discussing the cycling of Si in soils, highlighting the key drivers and dynamics that control its availability. The chapter then delves into the mechanisms by which Si alleviates both abiotic and biotic stresses in plants. Particular emphasis is placed on the benefits of Si in enhancing the legume-rhizobia symbiosis and improving nitrogen fixation. The chapter further explores the diverse strategies employed by SSB to dissolve insoluble Si minerals. The taxonomic distribution of SSB is discussed, along with techniques for their isolation and identification. Finally, the chapter highlights the beneficial effects of SSB application in soils, as well as emerging strategies, such as the use of silicon nanoparticles and the combined application of SSB with low-cost Si sources, to further advance sustainable agriculture and food security.

Keywords Sustainable agriculture · Legumes · Silicate-solubilizing bacteria (SSB) · Silicon (si) · Food security

7.1 Introduction

The world's population is increasing at a very fast rate, which is estimated to be about 80 million people annually (Borlaug and Dowswell 2003). This dramatic growth has resulted in significant pressure on food security worldwide, raising concerns about the ability to meet the nutritional needs of a growing population. Given that food security serves as a fundamental element of sustainable progress, the

A. H. Yadegari · H. Etesami (✉)
Department of Soil Science, University of Tehran, Tehran, Iran
e-mail: hassanetesami@ut.ac.ir

R. de Mello Prado et al. (eds.), *Silicon Advances for Sustainable Agriculture and Human Health*, Sustainable Plant Nutrition in a Changing World,
https://doi.org/10.1007/978-3-031-69876-7_7

identification of efficient strategies to tackle this urgent matter is of utmost importance (Godfray and Garnett 2014). Cultivable land is decreasing all over the world, and the climate pattern of the world is constantly changing (Zhang and Cai 2011); therefore, the only solution is to maximize production per unit area. One of the important strategies to achieve this goal is the use of new technologies and new tools to increase the fertility of soils and agricultural lands. The most important factor related to crop production and soil fertility is proper nutrition, which is necessary to increase the growth and yield of plants. On the other hand, plants face a multitude of biotic and abiotic stresses that can impede their growth and limit their overall yield potential (Etesami and Maheshwari 2018). Several studies have shown that mineral elements can effectively reduce the negative effects of different types of stress on different plants (Cakmak 2005; Nawaz et al. 2020; Kumari et al. 2022; Waraich et al. 2012; Etesami and Jeong 2018; Etesami and Adl 2020b). Among the various elements, the results of multiple investigations have repeatedly shown that silicon (Si) plays a beneficial role in increasing growth, plant function, and resistance to biotic and abiotic stresses (Ma 2004; Etesami and Jeong 2018, 2020; Etesami et al. 2022; Etesami and Jeong 2023). Plants that are abiotically and biotically stressed take advantage of silicon through various mechanisms, including effects on phytohormone signaling and secretion, metabolic processes, and gene expression or upregulation (Etesami and Jeong 2018; Frew et al. 2018; Etesami and Jeong 2023) (Fig. 7.1).

Silicon, which is the eighth most abundant element in the world, the second most abundant element in the earth's crust (~27%), and the most abundant element in soils, is not lacking in quantity (Epstein 1999). However, crop-available forms of silicon can be limited (Savant et al. 1997) due to the low solubility of most forms of

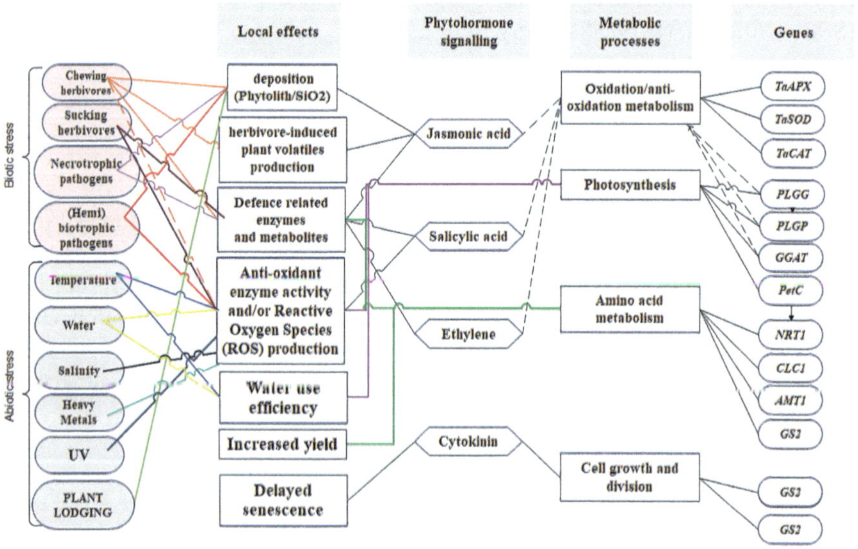

Fig. 7.1 Role of Si in various mechanisms including phytohormone signaling and secretion, metabolic processes, and gene expression or upregulation in plants under different stresses

silicon (Epstein 1999; Vasanthi et al. 2012). Since the early nineteenth century, the importance of silicon has been revealed to plant scientists (de Saussure 1804; Hall and Morison 1906; Guntzer et al. 2012). Although Si is currently considered a beneficial element, it is classified as a non-essential element (quasi-essential) (Richmond and Sussman 2003). Many studies have shown that the application of Si is not very significant in favorable and non-stressed conditions, and according to new articles, the benefits of silicon become apparent when the plant is under stress (Etesami and Jeong 2018, 2020; Etesami et al. 2022). Plants can only absorb silicon in the form of monosilicic acid (H_4SiO_4) when it is soluble. The production of monosilicic acid by rock weathering is primarily influenced by temperature, pH levels, redox potential, water content, and microbial activity (Schaller et al. 2021; Raturi et al. 2021). Silicon (Si) is mostly found in the form of insoluble silicates, including Mg-silicate, Al-silicate, Fe-silicate, Ca-silicate, Na-silicate, and K-silicate (Bist et al. 2020). In addition, several other types of silicon, such as kaolin, smectite, quartz, vermiculite, amorphous silica, and crystalline forms like orthoclase, feldspar, and plagioclase, are also found in the soil (Blume et al. 2016; Sahebi et al. 2015).

Silicon fertilizer has advantages in combating crop diseases and environmental challenges while also enhancing soil fertility through the augmentation of water retention, nutrient equilibrium, and preservation of the physiochemical characteristics of the soil (Meena et al. 2014; Rachappanavar et al. 2024; Etesami et al. 2022). Commercially available Si fertilizers include slag-based Ca-silicate fertilizer, fused Mg-phosphate silicate fertilizer, fused K-Mg silicate, porous hydrated-Ca silicate, soluble silicates like Na and K silicate, and natural sources like wollastonite and olivine (Liang et al. 2015b; Sharma et al. 2023; Etesami and Glick 2023). Sodium and potassium silicates are hydrophilic silicates, but their cost is prohibitive for conventional agricultural operations. Na and K silicates possess water-soluble characteristics, making them very suitable as foliar supplements that may be immediately sprayed as foliar spray (Liang et al. 2015b). The effectiveness of silicon through foliar use has not yet been proven because there is no evidence that silicon is absorbed through the leaves. Due to the high price of silicon fertilizers, nowadays relatively cheaper sources such as wollastonite, smelting by-products, iron ores, manganese, and silicon-rich crop residues are used as silicon fertilizers (Pang et al. 2024; Yang et al. 2022). Certain plants, such as rice, wheat, corn, barley, and sugarcane (graminaceous plants), are known to accumulate high amounts of silicon (Si) (Rao and Susmitha 2017; Rea et al. 2022; Haynes 2017; Souri et al. 2021). A recent study on Si-plant interactions estimated that plants across the world remove over 224 million tons of Si from the soil each year (Tubana et al. 2016). This can greatly reduce plant production in areas where there is mono-cropping of Si-accumulating crops, like rice.

The solubility of secondary and primary minerals is the main factor controlling the concentration of silicon in soil solution, and these minerals are quickly dissolved using biochemical and physicochemical interventions (Bin et al. 2008). Meanwhile, microbial activities are more important for biochemical performance than other factors (Haynes 2014; Tubaña and Heckman 2015; Sommer et al. 2006). There are many microorganisms in the soil whose effect on the solubility and mobility of elements in the soil is well known (Calvaruso et al. 2006; Sheng and He 2006;

Uroz et al. 2009) such as potassium-solubilizing bacteria (KSB) (Etesami et al. 2017; Ahmad et al. 2016; Zhang and Kong 2014), phosphorus-solubilizing bacteria (PSB) (Wan et al. 2020; Yu et al. 2022; Rawat et al. 2021; Li et al. 2023; Etesami 2020) and in the case of our topic silicate-solubilizing bacteria (SSB) (Raturi et al. 2021; Etesami et al. 2021; Bist et al. 2020; Rezakhani et al. 2020; Etesami and Jeong 2022). Considering the role of Si in biotic and abiotic stresses and increasing the yield and growth of plants, silicate-solubilizing bacteria (SSB), which convert the insoluble forms of silica minerals into plant-absorbable forms, have magnetized a lot of attention. In addition to converting insoluble forms of silica into soluble forms that can be absorbed by plants and reducing the effects of biotic and abiotic stresses, SSB have other side effects, such as improving potassium uptake by solubilizing potassium (Sheng 2005) and improving soil physical and biological properties (such as soil structure) (Chandrakala et al. 2019; Nagabovanalli et al. 2020). The aim of this book chapter is to provide a review on the critical role of silicon and silicon-solubilizing bacteria (SSB) in enhancing sustainable agriculture and improving food security in the face of a growing global population.

7.2 Silicon Cycling in Soils: Drivers and Dynamics

In the soil, silicon is generally in different phases, including the liquid phase, the solid phase, and the surface-adsorbed phase (Sauer et al. 2006). Nevertheless, the presence of Si is primarily affected by variables such as temperature, soil pH, soil organic matter, soil type, mineral weathering, and soil microbes (Raturi et al. 2021) (Fig. 7.2). Sources of silicon include feldspars, mica, olivine, and pyroxene (primary silicates), clay minerals (secondary silicates), quartz and irregular silica (silicate minerals), the remains of microorganisms, silicon-rich plants, and phytoliths (biogenic forms) pointed out (Tubaña and Heckman 2015; Laruelle et al. 2009). Si dissolution kinetics can be affected by redox potential, the amount of water, and sesquioxides (Savant et al. 1996). Adsorption or desorption of monosilicic acid (H_4SiO_4) depends on pH, which is one of the most important factors that affect the presence of monosilicic acid in soil.

The pH range of 9 to 10 affects the adsorption process of silicon, whereas pH levels outside of this range result in silicon desorption. This is why the polymerization and coagulation of H_4SiO_4 are more pronounced in saline soil. The pH levels of 9 and 11 can cause the dissociation of H_4SiO_4 into H_3SiO_4 and H_2SiO_4, respectively (Tubaña and Heckman 2015). Another factor that affects the solubility of silicon in soil is temperature. One of the reasons for producing rice in temperate regions is that silicon dissolved in temperate soils is 5–10 times higher than in tropical regions (Savant et al. 1996). Bacterial/fungal activity in soil plays a very important role in the cycle of silicon and the availability of silicon in soil. The biological weathering of minerals by bacteria causes the release of silicon in the soil and the growth and development of plants (Khan et al. 2021). Si can undergo leaching or accumulation in the soil during soil pedogenesis. Si loss is mostly caused by desilication, which is

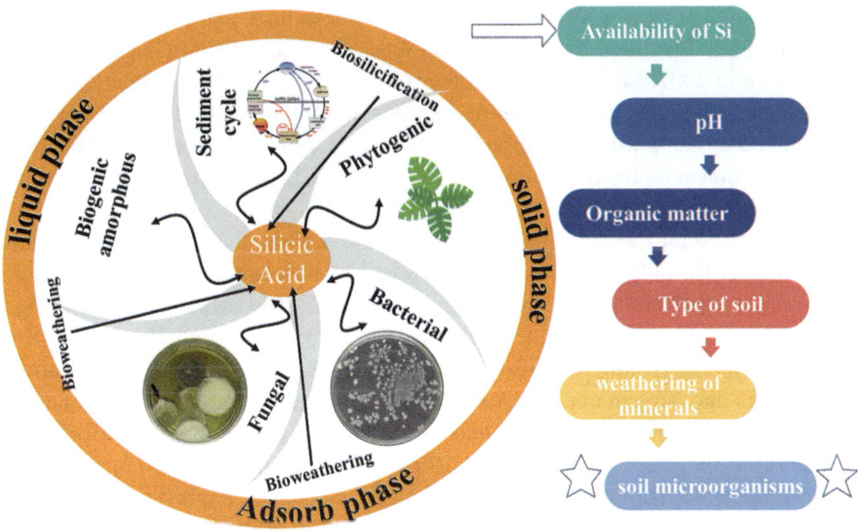

Fig. 7.2 Effects of Si presence on temperature, soil pH, soil organic matter, soil type, mineral weathering, and soil microbes

influenced by the degree of stratification, the Si concentration of the parent material, and the saturation of the profile (Haynes 2019; Sommer et al. 2006; White et al. 2012).

7.3 The Multifaceted Role of Silicon in Plant Stress Mitigation

The positive impacts of silicon have been recorded in various plants, such as sugarcane (Savant et al. 1999; Camargo and Keeping 2021; Majumdar and Prakash 2020), barley (Liang et al. 1996; Ali et al. 2013), and rice (Sakurai et al. 2015; Savant et al. 1997). In general, the concentration of Si in plants varies between 0.1% and 10% of the total dry matter (Epstein 1994). Plants are divided into several categories according to the presence of Si in their tissues: plants with a concentration of more than 1% are Si accumulators, plants with a concentration of less than 0.5% of Si are excluders, and plants with a concentration between these limits are known as intermediate (Frew et al. 2018; Zhang et al. 2017). Seven main plants, which include corn, wheat, sugarcane, rice, barley, soybean, and sugar beet, are known as Si accumulators (Guntzer et al. 2012). Also, many forages are considered Si accumulators because their leaves accumulate between 1 and 5% of Si (Baggio et al. 2021).

The first Si transporters were discovered in diatoms by which it was determined that the distribution of these transporters exists in various life forms (Hildebrand et al. 1997), but because their homologs were not found in plants, evidence for their existence in plants was not available until 2006 (Deshmukh and Bélanger 2016). In 2006, Ma et al. identified the first silicon transporter, named Lsi1 (Ma et al. 2006). Then again in 2007, Ma identified another transporter called Lsi2 (Ma et al. 2007). The Lsi1 is an influx Si transporter and Lsi2 is an efflux transporter. Lsi1 and Lsi2 are situated on the plasma membrane of the endodermis and exodermis. However, Lsi1 is positioned on the distal side of the cells, while Lsi2 is found on the proximal side. Table 7.1 summarizes the various Si transporters in different plants.

Some important mechanisms by which silicon mitigates abiotic stress (Etesami and Jeong 2018, 2023) include strengthening the plant structure (Meunier et al. 2017), changing the rate of photosynthesis (Perez et al. 2014; Mihaličová Malčovská et al. 2014; Sanglard et al. 2014; Rahman et al. 2015; Kang et al. 2016), adjusting stomatal behavior (Hattori et al. 2005), increasing the efficiency of water consumption (Fawaz and Mohammad 2013), osmotic adjustments to increase the potential and conductivity of water (Gong and Chen 2012; Ming et al. 2012), reducing oxidative stress (Shen et al. 2010; Ali et al. 2019; Kim et al. 2017), changing absorption and accumulation of other mineral elements (Pavlovic et al. 2016), changes in the concentration and amount of secretion of different phytohormones

Table 7.1 Influx and efflux transporters of silicon reported in different plants

Plant	(Scientific name)	Cotyledons	Transporter	References
Pepper	(*Capsicum annuum*)	Dicot	CaLsi1 CaLsi2	Gómez-Merino et al. (2020)
Tomato	(*Solanum lycopersicum*)	Dicot	SlLsi1 SlLsi2	Sun et al. (2020)
Cucumber	(*Cucumis sativus*)	Eudicots	CsLsi1 CsLsi2	Sun et al. (2017) and Sun et al. (2018)
Sorghum	(*Sorghum bicolor*)	Monocot	SbLsi1 SbLsi2 SbLsi6	(Soukup et al. 2017)
Strawberry	(*Fragaria ananassa*)	Dicot	FaLsi1 FaLsi2	Ouellette et al. (2017)
Horsetail	(*Equisetum arvense*)	–	EaNIP3; 1 EaNIP3; 3 EaNIP3; 4 EaLsi2-1 EaLsi2-2	Grégoire et al. (2012)
Barley	(*Hordeum vulgare*)	Monocot	HvLsi1 HvLsi2 HvLsi6	Yamaji et al. (2012)
Pumpkin	(*Cucurbita moschata*)	Eudicots	CmLsi1 CmLsi2-1 CmLsi2-2	Mitani et al. (2011) and Mitani-Ueno et al. (2011)
Maize	(*Zea mays*)	Monocot	ZmLsi1 ZmLsi2 ZmLsi6	Ma (2010)
Rice	(*Oryza sativa*)	Monocot	OsLsi1 OsLsi2 OsLsi6	Ma (2010)

(Hamayun et al. 2010; Kim et al. 2014) (e.g., the phytohormone ethylene plays a role in reducing salinity stress by reducing oxidative damage (Liang et al. 2015a)), reducing the stress of oxidative damages by reducing the production of ROS (such as O_2^- and H_2O_2 which boost when the plants are exposed to stress, they are very reactive and cause cell destruction (Van Bockhaven et al. 2013)), and/or increasing the performance of antioxidant metabolisms (Zhu et al. 2004; Liang et al. 2008; Miao et al. 2010; Ali et al. 2019; Kim et al. 2017). Also, the activity of photorespiratory enzymes and their gene expression are some of the other mechanisms by which silicon reduces abiotic stresses (Romero-Puertas et al. 2007; Rojas et al. 2012).

The silicon-induced metabolism of plant antioxidants at the molecular level and the expression of genes has also been investigated; for example, Si causes the expression of genes of antioxidant enzymes (*TaSOD* and *TaCAT*) and reduces the accumulation of H_2O_2 in wheat plants under drought stress (Ma et al. 2016). Also, the mechanisms by which Si alleviates biological stresses (Griffin et al. 2015; A Rodrigues et al. 2015; Reynolds et al. 2016) include physical defense through the accumulation of Si in plant tissues (McNaughton and Tarrants 1983; Katz 2015) (polymerization of monosilicic acid and turning it into phytolith by plants (Epstein 1994; Ma and Yamaji 2015)), and the formed phytolith in turn increases the tensile strength, hardness of the plant (Massey et al. 2007), and physical barrier for fungus penetration (Kim et al. 2002). Si deposition can also destroy the parts of the feeding mouth or lower jaw of insects (Kvedaras et al. 2009; Massey and Hartley 2009; Jeer et al. 2017) and diminish the digestibility of the plant against herbivorous insects and mammals (Massey et al. 2006; Massey and Hartley 2006; Frew et al. 2016). Si also impacts the concentration of a set of metabolites related to the crop defense system, which includes increasing the activity of chitinase, phenylalanine ammonia-lyase, β-1,3-glucanase, and polyphenol oxidase enzymes (Chérif et al. 1992; Rémus-Borel et al. 2005; Rahman et al. 2015; Debona et al. 2017). Also, the signaling pathway of plant phytohormones is one of the important effects of Si during biological stress control, for example, the secretion of a complex and special compound of salicylic acid against (hemi) biotrophic pathogens and the secretion of jasmonic acid against necrophytic pathogens and ethylene, which regulates the secretion of jasmonic acid during the defense (Glazebrook 2005; Wu and Baldwin 2010). Other mechanisms of Si in reducing biotic stresses include changing the composition of HIPVs (herbivore-induced plant volatiles) (Liu et al. 2017), identifying protected molecular patterns for plant recognition from biological threats, and stimulating the activation of plant phytohormones signaling to overcome the ability pathogens pointed to suppression of plant defense systems such as PAMPs (pathogen-associated molecular patterns), DAMPs (damage-associated molecular patterns), and HAMPs (herbivore-associated molecular patterns) (Erb et al. 2012).

In recent years, there has been an increased focus on the function of Si in boosting the growth of leguminous plants, both under environmental stress conditions and in normal conditions (Putra et al. 2020; Etesami and Adl 2020a). The legumes (Fabaceae) family includes about 20,000 species and 700 genera (Lewis et al. 2005). Legumes are considered the most valuable source of protein for plant and animal feed (Iqbal et al. 2006). Legumes form root nodules to accommodate rhizobia bacteria that fix atmospheric molecular nitrogen by the nitrogenase enzyme and convert it into ammonia (Vessey 1994). In contrast, host leguminous plants provide diverse resources, such as carbon-based compounds to their rhizobial partners (Checcucci et al. 2017). Bacteria associated with the rhizobia-legume symbiotic relationship, such as *Rhizobium leguminosarum*, are essential for the maintenance and productivity of drier agricultural ecosystems affected by salt, which is a global issue for agriculture (Al-Karaki 2006). Silicon has been able to improve the growth of legume plants mainly by improving the efficiency of the leguminous rhizobium symbiosis system (Putra et al. 2020; Etesami and Adl 2020a). Many studies showed the effect of silicon on nodulation and nitrogen fixation (Putra et al. 2020). It has been reported that feeding plants with silicon increases the generation of the endogenous hormone abscisic acid (ABA), which plays an essential role in nodulation (Ferguson and Mathesius 2003). It has also been reported that silicon increases the number of bacteroids and symbiosomes (Nelwamondo et al. 2001). In Putra et al.'s 2019 article (Putra et al. 2020), a hypothetical model for increasing root nodulation and nitrogen fixation in legumes by silicon is proposed. At first, Si is absorbed by plant roots under stress in the form of orthosilicic acid. The root secretes chemical cues (flavonoids) to attract free rhizobial bacteria, and the bacteria, in response to these chemicals, secrete nod factors to the root hairs that allow them to be recognized by the host. Following this identification, the root hairs begin to curl, and the rhizobia begin to induce the formation of the infection thread by injecting specific proteins that allow them to penetrate the cortex of roots. Subsequently, the induced crop proteins cause the cells of the root cortex to divide to initiate nodulation. At this stage, rhizobial bacteria begin to transform into N_2-fixing bacteroids. Finally, mature nodules are formed, and food exchange is carried out through the vascular tissues of the nodules. At the same time, Si starts to interfere with the release of chemical signals for rhizobial bacteria, which affects *Rhizobium*-legume communication and nodule development. In this stage, silicification occurs, and as a result, structural carbon may be replaced with silicon, increasing the permeability of the nodule and reducing the effects of environmental stress. Following these changes, compounds containing carbon are more supplied to rhizobial bacteria, and the rapid transfer of solutes and gases among nodules, root tissues, and soil occurs, and the host's resistance to stresses also increases.

7.4 Silicate-Solubilizing Bacteria and Their Role in Plant Nutrition and Soil Health

Silicate-solubilizing bacteria (SSB) are a group of bacteria that play a role in transforming silicates into soluble silica. They produce silicase, an enzyme that facilitates this conversion, making silicon accessible for plants to absorb (Vasanthi et al.

2018). Microorganisms compete with each other to alter minerals, as a means of increasing their ability to survive in a given environment (Brindavathy 2022; Fomina and Skorochod 2020). Decomposition of minerals provides mineral elements and terminal electron acceptor in cellular respiration (Li et al. 2019) that increases competition between microbial species (Brindavathy 2022). Many bacterial species are involved in the release and control of silicon via various steps such as formation, incorporation, or decomposition of minerals (Patrick and Holding 1985; Sheng et al. 2008a). Several mechanisms are adopted by element-solubilizing bacteria (including SSB) to dissolve insoluble forms of elements, which include chelation (ligand formation), pH change owing to the production of inorganic and organic acids, extracellular polysaccharides, and redox reactions. Solubilization based on organic acids is the most common method of bacteria for Si solubilization (Cama and Ganor 2006). Some bacterial strains release inorganic acids, for example, sulfuric acid by sulfur-oxidizing bacteria (SOB) (i.e., *Thiobacillus* genus) and nitric acid by nitrifying bacterial strains (Ranalli and Zanardini 2021). These acids enhance the weathering of silicate minerals. Organic acids such as pyruvic acids, citric acids, formic acids, acetic acids, gluconic acids, lactic acids, oxalic acids, and succinic acids are also produced by some bacterial strains, which increase the weathering of minerals. The pH changes caused by bacterial activities also increase the solubility of silicate minerals. These acids are released as by-products of C metabolic processes (Samuels et al. 2020; Zhu et al. 2014). The existence of carbonic anhydrase in aerobic respiration causes the release of CO_2, which leads to the production of carbonic acid. Carbonic acid production is a key process associated with acidification. The enzyme activities of many bacterial isolates (Smith and Ferry 2000) also cause silicon weathering (Xiao et al. 2014). Organic acids create acidic conditions in the environment surrounding minerals, and their deprotonated form creates a special type of chelate that increases the solubilization rate of insoluble silicates. Similarly, the related carboxylic groups of this organic acid mainly have a negative charge after dissociation of the hydrogen ion, and they act as a ligand site for cations. Di- and tricarboxylic acids are more efficient solvents compared to monocarboxylic acids (Adeleke et al. 2017). Siderophores, which are the most important chelating agents released by bacterial strains, increase the weathering of minerals (Torres et al. 2014). Solubilizing a mineral can also be done by altering the crystal structure through oxidation and reduction reactions (Kalinowski et al. 2000; Bonneville et al. 2004).

Although SSB are ubiquitous (in water, sediments, soil, and minerals), the number of SSB is relatively small compared to the total bacterial population (Vasanthi et al. 2018). The majority of the SSB are classified within the genera *Bacillus*, *Burkholderia*, *Pseudomonas*, *Janthinobacterium*, and *Aminobacter* (Vasanthi et al. 2018; Lee et al. 2019). According to a study based on 16S rRNA sequencing, *Pseudomonas* and *Bacillus* are the two main microbial genera responsible for Si dissolution (Bist et al. 2020). Also, this study showed the role of *Sphingobacterium* sp. in Si dissolution. Table 7.2 shows the taxonomic distribution of bacteria that have the ability to solubilize silicon.

Table 7.2 The taxonomical distribution of silicate-dissolving bacteria

Proteobacteria	Firmicutes
Enterobacter ludwigii	*Bacillus flexus*
Enterobacter hormaechei	*Bacillus megaterium*
Proteus mirabilis	*Bacillus circulans*
Pseudomonas stutzeri	*Bacillus cereus*
Pseudomonas fluorescens	*Bacillus amyloliquefaciens*
Aeromonas caviae	*[Bacillus] caldolyticus*
Burkholderia cenocepacia	*Paenibacillus mucilaginosus*
Burkholderia vietnamiensis	*Paenibacillus edaphicus*
Paraburkholderia eburnea	*Paenibacillus glucanolyticus*
Acidithiobacillus thiooxidans	*Paenibacillus* sp.
Acidithiobacillus ferrooxidans	*Sporosarcina globispora*
Rhizobium yantingense	
Rhizobium tropici	

For more details see Raturi et al. (2021)

Different effects of different strains of bacteria on Si solubilization have been reported (Etesami and Jeong 2022). For example, *B. mucilaginous*, *Bacillus globisporus*, and *Bacillus circulans* in liquid environments released silicon and potassium from silicate minerals (Liu et al. 2006). It has also been reported that *Bacillus mucilaginous* bacteria break down mica minerals and release Si and potassium from silicate minerals (Sumitha et al. 2018). Also, this bacterium secretes organic acids and polysaccharides (Shahbazi and Moosavy 2019). Many researches have shown that the bacterium *Pseudomonas syringae* enhances the effect of silica on the absorption of phosphorus and other elements in sorghum plants (Yin et al. 2016). The mesophilic bacterium *P. mirabilis* is known to make monomers of silica particles (Brindavathy 2022; Shivaraj et al. 2022). *B. caldolyticus* is a thermophilic bacterium that has caused positive effects on the plants that use a lot of Si under high Si conditions. The bacterium *Equisetum arvense* converts Si polymers into monomers (Currie and Perry 2007; Neethirajan et al. 2009). According to reports, *Rhizobia IIRR_1* has the potential to dissolve silicon from biogenic materials and silicate minerals, and this strain can increase silicate mineral weathering. Three strains, CCNWYC119 (*Rhizobium* sp.), H66T (*R. yantingense*), and Q34 (*Rh. tropici*), were found in the rhizosphere soil of legumes. They were able to break down silicate minerals and release Si (Chen et al. 2016; Wang et al. 2015; Yanni et al. 1997). Researchers investigated the silicon (Si)-solubilizing abilities of the bacterium Burkholderia eburnea CS4-2. They concluded that further research on this bacterium should explore the use of different soil types, silicate mineral sources, and pH levels to better understand its silicon dissolution capabilities (Ribeiro et al. 2020). In another study, it was concluded that the absorption of Si and P during the development and growth of the plant has been significantly affected by the *Enterobacter ludwigii* GAK2 strain. This strain has a natural capacity to dissolve phosphate and silicon and secretes hormones such as IAA and gibberellic acid (Gondil et al. 2017; Lee et al. 2019).

The isolation of SSB includes various steps such as sample collection, preparation of serial dilution, screening of microorganisms to dissolve insoluble silicate, and identification of bacterial strains. There are different culture media for isolating

Si-dissolving bacteria, which are supplemented with mineral-insoluble silicates such as magnesium trisilicate, illite, muscovite, biotite, talc, and feldspar (Kang et al. 2017; Sheng et al. 2008b; Vasanthi et al. 2018; Kannan and Raj 1998; Zhou et al. 2006; Archana 2007; Purushothaman et al. 1974). Magnesium trisilicate is commonly used as the insoluble inorganic mineral substrate for isolating silicon-solubilizing bacteria (SSB). When SSB are grown on media containing magnesium trisilicate, they should form a colorless halo around their colonies, indicating their ability to solubilize the mineral. Different media are used to isolate SSB bacteria including nutrient agar (NA) supplemented with magnesium trisilicate ($Mg_2O_8Si_3$, ~0.25%) (Vasanthi et al. 2018), modified Bunt and Rovira medium supplemented with $Mg_2O_8Si_3$ (0.25%) (Chandrakala et al. 2019), glucose agar medium having insoluble silicate (Lee et al. 2019), NBRISSM (to differentiate between SSB and PSB) (Lee et al. 2019), Zak-Alexandrov medium containing Na_2SiO_3 (Maleva et al. 2017), Aleksandrov's medium having $KAlSi_3O_8$ (Hu et al. 2018), and solid K-limited medium (KLM) supplemented feldspar and biotite (Wang et al. 2015).

The application of SSB to the soil causes re-mineralization; improvement of cation exchange capacity; acidification (pH change); release of acids, enzymes, and metabolites; chelation and availability of nutrients; change in reaction redox; dissolving minerals; and improving the physicochemical properties of soil. Also, SSB increases the surface area of the plant root, improves root exudates, improves the yield, protects the plant, reduces the consumption of expensive chemical fertilizers, increases the absorption of elements needed by the plant, and alleviates biotic and abiotic stresses in crops (Etesami and Jeong 2022; Chandrakala et al. 2019; Etesami and Glick 2023; Raturi et al. 2021).

Considering the multifaceted function of silicon in boosting the growth and performance of crops under environmental stress and, on the other hand, the affordability challenges associated with silicic fertilizers in agriculture, the use of silicate-dissolving bacteria appears to be a suitable alternative in the agricultural sector. Through the provision of silicon, these bacteria can help plants to leverage the multiple benefits of silicon, such as improved growth and resilience under environmental stresses.

7.5 Conclusions and Future Aspects

The chapter has highlighted the multifaceted and essential role of silicon (Si) and silicon-solubilizing bacteria (SSB) in advancing sustainable agriculture and addressing the challenges of food security. Silicon is a beneficial, quasi-essential element that plays a crucial role in enhancing plant growth, yield, and resistance to both biotic and abiotic stresses. The application of Si can significantly improve crop production, particularly in regions where Si is limited in the soil. Silicate-solubilizing bacteria (SSB) are pivotal in weathering and transforming insoluble Si minerals into plant-available forms, thereby improving Si nutrition and accessibility for crops. SSB employ diverse mechanisms, such as the production of inorganic and organic acids, to enhance Si solubilization, and their taxonomic distribution is dominated by genera such as *Bacillus*, *Burkholderia*, *Janthinobacterium*, and *Pseudomonas*, which have demonstrated remarkable abilities to dissolve various silicate minerals. The application of SSB in soils can lead to beneficial outcomes, including re-mineralization, improved cation exchange capacity, increased nutrient availability, and enhanced plant growth and stress tolerance while potentially reducing the need for expensive chemical fertilizers. Future research prospects include exploring the use of silicon nanoparticles (SINPs) as a novel approach to improving Si availability and uptake by plants, potentially in combination with SSB applications. Developing cost-effective and sustainable sources of Si, such as agricultural waste and by-products, for use as Si fertilizers, complementing the application of SSB to enhance Si cycling in soils, is another promising area. Investigating the specific mechanisms by which SSB modulate the legume-rhizobia symbiosis, with the aim of leveraging this beneficial relationship to improve nitrogen fixation and crop productivity, is also a key focus. Conducting large-scale field trials to validate the efficacy of SSB-based approaches in improving crop performance under diverse agro-ecological conditions, and expanding the understanding of the complex interactions between SSB, plant roots, and the rhizosphere, will further contribute to the advancement of Si-based and SSB-mediated strategies for sustainable agriculture and global food security.

Acknowledgments The author would like to express their gratitude to the University of Tehran, Tehran, Iran, for their support of this study.

References

Adeleke R, Nwangburuka C, Oboirien B (2017) Origins, roles and fate of organic acids in soils: a review. S Afr J Bot 108:393–406

Ahmad M, Nadeem SM, Naveed M, Zahir ZA (2016) Potassium-solubilizing bacteria and their application in agriculture. In: Potassium solubilizing microorganisms for sustainable agriculture, pp 293–313

Ali S, Farooq MA, Yasmeen T, Hussain S, Arif MS, Abbas F, Bharwana SA, Zhang G (2013) The influence of silicon on barley growth, photosynthesis and ultra-structure under chromium stress. Ecotoxicol Environ Saf 89:66–72

Ali A, Haq T, Mahmood R, Jaan M, Abbas MN (2019) Stimulating the anti-oxidative role and wheat growth improvement through silicon under salt stress. Silicon 11:2403–2406

Al-Karaki GN (2006) Nursery inoculation of tomato with arbuscular mycorrhizal fungi and subsequent performance under irrigation with saline water. Sci Hortic 109(1):1–7

Archana D (2007) Studies on potassium solubilizing bacteria. Department of Agricultural Microbiology, Dharwad

Baggio G, Dupas E, Galindo FS, Megda MM, Pereira NCM, Luchetta MO, Tritapepe CA, da Silva MR, Jalal A, Teixeira Filho MCM (2021) Silicon application induced alleviation of aluminum toxicity in Xaraés palisadegrass. Agronomy 11(10):1938

Bin L, Ye C, Lijun Z, Ruidong Y (2008) Effect of microbial weathering on carbonate rocks. Earth Sci Front 15(6):90–99

Bist V, Niranjan A, Ranjan M, Lehri A, Seem K, Srivastava S (2020) Silicon-solubilizing media and its implication for characterization of bacteria to mitigate biotic stress. Front Plant Sci 11:28

Blume H-P, Brümmer GW, Fleige H, Horn R, Kandeler E, Kögel-Knabner I, Kretzschmar R, Stahr K, Wilke B-M, Blume H-P (2016) Inorganic soil components—minerals and rocks. In: Scheffer/SchachtschabelSoil science, pp 7–53

Bonneville S, Van Cappellen P, Behrends T (2004) Microbial reduction of iron (III) oxyhydroxides: effects of mineral solubility and availability. Chem Geol 212(3–4):255–268

Borlaug NE, Dowswell CR (2003) Feeding a world of ten billion people: a 21st century challenge. In: Proceedings of the international congress in the wake of the double helix: From the green revolution to the gene revolution. Citeseer, pp 27–31

Brindavathy R (2022) Silicate minerals induced by microorganisms. In: Mineral formation by microorganisms: concepts and applications, pp 125–159

Cakmak I (2005) Role of mineral nutrients in tolerance of crop plants to environmental stress factors. In: Proceedings from the international symposium on fertigation–optimizing the utilization of water and nutrients, pp 35–48

Calvaruso C, Turpault M-P, Frey-Klett P (2006) Root-associated bacteria contribute to mineral weathering and to mineral nutrition in trees: a budgeting analysis. Appl Environ Microbiol 72(2):1258–1266

Cama J, Ganor J (2006) The effects of organic acids on the dissolution of silicate minerals: a case study of oxalate catalysis of kaolinite dissolution. Geochim Cosmochim Acta 70(9):2191–2209. https://doi.org/10.1016/j.gca.2006.01.028

Camargo MS, Keeping MG (2021) Silicon in sugarcane: availability in soil, fertilization, and uptake. Silicon 13:1–11

Chandrakala C, Voleti S, Bandeppa S, Sunil Kumar N, Latha P (2019) Silicate solubilization and plant growth promoting potential of Rhizobium sp. isolated from rice rhizosphere. Silicon 11(6):2895–2906. https://doi.org/10.1007/s12633-019-0079-2

Checcucci A, DiCenzo GC, Bazzicalupo M, Mengoni A (2017) Trade, diplomacy, and warfare: the quest for elite rhizobia inoculant strains. Front Microbiol 8:312157

Chen W, Luo L, He L-Y, Wang Q, Sheng X-F (2016) Distinct mineral weathering behaviors of the novel mineral-weathering strains Rhizobium yantingense H66 and Rhizobium etli CFN42. Appl Environ Microbiol 82(14):4090–4099

Chérif M, Benhamou N, Menzies JG, Bélanger R (1992) Silicon induced resistance in cucumber plants against Pythium ultimum. Physiol Mol Plant Pathol 41(6):411–425

Currie HA, Perry CC (2007) Silica in plants: biological, biochemical and chemical studies. Ann Bot 100(7):1383–1389

de Saussure NT (1804) Recherches chimiques sur la vegetation. Nyon

Debona D, Rodrigues FA, Datnoff LE (2017) Silicon's role in abiotic and biotic plant stresses. Annu Rev Phytopathol 55:85–107

Deshmukh R, Bélanger RR (2016) Molecular evolution of aquaporins and silicon influx in plants. Funct Ecol 30(8):1277–1285

Epstein E (1994) The anomaly of silicon in plant biology. Proc Natl Acad Sci 91(1):11–17

Epstein E (1999) Silicon. Annu Rev Plant Biol 50(1):641–664

Erb M, Meldau S, Howe GA (2012) Role of phytohormones in insect-specific plant reactions. Trends Plant Sci 17(5):250–259

Etesami H (2020) Enhanced phosphorus fertilizer use efficiency with microorganisms. In: Nutrient dynamics for sustainable crop production. Springer, pp 215–245

Etesami H, Adl SM (2020a) Can interaction between silicon and non–rhizobial bacteria help in improving nodulation and nitrogen fixation in salinity–stressed legumes? A review. Rhizosphere 15:100229

Etesami H, Adl SM (2020b) Plant growth-promoting Rhizobacteria (PGPR) and their action mechanisms in availability of nutrients to plants. In: Kumar M, Kumar V, Prasad R (eds) Phytomicrobiome in stress regulation. Springer Singapore, Singapore, pp 147–203. https://doi.org/10.1007/978-981-15-2576-6_9

Etesami H, Glick BR (2023) Exploring the potential: can mycorrhizal fungi and hyphosphere silicate-solubilizing bacteria synergistically alleviate cadmium stress in plants? Curr Res in Biotechnol 6:100158. https://doi.org/10.1016/j.crbiot.2023.100158

Etesami H, Jeong BR (2018) Silicon (Si): review and future prospects on the action mechanisms in alleviating biotic and abiotic stresses in plants. Ecotoxicol Environ Saf 147:881–896. https://doi.org/10.1016/j.ecoenv.2017.09.063

Etesami H, Jeong BR (2020) Importance of silicon in fruit nutrition: agronomic and physiological implications. In: Fruit crops. Elsevier, pp 255–277

Etesami H, Jeong BR (2022) Chapter 19: Biodissolution of silica by rhizospheric silicate-solubilizing bacteria. In: Etesami H, Al Saeedi AH, El-Ramady H, Fujita M, Pessarakli M, Anwar Hossain M (eds) Silicon and nano-silicon in environmental stress management and crop quality improvement. Academic, pp 265–276. https://doi.org/10.1016/B978-0-323-91225-9.00020-0

Etesami H, Jeong BR (2023) Chapter 22: How does silicon help alleviate biotic and abiotic stresses in plants? Mechanisms and future prospects. In: Ghorbanpour M, Adnan Shahid M (eds) Plant stress mitigators. Academic, pp 359–402. https://doi.org/10.1016/B978-0-323-89871-3.00031-8

Etesami H, Maheshwari DK (2018) Use of plant growth promoting rhizobacteria (PGPRs) with multiple plant growth promoting traits in stress agriculture: action mechanisms and future prospects. Ecotoxicol Environ Saf 156:225–246. https://doi.org/10.1016/j.ecoenv.2018.03.013

Etesami H, Emami S, Alikhani HA (2017) Potassium solubilizing bacteria (KSB): Mechanisms, promotion of plant growth, and future prospects-A review. J Soil Sci Plant Nutr 17(4):897–911

Etesami H, Jeong BR, Glick BR (2021) Contribution of arbuscular mycorrhizal fungi, phosphate–solubilizing bacteria, and silicon to P uptake by plant. Front Plant Sci 12:699618

Etesami H, Al Saeedi AH, El-Ramady H, Fujita M, Pessarakli M, Hossain MA (2022) Silicon and nano-silicon in environmental stress management and crop quality improvement: progress and prospects. Elsevier

Fawaz K, Mohammad A-C (2013) Growth and nitrogen fixation in silicon and/or potassium fed chickpeas grown under drought and well watered conditions. J Stress Physiol Biochem 9(3):385–406

Ferguson BJ, Mathesius U (2003) Signaling interactions during nodule development. J Plant Growth Regul 22:47–72

Fomina M, Skorochod I (2020) Microbial interaction with clay minerals and its environmental and biotechnological implications. Fortschr Mineral 10(10):861

Frew A, Powell JR, Sallam N, Allsopp PG, Johnson SN (2016) Trade-offs between silicon and phenolic defenses may explain enhanced performance of root herbivores on phenolic-rich plants. J Chem Ecol 42:768–771

Frew A, Weston LA, Reynolds OL, Gurr GM (2018) The role of silicon in plant biology: a paradigm shift in research approach. Ann Bot 121(7):1265–1273

Glazebrook J (2005) Contrasting mechanisms of defense against biotrophic and necrotrophic pathogens. Annu Rev Phytopathol 43:205–227

Godfray HCJ, Garnett T (2014) Food security and sustainable intensification. Philos Trans R Soc B: Biol Sci 369(1639):20120273

Gómez-Merino FC, Trejo-Téllez LI, García-Jiménez A, Escobar-Sepúlveda HF, Ramírez-Olvera SM (2020) Silicon flow from root to shoot in pepper: a comprehensive in silico analysis reveals a potential linkage between gene expression and hormone signaling that stimulates plant growth and metabolism. PeerJ 8:e10053

Gondil VS, Asif M, Bhalla TC (2017) Optimization of physicochemical parameters influencing the production of prodigiosin from Serratia nematodiphila RL2 and exploring its antibacterial activity. 3 Biotech 7:1–8

Gong H, Chen K (2012) The regulatory role of silicon on water relations, photosynthetic gas exchange, and carboxylation activities of wheat leaves in field drought conditions. Acta Physiol Plant 34:1589–1594

Grégoire C, Rémus-Borel W, Vivancos J, Labbé C, Belzile F, Bélanger RR (2012) Discovery of a multigene family of aquaporin silicon transporters in the primitive plant Equisetum arvense. Plant J 72(2):320–330

Griffin M, Hogan B, Schmidt O (2015) Silicon reduces slug feeding on wheat seedlings. J Pest Sci 88:17–24

Guntzer F, Keller C, Meunier J-D (2012) Benefits of plant silicon for crops: a review. Agron Sustain Dev 32:201–213
Hall AD, Morison C (1906) On the function of silica in the nutrition of cereals.—Part I. Proc R Soc London Ser B: Containing Papers Biol Character 77(520):455–477
Hamayun M, Sohn E-Y, Khan SA, Shinwari ZK, Khan AL, Lee I-J (2010) Silicon alleviates the adverse effects of salinity and drought stress on growth and endogenous plant growth hormones of soybean (*Glycine max* L.). Pak J Bot 42(3):1713–1722
Hattori T, Inanaga S, Araki H, An P, Morita S, Luxová M, Lux A (2005) Application of silicon enhanced drought tolerance in Sorghum bicolor. Physiol Plant 123(4):459–466
Haynes RJ (2014) A contemporary overview of silicon availability in agricultural soils. J Plant Nutr Soil Sci 177(6):831–844
Haynes RJ (2017) Significance and role of Si in crop production. Adv Agron 146:83–166
Haynes RJ (2019) What effect does liming have on silicon availability in agricultural soils? Geoderma 337:375–383
Hildebrand M, Volcani BE, Gassmann W, Schroeder JI (1997) A gene family of silicon transporters. Nature 385(6618):688–689
Hu L, Xia M, Lin X, Xu C, Li W, Wang J, Zeng R, Song Y (2018) Earthworm gut bacteria increase silicon bioavailability and acquisition by maize. Soil Biol Biochem 125:215–221
Iqbal A, Khalil IA, Ateeq N, Khan MS (2006) Nutritional quality of important food legumes. Food Chem 97(2):331–335
Jeer M, Telugu U, Voleti S, Padmakumari A (2017) Soil application of silicon reduces yellow stem borer, Scirpophaga incertulas (Walker) damage in rice. J Appl Entomol 141(3):189–201
Kalinowski BE, Liermann LJ, Givens S, Brantley SL (2000) Rates of bacteria-promoted solubilization of Fe from minerals: a review of problems and approaches. Chem Geol 169(3–4):357–370
Kang J, Zhao W, Zhu X (2016) Silicon improves photosynthesis and strengthens enzyme activities in the C3 succulent xerophyte Zygophyllum xanthoxylum under drought stress. J Plant Physiol 199:76–86
Kang S-M, Waqas M, Shahzad R, You Y-H, Asaf S, Khan MA, Lee K-E, Joo G-J, Kim S-J, Lee I-J (2017) Isolation and characterization of a novel silicate-solubilizing bacterial strain Burkholderia eburnea CS4-2 that promotes growth of japonica rice (Oryza sativa L. cv. Dongjin). Soil Sci Plant Nutr 63(3):233–241
Kannan NM, Raj SA (1998) Occurrence of silicate solubilizing bacteria in rice ecosystem. Madras Agric J 85:47
Katz O (2015) Silica phytoliths in angiosperms: phylogeny and early evolutionary history. New Phytol 208(3):642–646
Khan I, Awan SA, Rizwan M, Ali S, Hassan MJ, Brestic M, Zhang X, Huang L (2021) Effects of silicon on heavy metal uptake at the soil-plant interphase: a review. Ecotoxicol Environ Saf 222:112510
Kim SG, Kim KW, Park EW, Choi D (2002) Silicon-induced cell wall fortification of rice leaves: a possible cellular mechanism of enhanced host resistance to blast. Phytopathology 92(10):1095–1103
Kim Y-H, Khan AL, Kim D-H, Lee S-Y, Kim K-M, Waqas M, Jung H-Y, Shin J-H, Kim J-G, Lee I-J (2014) Silicon mitigates heavy metal stress by regulating P-type heavy metal ATPases, Oryza sativa low silicon genes, and endogenous phytohormones. BMC Plant Biol 14:1–13
Kim Y-H, Khan AL, Waqas M, Lee I-J (2017) Silicon regulates antioxidant activities of crop plants under abiotic-induced oxidative stress: a review. Front Plant Sci 8:256717
Kumari VV, Banerjee P, Verma VC, Sukumaran S, Chandran MAS, Gopinath KA, Venkatesh G, Yadav SK, Singh VK, Awasthi NK (2022) Plant nutrition: An effective way to alleviate abiotic stress in agricultural crops. Int J Mol Sci 23(15):8519
Kvedaras OL, Byrne MJ, Coombes NE, Keeping MG (2009) Influence of plant silicon and sugarcane cultivar on mandibular wear in the stalk borer Eldana saccharina. Agric For Entomol 11(3):301–306
Laruelle GG, Roubeix V, Sferratore A, Brodherr B, Ciuffa D, Conley D, Dürr H, Garnier J, Lancelot C, Le Thi PQ (2009) Anthropogenic perturbations of the silicon cycle at the global scale: key role of the land-ocean transition. Glob Biogeochem Cycles 23(4):GB4031

Lee K-E, Adhikari A, Kang S-M, You Y-H, Joo G-J, Kim J-H, Kim S-J, Lee I-J (2019) Isolation and characterization of the high silicate and phosphate solubilizing novel strain Enterobacter ludwigii GAK2 that promotes growth in rice plants. Agronomy 9(3):144

Lewis GP, Schrire B, Mackinder B, Lock M (2005) Legumes of the world, vol 577. Royal Botanic Gardens Kew

Li GL, Zhou CH, Fiore S, Yu WH (2019) Interactions between microorganisms and clay minerals: new insights and broader applications. Appl Clay Sci 177:91–113

Li Z, Wang Y, Liu Z, Han F, Chen S, Zhou W (2023) Integrated application of phosphorus-accumulating bacteria and phosphorus-solubilizing bacteria to achieve sustainable phosphorus management in saline soils. Sci Total Environ 885:163971

Liang Y, Shen Q, Shen Z, Ma T (1996) Effects of silicon on salinity tolerance of two barley cultivars. J Plant Nutr 19(1):173–183

Liang Y, Zhu J, Li Z, Chu G, Ding Y, Zhang J, Sun W (2008) Role of silicon in enhancing resistance to freezing stress in two contrasting winter wheat cultivars. Environ Exp Bot 64(3):286–294

Liang X, Wang H, Hu Y, Mao L, Sun L, Dong T, Nan W, Bi Y (2015a) Silicon does not mitigate cell death in cultured tobacco BY-2 cells subjected to salinity without ethylene emission. Plant Cell Rep 34:331–343

Liang Y, Nikolic M, Bélanger R, Gong H, Song A, Liang Y, Nikolic M, Bélanger R, Gong H, Song A (2015b) Silicon sources for agriculture. In: Silicon in agriculture: from theory to practice, pp 225–232

Liu W, Xu X, Wu X, Yang Q, Luo Y, Christie P (2006) Decomposition of silicate minerals by Bacillus mucilaginosus in liquid culture. Environ Geochem Health 28:133–140

Liu J, Zhu J, Zhang P, Han L, Reynolds OL, Zeng R, Wu J, Shao Y, You M, Gurr GM (2017) Silicon supplementation alters the composition of herbivore induced plant volatiles and enhances attraction of parasitoids to infested rice plants. Front Plant Sci 8:1265

Ma JF (2004) Role of silicon in enhancing the resistance of plants to biotic and abiotic stresses. Soil Sci Plant Nutr 50(1):11–18

Ma JF (2010) Silicon transporters in higher plants. In: MIPs and their role in the exchange of metalloids, pp 99–109

Ma JF, Yamaji N (2015) A cooperative system of silicon transport in plants. Trends Plant Sci 20(7):435–442

Ma JF, Tamai K, Yamaji N, Mitani N, Konishi S, Katsuhara M, Ishiguro M, Murata Y, Yano M (2006) A silicon transporter in rice. Nature 440(7084):688–691

Ma JF, Yamaji N, Mitani N, Tamai K, Konishi S, Fujiwara T, Katsuhara M, Yano M (2007) An efflux transporter of silicon in rice. Nature 448(7150):209–212

Ma D, Sun D, Wang C, Qin H, Ding H, Li Y, Guo T (2016) Silicon application alleviates drought stress in wheat through transcriptional regulation of multiple antioxidant defense pathways. J Plant Growth Regul 35:1–10

Majumdar S, Prakash NB (2020) An overview on the potential of silicon in promoting defence against biotic and abiotic stresses in sugarcane. J Soil Sci Plant Nutr 20(4):1969–1998

Maleva M, Borisova G, Koshcheeva O, Sinenko O (2017) Biofertilizer based on silicate solubilizing bacteria improves photosynthetic function of Brassica juncea. AGROFOR 2(3). https://doi.org/10.7251/AGRENG1703013M

Massey FP, Hartley SE (2006) Experimental demonstration of the antiherbivore effects of silica in grasses: impacts on foliage digestibility and vole growth rates. Proc R Soc B Biol Sci 273(1599):2299–2304

Massey FP, Hartley SE (2009) Physical defences wear you down: progressive and irreversible impacts of silica on insect herbivores. J Anim Ecol 78(1):281–291

Massey FP, Ennos AR, Hartley SE (2006) Silica in grasses as a defence against insect herbivores: contrasting effects on folivores and a phloem feeder. J Anim Ecol 75(2):595–603

Massey FP, Ennos AR, Hartley SE (2007) Grasses and the resource availability hypothesis: the importance of silica-based defences. J Ecol 95:414–424

McNaughton S, Tarrants J (1983) Grass leaf silicification: natural selection for an inducible defense against herbivores. Proc Natl Acad Sci 80(3):790–791

Meena V, Dotaniya M, Coumar V, Rajendiran S, Ajay KS, Subba Rao A (2014) A case for silicon fertilization to improve crop yields in tropical soils. Proc Natl Acad Sci India Sect B: Biol Sci 84:505–518

Meunier JD, Barboni D, Anwar-ul-Haq M, Levard C, Chaurand P, Vidal V, Grauby O, Huc R, Laffont-Schwob I, Rabier J (2017) Effect of phytoliths for mitigating water stress in durum wheat. New Phytol 215(1):229–239

Miao B-H, Han X-G, Zhang W-H (2010) The ameliorative effect of silicon on soybean seedlings grown in potassium-deficient medium. Ann Bot 105(6):967–973

Mihaličová Malčovská S, Dučaiová Z, Maslaňáková I, Bačkor M (2014) Effect of silicon on growth, photosynthesis, oxidative status and phenolic compounds of maize (*Zea mays* L.) grown in cadmium excess. Water Air Soil Pollut 225(8):2056

Ming D, Pei Z, Naeem M, Gong H, Zhou W (2012) Silicon alleviates PEG-induced water-deficit stress in upland rice seedlings by enhancing osmotic adjustment. J Agron Crop Sci 198(1):14–26

Mitani N, Yamaji N, Ago Y, Iwasaki K, Ma JF (2011) Isolation and functional characterization of an influx silicon transporter in two pumpkin cultivars contrasting in silicon accumulation. Plant J 66(2):231–240

Mitani-Ueno N, Yamaji N, Ma JF (2011) Silicon efflux transporters isolated from two pumpkin cultivars contrasting in Si uptake. Plant Signal Behav 6(7):991–994

Nagabovanalli PB, Majumdar S, Kollalu S (2020) Status of silicon in ecosystem, silicon solubilization by rhizospheric microorganisms and their impact on crop productivity. In: Rhizosphere microbes: soil and plant functions, pp 409–429

Nawaz F, Shehzad MA, Majeed S, Ahmad KS, Aqib M, Usmani MM, Shabbir RN (2020) Role of mineral nutrition in improving drought and salinity tolerance in field crops. In: Agronomic crops: volume 3: stress responses and tolerance, pp 129–147

Neethirajan S, Gordon R, Wang L (2009) Potential of silica bodies (phytoliths) for nanotechnology. Trends Biotechnol 27(8):461–467

Nelwamondo A, Jaffer MA, Dakora FD (2001) Subcellular organization of N 2-fixing nodules of cowpea (Vigna unguiculata) supplied with silicon. Protoplasma 216:94–100

Ouellette S, Goyette M-H, Labbé C, Laur J, Gaudreau L, Gosselin A, Dorais M, Deshmukh RK, Bélanger RR (2017) Silicon transporters and effects of silicon amendments in strawberry under high tunnel and field conditions. Front Plant Sci 8:949

Pang Z, Peng H, Lin S, Liang Y (2024) Theory and application of a Si based defense barrier for plants: implications for soil-plant-atmosphere system health. Crit Rev Environ Sci Technol 54(9):722–746

Patrick S, Holding A (1985) The effect of bacteria on the solubilization of silica in diatom frustules. J Appl Bacteriol 59(1):7–16

Pavlovic J, Samardzic J, Kostic L, Laursen KH, Natic M, Timotijevic G, Schjoerring JK, Nikolic M (2016) Silicon enhances leaf remobilization of iron in cucumber under limited iron conditions. Ann Bot 118(2):271–280

Perez CEA, Rodrigues FÁ, Moreira WR, DaMatta FM (2014) Leaf gas exchange and chlorophyll a fluorescence in wheat plants supplied with silicon and infected with Pyricularia oryzae. Phytopathology 104(2):143–149

Purushothaman A, Chandramohan D, Natarajan R (1974) Distribution of silicate dissolving bacteria in vellar estuary. Curr Sci 43:282–283

Putra R, Powell JR, Hartley SE, Johnson SN (2020) Is it time to include legumes in plant silicon research? Funct Ecol 34(6):1142–1157

Rachappanavar V, Kumar M, Negi N, Chowdhury S, Kapoor M, Singh S, Rustagi S, Rai AK, Shreaz S, Negi R (2024) Silicon derived benefits to combat biotic and abiotic stresses in fruit crops: current research and future challenges. Plant Physiol Biochem 211:108680

Rahman A, Wallis CM, Uddin W (2015) Silicon-induced systemic defense responses in perennial ryegrass against infection by Magnaporthe oryzae. Phytopathology 105(6):748–757

Ranalli G, Zanardini E (2021) Biocleaning on cultural heritage: new frontiers of microbial biotechnologies. J Appl Microbiol 131(2):583–603

Rao GB, Susmitha P (2017) Silicon uptake, transportation and accumulation in rice. J Pharmacogn Phytother 6(6):290–293

Raturi G, Sharma Y, Rana V, Thakral V, Myaka B, Salvi P, Singh M, Dhar H, Deshmukh R (2021) Exploration of silicate solubilizing bacteria for sustainable agriculture and silicon biogeochemical cycle. Plant Physiol Biochem 166:827–838

Rawat P, Das S, Shankhdhar D, Shankhdhar S (2021) Phosphate-solubilizing microorganisms: mechanism and their role in phosphate solubilization and uptake. J Soil Sci Plant Nutr 21(1):49–68

Rea RS, Islam MR, Rahman MM, Nath B, Mix K (2022) Growth, nutrient accumulation, and drought tolerance in crop plants with silicon application: a review. Sustain For 14(8):4525

Rémus-Borel W, Menzies JG, Bélanger RR (2005) Silicon induces antifungal compounds in powdery mildew-infected wheat. Physiol Mol Plant Pathol 66(3):108–115

Reynolds OL, Padula MP, Zeng R, Gurr GM (2016) Silicon: potential to promote direct and indirect effects on plant defense against arthropod pests in agriculture. Front Plant Sci 7:188037

Rezakhani L, Motesharezadeh B, Tehrani MM, Etesami H, Mirseyed Hosseini H (2020) Effect of silicon and phosphate-solubilizing bacteria on improved phosphorus (P) uptake is not specific to insoluble P-fertilized sorghum (Sorghum bicolor L.) plants. J Plant Growth Regul 39:239–253

Ribeiro IDA, Volpiano CG, Vargas LK, Granada CE, Lisboa BB, Passaglia LMP (2020) Use of mineral weathering bacteria to enhance nutrient availability in crops: a review. Front Plant Sci 11:590774

Richmond KE, Sussman M (2003) Got silicon? The non-essential beneficial plant nutrient. Curr Opin Plant Biol 6(3):268–272

Rodrigues FA, Dallagnol LJ, Duarte HSS, Datnoff LE (2015) Silicon control of foliar diseases in monocots and dicots. In: Silicon and plant diseases, pp 67–108

Rojas CM, Senthil-Kumar M, Wang K, Ryu C-M, Kaundal A, Mysore KS (2012) Glycolate oxidase modulates reactive oxygen species–mediated signal transduction during nonhost resistance in Nicotiana benthamiana and Arabidopsis. Plant Cell 24(1):336–352

Romero-Puertas MC, Corpas FJ, Rodríguez-Serrano M, Gómez M, Del Río LA, Sandalio LM (2007) Differential expression and regulation of antioxidative enzymes by cadmium in pea plants. J Plant Physiol 164(10):1346–1357

Sahebi M, Hanafi MM, Siti Nor Akmar A, Rafii MY, Azizi P, Tengoua F, Nurul Mayzaitul Azwa J, Shabanimofrad M (2015) Importance of silicon and mechanisms of biosilica formation in plants. BioMed Res Int 2015:396010

Sakurai G, Satake A, Yamaji N, Mitani-Ueno N, Yokozawa M, Feugier FG, Ma JF (2015) In silico simulation modeling reveals the importance of the Casparian strip for efficient silicon uptake in rice roots. Plant Cell Physiol 56(4):631–639

Samuels T, Bryce C, Landenmark H, Marie-Loudon C, Nicholson N, Stevens AH, Cockell C (2020) Microbial weathering of minerals and rocks in natural environments. In: Biogeochemical cycles: ecological drivers and environmental impact, pp 59–79

Sanglard LM, Martins SC, Detmann KC, Silva PE, Lavinsky AO, Silva MM, Detmann E, Araújo WL, DaMatta FM (2014) Silicon nutrition alleviates the negative impacts of arsenic on the photosynthetic apparatus of rice leaves: an analysis of the key limitations of photosynthesis. Physiol Plant 152(2):355–366

Sauer D, Saccone L, Conley DJ, Herrmann L, Sommer M (2006) Review of methodologies for extracting plant-available and amorphous Si from soils and aquatic sediments. Biogeochemistry 80:89–108

Savant N, Snyder G, Datnoff L (1996) Silicon management and sustainable rice production. Adv Agron 58:151–199

Savant NK, Datnoff LE, Snyder GH (1997) Depletion of plant-available silicon in soils: a possible cause of declining rice yields. Commun Soil Sci Plant Anal 28(13–14):1245–1252

Savant NK, Korndörfer GH, Datnoff LE, Snyder GH (1999) Silicon nutrition and sugarcane production: a review. J Plant Nutr 22(12):1853–1903

Schaller J, Puppe D, Kaczorek D, Ellerbrock R, Sommer M (2021) Silicon cycling in soils revisited. Plants 10(2):295

Shahbazi Y, Moosavy M-H (2019) Physico-mechanical and antimicrobial properties of quince seed mucilage supplemented with titanium dioxide and silicon oxide nanoparticles. Nanomed Res J 4(3):157–163

Sharma B, Kumawat KC, Tiwari S, Kumar A, Dar RA, Singh U, Cardinale M (2023) Silicon and plant nutrition—dynamics, mechanisms of transport and role of silicon solubilizer microbiomes in sustainable agriculture: a review. Pedosphere 33(4):534–555

Shen X, Zhou Y, Duan L, Li Z, Eneji AE, Li J (2010) Silicon effects on photosynthesis and antioxidant parameters of soybean seedlings under drought and ultraviolet-B radiation. J Plant Physiol 167(15):1248–1252

Sheng X (2005) Growth promotion and increased potassium uptake of cotton and rape by a potassium releasing strain of Bacillus edaphicus. Soil Biol Biochem 37(10):1918–1922

Sheng XF, He LY (2006) Solubilization of potassium-bearing minerals by a wild-type strain of Bacillus edaphicus and its mutants and increased potassium uptake by wheat. Can J Microbiol 52(1):66–72

Sheng M, Tang M, Chen H, Yang B, Zhang F, Huang Y (2008a) Influence of arbuscular mycorrhizae on photosynthesis and water status of maize plants under salt stress. Mycorrhiza 18:287–296

Sheng XF, Zhao F, He LY, Qiu G, Chen L (2008b) Isolation and characterization of silicate mineral solubilizing Bacillus globisporus Q12 from the surfaces of weathered feldspar. Can J Microbiol 54(12):1064–1068

Shivaraj S, Mandlik R, Bhat JA, Raturi G, Elbaum R, Alexander L, Tripathi DK, Deshmukh R, Sonah H (2022) Outstanding questions on the beneficial role of silicon in crop plants. Plant Cell Physiol 63(1):4–18

Smith KS, Ferry JG (2000) Prokaryotic carbonic anhydrases. FEMS Microbiol Rev 24(4):335–366

Sommer M, Kaczorek D, Kuzyakov Y, Breuer J (2006) Silicon pools and fluxes in soils and landscapes—a review. J Plant Nutr Soil Sci 169(3):310–329

Soukup M, Martinka M, Bosnić D, Čaplovičová M, Elbaum R, Lux A (2017) Formation of silica aggregates in sorghum root endodermis is predetermined by cell wall architecture and development. Ann Bot 120(5):739–753

Souri Z, Khanna K, Karimi N, Ahmad P (2021) Silicon and plants: current knowledge and future prospects. J Plant Growth Regul 40:906–925

Sumitha S, Vasanthi S, Shalini S, Chinni SV, Gopinath SC, Anbu P, Bahari MB, Harish R, Kathiresan S, Ravichandran V (2018) Phyto-mediated photo catalysed green synthesis of silver nanoparticles using Durio zibethinus seed extract: antimicrobial and cytotoxic activity and photocatalytic applications. Molecules 23(12):3311

Sun H, Guo J, Duan Y, Zhang T, Huo H, Gong H (2017) Isolation and functional characterization of CsLsi1, a silicon transporter gene in Cucumis sativus. Physiol Plant 159(2):201–214

Sun H, Duan Y, Qi X, Zhang L, Huo H, Gong H (2018) Isolation and functional characterization of CsLsi2, a cucumber silicon efflux transporter gene. Ann Bot 122(4):641–648

Sun H, Duan Y, Mitani-Ueno N, Che J, Jia J, Liu J, Guo J, Ma JF, Gong H (2020) Tomato roots have a functional silicon influx transporter but not a functional silicon efflux transporter. Plant Cell Environ 43(3):732–744

Torres MA, West AJ, Nealson K (2014) Microbial acceleration of olivine dissolution via siderophore production. Procedia Earth Planet Sci 10:118–122

Tubaña BS, Heckman JR (2015) Silicon in soils and plants. In: Silicon and plant diseases, pp 7–51

Tubana BS, Babu T, Datnoff LE (2016) A review of silicon in soils and plants and its role in US agriculture: history and future perspectives. Soil Sci 181(9/10):393–411

Uroz S, Calvaruso C, Turpault M-P, Frey-Klett P (2009) Mineral weathering by bacteria: ecology, actors and mechanisms. Trends Microbiol 17(8):378–387

Van Bockhaven J, De Vleesschauwer D, Höfte M (2013) Towards establishing broad-spectrum disease resistance in plants: silicon leads the way. J Exp Bot 64(5):1281–1293

Vasanthi N, Saleena LM, Raj SA (2012) Silicon in day today life. World Appl Sci J 17(11):1425–1440

Vasanthi N, Saleena L, Raj SA (2018) Silica solubilization potential of certain bacterial species in the presence of different silicate minerals. Silicon 10:267–275

Vessey JK (1994) Measurement of nitrogenase activity in legume root nodules: in defense of the acetylene reduction assay. Plant Soil 158:151–162

Wan W, Qin Y, Wu H, Zuo W, He H, Tan J, Wang Y, He D (2020) Isolation and characterization of phosphorus solubilizing bacteria with multiple phosphorus sources utilizing capability and their potential for lead immobilization in soil. Front Microbiol 11:752

Wang RR, Wang Q, He LY, Qiu G, Sheng XF (2015) Isolation and the interaction between a mineral-weathering Rhizobium tropici Q34 and silicate minerals. World J Microbiol Biotechnol 31:747–753

Waraich E, Ahmad R, Halim A, Aziz T (2012) Alleviation of temperature stress by nutrient management in crop plants: a review. J Soil Sci Plant Nutr 12(2):221–244

White AF, Vivit DV, Schulz MS, Bullen TD, Evett RR, Aagarwal J (2012) Biogenic and pedogenic controls on Si distributions and cycling in grasslands of the Santa Cruz soil chronosequence, California. Geochim Cosmochim Acta 94:72–94

Wu J, Baldwin IT (2010) New insights into plant responses to the attack from insect herbivores. Annu Rev Genet 44:1–24

Xiao L, Hao J, Wang W, Lian B, Shang G, Yang Y, Liu C, Wang S (2014) The up-regulation of carbonic anhydrase genes of Bacillus mucilaginosus under soluble Ca2+ deficiency and the heterologously expressed enzyme promotes calcite dissolution. Geomicrobiol J 31(7):632–641

Yamaji N, Chiba Y, Mitani-Ueno N, Feng Ma J (2012) Functional characterization of a silicon transporter gene implicated in silicon distribution in barley. Plant Physiol 160(3):1491–1497

Yang J, Firsbach F, Sohn I (2022) Pyrometallurgical processing of ferrous slag "co-product" zero waste full utilization: a critical review. Resour Conserv Recycl 178:106021

Yanni YG, Rizk R, Corich V, Squartini A, Ninke K, Philip-Hollingsworth S, Orgambide G, de Bruijn FD, Stoltzfus J, Buckley D (1997) Natural endophytic association between Rhizobium leguminosarum bv. Trifolii and rice roots and assessment of its potential to promote rice growth. In: Opportunities for biological nitrogen fixation in rice and other non-legumes: papers presented at the second working group meeting of the frontier project on nitrogen fixation in rice held at the National Institute for Biotechnology and Genetic Engineering (NIBGE), Faisalabad, Pakistan, 13–15 October 1996. Springer, pp 99–114

Yin L, Wang S, Tanaka K, Fujihara S, Itai A, Den X, Zhang S (2016) Silicon-mediated changes in polyamines participate in silicon-induced salt tolerance in *Sorghum bicolor* L. Plant Cell Environ 39(2):245–258

Yu H, Wu X, Zhang G, Zhou F, Harvey PR, Wang L, Fan S, Xie X, Li F, Zhou H (2022) Identification of the phosphorus-solubilizing bacteria strain JP233 and its effects on soil phosphorus leaching loss and crop growth. Front Microbiol 13:892533

Zhang X, Cai X (2011) Climate change impacts on global agricultural land availability. Environ Res Lett 6(1):014014

Zhang C, Kong F (2014) Isolation and identification of potassium-solubilizing bacteria from tobacco rhizospheric soil and their effect on tobacco plants. Appl Soil Ecol 82:18–25

Zhang W, Xie Z, Lang D, Cui J, Zhang X (2017) Beneficial effects of silicon on abiotic stress tolerance in legumes. J Plant Nutr 40(15):2224–2236

Zhou H-b, Zeng X-x, Liu F-f, Qiu G-z, Hu Y-h (2006) Screening, identification and desilication of a silicate bacterium. J Cent S Univ Technol 13(4):337–341

Zhu Z, Wei G, Li J, Qian Q, Yu J (2004) Silicon alleviates salt stress and increases antioxidant enzymes activity in leaves of salt-stressed cucumber (Cucumis sativus L.). Plant Sci 167(3):527–533

Zhu Y, Duan G, Chen B, Peng X, Chen Z, Sun G (2014) Mineral weathering and element cycling in soil-microorganism-plant system. Sci China Earth Sci 57:888–896

Chapter 8
Role of Silicon in the Diversity of the Rhizospheric Microbial Community of Crops

Marcos Renan Lima Leite, Romário Martins Costa, Sandra Mara Barbosa Rocha, Francisco de Alcântara Neto, Ademir Sérgio Ferreira de Araújo, Rafael de Souza Miranda, and Renato de Mello Prado

Abstract In recent years, silicon (Si) has emerged as an alternative for agricultural production, impacting plant rhizosphere dynamics by promoting plant-microorganism interactions and improving crop performance under stressful conditions and different production systems. Here, we provide evidence into how Si modulates soil pH, enzymatic activity, nutrient availability, and the structure, composition, and functional potential of soil microbial communities. The mechanisms behind these changes are not fully understood but might involve direct and indirect associations from Si-soil-microorganism-plant dynamics. Understanding these interactions is crucial for promoting sustainable and resilient agronomic practices.

Keywords Plant nutrition · Beneficial element · Soil microbiology · Plant-microorganism interaction · Agricultural sustainability

8.1 Introduction

In contemporary agriculture, the strategic use of silicon (Si) is widely recognized for its ability to promote plant growth and play an essential role in reducing biotic and abiotic stresses (Mir et al. 2022; Bhardwaj et al. 2023; Misra et al. 2023). This

M. R. L. Leite · R. M. Costa
Federal University of Piaui, Bom Jesus, Brazil

S. M. B. Rocha · F. de Alcântara Neto (✉) · A. S. F. de Araújo · R. de Souza Miranda
Federal University of Piauí, Teresina, Brazil
e-mail: fneto@ufpi.edu.br

R. de Mello Prado
Department of Soil Science, São Paulo State University, Jaboticabal, São Paulo, Brazil

R. de Mello Prado et al. (eds.), *Silicon Advances for Sustainable Agriculture and Human Health*, Sustainable Plant Nutrition in a Changing World,
https://doi.org/10.1007/978-3-031-69876-7_8

process enhances the resistance and resilience of cultivated plants, resulting in increased productivity and stability in agricultural systems (Du et al. 2022; Etesami and Jeong 2023). However, these benefits are not limited solely to the direct effects of Si on plants but also to the complex interactions among the element, plants, and microorganisms. Some research suggests that Si may play an active role in plant-microbe interactions (Song et al. 2021a; Wang et al. 2022; Deng et al. 2023a; Gao et al. 2024).

In the soil, microorganisms maintain constant interactions with each other and with plants (Zhang and Guan 2022). The rhizosphere region influenced by plant roots acts as a focal point for these interactions, hosting a wide variety of microbial groups (Prashar et al. 2014; Pathan et al. 2020; Leite et al. 2023, 2024b). This increase in microbial diversity in the rhizosphere is a result of the continuous release of organic compounds, such as enzymes, sugars, and organic acids, among others, which can promote microbial growth in the rhizosphere (Chukwuneme et al. 2021; Wang et al. 2023a).

Silicon can influence interactions between plants and soil microorganisms, affecting rhizosphere dynamics, as it exerts influence on various factors such as soil pH (Epstein 1994), enzyme activity (Teixeira et al. 2022), and nutrient availability (Dutra et al. 2023). By affecting nutrient availability in the soil, Si can induce changes in the competitive interactions between plants and soil microorganisms for nutrient acquisition. This can lead to a more diverse and balanced microbial community, contributing to soil health and plant growth (Sharma et al. 2023). Consequently, this dynamic can impact the composition and activity of microbial communities in the rhizosphere, as well as regulate the abundance and diversity of microorganisms in this environment.

Given the current scenario, with the increasing environmental stress, climate instability, and the growing process of soil degradation, both due to erosion and desertification resulting from intensive agricultural practices, it becomes essential to adopt reliable and environmentally sustainable techniques to meet the growing demands of agricultural productivity globally. Therefore, in agricultural systems, the interaction between Si, microorganisms, and plants is particularly relevant, and understanding this relationship is fundamental to optimizing management practices and promoting more sustainable and resilient agricultural systems.

The next sections of this chapter address fundamental aspects, such as the beneficial role of Si in agriculture, the interaction of the element with soil microbiota and its effects on plant resistance to environmental stresses, as well as its influence on the composition of microbial communities in the rhizosphere of cultivated plants. Additionally, future perspectives are presented that may guide further research on the subject.

8.2 The Beneficial Role of Silicon in Agriculture

Silicon is a widely distributed element in nature and plays a crucial role in various biological and environmental processes (Souri et al. 2021). Si can be found in the soil in soluble and insoluble forms, and its availability and mobility are influenced by several factors such as soil pH, organic matter content, types of minerals present, and microbial activity. In soil solution, Si predominantly appears in the form of silicic acid or monosilicic acid [$Si(OH)_4$], which is readily absorbed by plants (Epstein 1994; Raturi et al. 2021). Within the natural ecosystem, Si is reintroduced into the soil through the decomposition of leaf litter and plant debris, forming biogenic Si deposits that reintegrate the element cycle in the soil (Nakamura et al. 2020).

In recent years, the strategic use of Si has been widely recognized in agriculture as it can stimulate plant growth and play an important role in mitigating biotic and abiotic stresses in plants. Studies have shown that Si contributes to increasing the resistance and resilience of agricultural crops, resulting in higher productivity and stability of agricultural systems (Mir et al. 2022; Du et al. 2022; Ahmed et al. 2023b; Bhardwaj et al. 2023; Etesami and Jeong 2023; Misra Misra et al. 2023). In the soil, monosilicic acid is absorbed by plant roots, transported through the xylem, and deposited in cells in the form of silica phytoliths that strengthen the cell wall, making them more resistant to penetration by phytopathogens (Epstein 1994; Barreto and Barão 2023; Tubana 2023). Additionally, Si induces defense responses in plants such as the production of phenolic compounds and the activation of defense enzymes, which help combat pests and diseases (Deng et al. 2023b; Gulzar et al. 2024; Qi et al. 2024). Si can also increase plants' ability to tolerate drought and water stress by reducing transpiration and improving water use efficiency. Plants supplemented with Si tend to maintain cell turgor and growth rate even under low water availability conditions (Rea et al. 2022; de Souza Júnior et al. 2022a; Raza et al. 2023; Singhal et al. 2023; Saudy et al. 2023).

Salinity is identified as one of the main abiotic stresses affecting the production of most cultivated plants (Gul et al. 2022). This issue is particularly alarming in arid and semi-arid regions, typically characterized by higher evapotranspiration rates compared to precipitation, where soil salinization is relatively high throughout the year. This occurs due to low rainfall, resulting in the accumulation of soluble salts in the soil, affecting everything from seed germination to the growth, development, and productivity of crops (Munns and Tester 2008). In this context, the use of Si has been reported as a mitigator of the adverse effects of saline stress on plants (Venâncio et al. 2022; Alayafi et al. 2022; Thorne et al. 2022), as the element induces a reduction in the absorption of toxic ions, such as sodium (Na^+) and chloride (Cl^-), maintaining ionic balance within cells. This helps plants maintain osmotic homeostasis and avoid damage caused by salt accumulation (Naz et al. 2022; Muhammad et al. 2022).

In soils contaminated with heavy metals, Si can help reduce the absorption and translocation of these metals by plants, minimizing the toxic effects on physiological processes and plant growth (Ngugi et al. 2022; Ahmed et al. 2023a; Wang et al. 2023a; Asgher et al. 2024). Additionally, Si can induce responses in plants to tolerate extreme temperatures by regulating the expression of genes related to heat stress and protecting cell membranes from damage caused by abrupt temperature changes (Bocharnikova and Matichenkov 2022; Parveen et al. 2022; Basirat and Mousavi 2022; Bilal et al. 2023).

The use of Si in plant nutrition has proven to be a promising strategy due to the various benefits it provides, as the element enhances nutrient use efficiency (Seyed Hajizadeh et al. 2023; Barão 2023; Khan et al. 2023; Dutra et al. 2023), directly impacting crop productivity and promoting greater sustainability in agricultural systems. In crops such as rice, sugarcane, wheat, and corn, Si supplementation can result in significant increases in production and product quality (Flores et al. 2022; de Souza Júnior et al. 2022b; Salem et al. 2022; Gou et al. 2023). Furthermore, incorporating Si into agricultural management can reduce the need for fertilizers and insecticides, contributing to more economical and environmentally sustainable agricultural practices (Kovács et al. 2022).

The benefits of Si are not limited to just the direct effects on plants but also to the complex interactions between the element, plants, and microorganisms. In this context, in recent years, with the advancement of biotechnology, several studies have been developed to understand how Si affects the soil microbial community. These results can provide insights into management strategies that promote soil and plant health, increasing nutrient use efficiency and agricultural sustainability.

8.3 Microorganism-Plant Interaction

8.3.1 Plant Growth Promoters

Soil microorganisms play a fundamental role in the biogeochemical cycles of nutrients, performing processes essential to ecosystem dynamics (del Orozco-Mosqueda Ma et al. 2021). These microorganisms, including bacteria, fungi, and other microbes, are responsible for the decomposition of organic matter and the release of nutrients such as nitrogen, phosphorus, and sulfur into the soil (Basu et al. 2021). The continuous cycling of nutrients, in turn, contributes to the ecological balance and sustainable productivity of ecosystems (Basu et al. 2021). In addition to influencing nutrient availability for plants, the active participation of microorganisms in the soil plays an essential role in maintaining soil quality and environmental quality due to their central role in carbon dynamics (Dubey et al. 2019).

Soil microorganisms continuously interact with each other (Zhang and Guan 2022) and with plants (Andreote and de Pereira e Silva 2017). These interactions between microorganisms and plants can be either positive or negative (Chhetri et al.

2022; Farhaoui et al. 2023). The rhizosphere, a narrow region of soil with physical, chemical, and biological properties altered by plant roots (Prashar et al. 2014), acts as a hotspot for microbe-plant interactions (Pathan et al. 2020) and hosts a vast array of microbial taxa, as reported in recent studies (Yan et al. 2017; Ling et al. 2022; Leite et al. 2023, 2024a). This enrichment of microbial taxa in the rhizosphere results from the continuous release of organic compounds (including enzymes, sugars, organic acids, among others) (Chukwuneme et al. 2021), which can favor the development of microorganisms in the rhizosphere, especially those of a heterotrophic nature (Ling et al. 2022; Wang et al. 2023b). Plants selectively recruit specific taxa in the rhizosphere based on functional characteristics (Yan et al. 2017), with potential in plant growth promotion, such as plant growth-promoting bacteria (PGPB).

PGPBs are a diverse class of soil microorganisms widely studied in the literature, performing various important functions for plant growth and development. These functions include phosphate solubilization (Gupta et al. 2022), biological nitrogen fixation (Solanki et al. 2019; Asiloglu et al. 2020), protection against phytopathogens (Viljoen et al. 2019; Farhaoui et al. 2023), production of phytohormones (Majeed et al. 2018; Zerrouk et al. 2019), and siderophore production (Delaporte-Quintana et al. 2020), as well as pollutant degradation (Mishra et al. 2021) and other plant growth-promoting functions. In this topic, we will focus solely on the beneficial interactions between microorganisms and plants, especially interactions with PGPBs. Figure 8.1 provides an overview of the main mechanisms of action of PGPBs.

During plant development, bacterial communities in the rhizosphere undergo several changes, especially a significant increase in rhizobacteria with the potential to stimulate plant growth (Sugiyama et al. 2014). In previous studies, Arruda et al. (2013) found the presence of rhizobacterial taxa isolated from maize plants (*Zea mays* L.) with the ability to solubilize phosphates and produce siderophores and indoleacetic acid (IAA), including the genera *Burkholderia*, *Pseudomonas*, and *Chryseobacterium*. Delaporte-Quintana et al. (2020) further verified that the plant growth-promoting bacteria *Gluconacetobacter diazotrophicus* and *Azospirillum brasilense* play a crucial role in the iron nutrition of strawberry plants due to siderophore production. Sunera et al. (2020), in a study characterizing and assessing the phytostimulating activity of bacteria isolated from tomato rhizosphere (*Lycopersicon esculentum* Mill.), identified strains of *Bacillus cereus* producing IAA, gibberellic acid (GA3), and kinetin. These substances demonstrated positive effects on the growth of crops such as tomato and mung bean cultivated in a greenhouse environment. Other studies employing the 16S rRNA approach also identified potential growth promoters in the rhizosphere of various crops, including sugarcane (Leite et al. 2024b), cucumber (Zhou et al. 2018), and rice (Chhetri et al. 2022), among others. Further experimental evidence is described in Table 8.1, highlighting the mechanisms of action and observed effects on agricultural crops.

In general, PGPB stimulate plant growth and significantly mitigate the deleterious effects of biotic and abiotic stresses on plants through various mechanisms,

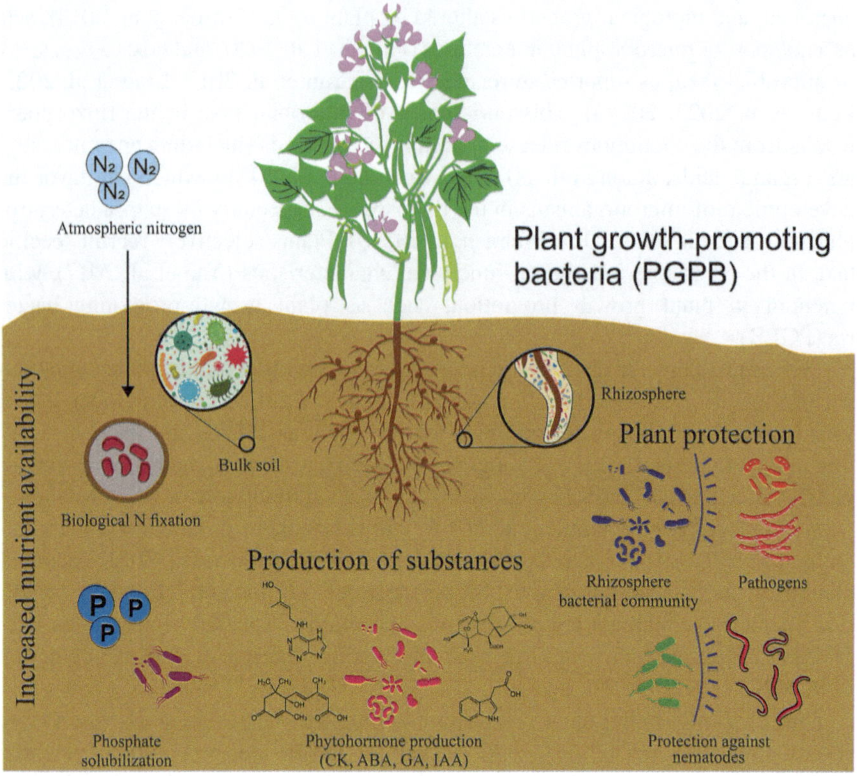

Fig. 8.1 Summary of some of the main mechanisms of action of PGPR. Plant growth-promoting bacteria can increase nutrient availability through biological nitrogen fixation and phosphorus (P) solubilization; they can produce phytohormones such as cytokinin (CK), abscisic acid (ABA), gibberellins (GA), and indole-3-acetic acid (IAA) and act in plant protection against pathogenic bacteria, fungi, and nematodes. (Source: Authors' elaboration)

such as the production of phytohormones and antibiotics, among others (Cao et al. 2018; Verma et al. 2023; Yadav et al. 2023). These mechanisms are well documented in the literature, and the knowledge gained in recent years has been applied in agriculture through biotechnologies, such as bioinoculants, bionematicides, and bioinsecticides, among others (Ruiu 2020; del Orozco-Mosqueda Ma et al. 2021). However, numerous gaps and important questions still remain unknown. Therefore, the use of multidisciplinary techniques and approaches will be crucial to increase agricultural productivity and environmental sustainability while reducing dependence on external inputs, which are becoming increasingly scarce.

Table 8.1 PGPB genres described in literature, mechanisms of action, and beneficial effects observed in various agricultural crops

PGPB	PGPB mechanisms	Crops	Highlights	Ref.
Azospirillum *Gluconacetobacter*	Siderophore production	Strawberry	Increased iron absorption	Delaporte-Quintana et al. (2020)
Bacillus *Paenibacillus*	Production of secondary metabolites	Tomato Carrot	Nematode gall control	Viljoen et al. (2019)
Pseudomonas	Gene expression regulation	Chickpea	Mitigation of water stress	Yadav et al. (2023)
Enterobacter	Immobilization of Cd in soil	Pea	Reduces the toxic effect of cadmium on plants	Naveed et al. (2020)
Pseudomonas	ACC deaminase and IAA	Maize	Increases tolerance to salt and aluminum toxicity	Zerrouk et al. (2019)
Bacillus	Antifungal metabolites synthesis	Sugar beet	Fungal biocontrol against Rhizoctonia	Farhaoui et al. (2023)
Pseudomonas	Secretion systems	Tomato	Ability to combat rhizosphere microbiome competitors	Durán et al. (2021)
Bacillus *Pseudomonas*	Phosphate solubilization	Rice	Increased availability of phosphorus in the soil	Gupta et al. (2022)
Azospirillum	Biological nitrogen fixation	Rice	Increased nitrogen absorption	Asiloglu et al. (2020)
Bacillus *Streptomyces* *Pantoea*	Biological nitrogen fixation	Sugarcane	Potential increase in availability of nitrogen	Solanki et al. (2019)
Pantoea *Burkholderia* *Gluconacetobacter* *Enterobacter*	Phosphate solubilization	Sugarcane	Potential increase in phosphorus availability	Solanki et al. (2019)
Arthrobacter	IAA synthesis; production of antimicrobial secondary metabolites	Rice	Increase in plant growth	Chhetri et al. (2022)
Bacillus	Antibiotic Production	Tomato Banana	Antagonistic activity against pathogens	Cao et al. (2018)

ACC 1-aminocyclopropane-1-carboxylic acid; *IAA* indole-3-acetic acid

8.4 Silicon-Microbiota-Plant Interaction

8.4.1 The Interaction Between Silicon and Microbiota Affects the Bioavailability of Heavy Metals in the Soil

Studies on understanding the interaction between Si and soil microorganisms have revealed important insights into the bioavailability of heavy metals in this environment. Soils contaminated with heavy metals can severely affect the diversity, distribution, quantity, and structure of soil microorganisms (Wang et al. 2020a; Peng et al. 2023). Conversely, stress caused by heavy metals can increase the abundance of microorganisms that tolerate this condition (Huang et al. 2024) through different direct and indirect mechanisms. In this context, the application of Si has been shown to be effective in alleviating heavy metal toxicity in various crops such as rice (Peng et al. 2023), corn (Yasin et al. 2024), wheat (Song et al. 2021b), and tobacco (*Nicotiana tabacum* L.) (Wang et al. 2023a), through physiological and biochemical processes. However, it is not yet fully understood what the mechanisms of the synergistic effect of applying silicate fertilizers and the soil microbial community are that can mitigate the effect of heavy metals on plants.

The positive effects of Si on mitigating heavy metal toxicity have sparked interest in the scientific community. Studies have shown that the use of silicate fertilizers promotes microbial interactions (Gao et al. 2024) and can play a significant role in reducing the bioavailability of heavy metals in the soil, such as cadmium (Cd) toxicity (del Orozco-Mosqueda Ma et al. 2021; Song et al. 2021b; Peng et al. 2023; Pan et al. 2023), arsenic (As) (Das et al. 2021; Gao et al. 2022, 2024), lead (Pb) (Wang et al. 2023a), and copper (Cu) (Kim et al. 2014; Wang et al. 2020a). These studies can be observed in more detail in Table 8.2. Improvement in microbial interactions involves the formation of insoluble complexes and the induction of resistance in plants. Recently, it has been reported that Si interacts with the growth of beneficial soil microorganisms and alters their structure and diversity (Song et al. 2021b; Peng et al. 2023), providing an increase in the abundance of microorganisms capable of significantly reducing the bioavailability of metals in the soil for plant roots (Song et al. 2021b).

The presence and activity of soil microorganisms mediated by Si influence the availability of heavy metals (Gao et al. 2024), both directly through sorption and complexation processes and indirectly by modifying soil properties and nutrient availability for plants (del Orozco-Mosqueda Ma et al. 2021; Das et al. 2021). Si also aids in the survival of microorganisms in stressed environments, making them capable of adsorbing metals through biological colloids, exopolysaccharides, and other substances (Ren et al. 2024). Additionally, recent studies have shown that groups of Si-stimulated filamentous bacteria with biological resistance capacity can adsorb, precipitate, degrade, or reoxidize heavy metals (Yu et al. 2023; Shankar et al. 2023). Gao et al. (2024) observed that Si can regulate arsenic levels in rice roots through microbial interactions or molecular signaling.

Table 8.2 Silicon-mediated microbial community response to heavy metal application

Heavy metals	Enriched microorganisms	Crops	Highlights	Ref.
Cadmium (Cd)	*Acidobacteria* *Thaumarchaeota*	Wheat	The supply of silicon shapes the abundance of tolerant bacterial communities, reducing the availability of Cd in the soil for plant roots	Song et al. (2021b)
Cadmium and lead (Pb)	*Actinobacteria* *Bacteroidetes*	Tobacco	Reduction of oxidative damage, increase in peroxidase and ascorbate content; increase in photosynthesis, chlorophyll, and carotenoids	Ren et al. (2024), Wang et al. (2023a)
Arsenic (As)	*Mycobacterium, Streptomyces, Anaeromyxobacter, Geotale*a	Rice	Si can regulate arsenic levels in roots, signaling molecules or microbial interlaces	Das et al. (2021), Gao et al. (2022, 2024)
Cadmium		Rice	Reduce the absorption and accumulation of Cd in rice	del Orozco-Mosqueda Ma et al. (2021), Peng et al. (2023), Pan et al. (2023)
Multiple metals: Pb, Cd, As, Cu, Hg e Cr	*Proteobacteria* *Actinobacteria*	Pakchoi (*Brassica chinensis*)	Selection of tolerant bacterial groups that provide a physicochemical improvement in soil for pakchoi growth	Wang et al. (2020a)

Song et al. (2021a, b) demonstrated a significant effect of the application of silicate fertilizer resulting in an increase in soil available Si and microbial community, mainly *Acidobacteria* and *Thaumarchaeota* phyla, capable of reducing Cd availability to wheat roots. Similarly, Das et al. (2021) concluded that silicate fertilizers were responsible for stimulating microbial functions beneficial for rice growth while controlling As availability. Peng et al. (2023) observed that the accumulation of Cd in soil and rice plants decreased in treatments with silicate fertilizers. Additionally, they noted an increase in microbial richness and diversity, which may contribute to an improvement in soil biochemistry. The research emphasizes an enrichment of specific bacterial groups such as *Proteobacteria*, *Nitrospirae*, and Gemmatimonadetes, suggesting that these groups may participate directly or indirectly in the transformation of cadmium in the soil-plant system and in reducing the accumulation of this metal. In another study, Wang et al. (2023a) found that organic and mineral silicate fertilizers altered the microbial community and mitigated Cd and Pb stress in tobacco plants, promoting growth through improved photosynthesis processes and reducing oxidative damage caused by these metals.

The application of silicate fertilizers can increase the density and richness of the soil bacterial community, providing a more stable and abundant environment (Wang et al. 2020a). This helps to minimize the effect that the presence of heavy metals can cause, such as significant loss in diversity and some key functions of the soil microbial community. Although it is evident that the presence of metals alters the composition and diversity of soil microorganisms (Song et al. 2021b; Peng et al. 2023) and that soil functions are directly linked to community composition, it is still not fully understood which groups are favored under these conditions. When evaluating the effect of silicate fertilizers on *Brassica chinensis* plants in soils with multiple metals (Pb, Cd, As, Cu, Hg, and Cr), Wang et al. (2020a, b) observed a higher relative abundance of *Proteobacteria* and *Actinobacteria*, suggesting they are more active in metal-contaminated soils. However, Ren et al. (2024) observed a predominance of *Actinobacteria* and *Bacteroidetes* as important microorganisms composing an ecological network in soils with Cd and Pb.

Silicon also plays an important role in establishing a complex, efficient, and interdependent molecular ecological network among soil microorganisms in the presence of metals (Cd and Pb) (Ren et al. 2024). Additionally, it promotes an increase in microbial capacity and provides a positive impact on the metabolic functions and organic systems of soil microorganisms, such as enzymatic activity (Ren et al. 2024). When evaluating the potential of Si in mitigating the effects of Cd and Pb on tobacco cultivation, Ren et al. (2024) found that Si is capable of improving agronomic characteristics, such as biomass accumulation in this crop. Furthermore, it was discovered that organosilicate fertilizers are more beneficial in reducing Cd and Pb in the soil and improving agronomic characteristics in tobacco compared to mineral silicate fertilizers. According to Huang et al. (2019), this can be attributed to the effective increase in antagonism between Pb and Cd promoted by organosilicates, leading to a reduction in the transfer of heavy metals in the soil and consequently decreasing toxicity in plants.

The interaction between silicon, soil microorganisms, and heavy metals has emerged as a promising field of research, yielding significant results recently. Further studies aiming to elucidate the functional mechanisms by which microorganisms act to mitigate the effects of heavy metals in the environment are of utmost importance to broaden our understanding of this complex interaction.

8.4.1.1 Functional Microbial Activity

Understanding microbial functional activity mediated by silicon has been important in unraveling the mechanisms of bacterial responses to heavy metal toxicity. Silicon plays a crucial role in enhancing microbial community resilience against oxidative and osmotic stresses, carbon and nitrogen fixation, as well as methane (CH_4) consumption (Das et al. 2021). These benefits are particularly relevant in metal-contaminated soil environments, where silicon increases microbial diversity, expanding their genetic potential (Das et al. 2021; Peng et al. 2023). Silicon application influences the composition and functional adaptation of soil microorganisms, resulting in the bioavailability and transformation of heavy metals in the soil-plant system while also contributing to soil fertility enhancement (Peng et al. 2023). The use of silicon further promotes increased soil enzymatic activity, especially related to carbon, nitrogen, and phosphorus cycles (Liang et al. 2021; Ren et al. 2024).

Previous studies have shown that reducing microbial diversity leads to a decrease in functional redundancy, especially when considering functions performed by specific groups such as symbiotic nitrogen fixation and xenobiotic degradation, while functions related to organic matter decomposition carried out by many microorganisms remain constant (Xun et al. 2021). On the other hand, Gao et al. (2024) found that soil-native microorganisms involved in specialized functions, such as the effect of silicon on arsenic (As) levels in roots, are primarily responsible for silicon-regulated processes.

Currently, scientific research has been dedicated to studying the role of silicon in mediating microbial functional activity in plants (Das et al. 2021; Peng et al. 2023), aiming to understand the involved mechanisms and impacts on sustainable agricultural production. Peng et al. (2023) observed that Si application induced functional responses of soil microorganisms, including membrane transport, and promoted resistance to cadmium (Cd) pollution. These findings suggest that microorganisms present in Si-rich soils may produce heavy metal-chelating agents, such as siderophores, organic acids, and biosurfactants, which reduce the bioavailability of ions of these metals.

Silicon in the form of organosilicate fertilizers provides improvements in functional metabolic pathways such as terpenoid and polyketide metabolism, lipids, xenobiotics, amino acids, and biodegradation (Ren et al. 2024). Wang et al. (2023a) found that the application of silicate fertilizers (organic and mineral) altered bacterial communities and consequently the functional prediction of these microorganisms in the presence of cadmium (Cd) and lead (Pb). This primarily contributed to the increase in enzymes related to amino acids, thiamine metabolism, and arachidonic acid and the biosynthesis of valine, leucine, and isoleucine.

In a study by Das et al. (2021), it was observed that the application of Si fertilizer increases the abundance of genes associated with the carbon/nitrogen cycle and the metal detoxification process in rice-cultivated soils. Additionally, the cultivation of this crop in soils contaminated with arsenic (As) results in the modification of the composition and diversity of the microbial community (Shao et al. 2017). Ren et al. (2024), when assessing the potential of Si in mitigating the effects of heavy metals such as Cd and Pb through changes in bacterial functional profiles, found that silicate fertilizers improved the metabolic activities of microorganisms and the overall functioning of organic systems. Adhikari et al. (2020) demonstrated that silicate ($Mg_2O_8Si_3$), besides improving the growth characteristics of rice, reduced the expression of the cadmium transporter gene (OsHMA2) and increased the expression of the silicon transporter gene (OsLsi1).

In general, advances in scientific production in this area have significant practical implications for agriculture, offering innovative strategies to improve crop productivity, reduce dependency on chemicals, and strengthen the resilience of agricultural systems in the face of environmental challenges. The silicon-centered approach and microbial activity represent a promising perspective for the development of more sustainable and efficient agricultural practices.

8.4.2 The Silicon Alters the Composition of Microbial Communities in the Rhizosphere of Cultivated Plants

Silicon can have significant effects on the composition and activity of microbial communities in plant rhizospheres, affecting plant-microorganism interactions and, consequently, plant health and performance in adverse environmental conditions and different production systems (Rea et al. 2022; Yuan et al. 2022; Jin et al. 2023; Ahmed et al. 2024; Leite et al. 2024b; Lewin et al. 2024). As previously reported, this influence of Si encompasses various factors such as soil pH, nutrient availability, and microbial activity, which can regulate the abundance and diversity of microorganisms in the rhizosphere (Epstein 1994; Deng et al. 2023a). By affecting nutrient availability in the soil, Si can induce changes in competitive interactions between plants and soil microorganisms for nutrient acquisition. This, in turn, can influence the composition and activity of microbial communities in the rhizosphere (Verma et al. 2022; Sharma et al. 2023). The presence of Si in the soil can also influence enzymatic activity, including enzymes involved in organic matter decomposition and nutrient cycling (Teixeira et al. 2022). Such alterations in enzymatic activity can influence nutrient availability for plants and soil microorganisms, shaping the microbial community in the rhizosphere of that environment (Ndabankulu et al. 2022).

Silicon can induce plant resistance to pathogens by activating systemic defense responses, which may include the production of secondary metabolites and the activation of defense-related genes (Sun et al. 2022), which can also affect the microbial community in crop rhizosphere (Verma et al. 2022). The rhizosphere is considered a dominant factor in shaping the composition of the bacterial community, explaining up to 38% of the total variation (Nuccio et al. 2016). However, variables such as environmental conditions (Nuccio et al. 2016; Pascual et al. 2018), soil characteristics (Wang et al. 2018; Wan et al. 2020), soil management practices (Wang et al. 2020b), as well as the specific soil and area history (Schmid et al. 2021) play a significant role in structuring and differentiating microbial communities present in the rhizosphere.

In recent years, studies focusing on the relationship between Si and microorganisms have been intensified, especially in modulating the composition, structure, diversity, and activity of microbial communities in the rhizosphere of various agricultural crops (Table 8.3). In a study on cucumber (*Cucumis sativus* L.) cultivation, fertilization with sodium silicate increased the abundance and diversity of soil bacterial communities, as well as enhancing plant resistance to *Fusarium* wilt (Zhou et al. 2018). According to the authors, increased disease resistance was more associated with changes in microbial communities mediated by Si than with direct

Table 8.3 Responses of applying different sources of Si on the microbial community

Cultura	Fonte de Si	Destaques	Ref.
Cucumber (*Cucumis sativus* L.)	Sodium silicate	Increased abundance and diversity of soil bacterial communities	Zhou et al. (2018)
Pakchoi (*Brassica chinensis* L.)	By-product of steelmaking, containing 35% available SiO_2	Increase in the abundance of the *Nitrospirae*, *Flavobacteria*, *Anaerolineae*, and *Bacillales* genera	Wang et al. (2020a)
Wheat (*Triticale* spp.)	Silica gel	The Si application altered soil microbial community (*Acidobacteria* and *Thaumarchaeota*), reducing soil Cd bioavailability toward plant parts	Song et al. (2021b)
Sugarcane (*Saccharum* spp.)	Sodium metasilicate	Enrichment of the rhizosphere bacterial network, specifically Proteobacteria; positive impacts on crop growth	Deng et al. (2021)
Peanut (*Arachis hypogaea* L.)	Sodium silicate	Increase in the abundance of *Planctomycetes* and *Firmicutes* phyla under *Ralstonia solanacearum* inoculation	Deng et al. (2023a)
Sugarcane (*Saccharum* spp.)	Sodium silicate	Changed the abundance of specific microbial groups, with variations among genotypes; enrichment of microbial groups, particularly some N-fixers	Leite et al. (2023)
Sugarcane (*Saccharum* spp.)	Sodium silicate	Changes the complexity of the growth-promoting rhizobacteria network	Leite et al. (2024b)

inhibition of pathogen growth by Si. In pakchoi cabbage (*Brassica chinensis* L.) cultivation, Wang et al. (2020a) evaluated the effects of Si application on the bacterial community in soil contaminated with heavy metals. The results demonstrated significant changes in the structure of the bacterial community, especially at higher doses of Si fertilizer. An interesting aspect was the observation of an increase in the abundance of bacteria associated with soil metabolic function, specifically those involved in carbon/nitrogen cycles. Genera such as *Nitrospirae*, *Flavobacteriales*, *Anaerolineae*, and *Bacillales* were more abundant in treatments receiving Si fertilization.

In the study conducted by Song et al. (2021a, b) on wheat cultivation in cadmium (Cd)-contaminated soil, it was observed that the addition of Si altered the microbial community present in the soil and substantially contributed to increasing wheat productivity. Si fertilization also promoted an increase in the abundance of *Acidobacteria* and *Thaumarchaeota*, which contributed to the reduction of Cd availability in the soil. In another study, Song et al. (2021a) observed that the return of rice straw to the soil had a significant impact on increasing bioavailable Si and altering the abundance of the soil microbial community. Redundancy analysis showed that Si status was responsible for explaining 12% of the variation in the soil bacterial community. These results suggest that the interaction between soil bacterial community diversity and mobility of Si contributes to balancing various nutrient sources, thus driving the biological cycle of Si in agroecosystems.

In a study conducted on sugarcane during the early stage of stalk growth, Deng et al. (2021) used anhydrous sodium metasilicate (Na_2SiO_3) as a source of silicon and demonstrated, through a network association analysis, that Si application enriched the bacterial network in the rhizosphere, resulting in positive impacts on crop growth. Si promoted a significant increase in the relative abundance of *Proteobacteria*, suggesting that this group may be dominant in sugarcane growth when fertilized with Si. Deng et al. (2023a) investigated the effects of silicon application on modulating the bacterial community in the rhizosphere of soil cultivated with peanuts and inoculated with *Ralstonia solanacearum*, the causative agent of bacterial wilt. Si altered the soil rhizosphere bacterial community structures and metabolic profiles. According to the authors, Si can generally act on the microbial community in the rhizosphere, altering plant exudation profile modulation and microbial growth. These changes result from alterations in the physical-chemical properties and soil enzymes induced by Si, leading to the establishment of new microbial communities.

When evaluating the effect of silicon on the diversity and composition of the prokaryotic community in the rhizosphere of sugarcane genotypes over 2 years of cultivation (plant cane and ratoon cane), Dutra et al. (2023) found that Si application caused changes in the abundance of specific microbial groups, with variations among the genotypes studied. Additionally, they observed that the differential plant exudation profile depended on its stage of development, influencing the interaction between microorganisms and plants, including plant growth promoters. In Si treatments, there was an enrichment of microbial groups, especially some nitrogen (N) and phosphate (P) fixers. In a more recent study, Leite et al. (2024b) found that the application of sodium silicate alters the complexity of the network of plant growth-promoting rhizobacteria in the rhizosphere of sugarcane genotypes. Additionally, these authors also found that the most abundant genera in this culture were *Bacillus* (~42%), followed by *Sphingomonas* (~16%), *Paenibacillus* (~10%), *Streptomyces* (~7%), and *Mycobacterium* (~6%).

8.4.3 *The Silicon-Microbiota Interaction in Plant Resistance to Environmental Stresses*

As previously demonstrated, Si acts as a mitigator of biotic and abiotic stresses, resulting in increased resistance and yield of cultivated plants (Kidd et al. 2001; Fan et al. 2016; Etesami and Jeong 2018). Similarly, there is ample documentation on the action of beneficial microorganisms in promoting growth and potential effects on stress reduction in plants under adverse conditions (Majeed et al. 2018; Ruiu 2020; Porter et al. 2020; Ngalimat et al. 2021). However, it is still not fully understood how the interaction between silicon-microorganisms affects plant resistance to biotic and abiotic stresses. In this topic, we explore recent literature to clarify important issues related to the interaction between Si and microorganisms in plant resistance to stress. Some of the addressed questions include the following: How does the alteration of the microbial community mediated by Si affect plant resistance to stress? What is the role of microorganisms in mediating the silicon-plant relationship to increase resistance to biotic and abiotic stresses? Understanding these interactions is crucial for the development of agricultural management strategies that optimize the beneficial role of Si in promoting plant health and improving stress resistance. Ongoing research in this field aims to broaden our understanding and explore practical applications for more sustainable agriculture.

The exogenous application of Si has been recognized to increase plant resistance to environmental stresses (Das et al. 2021; Peng et al. 2023). However, the enhanced plant resistance seems not to be merely a direct response to Si application, but rather a result of indirect responses due to Si-microorganism-plant interactions (Wang et al. 2013; Zhou et al. 2018; Yu et al. 2022). Wang et al. (2013) observed that Si application increases tomato resistance against *Ralstonia solanacearum*, the causal agent of bacterial wilt, due to modification of soil microbial community structure and activity by Si. Additionally, Zhou et al. (2018) also found that Si application (sodium silicate) alters soil microbial communities and consequently enhances cucumber resistance to *Fusarium* wilt caused by the phytopathogen *Fusarium oxysporum*. In the same study, Zhou et al. (2018) noted that Si contributed to reducing the relative abundance of microbial taxa containing pathogens and increased those with potential beneficial effects for plants, including *Streptomyces* DHV3–2, an antagonist bacterium of *Fusarium*. In a short-term study, Li et al. (2019) reported that Si application and related changes in soil bacterial community dynamics can reduce the incidence of black spot disease in *Panax ginseng* (medicinal plant). Peng et al. (2023) also found that Si application (steel slag) can reduce cadmium (Cd) uptake by rice crops in polluted soils, with alterations in the microbial community.

Biotic and abiotic stresses pose significant challenges to agriculture, leading to substantial productivity losses that have the potential to compromise food security (Tyczewska et al. 2018; Verma et al. 2020). Beneficial microorganisms present in the soil demonstrate great potential in mitigating both abiotic and biotic stresses, especially when used in conjunction with Si application (Verma et al. 2020; Alharbi et al. 2022; Chaganti et al. 2023). Akhtar et al. (2021) found that the combined application of Si oxide nanoparticles and PGPR promoted drought tolerance in wheat by modulating various physiological and metabolic processes in plants, resulting in significant improvements in wheat growth and productivity under water stress conditions.

Silicon also acts in mitigating salinity stress. In a study conducted by Alharbi et al. (2022), it was observed that the combination of rhizobacteria and silica nanoparticles has a promotive effect on plant growth, besides stimulating the resilience of sugar beet irrigated with saline water in soils impacted by salinity. Additionally, microorganisms can also influence the bioavailability of Si in the soil. In a study conducted by Chaganti et al. (2023), it was observed that silicate-solubilizing and plant growth-promoting bacteria interact with biogenic silica, conferring tolerance to thermal stress in rice, modulating physiology and gene expression. Das et al. (2021), when evaluating silicate fertilization in arsenic (As)-contaminated rice cultivation systems, observed improvement in microbial functional potential, evidenced by an increase in functional genes related to stress response. Furthermore, the study also showed that silicate fertilization promotes increased soil enzyme activity related to nutrient cycling (carbon, nitrogen, and phosphorus), raises plant photosynthetic rate, and strengthens microorganism resistance under stress conditions. Such information is relevant for understanding how the Si-microorganism relationship enhances plant resistance. In Table 8.4, we present some studies evaluating the silicon-microorganism relationship in enhancing stress resistance in cultivated plants.

Stusdies show that Si application can indeed modify the structure, composition, and functional potential of microbial communities (Wang et al. 2013; Das et al. 2021; Kang et al. 2022). However, the exact mechanisms behind these changes are not yet fully understood. Considering that plants play a crucial role in recruiting specific taxa to the rhizosphere (Deng et al. 2021; Yin et al. 2021), a widely accepted view in the scientific community, it is likely that observed changes in the microbial community due to Si occur through direct and indirect effects mediated by plant-microbe interactions. Si can directly alter the microbial community by being incorporated into microbial cell walls, influencing their survival capacity under stress conditions (Zhang and Guan 2022). Additionally, Si can have indirect effects by modifying plant metabolism and consequently the pattern of organic compound exudation in the rhizosphere (Fan et al. 2016). Kidd et al. (2001), evaluating the role of root exudates in plant resistance to aluminum and mitigation of silicon-induced toxicity in corn, found that Si-treated plants produced more phenolic compounds, including flavonoids, compared to untreated plants. In this context, it is possible to assert that changes in the chemical composition of root exudates can regulate the composition and activity of the microbial community (Wen et al. 2022). Microorganisms play a crucial role in organic matter degradation and Si cycling in the soil, directly influencing its availability to plants (Song et al. 2021b). According to these authors, microorganisms interact with Si in various ways, influencing its form, mobility, and solubility in the soil. A summary of the main known mechanisms involving the interaction silicon-microorganisms-plant is shown in Fig. 8.2.

8.5 Conclusions and Future Perspectives

The data presented in this chapter provide solid evidence that silicon can exert significant effects on the composition and activity of microbial communities in plant rhizospheres. This element affects plant-microorganism interactions and, consequently, the metabolism and performance of plants under different environmental conditions. The mechanisms activated during the Si-plant interaction in mitigating biotic and abiotic stresses have been extensively studied and reported in the scientific community. The interaction of a third factor makes the process even more dynamic, with the relationship between Si-microorganisms-plants demonstrating

Table 8.4 Silicon-microorganism relationship in enhancing stress resistance in cultivated plants

Silicon form	Applied or enriched microorganisms	Crops	Highlights	Ref.
Silicon dioxide nanoparticles	*Bacillus* sp. *Azospirillum lipoferum* *Azospirillum brasilense*	Wheat	Increased drought tolerance; Increase in yield; enhanced nutrient uptake	Akhtar et al. (2021)
Water glass	Increased in *Acidobacteria* and *Thaumarchaeota*	Wheat	Decreased Cd bioavailability in soil and plant uptake	Song et al. (2021b)
Amorphous silica	Increased in *Arthrobacter e Sphingomonas*	Wheat	Mitigated the drought stress, enhance the availability of plant nutrients	Lewin et al. (2024)
Potassium silicate	*Bacillus* sp.	Cucumber	Reduction of salt stress in seedlings	Kaloterakis et al. (2021)
Silicic acid	Increased in *Proteobacteria*, *Bacteroidota*, *Ascomycota*	Soybean	Cold tolerance in early growth stages	Ahmad et al. (2024)
Calcium silicate	–	Rice	Increased plant photosynthetic rate, transpiration efficiency, and stomatal conductance, while decreased the transpiration rate in soil enriched with As	Das et al. (2021)
Potassium silicate	Increased the abundance of *Sandaracinus*, *Polycyclovorans*, *Hirschia*, *Haliangium*, *Nitrospira*, *Saccharothrix*, *Aeromicrobium*, *Luteimonas*, and *Rubellimicrobium*	*Panax ginseng*	Prevented and controlled the occurrence of ginseng black spot	Li et al. (2019)
Potassium silicate	*Bacillus drentensis*	Mung bean	Improving physiology, growth and productivity under salt stress	Mahmood et al. (2016)
Sodium silicate	Increased the abundance of *Streptomyces* DHV3-2 in soil	Cucumber	Increased cucumber seedling resistance to Fusarium	Zhou et al. (2018)
Silica nanoparticles	*Pseudomonas koreensis* and *Bacillus coagulans*	Sugar beet	Reduction of salt stress and improvement of morphophysiological characteristics and productivity	Alharbi et al. (2022)

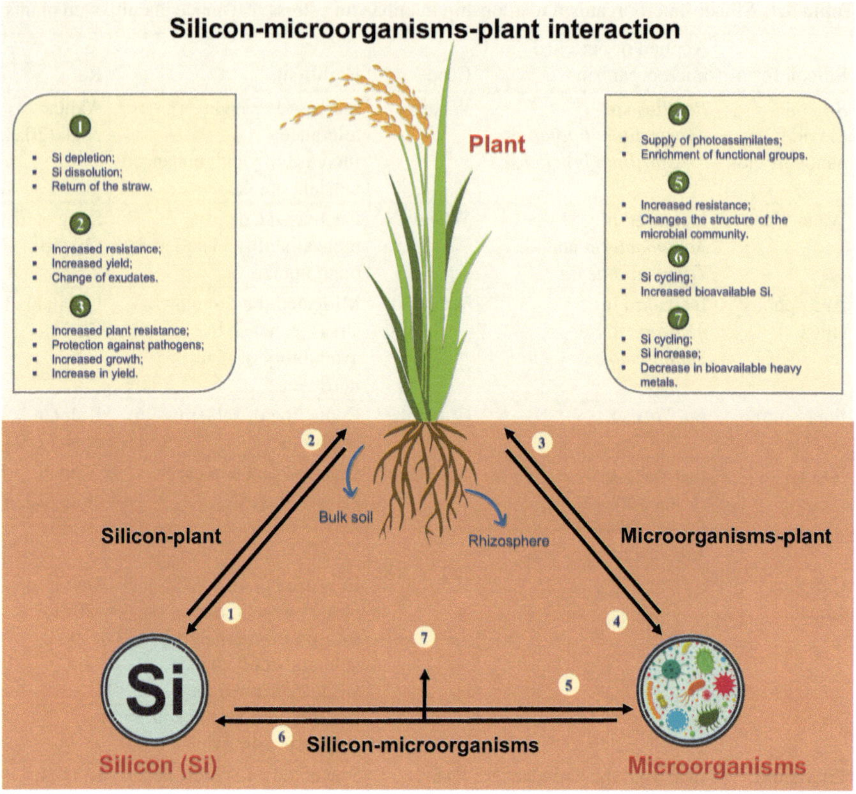

Fig. 8.2 Summary of main responses resulting from silicon-microorganism-plant interaction. Arrows indicate the direction of effect of one component on another component: (1) effect of the plant on Si; (2) effect of Si on the plant; (3) effect of microorganisms on the plant; (4) effect of the plant on microorganisms; (5) effect of Si on microorganisms; (6) effect of microorganisms on the plant; (7) effects of Si-microorganism interaction on the plant. (Source: developed by the authors)

how silicon influences soil microbial communities, including bulk soil and the rhizosphere, to exert beneficial effects on crop yields.

Even in the face of extensive knowledge, this is a research field that still requires further detailed exploration. There remains a gap in understanding the molecular mechanisms governing the relationship between silicon, microorganisms, and plants, as well as in quantifying the impact of each individual component on the others. Additional studies are essential and should focus on the complex interactions employing omics approaches such as metagenomics, metataxonomics, and metabolomics, along with biochemical analyses. A multidisciplinary approach, combining different techniques, will allow for a more comprehensive understanding of the interactions and their potential for the development of more sustainable agricultural practices. Such studies should aim to improve soil health and quality while simultaneously increasing crop productivity, thus contributing to more efficient and sustainable agriculture.

References

Adhikari A, Lee K-E, Khan MA et al (2020) Effect of silicate and phosphate solubilizing Rhizobacterium *Enterobacter ludwigii* GAK2 on *Oryza sativa* L. under cadmium stress. J Microbiol Biotechnol 30:118–126. https://doi.org/10.4014/jmb.1906.06010

Ahmad W, Coffman L, Weerasooriya AD et al (2024) The silicon regulates microbiome diversity and plant defenses during cold stress in *Glycine max* L. Front Plant Sci 14. https://doi.org/10.3389/fpls.2023.1280251

Ahmed S, Iqbal M, Ahmad Z et al (2023a) Foliar application of silicon-based nanoparticles improve the adaptability of maize (*Zea mays* L.) in cadmium contaminated soils. Environ Sci Pollut Res 30:41002–41013. https://doi.org/10.1007/s11356-023-25189-0

Ahmed SR, Anwar Z, Shahbaz U et al (2023b) Potential role of silicon in plants against biotic and abiotic stresses. SILICON 15:3283–3303. https://doi.org/10.1007/s12633-022-02254-w

Ahmed T, Guo J, Noman M et al (2024) Metagenomic and biochemical analyses reveal the potential of silicon to alleviate arsenic toxicity in rice (*Oryza sativa* L.). Environ Pollut 345:123537. https://doi.org/10.1016/j.envpol.2024.123537

Akhtar N, Ilyas N, Mashwani Z-R et al (2021) Synergistic effects of plant growth promoting rhizobacteria and silicon dioxide nano-particles for amelioration of drought stress in wheat. Plant Physiol Biochem 166:160–176. https://doi.org/10.1016/j.plaphy.2021.05.039

Alayafi AH, Al-Solaimani SGM, Abd El-Wahed MH et al (2022) Silicon supplementation enhances productivity, water use efficiency and salinity tolerance in maize. Front Plant Sci 13. https://doi.org/10.3389/fpls.2022.953451

Alharbi K, Hafez E, Omara AE-D et al (2022) Plant growth promoting rhizobacteria and silica nanoparticles stimulate sugar beet resilience to irrigation with saline water in salt-affected soils. Plan Theory 11:3117. https://doi.org/10.3390/plants11223117

Andreote FD, de Pereira e Silva MC (2017) Microbial communities associated with plants: learning from nature to apply it in agriculture. Curr Opin Microbiol 37:29–34. https://doi.org/10.1016/j.mib.2017.03.011

Arruda L, Beneduzi A, Martins A et al (2013) Screening of rhizobacteria isolated from maize (*Zea mays* L.) in Rio Grande do Sul State (South Brazil) and analysis of their potential to improve plant growth. Appl Soil Ecol 63:15–22. https://doi.org/10.1016/j.apsoil.2012.09.001

Asgher M, Rehaman A, Nazar ul Islam S, Khan NA (2024) Multifaceted roles of silicon nano particles in heavy metals-stressed plants. Environ Pollut 341:122886. https://doi.org/10.1016/j.envpol.2023.122886

Asiloglu R, Shiroishi K, Suzuki K et al (2020) Protist-enhanced survival of a plant growth promoting rhizobacteria, *Azospirillum* sp. B510, and the growth of rice (*Oryza sativa* L.) plants. Appl Soil Ecol 154:103599. https://doi.org/10.1016/j.apsoil.2020.103599

Barão L (2023) The use of Si-based fertilization to improve agricultural performance. J Soil Sci Plant Nutr 23:1096–1108. https://doi.org/10.1007/s42729-022-01106-1

Barreto RF, Barão L (2023) Silicon: transcellular and apoplastic absorption and transport in the xylem. In: Benefits of silicon in the nutrition of plants. Springer International Publishing, Cham, pp 17–25

Basirat M, Mousavi SM (2022) Effect of foliar application of silicon and salicylic acid on regulation of yield and nutritional responses of greenhouse cucumber under high temperature. J Plant Growth Regul 41:1978–1988. https://doi.org/10.1007/s00344-021-10562-5

Basu S, Kumar G, Chhabra S, Prasad R (2021) Role of soil microbes in biogeochemical cycle for enhancing soil fertility. In: Verma JP, Gupta VK, Podile AR, Macdonald CA (eds) New and future developments in microbial biotechnology and bioengineering. Elsevier, pp 149–157

Bhardwaj S, Sharma D, Singh S et al (2023) Physiological and molecular insights into the role of silicon in improving plant performance under abiotic stresses. Plant Soil 486:25–43. https://doi.org/10.1007/s11104-022-05395-4

Bilal S, Khan T, Asaf S et al (2023) Silicon induced morphological, biochemical and molecular regulation in *Phoenix dactylifera* L. under low-temperature stress. Int J Mol Sci 24:6036. https://doi.org/10.3390/ijms24076036

Bocharnikova E, Matichenkov V (2022) Silicon induced mitigation of low-temperature stress in sugarcane. In: Agro-industrial perspectives on sugarcane production under environmental stress. Springer Nature Singapore, Singapore, pp 215–229

Cao Y, Pi H, Chandrangsu P et al (2018) Antagonism of two plant-growth promoting *Bacillus velezensis* isolates against *Ralstonia solanacearum* and *Fusarium oxysporum*. Sci Rep 8:4360. https://doi.org/10.1038/s41598-018-22782-z

Chaganti C, Phule AS, Chandran LP et al (2023) Silicate solubilizing and plant growth promoting bacteria interact with biogenic silica to impart heat stress tolerance in rice by modulating physiology and gene expression. Front Microbiol 14. https://doi.org/10.3389/fmicb.2023.1168415

Chhetri G, Kim I, Kang M et al (2022) An isolated *Arthrobacter* sp. enhances rice (*Oryza sativa* L.) plant growth. Microorganisms 10:1187. https://doi.org/10.3390/microorganisms10061187

Chukwuneme CF, Ayangbenro AS, Babalola OO, Raphael Kutu F (2021) Functional diversity of microbial communities in two contrasting maize rhizosphere soils. Rhizosphere 17:100282. https://doi.org/10.1016/j.rhisph.2020.100282

Das S, Kim GW, Lee JG et al (2021) Silicate fertilization improves microbial functional potentials for stress tolerance in arsenic-enriched rice cropping systems. J Hazard Mater 417:125953. https://doi.org/10.1016/j.jhazmat.2021.125953

de Souza Júnior JP, de Mello PR, Campos CNS et al (2022a) Silicon foliar spraying in the reproductive stage of cotton plays an equivalent role to boron in increasing yield, and combined boron-silicon application, without polymerization, increases fiber quality. Ind Crop Prod 182:114888. https://doi.org/10.1016/j.indcrop.2022.114888

de Souza Júnior JP, de Oliveira TL, de Mello PR et al (2022b) Analyzing the role of silicon in leaf C:N:P stoichiometry and its effects on nutritional efficiency and dry weight production in two sugarcane cultivars. J Soil Sci Plant Nutr 22:2687–2694. https://doi.org/10.1007/s42729-022-00836-6

del Orozco-Mosqueda Ma C, Flores A, Rojas-Sánchez B et al (2021) Plant growth-promoting bacteria as bioinoculants: attributes and challenges for sustainable crop improvement. Agronomy 11:1167. https://doi.org/10.3390/agronomy11061167

Delaporte-Quintana P, Lovaisa NC, Rapisarda VA, Pedraza RO (2020) The plant growth promoting bacteria *Gluconacetobacter diazotrophicus* and *Azospirillum brasilense* contribute to the iron nutrition of strawberry plants through siderophores production. Plant Growth Regul 91:185–199. https://doi.org/10.1007/s10725-020-00598-0

Deng Q, Yu T, Zeng Z et al (2021) Silicon application modulates the growth, rhizosphere soil characteristics, and bacterial community structure in sugarcane. Front Plant Sci 12. https://doi.org/10.3389/fpls.2021.710139

Deng Q, Liu H, Lu Q et al (2023a) Silicon application for the modulation of rhizosphere soil bacterial community structures and metabolite profiles in peanut under ralstonia solanacearum inoculation. Int J Mol Sci 24:3268. https://doi.org/10.3390/ijms24043268

Deng Y, Yang M, Li T et al (2023b) Silicon supplementation improves biomass and direct defense of ryegrass: a multi-omics study. Ind Crop Prod 204:117357. https://doi.org/10.1016/j.indcrop.2023.117357

Du J, Liu B, Zhao T et al (2022) Silica nanoparticles protect rice against biotic and abiotic stresses. J Nanobiotechnol 20:197. https://doi.org/10.1186/s12951-022-01420-x

Dubey A, Malla MA, Khan F et al (2019) Soil microbiome: a key player for conservation of soil health under changing climate. Biodivers Conserv 28:2405–2429. https://doi.org/10.1007/s10531-019-01760-5

Durán D, Bernal P, Vazquez-Arias D et al (2021) *Pseudomonas fluorescens* F113 type VI secretion systems mediate bacterial killing and adaption to the rhizosphere microbiome. Sci Rep 11:5772. https://doi.org/10.1038/s41598-021-85218-1

Dutra AF, Leite MRL, de Melo CCF, et al (2023) Soil and foliar Si fertilization alters elemental stoichiometry and increases yield of sugarcane cultivars. Sci Rep 13:16040. https://doi.org/10.1038/s41598-023-43351-z

Epstein E (1994) The anomaly of silicon in plant biology. Proc Natl Acad Sci 91:11–17. https://doi.org/10.1073/pnas.91.1.11

Etesami H, Jeong BR (2018) Silicon (Si): review and future prospects on the action mechanisms in alleviating biotic and abiotic stresses in plants. Ecotoxicol Environ Saf 147:881–896. https://doi.org/10.1016/j.ecoenv.2017.09.063

Etesami H, Jeong BR (2023) How does silicon help alleviate biotic and abiotic stresses in plants? Mechanisms and future prospects. In: Plant Stress Mitigators. Elsevier, pp 359–402

Fan X, Wen X, Huang F et al (2016) Effects of silicon on morphology, ultrastructure and exudates of rice root under heavy metal stress. Acta Physiol Plant 38:197. https://doi.org/10.1007/s11738-016-2221-8

Farhaoui A, El Alami N, Khadiri M et al (2023) Biological control of diseases caused by *Rhizoctonia solani* AG-2-2 in sugar beet (*Beta vulgaris* L.) using plant growth-promoting rhizobacteria (PGPR). Physiol Mol Plant Pathol 124:101966. https://doi.org/10.1016/j.pmpp.2023.101966

Flores RA, Pessoa-de-Souza MA, de Andrade AF et al (2022) Does foliar application of silicon under natural water stress conditions increase rice yield in subtropical dry regions? Silicon 14:3591–3600. https://doi.org/10.1007/s12633-021-01109-0

Gao Z, Chen H, Zhang X et al (2022) Silicon enhances abundances of reducing microbes in rhizoplane and decreases arsenite uptake by rice (*Oryza sativa* L.). Environ Pollut 306:119405. https://doi.org/10.1016/j.envpol.2022.119405

Gao Z, Jiang Y, Li W et al (2024) Evidence for the role of microbes in the silicon-regulated arsenic concentrations of rice roots in the soil environment. Sci Total Environ 908:168321. https://doi.org/10.1016/j.scitotenv.2023.168321

Gou C, Huang Q, Rady MM et al (2023) Integrative application of silicon and/or proline improves sweet corn (*Zea mays* L. *saccharata*) production and antioxidant defense system under salt stress condition. Sci Rep 13:18315. https://doi.org/10.1038/s41598-023-45003-8

Gul Z, Tang Z-H, Arif M, Ye Z (2022) An insight into abiotic stress and influx tolerance mechanisms in plants to cope in saline environments. Biology (Basel) 11:597. https://doi.org/10.3390/biology11040597

Gulzar N, Lone R, Al-Sadi AM, Azeez A (2024) Interactive role of silicon and phenolics in biotic stress regulation in plants and expression of phenylpropanoid pathway genes. In: Plant phenolics in biotic stress management. Springer Nature Singapore, Singapore, pp 305–324

Gupta R, Kumari A, Sharma S et al (2022) Identification, characterization and optimization of phosphate solubilizing rhizobacteria (PSRB) from rice rhizosphere. Saudi J Biol Sci 29:35–42. https://doi.org/10.1016/j.sjbs.2021.09.075

Huang H, Rizwan M, Li M et al (2019) Comparative efficacy of organic and inorganic silicon fertilizers on antioxidant response, Cd/Pb accumulation and health risk assessment in wheat (*Triticum aestivum* L.). Environ Pollut 255:113146. https://doi.org/10.1016/j.envpol.2019.113146

Huang F, Li Z, Yang X et al (2024) Silicon reduces toxicity and accumulation of arsenic and cadmium in cereal crops: a meta-analysis, mechanism, and perspective study. Sci Total Environ 918:170663. https://doi.org/10.1016/j.scitotenv.2024.170663

Jin X, Rahman MK, Ma C et al (2023) Silicon modification improves biochar's ability to mitigate cadmium toxicity in tomato by enhancing root colonization of plant-beneficial bacteria. Ecotoxicol Environ Saf 249:114407. https://doi.org/10.1016/j.ecoenv.2022.114407

Kaloterakis N, van Delden SH, Hartley S, De Deyn GB (2021) Silicon application and plant growth promoting rhizobacteria consisting of six pure *Bacillus* species alleviate salinity stress in cucumber (*Cucumis sativus* L). Sci Hortic 288:110383. https://doi.org/10.1016/j.scienta.2021.110383

Kang Y, Ma Y, Wu W et al (2022) Bioorganic and silicon amendments alleviate early defoliation of pear trees by improving the soil nutrient bioavailability, microbial activity, and reshap-

ing the soil microbiome network. Appl Soil Ecol 173:104383. https://doi.org/10.1016/j.apsoil.2021.104383
Khan I, Awan SA, Rizwan M et al (2023) Silicon: an essential element for plant nutrition and phytohormones signaling mechanism under stressful conditions. Plant Growth Regul 100:301–319. https://doi.org/10.1007/s10725-022-00872-3
Kidd PS, Llugany M, Poschenrieder C et al (2001) The role of root exudates in aluminium resistance and silicon-induced amelioration of aluminium toxicity in three varieties of maize (*Zea mays* L.). J Exp Bot 52:1339–1352. https://doi.org/10.1093/jexbot/52.359.1339
Kim Y-H, Khan AL, Kim D-H et al (2014) Silicon mitigates heavy metal stress by regulating P-type heavy metal ATPases, *Oryza sativa* low silicon genes, and endogenous phytohormones. BMC Plant Biol 14:13. https://doi.org/10.1186/1471-2229-14-13
Kovács S, Kutasy E, Csajbók J (2022) The multiple role of silicon nutrition in alleviating environmental stresses in sustainable crop production. Plan Theory 11:1223. https://doi.org/10.3390/plants11091223
Leite MRL, Alcantara Neto F, Dutra AF et al (2023) Silicon application influences the prokaryotic communities in the rhizosphere of sugarcane genotypes. Appl Soil Ecol 187:104818. https://doi.org/10.1016/j.apsoil.2023.104818
Leite MRL, Alcântara Neto F, Dutra AF et al (2024a) Distinct sources of silicon shape differently the rhizospheric microbial community in sugarcane. Appl Soil Ecol 193:105131. https://doi.org/10.1016/j.apsoil.2023.105131
Leite MRL, Costa RM, Costa MKL et al (2024b) Application of silicate changes the network complexity of plant growth-promoting rhizobacteria in the rhizosphere of sugarcane genotypes. Rhizosphere 29:100855. https://doi.org/10.1016/j.rhisph.2024.100855
Lewin S, Schaller J, Kolb S, Francioli D (2024) Amorphous silica fertilization ameliorated soil properties and promoted putative soil beneficial microbial taxa in a wheat field under drought. Appl Soil Ecol 196:105286. https://doi.org/10.1016/j.apsoil.2024.105286
Li M, Wang Q, Liu Z et al (2019) Silicon application and related changes in soil bacterial community dynamics reduced ginseng black spot incidence in *Panax ginseng* in a short-term study. BMC Microbiol 19:263. https://doi.org/10.1186/s12866-019-1627-z
Liang Y, Liao M, Fang Z et al (2021) How silicon fertilizer improves nitrogen and phosphorus nutrient availability in paddy soil? J Zhejiang Univ-Sci B 22:521–532. https://doi.org/10.1631/jzus.B2000708
Ling N, Wang T, Kuzyakov Y (2022) Rhizosphere bacteriome structure and functions. Nat Commun 13:836. https://doi.org/10.1038/s41467-022-28448-9
Mahmood S, Daur I, Al-Solaimani SG et al (2016) Plant growth promoting rhizobacteria and silicon synergistically enhance salinity tolerance of mung bean. Front Plant Sci 7. https://doi.org/10.3389/fpls.2016.00876
Majeed A, Muhammad Z, Ahmad H (2018) Plant growth promoting bacteria: role in soil improvement, abiotic and biotic stress management of crops. Plant Cell Rep 37:1599–1609. https://doi.org/10.1007/s00299-018-2341-2
Mir RA, Bhat BA, Yousuf H et al (2022) Multidimensional role of silicon to activate resilient plant growth and to mitigate abiotic stress. Front Plant Sci 13. https://doi.org/10.3389/fpls.2022.819658
Mishra S, Lin Z, Pang S et al (2021) Recent advanced technologies for the characterization of xenobiotic-degrading microorganisms and microbial communities. Front Bioeng Biotechnol 9. https://doi.org/10.3389/fbioe.2021.632059
Misra V, Mall AK, Ansari SA et al (2023) Silicon as a beneficial nutrient for productivity augmentation and abiotic/biotic stress tolerance in sugarcane. Biocatal Agric Biotechnol 54:102944. https://doi.org/10.1016/j.bcab.2023.102944
Muhammad HMD, Abbas A, Ahmad R (2022) Fascinating role of silicon nanoparticles to mitigate adverse effects of salinity in fruit trees: a mechanistic approach. SILICON 14:8319–8326. https://doi.org/10.1007/s12633-021-01604-4

Munns R, Tester M (2008) Mechanisms of salinity tolerance. Annu Rev Plant Biol 59:651–681. https://doi.org/10.1146/annurev.arplant.59.032607.092911

Nakamura R, Cornelis J-T, de Tombeur F et al (2020) Diversity of silicon release rates among tropical tree species during leaf-litter decomposition. Geoderma 368:114288. https://doi.org/10.1016/j.geoderma.2020.114288

Naveed M, Mustafa A, Majeed S et al (2020) Enhancing cadmium tolerance and pea plant health through *Enterobacter* sp. MN17 inoculation together with biochar and gravel sand. Plan Theory 9:530. https://doi.org/10.3390/plants9040530

Naz R, Zaman Q, Nazir S et al (2022) Silicon fertilization counteracts salinity-induced damages associated with changes in physio-biochemical modulations in spinach. PLoS One 17:e0267939. https://doi.org/10.1371/journal.pone.0267939

Ndabankulu K, Egbewale SO, Tsvuura Z, Magadlela A (2022) Soil microbes and associated extracellular enzymes largely impact nutrient bioavailability in acidic and nutrient poor grassland ecosystem soils. Sci Rep 12:12601. https://doi.org/10.1038/s41598-022-16949-y

Ngalimat MS, Mohd Hata E, Zulperi D et al (2021) Plant growth-promoting bacteria as an emerging tool to manage bacterial rice pathogens. Microorganisms 9:682. https://doi.org/10.3390/microorganisms9040682

Ngugi MM, Gitari HI, Muui CW, Gweyi-Onyango JP (2022) Growth tolerance, concentration, and uptake of heavy metals as ameliorated by silicon application in vegetables. Int J Phytoremediation 24:1543–1556. https://doi.org/10.1080/15226514.2022.2045251

Nuccio EE, Anderson-Furgeson J, Estera KY et al (2016) Climate and edaphic controllers influence rhizosphere community assembly for a wild annual grass. Ecology 97:1307–1318. https://doi.org/10.1890/15-0882.1

Pan B, Wang W, Liu B et al (2023) Significant difference in the efficacies of silicon application regimes on cadmium species and environmental risks in rice rhizosphere. Environ Pollut 327:121521. https://doi.org/10.1016/j.envpol.2023.121521

Parveen A, Mumtaz S, Saleem MH et al (2022) Silicon and nanosilicon mediated heat stress tolerance in plants. In: Silicon and nano-silicon in environmental stress management and crop quality improvement. Elsevier, pp 153–159

Pascual J, Blanco S, Ramos JL, van Dillewijn P (2018) Responses of bulk and rhizosphere soil microbial communities to thermoclimatic changes in a Mediterranean ecosystem. Soil Biol Biochem 118:130–144. https://doi.org/10.1016/j.soilbio.2017.12.013

Pathan SI, Ceccherini MT, Sunseri F, Lupini A (2020) Rhizosphere as hotspot for plant-soil-microbe interaction. In: Datta R, Meena R, Pathan S, Ceccherini M (eds) Carbon and nitrogen cycling in soil. Springer Singapore, Singapore, pp 17–43

Peng H, Deng K, Shi Y et al (2023) Alleviation of Cd-polluted paddy soils through Si fertilizer application and its effects on the soil microbial community. Sci Total Environ 855:158735. https://doi.org/10.1016/j.scitotenv.2022.158735

Porter SS, Bantay R, Friel CA et al (2020) Beneficial microbes ameliorate abiotic and biotic sources of stress on plants. Funct Ecol 34:2075–2086. https://doi.org/10.1111/1365-2435.13499

Prashar P, Kapoor N, Sachdeva S (2014) Rhizosphere: its structure, bacterial diversity and significance. Rev Environ Sci Biotechnol 13:63–77. https://doi.org/10.1007/s11157-013-9317-z

Qi X, Xue X, Wang Z et al (2024) Silicon application enhances wheat defence against Sitobion avenae F. By regulating plant physiological-biochemical responses. Basic Appl Ecol 74:13–23. https://doi.org/10.1016/j.baae.2023.11.003

Raturi G, Sharma Y, Rana V et al (2021) Exploration of silicate solubilizing bacteria for sustainable agriculture and silicon biogeochemical cycle. Plant Physiol Biochem 166:827–838. https://doi.org/10.1016/j.plaphy.2021.06.039

Raza MAS, Zulfiqar B, Iqbal R et al (2023) Morpho-physiological and biochemical response of wheat to various treatments of silicon nano-particles under drought stress conditions. Sci Rep 13:2700. https://doi.org/10.1038/s41598-023-29784-6

Rea RS, Islam MR, Rahman MM et al (2022) Growth, nutrient accumulation, and drought tolerance in crop plants with silicon application: a review. Sustain For 14:4525. https://doi.org/10.3390/su14084525

Ren K, Yang X, Li J et al (2024) Alleviating the adverse effects of Cd–Pb contamination through the application of silicon fertilizer: enhancing soil microbial diversity and mitigating heavy metal contamination. Chemosphere 352:141414. https://doi.org/10.1016/j.chemosphere.2024.141414

Ruiu L (2020) Plant-growth-promoting bacteria (PGPB) against insects and other agricultural pests. Agronomy 10:861. https://doi.org/10.3390/agronomy10060861

Salem EMM, Kenawey MKM, Saudy HS, Mubarak M (2022) Influence of silicon forms on nutrients accumulation and grain yield of wheat under water deficit conditions. Gesunde Pflanzen 74:539–548. https://doi.org/10.1007/s10343-022-00629-y

Saudy HS, Salem EMM, Abd El-Momen WR (2023) Effect of potassium silicate and irrigation on grain nutrient uptake and water use efficiency of wheat under calcareous soils. Gesunde Pflanzen 75:647–654. https://doi.org/10.1007/s10343-022-00729-9

Schmid MW, van Moorsel SJ, Hahl T et al (2021) Effects of plant community history, soil legacy and plant diversity on soil microbial communities. J Ecol 109:3007–3023. https://doi.org/10.1111/1365-2745.13714

Seyed Hajizadeh H, Azizi S, Rasouli F, Kaya O (2023) Evaluation of nano-silicon efficiency on compatible solutes and nutrient status of Damask rose affected by in vitro simulated drought stress. Chem Biol Technol Agric 10:22. https://doi.org/10.1186/s40538-023-00397-5

Shankar S, Joshi S, Srivastava RK (2023) A review on heavy metal biosorption utilizing modified chitosan. Environ Monit Assess 195:1350. https://doi.org/10.1007/s10661-023-11963-7

Shao JF, Che J, Yamaji N et al (2017) Silicon reduces cadmium accumulation by suppressing expression of transporter genes involved in cadmium uptake and translocation in rice. J Exp Bot 68:5641–5651. https://doi.org/10.1093/jxb/erx364

Sharma B, Kumawat KC, Tiwari S et al (2023) Silicon and plant nutrition—dynamics, mechanisms of transport and role of silicon solubilizer microbiomes in sustainable agriculture: a review. Pedosphere 33:534–555. https://doi.org/10.1016/j.pedsph.2022.11.004

Singhal RK, Fahad S, Kumar P et al (2023) Beneficial elements: new players in improving nutrient use efficiency and abiotic stress tolerance. Plant Growth Regul 100:237–265. https://doi.org/10.1007/s10725-022-00843-8

Solanki MK, Wang F-Y, Wang Z et al (2019) Rhizospheric and endospheric diazotrophs mediated soil fertility intensification in sugarcane-legume intercropping systems. J Soils Sediments 19:1911–1927. https://doi.org/10.1007/s11368-018-2156-3

Song A, Li Z, Liao Y et al (2021a) Soil bacterial communities interact with silicon fraction transformation and promote rice yield after long-term straw return. Soil Ecol Lett 3:395–408. https://doi.org/10.1007/s42832-021-0076-4

Song A, Li Z, Wang E et al (2021b) Supplying silicon alters microbial community and reduces soil cadmium bioavailability to promote health wheat growth and yield. Sci Total Environ 796:148797. https://doi.org/10.1016/j.scitotenv.2021.148797

Souri Z, Khanna K, Karimi N, Ahmad P (2021) Silicon and plants: current knowledge and future prospects. J Plant Growth Regul 40:906–925. https://doi.org/10.1007/s00344-020-10172-7

Sugiyama A, Ueda Y, Zushi T et al (2014) Changes in the bacterial community of soybean rhizospheres during growth in the field. PLoS One 9:e100709. https://doi.org/10.1371/journal.pone.0100709

Sun S, Yang Z, Song Z et al (2022) Silicon enhances plant resistance to Fusarium wilt by promoting antioxidant potential and photosynthetic capacity in cucumber (*Cucumis sativus* L.). Front Plant Sci 13:10.3389/fpls.2022.1011859

Sunera A, Saqib S et al (2020) Characterization and phytostimulatory activity of bacteria isolated from tomato (*Lycopersicon esculentum* Mill.) rhizosphere. Microb Pathog 140:103966. https://doi.org/10.1016/j.micpath.2020.103966

Teixeira GCM, de Prado RM, Rocha AMS et al (2022) Action of silicon on the activity of antioxidant enzymes and on physiological mechanisms mitigates water deficit in sugarcane and energy cane plants. Sci Rep 12:17487. https://doi.org/10.1038/s41598-022-21680-9

Thorne SJ, Stirnberg PM, Hartley SE, Maathuis FJM (2022) The ability of silicon fertilisation to alleviate salinity stress in rice is critically dependent on cultivar. Rice 15:8. https://doi.org/10.1186/s12284-022-00555-7

Tubana BS (2023) Dynamics of silicon in soil and plant to establish silicate fertilization. In: Benefits of silicon in the nutrition of plants. Springer International Publishing, Cham, pp 57–73

Tyczewska A, Woźniak E, Gracz J et al (2018) Towards food security: current state and future prospects of agrobiotechnology. Trends Biotechnol 36:1219–1229. https://doi.org/10.1016/j.tibtech.2018.07.008

Venâncio JB, da Dias NS, de Medeiros JF et al (2022) Yield and morphophysiology of onion grown under salinity and fertilization with silicon. Sci Hortic 301:111095. https://doi.org/10.1016/j.scienta.2022.111095

Verma KK, Song X-P, Li D-M et al (2020) Interactive role of silicon and plant–rhizobacteria mitigating abiotic stresses: a new approach for sustainable agriculture and climate change. Plan Theory 9:1055. https://doi.org/10.3390/plants9091055

Verma KK, Song X-P, Li D-M et al (2022) Silicon and soil microorganisms improve rhizospheric soil health with bacterial community, plant growth, performance and yield. Plant Signal Behav 17. https://doi.org/10.1080/15592324.2022.2104004

Verma KK, Song X-P, Singh M et al (2023) Association of silicon and soil microorganisms induces stress mitigation, increasing plant productivity. In: Benefits of silicon in the nutrition of plants. Springer International Publishing, Cham, pp 299–328

Viljoen JJF, Labuschagne N, Fourie H, Sikora RA (2019) Biological control of the root-knot nematode *Meloidogyne incognita* on tomatoes and carrots by plant growth-promoting rhizobacteria. Trop Plant Pathol 44:284–291. https://doi.org/10.1007/s40858-019-00283-2

Wan W, Tan J, Wang Y et al (2020) Responses of the rhizosphere bacterial community in acidic crop soil to pH: changes in diversity, composition, interaction, and function. Sci Total Environ 700:134418. https://doi.org/10.1016/j.scitotenv.2019.134418

Wang L, Cai K, Chen Y, Wang G (2013) Silicon-mediated tomato resistance against *Ralstonia solanacearum* is associated with modification of soil microbial community structure and activity. Biol Trace Elem Res 152:275–283. https://doi.org/10.1007/s12011-013-9611-1

Wang Q, Jiang X, Guan D et al (2018) Long-term fertilization changes bacterial diversity and bacterial communities in the maize rhizosphere of Chinese Mollisols. Appl Soil Ecol 125:88–96. https://doi.org/10.1016/j.apsoil.2017.12.007

Wang B, Chu C, Wei H et al (2020a) Ameliorative effects of silicon fertilizer on soil bacterial community and pakchoi (*Brassica chinensis* L.) grown on soil contaminated with multiple heavy metals. Environ Pollut 267:115411. https://doi.org/10.1016/j.envpol.2020.115411

Wang Z, Li Y, Li T et al (2020b) Tillage practices with different soil disturbance shape the rhizosphere bacterial community throughout crop growth. Soil Tillage Res 197:104501. https://doi.org/10.1016/j.still.2019.104501

Wang L, Gao Y, Jiang N et al (2022) Silicon controls bacterial wilt disease in tomato plants and inhibits the virulence-related gene expression of *Ralstonia solanacearum*. Int J Mol Sci 23:6965. https://doi.org/10.3390/ijms23136965

Wang B, Xiao L, Xu A et al (2023a) Silicon fertilization enhances the resistance of tobacco plants to combined cd and Pb contamination: physiological and microbial mechanisms. Ecotoxicol Environ Saf 255:114816. https://doi.org/10.1016/j.ecoenv.2023.114816

Wang X-L, Tian S-S, Yu H et al (2023b) Enhanced post-drought compensatory growth and water utilization in maize via rhizosphere soil nitrification by heterotrophic ammonia-oxidizing bacteria. Water (Basel) 15:3933. https://doi.org/10.3390/w15223933

Wen T, Yu G-H, Hong W-D et al (2022) Root exudate chemistry affects soil carbon mobilization via microbial community reassembly. Fundament Res 2:697–707. https://doi.org/10.1016/j.fmre.2021.12.016

Xun W, Liu Y, Li W et al (2021) Specialized metabolic functions of keystone taxa sustain soil microbiome stability. Microbiome 9:35. https://doi.org/10.1186/s40168-020-00985-9

Yadav A, Kumar S, Verma R et al (2023) Overexpression of PGPR responsive chickpea miRNA166 targeting ATHB15 for drought stress mitigation. Plant Cell Tissue Organ Cult 154:381–398. https://doi.org/10.1007/s11240-023-02458-x

Yan Y, Kuramae EE, de Hollander M et al (2017) Functional traits dominate the diversity-related selection of bacterial communities in the rhizosphere. ISME J 11:56–66. https://doi.org/10.1038/ismej.2016.108

Yasin MU, Hannan F, Munir R et al (2024) Interactive mode of biochar-based silicon and iron nanoparticles mitigated Cd-toxicity in maize. Sci Total Environ 912:169288. https://doi.org/10.1016/j.scitotenv.2023.169288

Yin C, Casa Vargas JM, Schlatter DC et al (2021) Rhizosphere community selection reveals bacteria associated with reduced root disease. Microbiome 9:86. https://doi.org/10.1186/s40168-020-00997-5

Yu Y, Zhang L, Li Y et al (2022) Silicon fertilizer and microbial agents changed the bacterial community in the consecutive replant soil of lilies. Agronomy 12:1530. https://doi.org/10.3390/agronomy12071530

Yu H, Li C, Yan J et al (2023) A review on adsorption characteristics and influencing mechanism of heavy metals in farmland soil. RSC Adv 13:3505–3519. https://doi.org/10.1039/D2RA07095B

Yuan Z, Pang Z, Fallah N et al (2022) Silicon fertilizer mediated structural variation and niche differentiation in the rhizosphere and endosphere bacterial microbiome and metabolites of sugarcane. Front Microbiol 13. https://doi.org/10.3389/fmicb.2022.1009505

Zerrouk IZ, Rahmoune B, Khelifi L et al (2019) Algerian Sahara PGPR confers maize root tolerance to salt and aluminum toxicity via ACC deaminase and IAA. Acta Physiol Plant 41:91. https://doi.org/10.1007/s11738-019-2881-2

Zhang L, Guan Y (2022) Consistent responses of soil bacterial communities to bioavailable silicon deficiency in croplands. Geoderma 408:115587. https://doi.org/10.1016/j.geoderma.2021.115587

Zhou X, Shen Y, Fu X, Wu F (2018) Application of sodium silicate enhances cucumber resistance to Fusarium wilt and alters soil microbial communities. Front Plant Sci 9. https://doi.org/10.3389/fpls.2018.00624

Chapter 9
Silica and Silica Nanoparticles: An Approach to Biogenic Synthesis and Their Main Applications

Liliam Katsue Harada, Mariana Guilger-Casagrande, Tais Germano-Costa, Natália Bilesky-José, Leonardo F. Fraceto, and Renata Lima

Abstract Silica is an inert material, and it is interesting due to its applications as a nanomaterial in many areas. Silica nanoparticles (SiO_2-NPs) have silanol (Si-OH) surface groups. These groups enable functionalization and the development of several modified and hybrid materials with excellent performance. Although silica is a versatile and relatively safe material, SiO_2-NPs obtained through synthetic processes may be considered high cost and environmentally damaging. In contrast, green methods and biogenic synthesis routes are promising options to replace synthetic methods, showing countless advantages ranging from low cost to environmental sustainability. Due to the growing interest in silica nanoparticles, the methods of synthesis, and the optimization of existing processes, studies on this nanomaterial are essential for better comprehending its potential and exploring alternative methods for the green synthesis of these nanoparticles.

Keywords Silicon · Nanosilica · Nanoparticle · Synthetic processes · Biogenic synthesis

---

L. K. Harada · T. Germano-Costa · N. Bilesky-José · R. Lima (✉)
Laboratory for Evaluation of the Bioactivity and Toxicology of Nanomaterials, University of Sorocaba (UNISO), Sorocaba, São Paulo, Brazil

M. Guilger-Casagrande
Laboratory of Environmental Nanotechnology, São Paulo State University (UNESP), Sorocaba, São Paulo, Brazil

Laboratory for Evaluation of the Bioactivity and Toxicology of Nanomaterials, University of Sorocaba (UNISO), Sorocaba, São Paulo, Brazil

L. F. Fraceto
Laboratory of Environmental Nanotechnology, São Paulo State University (UNESP), Sorocaba, São Paulo, Brazil

R. de Mello Prado et al. (eds.), *Silicon Advances for Sustainable Agriculture and Human Health*, Sustainable Plant Nutrition in a Changing World,
https://doi.org/10.1007/978-3-031-69876-7_9

9.1 Introduction

Silicon (Si) is the second most abundant element in nature. It occurs in combination with oxygen and other elements, such as silicas (SiO_2) and silicates (compounds which contain oxygen and metals), along with oxygen composing 50–70% of soil mass (Epstein 1994; Sommer et al. 2006; Ma and Yamaji 2006; Gomes et al. 2018). It mainly occurs in the soils as silicate minerals (SiO_3^{-2}) and silicon dioxide (SiO_2). It occupies, on average, more than 25% of soil mass in a range of 0.5 and 47% (Tubana et al. 2016). In the environment, Si is bound to oxygen (O_2), forming the silicon dioxide compound (or silica), which, when solubilized in water, forms monosilicic or orthosilicic acid (H_4SiO_4). This acid is absorbable and potentially bioavailable for organisms showing biological activity. Silicates (SiO_4^{2-}) mainly occur in sand or quartz and show an ionic group as a main anion commonly associated with a metal (Ma et al. 2007).

Silica is released into the soil through chemical and biological processes. Plants have a significant contribution to the biogeochemical cycle of silica since they absorb silicon from the soil in the form of silicic acid (hydrosoluble form), which is polymerized and precipitated as amorphous silica and returns to the soil as these plants die. Microorganisms such as actinomycetes, bacteria, and fungi also contribute to the global cycling of silicon (Bansal et al. 2006). As a material, silica is an essential source of solid mesoporous matrices and hollow microspheres (Bansal et al. 2005).

Orthosilicic acid (H_4SiO_4) may be employed in food supplementation. According to De Medeiros et al. (2020), there is no minimum or maximum recommended concentration. A daily intake of 10–25 mg is suggested, with possible variations according to population, age, and others. Regulatory organizations such as the FDA (Food and Drug Administration) and EFSA (European Food Safety Authority) consider the daily intake of 1.500 mg as a safe oral administration dosage of silicates and amorphous silica (Diab et al. 2017). It is worth noting that Si is the human body's third most abundant trace element. It comprises different tissues, such as nails and hair. It is actively involved in bone mineralization, prevention of osteoporosis, collagen synthesis, prevention of skin aging, reduction of the risk of atherosclerosis and Alzheimer's disease, and other biological effects (Russell 2010; Nielsen 2014).

Reffitt et al. (2003) showed that orthosilicic acid stimulated the synthesis of type I collagen in human osteoblast-like cells and promoted osteoblastic differentiation, suggesting a possible link with the regulation of prolyl hydroxylase activity. Years later, Dong et al. (2016) concluded that silicon stimulates the synthesis of COL-1 and osteocalcin in human osteoblast-like cells through the BMP-2/Smad1/5/RUNX2 signaling pathway. Accordingly, orthosilicic acid may be a promising therapeutical agent for human healthcare since its potential biological and therapeutical effects have been reported, as in the conditions above (Jurkić et al. 2013).

Silica is a relatively safe material with many applications and may be used in several sectors. Therefore, the importance of the studies involving silica can be related to its applications and the optimization of the obtaining processes. In this context, this study briefly reviews the obtaining methods and the use of silica and silica nanoparticles, emphasizing the biogenic synthesis of these materials.

9.2 Silicon and Plants

All plants have Si in their composition. Although it is not an essential element, it occurs in the range of 0.1 and 10% of the dry weight (Ma and Yamaji 2006; Cornelis et al. 2010; Liang et al. 2015; Sahebi et al. 2015), varying according to the plant species (Currie and Perry 2007). The roots absorb silicon in the form of silicic acid and mainly transport it to the leaves. In plant transpiration, water flows out, and silicon is deposited on the outer wall of epidermal cells in silica gel (Ma and Yamaji 2006; Carneiro et al. 2015).

Throughout history, there have been several evidences that the presence of Si enhances plant growth and contributes to the control of pathogens (Sommer 1926; Raleigh 1939; Jones and Handreck 1967; Epstein 1994; Hattori et al. 2005; Fauteux et al. 2006). High levels of Si in plants promote a faster and more efficient response of hypersensitivity to pathogens (Fauteux et al. 2006; Bélanger et al. 2003). Silicon also contributes to plant physiology in the tolerance of abiotic stress (Liang et al. 2015), absorptions that promote the increase of plant resistance (suberization and lignification), and the reduction of disease proliferation and attacks of pests (Fleck et al. 2011; Gomes et al. 2018). Hence, silicic acid has an important role in stress responses (Van Bockhaven et al. 2013). This effect may be triggered by free and non-bonded silicic acid (Bauer et al. 2011). However, this element was only recently officially considered a beneficial element. It provides a higher resiliency to many plants against biotic and abiotic environmental stresses (Coskun et al. 2019).

Orthosilicic acid (H_4SiO_4) is the plant-available form of Si that is absorbed through the Si transporters LSi1(influx) and LSi2 (efflux) from the roots to the xylem, where it accumulates (Fig. 9.1) (Raven 1983; Ma et al. 2007; Ma 2010). Orthosilicic acid (H_4SiO_4) is transported out of the xylem via LSi6, polymerized, and transformed into amorphous silica ($Si_2.nH_2O$), which accumulates in the cell wall, specific cells, cell lumens, and intercellular spaces (Sangster 1978; Mukarram et al. 2022). This deposition of $Si_2.nH_2O$ in plant tissues is known as silica bodies or phytoliths (Ma et al. 2001, 2011; Ma and Yamaji 2015; Mandlik et al. 2020). According to Zexer and Elbaum (2020), polymerization only occurs in specific regions of the plant. Kumar et al. (2020) reported a single protein in sorghum leaves, the Siliplant1 (Slp1), which acts in silica precipitation.

Deshmukh et al. (2020) recently identified several transporter proteins, including the group III aquaporins, such as nodulin 26 (NIP-III). The NIPs are transporter proteins from the plant kingdom responsible for the uptake, translocation, or extrusion of several metalloid compounds. They are a transmembrane transport pathway for metalloids such as Si. These transporters are also studied due to the possibility of absorption of toxic metalloids, contributing to the bioremediation of metals. Among the NIPs, NIP-III transports the widest range of substrates (Pommerrenig et al. 2015).

Studies conducted with plants aimed at expanding knowledge on Lsi1 and Lsi2 contribute to the comprehension of transport mechanisms, Si accumulation in plants, and its effects on different activities. The Lsi2 seems to act in combination with Lsi1, as shown in studies performed with corresponding proteins in rice

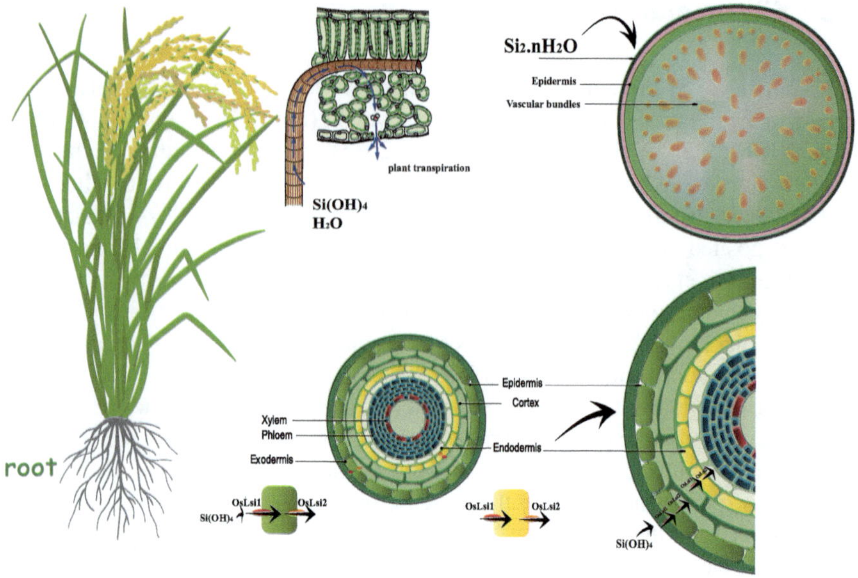

Fig. 9.1 Schematic representation showing silica transport inside the plant. (Adapted from Mitani-Ueno and Ma 2021; Coskun et al. 2021)

(OsLsi2 and OsLsi1) (Deshmukh et al. 2020; Sun et al. 2018, 2020). In these studies, OsLsi2 acts as an efflux transporter in combination with OsLsi1, transporting Si from the roots to the stem. Although more studies are required, there is a hypothesis that the "coordinated system" of OsLsi1 and OsLsi2 may be an exclusive mechanism of rice, explaining the higher content of Si in the tissues compared to other plant species (Ma and Yamaji 2015; Barberon and Geldner 2014; Raggi et al. 2020). Silicon also demonstrated an influence on the expression of the *OsZEP* gene, which is upregulated when the plant is exposed to salinity stress. There was downregulation of *OsZE* 24 h after plant exposure to Si, indicating that Si may mitigate the negative effects of salinity stress over time. Another observation pointed out that other genes involved in the biosynthesis of ABA, such as 9-cis-epoxycarotenoid dioxygenase (NCED1, NCED3, and NCED4), showed an expression pattern similar to *OsZEP and a downregulation with the increase of Si concentration* (Abdel-Haliem et al. 2017).

Due to the presence of Si in all the plants and the knowledge that this element is obtained from the environment, there is a classification based on Si content. There are three different groups of plants based on their absorption and tissue accumulation capacity: Si accumulators (active absorption of Si that triggers a decrease in the concentration of Si in soil solution; rice, wheat, millet, sugar cane), Si intermediates or non-accumulators (Si passive uptake), and Si excluders (discard Si, soybean) (Takahashi et al. 1990; Cornelis et al. 2010). Therefore, there are passive accumulator plants (which accumulate silica according to the volume of water in the transpiration flow) and those plants that uptake silica actively from plant tissue (Raven 1983; Mayland et al. 1991).

The accumulation of Si in the plants may improve its capacity to respond to hydric stress. Predictions of global environmental changes point to an increase in both the severity and frequency of water stress in the near future. Therefore, the application of Si and its performance in field conditions may be one of the solutions to overcome the problem. However, further extensive studies are required (Chen et al. 2018).

The resistance to diseases induced by Si may occur due to the physical barrier formed by the deposition of Si in plant cuticles, which mechanically avoids the penetration of fungi and interrupts the infection process. Silicon also stimulates chitinase activity and the fast activation of peroxidases and polyphenoloxidases after fungal infection (Chérif et al. 1994). Phenolic glycosides extracted from plants supplemented with Si have shown potential antifungal activity when submitted to acid hydrolysis or ß-glucosidase. Dann and Muir (2002) reported that pea plantlets (*Pisum sativum* L.) supplemented with potassium silicate had shown an increase in chitinase and ß-1.3-glucanase activity against the fungal disease caused by *Mycosphaerella pinodes*. Furthermore, the leaves of pea plantlets supplemented with silicon showed fewer injuries than non-exposed plantlets. The deposition of Si on plant epidermal layers also acts as a physical barrier against insects, given that leaves with tissues containing Si are more difficult to be attacked by sucking pests and caterpillars.

Jinger et al. (2017) showed that using Si fertilizers may reduce leaf lesions caused by foliar bacterial pests by 5–22%, decreasing the soluble sugar in the leaves due to the presence of silicon. However, the mechanisms involved in the beneficial effects of Si are not well known yet, and more investigations are required, mainly regarding the molecular and biochemical mechanisms involved in the regulation of water absorption by Si.

9.3 Structure of Silica or Silica Dioxide (SiO_2)

Silica presents two possible forms: crystalline and amorphous. Crystalline forms commonly occur in sands and quartz. Meanwhile, amorphous forms occur in living organisms. These two forms are very different, and it is not possible to make a practical comparison, only in theory. In crystalline silica, the silicon atom is bonded to four adjacent atoms in a tetrahedral molecular geometry (Fig. 9.2), forming a well-defined tetrahedral structure. In turn, amorphous silica does not present an organized form (Fig. 9.2) (Kang et al. 2021), and, in addition to its presence in plants such as rice and sugar cane, it may occur in other organisms such as diatoms, which are formed by soluble silica. In general, the crystalline form shows toxicity and results from the exposure of amorphous silica to high temperatures (Waddell and Evans 2000; Gomes et al. 2018).

Amorphous silica is formed through the condensation of H_4SiO_4 monomers into Si-O-Si bonds through polymerization in bacteria or plants, where the silica nuclei reach a critical size (0.5–2.0 nm). Amorphous silica can be removed and used for different purposes.

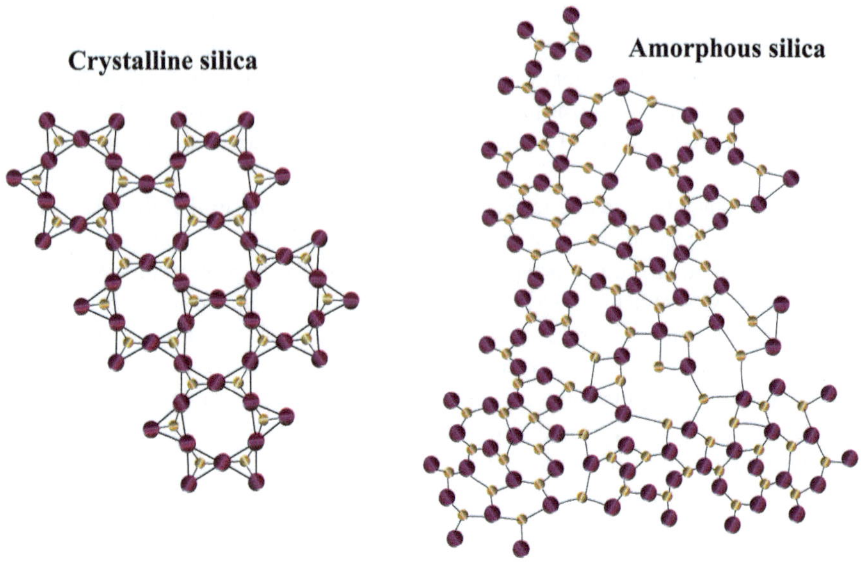

Fig. 9.2 Schematic representation showing the main arrangements of Si

There are different ways to remove amorphous silica from plants. Biosilicification by living organisms is one way of doing so. It occurs in mild physiological conditions and produces a variety of nanostructures. The biosilicification mediated by fungi, such as *Aspergillus parasiticus* or *Fusarium oxysporum*, is an efficient and sustainable extraction process. However, the obtained nanosilica includes proteins associated with silica structures, and its final form is crystalline (Zielonka et al. 2018; Bansal et al. 2006).

Mesoporous silica (MS) is formed by highly ordered silica junctions in different structures, which provide a wide geometric range of pores with diameters from 2 to 50 nm and may be mainly hexagonal, spherical, and cubic. The different arrangements of the pores, which depend on the interaction of surfactant and silica, determine the difference between mesoporous silicas. For example, MCM-41 has spherical pores due to the interaction of cationic surfactants. Meanwhile, SBA-15 has hexagonal pores due to amphiphilic copolymers acting as a directing agent for the structure's organic portion (Shamim et al. 2022).

In recent years, mesoporous silica-based materials have shown great potential for different applications due to their single properties, such as biocompatibility, mechanical stability because of the presence of siloxane groups (Si-O-Si), the possibility of adjusting pore size, high surface area, and surface reactivity (Gartmann et al. 2010; Sattary et al. 2020). These NPs may be obtained through green synthesis using plants, agricultural waste, industrial waste, fungi, bacteria, yeasts, clay minerals, and actinomycetes, among others (Jeelani et al. 2020).

9.4 Obtaining and Using Silica

Synthetic silicas may be obtained using aqueous solutions of dissociated silicic acid or from the vaporization of silicon compounds, such as silicon tetrachloride ($SiCl_4$), tetraethyl orthosilicate (TEOS), tetramethyl orthosilicate (TMOS), and sodium silicate (Na_2SiO_3) (Gomes et al. 2018; Jeelani et al. 2020).

Synthetic methods can obtain three classes of amorphous silica: pyrogenic silica, precipitated silica, and silica gel. The pyrogenic silica is obtained through the pyrolysis of silicon tetrachloride in the presence of hydrogen and oxygen. They are commonly used as anti-caking agents for powder products and are considered environmentally friendly and safe for human health. Precipitated silica is obtained through the combination of sand and sodium carbonate or caustic soda, forming sodium silicate, which is then treated with sulfuric acid to precipitate the silica and eliminate products by filtration. The obtained silica is submitted to a washing, drying, and storage process. In the food industry, silicas are used as anti-caking and binder agents for concrete and ceramic materials and as an adsorbent of impurities in liquid and gaseous systems. Silica gel is synthesized by the acidification of aqueous solutions of sodium silicate, producing a gelatinous precipitate that is washed and dehydrated. It has a highly porous structure and is usually used as a drying agent and anti-caking agent in the food industry (Gomes et al. 2018). However, synthetic silicas show low biocompatibility (Alshatwi et al. 2015).

Biogenic silicas are produced in living organisms such as rice, cacti, algae, sugarcane, and the exoskeleton of some animals. Furthermore, they present silica gel, which is formed through the aggregation of small particles deposited on epidermal cells of vegetal species (Gomes et al. 2018). This type of silica may be obtained through the biomass extraction method. For example, the silica obtained from rice husk is usually extracted through chemical treatment by acid leaching with 10% chloridric acid to remove metallic impurities, followed by heat treatment. Eliminating other compounds makes it possible to obtain a high amorphous silica content (Della et al. 2006; Sharma et al. 2022). Moreover, depending on the applied thermic cycle, amorphous and crystalline silica may be obtained (Zou and Yang 2019).

As an example of the variety of applications of silica, Damavandi et al. (2021) developed a capturing and enhancing system in nucleic acid probes to optimize the capture of low detection targets for TaqMan qPCR. This system comprises iron magnetic nanoparticles capped with a layer of inert silica, which protects the nucleus and enables the functionalization of the probe-binding surface. According to the authors, the capping of the nanoparticles with inert silica reduces the non-specific interactions, making it the appropriate capping. Garcia-Rios et al. (2018) investigated the immobilization of essential oils on the surface of different commercial supports, including the amorphous silica microparticles. They obtained a possible antimicrobial agent used as an adjuvant in the vinification process, which contributes to wine preservation. It prevents and optimizes the inhibitory activity of the bioactive agents (oils), showing superior results against known bacteria involved in wine deterioration compared to free bioactive compounds.

9.5 Advantages of Using Nanoscale Silica

Studies on nanoscale materials have increased, mainly due to their chemical stability properties, surface adsorption, and varied sizes, types, and morphologies (Goswami and Mathur 2019; Mitra et al. 2023). According to Mitra et al. (2023), due to their excellent physical and chemical characteristics, these nanoparticles show remarkable properties in the environment. Nevertheless, relatively few research laboratories consider these nanoparticles sustainable for agricultural applications.

The sol-gel method, proposed by Stöber et al. (1968), which employs TEOS as a raw material, triggered the formation of silica microparticles with uniform size. From there on, different methods and precursors were used to prepare nanosilica (Azlina et al. 2016).

Silica nanoparticles show unique characteristics, such as easy obtainment (synthesis), easy surface modifications, porosity adjustment, and great biocompatibility, which make them extremely relevant for the research and development of new materials with applications in several areas (Li et al. 2021; Yaakov et al. 2022). Tetraethyl orthosilicate (TEOS) and tetramethyl orthosilicate (TMOS) are usually employed for the chemical obtainment of nanoparticles by the sol-gel method (Sharma et al. 2022). Microemulsion, precipitation, and gaseous phase methods, which use silicon tetrachloride as raw material, are also used to produce the nanoparticles and show excellent reinforcing properties.

The microemulsion method enables the obtainment of nanoparticles lower than 100 nm, allowing size changes according to alterations of the parameters applied in the production process.

Regarding the precipitation method, the nanoparticles may be obtained after drying or calcinating the resulting product (Li et al. 2021). These nanoparticles may also be obtained from sand through treatment and calcination (Aisvarya et al. 2023).

Mesoporous silica (MS) is formed by silica junctions highly organized in different structures, which provides a wide geometric range of pores. The TEOS compound is commonly used as raw material in these nanomaterials. The MS is a promising alternative for the composition of nanotransporters of beneficial and interesting molecules due to its properties, such as the high surface area and the number of pores, which contribute to the accessibility, the loading, and the homogeneous release of these active compounds (Venkatesan et al. 2003; Ahern et al. 2013).

Elmotasem et al. (2023) encapsulated benzocaine in mesoporous silica nanoparticles and incorporated them in hydrogels based on polysaccharides to obtain a delivery system of topical anesthetic that prolonged the release and the action of the active compound.

The scientific literature has already reported the benefits of silicon nanoparticles for agricultural applications, such as nanoherbicides and pesticides, nanofertilizers, and nanosensors (Mukarram et al. 2022). Sattary et al. (2020) synthesized mesoporous silica nanoparticles using cetyl trimethyl ammonium bromide (CTAB) and TEOS and encapsulated lemon grass and clove essential oils for the application as nanopesticide against *Gaeumannomyces graminis* var. *tritici*, which causes damages to wheat plants.

Delaney et al. (2011) studied the application of mesoporous silica doped with titanium, iron, aluminum, and zirconium to remove phosphate ions from contaminated water by adsorption.

The possibility of surface modification shown by these NPs enables its potential applications in the construction sector, in bitumen mixtures to optimize thermodynamic properties to produce asphalt (Mousavi and Fini 2020).

In general, the synthetic methods to obtain SiO_2-NPs may be considered relatively expensive due to the high costs of raw materials. They may cause environmental damage and a certain concern regarding toxicity (Sharma et al. 2022). Although microemulsion and precipitation are considered simpler and easier to handle, these methods result in the production of NPs with lower purity when compared to other methods due to the presence of organic compounds generated during the process and a higher aggregation (Li et al. 2021). Accordingly, the methods that propose obtaining green and sustainable SiO_2-NPs provide encouraging characteristics to produce nanomaterials with target properties (Karande et al. 2021).

9.6 Biogenic Synthesis of Nanoparticles

Biological systems for the synthesis of nanoparticles offer a sustainable, efficient, and low-cost method using living microorganisms, such as fungi, bacteria, and plants. Table 9.1 shows some advantages of using these agents.

Regarding metallic particles, the synthesis may occur by bioreduction, where the metal ion is chemically reduced into its most biologically stable form. The synthesis may also occur by biosorption, where the metallic ions of a sample are bound to the microorganism (e.g., cell wall). The binding between the metal ion and the modified cell wall, or the peptides expressed by some fungi, bacteria, or plants, may generate stable complexes in the form of nanoparticles (Pantidos and Horsfall 2014). Regarding oxides such as silica, the process of formation of the nanoparticles may occur in two steps: silica extraction (biolixiviation) in the form of silicic acid mediated by proteins from the microorganism and the hydrolysis of the silicate complex by other specific proteins, giving rise to stable particles which are capped with the same proteins (Bansal et al. 2005, 2006). Regarding the use of plants, bioactive compounds play an important role in the reduction process for the formation and stabilization of SiO_2NPs (Sankareswaran et al. 2022). Chakraborty et al. (2022) also reported the synthesis of metallic nanoparticles using yeasts and viruses.

9.7 Synthetic Nanoparticles x Biogenic Nanoparticles

The reactions for the production of nanoparticles by physical and chemical methods, in general, require high temperatures, resulting in excessive energy expenditure and the use of chemical substances, which may present toxicity (Pantidos and

Table 9.1 Biological systems used in the biogenic synthesis of nanoparticles and main advantages (Pantidos and Horsfall 2014)

Agent	Advantage
Bacteria	Easy and low-cost manipulation High growth rate compared to other agents
Fungi	Higher protein secretion which may increase the nanoparticle concentration Better performance for large-scale production in reactors when compared to bacteria
Plants	Isolated extracts or the whole plant may be used Absence of pathogenicity

Horsfall 2014). The synthesis of SiO_2NPs by chemical methods may also cause environmental damage and depend on relatively high production costs, in addition to the requirement of extreme conditions of temperature, pressure, and pH (Bansal et al. 2005; Sharma et al. 2022). Despite its wide application, silica nanoparticles still present some problems related to biosecurity. Cytotoxicity and genotoxicity studies have shown adverse results for these materials (Lin et al. 2006; Tang et al. 2012; Alshatwi et al. 2015; Li et al. 2021). Silicas obtained from synthetic materials and natural quartz have higher toxicity risks for human health (Napierska et al. 2010).

In contrast, biosilicification by the activity of living organisms occurs in physiological conditions, producing complex nanostructures of biogenic silica and diversified structures (Bansal et al. 2005), resulting in a safe, environmentally friendly, and low-cost synthesis. Furthermore, these biogenic SiO_2-NPs may be obtained from wastes with high content of silica, such as rice husk, which are an alternative to synthetic SiO_2-NPs, giving rise to porous and polar materials with high adsorption capacity (Liou et al. 2022; Shrestha et al. 2023).

Porrang et al. (2021) found promising results associated with the cytotoxicity of biogenic silica nanoparticles, highlighting the biocompatibility of these nanomaterials. Therefore, as the main advantage, biogenic silica nanoparticles show biocompatibility for the interaction with the targets, being potentially non-toxic for the environment and cells/non-target organisms, besides its production by a sustainable method.

Moreover, Guilger-Casagrande et al. (2021) found that the capping formed during the biogenic synthesis of metallic nanoparticles using fungi (e.g., *Trichoderma harzianum*) may contribute to this nanomaterial's biological activity.

9.8 Methods Used in the Biogenic Synthesis of Silica Nanoparticles

There are three approaches regarding the association between the obtainment of silica nanoparticles and the biogenic methods: synthesis of nanoparticles from biogenic silica (Umeda and Kondoh 2010; Alshatwi et al. 2015; Schneider et al. 2019; Ariaeenejad et al. 2020), biogenic synthesis of nanoparticles from materials which provide non-biogenic silica (Bansal et al. 2005), and biogenic synthesis of nanoparticles using biogenic silica as a precursor (Bansal et al. 2006).

The synthesis of SiO_2-NPs using biogenic silica shows as an advantage the possibility of biological extraction of these NPs from agriculture wastes with high content of silica, such as rice husk and straw, sugarcane bagasse, and oat husk (Schneider et al. 2020; Yadav et al. 2022). Alshatwi et al. (2015) performed the synthesis of silica NPs through the acid digestion of rice husk using chloridric acid at 120 °C, followed by calcination, and it was possible to observe that the applied temperature influenced the formation of the nanoparticles since the size decreased with the temperature increase. The use of different concentrations of acid may reduce the metallic impurities present in the extraction by acid lixiviation (Umeda and Kondoh 2010). Shrestha et al. (2023) successfully obtained mesoporous SiO_2-NPs by a technique similar to acid lixiviation followed by pyrolysis at 600 °C for the application as a stationary phase in column chromatography. Porrang et al. (2021) also obtained NPs from rice and wheat husks, which showed high potential for application as drug carriers.

Bansal et al. (2005) synthesized silica NPs using white sand as a precursor and the fungus *Fusarium oxysporum* as an agent of biotransformation. The study suggested that the silica complex is hydrolyzed by the action of specific proteins or enzymes secreted by the fungi, resulting in crystalline nanoparticles capped with stabilizing proteins. In the same study, a calcination process was performed at 400 °C for 2 h to eliminate proteins from the structure, obtaining hollow particles. Then, a new study addressed the same fungus species. However, it used rice husk as a silica precursor, which provides amorphous silica due to its natural origin. The studies indicated that some specific biomolecules released by the fungus initially act in the biolixiviation of amorphous silica from the rice husk and, subsequently, promote their biotransformation in silica crystalline particles (Bansal et al. 2006). Changes in the forms of silica particles may be associated with the proteins secreted by the fungus, which interact with the surface of rice husk (Pineda-Vasquez et al. 2014).

9.9 Potential Applications of Nanosilica

Both rice husk and rice ash show great potential for obtaining silica gel and aerogel (Zou and Yang 2019). Silicas obtained through the biosynthesis process have a wide range of applications in biomedicine, agriculture, environment, photocatalysis, and the development of sensors, among others (Jeelani et al. 2020; Yadav et al. 2022).

Alshatwi et al. (2015) obtained encouraging results regarding the toxicity of biogenic SiO_2-NPs. They evaluated the cell viability, morphological alterations, presence of intracellular ROS, and mitochondrial transmembrane potential of human pulmonary fibroblasts exposed to the nanoparticles. Furthermore, they confirmed the biocompatibility of this nanomaterial. This nanomaterial may be an excellent alternative for biomedical applications and as food additives, replacing synthetic silicas.

The presence of pores in mesoporous silicas is an important parameter related to release rates and the quantity of drugs delivered. Due to their porosity, SiO_2-NPs enable drug loading both in the inner and outer surfaces of the nanoparticles (Porrang et al. 2021; Diab et al. 2017).

In a study for the development of a biocompatible system for the delivery of an anticancer drug (Doxorubicin), Porrang et al. (2021) promoted the synthesis of biogenic silica nanoparticles from rice husk and wheat by the acid lixiviation method. Silica from diatoms is a biosilica that shows highly ordered pores. They are synthesized in the cell walls of eukaryotic unicellular algae. Some studies report investigations on its application, combined with orally administered drugs, as complex oral carriers (Diab et al. 2017). Ariaeenejad et al. (2020) developed a nanobiocatalyst based on biogenic mesoporous SiO_2-NPs to immobilize thermostable xylanase to optimize its stability and reuse. It was shown as a potential candidate for application in animal feeding and the hydrolysis of biomass of agricultural wastes.

9.10 Environment

The application of biogenic silica nanoparticles is a viable alternative to remove water pollutants and refine vegetal oils. As low-cost adsorbents, these nanoparticles promote the efficient removal of cadmium, lead, copper, nickel, mercury, cobalt, and zinc from residual waters and polluted air (Zou and Yang 2019; Yadav et al. 2022). Sharma et al. (2022) synthesized silica NPs from bamboo leaves. Then, they functionalized the surface with amines and carboxylic groups using APTES (3-aminopropyl triethoxysilane) and CPTES (3-cianopropyl trietoxysilane), respectively, for the application in the removal of hexavalent chromium (Cr(VI)) by adsorption. Liou et al. (2022) studied the capacity of mesoporous functionalized SiO_2-NPs obtained from rice husk ashes in the adsorption of tannic acid from water. It has shown a potential application for removing organic contaminants from residual waters.

9.11 Agriculture

Regarding the absorption of silicon by plants, studies have shown that this element may increase vegetal resistance, contributing to the decrease of disease proliferation and attacks of pests, improving the photosynthetic process, activating defense enzymes, and increasing the nutrient uptake by the plants in adverse conditions (Gomes et al. 2018; Wang et al. 2022).

Silica nanoparticles stimulate plant growth and may be applied as nanoherbicides, nanofertilizers, and nanopesticides (Wang et al. 2022). Sankareswaran et al. (2022) synthesized SiO_2-NPs using the aqueous extract of *Bryophyllum pinnatum* leaves as a reducing agent and TEOS as a silica precursor. The obtained nanoparticles were applied in a seed germination assay with *Vigna radiata.* The results showed a positive germination potential of the seeds treated with these nanoparticles, promoting an enhanced growth of the aerial part and an increase in the development of roots, even in low concentrations.

Food production has a considerable environmental impact, occasioned by forming solid residues and using agrochemicals, which present problems such as high solubility and poor soil retention. Considering both problems, Dorairaj et al. (2022) studied the synthesis of mesoporous silica nanoparticles obtained from rice husk without using surfactants to control the formation of the pores. According to the authors, these NPs may be applied to deliver target agrochemicals, decreasing possible environmental impacts.

9.12 Conclusion

The green synthesis of silica nanoparticles is a relatively simple and low-cost process that reduces the use of toxic solvents. Furthermore, it has an added value due to the possibility of reusing a waste material, such as rice husk, as a substrate for the synthesis. Reducing the use of chemical substances reduces the generation of residues and the use of high-cost and complex equipment and procedures.

Despite their great potential for applications in several areas, most studies on the green synthesis of SiO_2-NPs are performed on a laboratory scale. More investigations regarding large-scale applications and their consequences are necessary. However, the advantages regarding their biocompatibility and the possibility of combination with other materials are well-known. It may be one of the most versatile materials for future applications.

Acknowledgments This work was financed in part by the São Paulo State Research Support Foundation (FAPESP, grants, #2024/01693-4, #2022/03399-0, #2024/12652-7, #CEPID-CBioClima #2021/10639-5 and #2017/21004-5), Brazilian National Council for Scientific and Technological Development (CNPq - INCT NanoAgro #405924/2022-4) and Coordenação de Aperfeiçoamento de Pessoal de Nível Superior – Brasil (MEC-CAPES INCTNanoAgro #88887.953443/2024-00).

References

Abdel-Haliem ME, Hegazy HS, Hassan NS, Naguib DM (2017) Effect of silica ions and nano-silica on rice plants under salinity stress. Ecol Eng 99:282–289. https://doi.org/10.1016/j.ecoleng.2016.11.060

Ahern RJ, Hanrahan JP, Tobin JM, Ryan KB, Crean AM (2013) Comparison of fenofibrate–mesoporous silica drug-loading processes for enhanced drug delivery. Eur J Pharm Sci 50:400–409. https://doi.org/10.1016/j.ejps.2013.08.026

Aisvarya S, Kalyanasundaram M, Kannan M, Preetha S, Elango K, Srinivasan T, Govindaraju K (2023) Preparation and characterization of silica nanoparticles (sol–gel & river sand): a comparative toxicity studies against seed weevil (Sitophilus oryzae L). and effect on agro-morphological characteristics of maize seeds. Mater Sci Eng B 293:116503. https://doi.org/10.1016/j.mseb.2023.116503

Alshatwi AA, Athinarayanan J, Periasamy VS (2015) Biocompatibility assessment of rice husk-derived biogenic silica nanoparticles for biomedical applications. Mater Sci Eng 47:8–16. https://doi.org/10.1016/j.msec.2014.11.005

Ariaeenejad S, Jokar F, Hadian P, Ma'mani L, Gharaghani S, Fereidoonnezhad M, Salekdeh GH (2020) An efficient nano-biocatalyst for lignocellulosic biomass hydrolysis: Xylanase immobilization on organically modified biogenic mesoporous silica nanoparticles. Int J Biol Macromol 164:3462–3473. https://doi.org/10.1016/j.ijbiomac.2020.08.211

Azlina H, Hasnidawani J, Norita H, Surip S (2016) Synthesis of SiO_2 nanostructures using Sol-Gel method. Acta Phys Pol A 129:842–844. https://doi.org/10.12693/APhysPolA.129.842

Bansal V, Sanyal A, Rautaray D, Ahmad A, Sastry M (2005) Bioleaching of sand by the fungus Fusarium oxysporum as a means of producing extracellular silica nanoparticles. Adv Mater 17:889–892. https://doi.org/10.1002/adma.200401176

Bansal V, Ahmad A, Sastry M (2006) Fungus-mediated biotransformation of amorphous silica in rice husk to nanocrystalline silica. J Am Chem Soc 128:14059–14066. https://doi.org/10.1021/ja062113

Barberon M, Geldner N (2014) Radial transport of nutrients: the plant root as a polarized epithelium. Plant Physiol 166:528–537. https://doi.org/10.1104/pp.114.246124

Bauer P, Elbaum R, Weiss IM (2011) Calcium and silicon mineralization in land plants: transport, structure and function. Plant Sci 180:746–756. https://doi.org/10.1016/j.plantsci.2011.01.019

Bélanger RR, Benhamou N, Menzies JG (2003) Cytological evidence of an active role of silicon in wheat resistance to powdery mildew (Blumeria graminis f. sp. tritici). Phytopathology 93:402–412. https://doi.org/10.1094/PHYTO.2003.93.4.402

Carneiro ME, Magalhães WL, Bolzon MGI, Nisgoski S, Satyanarayana KG (2015) Preparation and characterization of nano silica from Equisetum arvense. J Bioprocess Biotech 5:205. https://doi.org/10.4172/2155-9821.1000205

Chakraborty S, Singh A, Roychoudhury A (2022) Biogenic nanoparticles and generation of abiotic stress-resilient plants: a new approach for sustainable agriculture. Plant Stress 6:100117. https://doi.org/10.1016/j.stress.2022.100117

Chen D, Wang S, Yin L, Deng X (2018) How does silicon mediate plant water uptake and loss under water deficiency? Front Plant Sci 9:340168. https://doi.org/10.3389/fpls.2018.00281

Chérif M, Asselin A, Bélanger RR (1994) Defense responses induced by soluble silicon in cucumber roots infected by Pythium spp. Phytopathology 84:236–242

Cornelis JT, Delvaux B, Titeux H (2010) Contrasting silicon uptakes by coniferous trees: a hydroponic experiment on young seedlings. Plant Soil 336:99–106. https://doi.org/10.1007/s11104-010-0451-x

Coskun D, Deshmukh R, Sonah H, Menzies JG, Reynolds O, Ma JF, Kronzucker HJ, Bélanger RR (2019) The controversies of silicon's role in plant biology. New Phytol 221:67–85. https://doi.org/10.1111/nph.15343

Coskun D, Deshmukh R, Shivaraj SM, Isenring P, Bélanger RR (2021) Lsi2: a black box in plant silicon transport. Plant Soil 466:1–20. https://doi.org/10.1007/s11104-021-05061-1

Currie HA, Perry CC (2007) Silica in plants: biological, biochemical and chemical studies. Ann Bot 100:1383–1389. https://doi.org/10.1093/aob/mcm247

Damavandi F, Wang W, Shen WZ, Cetinel S, Jordan T et al (2021) Enrichment of low abundance DNA/RNA by oligonucleotide-clicked iron oxide nanoparticles. Sci Rep 11:1–10. https://doi.org/10.1038/s41598-021-92376-9

Dann EK, Muir S (2002) Peas grown in media with elevated plant-available silicon levels have higher activities of chitinase and β-1,3- glucanase, are less susceptible to a fungal leaf spot pathogen and accumulate more foliar silicon. Australas Plant Pathol 31:9–13. https://doi.org/10.1071/AP01047

De Medeiros TA, Suzuki VY, Leite JATM, Fernandes PS, Santana VC, Oliveira CR, Nicodemo D, Ferreira LM (2020) Silício nos alimentos: uma revisão. Adv Nutr Sci 1:41–48. https://doi.org/10.47693/ans.v1i1.7

Delaney P, McManamon C, Hanrahan P, Copley MP, Holmes JD, Morris MA (2011) Development of chemically engineered porous metal oxides for phosphate removal. J Hazard Mater 185:382–391. https://doi.org/10.1016/j.jhazmat.2010.08.128

Della VP, Hotza D, Junkes JA, Oliveira APND (2006) Comparative study of silica obtained from acid leaching of rice husk and the silica obtained by thermal treatment of rice husk ash. Quim Nova 29:1175–1179. https://doi.org/10.1590/S0100-40422006000600005

Deshmukh R, Sonah H, Belanger RR (2020) New evidence defining the evolutionary path of aquaporins regulating silicon uptake in land plants. J Exp Bot 71:6775–6788. https://doi.org/10.1093/jxb/eraa342

Diab R, Canilho N, Pavel IA, Haffner FB, Girardon M, Pasc A (2017) Silica-based systems for oral delivery of drugs, macromolecules and cells. Adv Colloid Interf Sci 249:346–362. https://doi.org/10.1016/j.cis.2017.04.005

Dong M, Jiao G, Liu H, Wu W, Li S, Wang Q, Xu D, Li X, Liu H, Chen Y (2016) Biological silicon stimulates collagen type 1 and osteocalcin synthesis in human osteoblast-like cells through the BMP-2/Smad/RUNX2 signaling pathway. Biol Trace Elem Res 173:306–315. https://doi.org/10.1007/s12011-016-0686-3

Dorairaj D, Govender N, Zakaria S, Wickneswari R (2022) Green synthesis and characterization of UKMRC-8 rice husk-derived mesoporous silica nanoparticle for agricultural application. Sci Rep 12:20162. https://doi.org/10.1038/s41598-022-24484-z

Elmotasem H, El Marasy SA, Mohamed AL (2023) Benzocaine mesoporous silica nanoparticles/bio polysaccharides-based hydrogels loaded cotton bandage as a platform for topical anesthesia. Results Chem 5:100830. https://doi.org/10.1016/j.rechem.2023.100830

Epstein E (1994) The anomaly of silicon in plant biology. Proc Natl Acad Sci 91(1):11–17. https://doi.org/10.1073/pnas.91.1.11

Fauteux F, Chain F, Belzile F, Menzies JG, Bélanger RR (2006) The protective role of silicon in the Arabidopsis–powdery mildew pathosystem. Proc Natl Acad Sci 103:17554–17559. https://doi.org/10.1073/pnas.0606330103

Fleck AT, Nye T, Repenning C, Stahl F, Zahn M, Schenk MK (2011) Silicon enhances suberization and lignification in roots of rice (Oryza sativa). J Exp Bot 62:2001–2011. https://doi.org/10.1093/jxb/erq392

García-Ríos E, Ruiz-Rico M, Guillamón JM, Pérez-Esteve É, Barat JM (2018) Improved antimicrobial activity of immobilised essential oil components against representative spoilage wine microorganisms. Food Control 94:177–186. https://doi.org/10.1016/j.foodcont.2018.07.005

Gartmann N, Schütze C, Ritter H, Brühwiler D (2010) The effect of water on the functionalization of mesoporous silica with 3-aminopropyltriethoxysilane. J Phys Chem Lett 1:379–382. https://doi.org/10.1021/jz9002795

Gomes LS, de Souza MC, Furtado ACR (2018) A sílica e suas particularidades. Rev Virtual Quím 10(4). https://doi.org/10.21577/1984-6835.20180072

Goswami P, Mathur J (2019) Positive and negative effects of nanoparticles on plants and their applic-Fations in agriculture. Plant Sci Today 6:232–242. https://doi.org/10.14719/pst.2019.6.2.502

Guilger-Casagrande M, Germano-Costa T, Bilesky-José N, Pasquoto-Stigliani T, Carvalho L, Fraceto LF, De Lima R (2021) Influence of the capping of biogenic silver nanoparticles on their toxicity and mechanism of action towards Sclerotinia sclerotiorum. J Nanobiotechnol 19:1–18. https://doi.org/10.1186/s12951-021-00797-5

Hattori T, Inanaga S, Araki H, An P, Morita S, Luxová M, Lux A (2005) Application of silicon enhanced drought tolerance in Sorghum bicolor. Physiol Plant 123:459–466. https://doi.org/10.1111/j.1399-3054.2005.00481.x

Jeelani PG, Mulay P, Venkat R, Ramalingam C (2020) Multifaceted application of silica nanoparticles. A review. Silicon 12:1337–1354. https://doi.org/10.1007/s12633-019-00229-y

Jinger D, Devi MT, Dhar S, Dass A, Rajanna GA, Upadhaya P, Raj R (2017) Silicon in mitigating biotic stresses in rice (Oryza sativa L.) – a review. Ann Agric Res 38:1–14

Jones LHP, Handreck KA (1967) Silica in soils, plants, and animals. Adv Agron 19:107–149. https://doi.org/10.1016/S0065-2113(08)60734-8

Jurkić LM, Cepanec I, Pavelić SK, Pavelić K (2013) Biological and therapeutic effects of ortho-silicic acid and some ortho-silicic acid-releasing compounds: new perspectives for therapy. Nutr Metab 10:1–12. https://doi.org/10.1186/1743-7075-10-2

Kang H, Elmer W, Shen Y, Zuverza-Mena N, Ma C, Botella P, White JC, Haynes CL (2021) Silica nanoparticle dissolution rate controls the suppression of Fusarium Wilt of watermelon (Citrullus lanatus). Environ Sci Technol 55:13513–13522. https://doi.org/10.1021/acs.est.0c07126

Karande SD, Jadhav SA, Garud HB, Kalantre VA, Burungale SH, Patil PS (2021) Green and sustainable synthesis of silica nanoparticles. Nanotechnol Environ Eng 6:29. https://doi.org/10.1007/s41204-021-00124-1

Kumar S, Adiram-Filiba N, Blum S, Sanchez-Lopez JA, Tzfadia O et al (2020) Siliplant1 protein precipitates silica in sorghum silica cells. J Exp Bot 71:6830–6843. https://doi.org/10.1093/jxb/eraa258

Li Z, Mu Y, Peng C, Lavin MF, Shao H, Du Z (2021) Understanding the mechanisms of silica nanoparticles for nanomedicine. Wiley Interdiscip Rev Nanomed Nanobiotechnol 13:e1658. https://doi.org/10.1002/wnan.1658

Liang Y, Nikolic M, Belanger R, Gong H, Song A (2015) Silicon in agriculture: from theory to practice. Springer Publishing, New York. https://doi.org/10.1007/978-94-017-9978-2

Lin W, Huang YW, Zhou XD, Ma Y (2006) In vitro toxicity of silica nanoparticles in human lung cancer cells. Toxicol Appl Pharmacol 217:252–259. https://doi.org/10.1016/j.taap.2006.10.004

Liou TH, Wang SY, Lin YT, Yang S (2022) Sustainable utilization of rice husk waste for preparation of ordered nanostructured mesoporous silica and mesoporous carbon: characterization and adsorption performance. Colloids Surf A Physicochem Eng Asp 636:128150. https://doi.org/10.1016/j.colsurfa.2021.128150

Ma JF (2010) Silicon transporters in higher plants. In: Jahn TP, Bienert GP (eds) MIPs and their role in the exchange of metalloids. Springer, New York, pp 99–106. https://doi.org/10.1007/978-1-4419-6315-4_8

Ma JF, Yamaji N (2006) Silicon uptake and accumulation in higher plants. Trends Plant Sci 11:392–397. https://doi.org/10.1016/j.tplants.2006.06.007

Ma JF, Yamaji N (2015) A cooperative system of silicon transport in plants. Trends Plant Sci 20:435–442. https://doi.org/10.1016/j.tplants.2015.04.007

Ma JF, Miyake Y, Takahashi E (2001) Chapter 2: Silicon as a beneficial element for crop plants. Stud Plant Sci 8:17–39. https://doi.org/10.1016/S0928-3420(01)80006-9

Ma JF, Yamaji N, Mitani N, Tamai K, Konishi S, Fujiwara T, Katsuhara M, Yano M (2007) An efflux transporter of silicon in rice. Nature 448:209–212. https://doi.org/10.1038/nature05964

Ma JF, Yamaji N, Mitani-Ueno N (2011) Transport of silicon from roots to panicles in plants. Proc Jpn Acad Ser B Phys Biol Sci 87:377–385. https://doi.org/10.2183/pjab.87.377

Mandlik R, Thakral V, Raturi G, Shinde S, Nikolić M, Tripathi DK, Sonah H, Deshmukh R (2020) Significance of silicon uptake, transport, and deposition in plants. J Exp Bot 71:6703–6718. https://doi.org/10.1093/jxb/eraa301

Mayland HF, Wright JL, Sojka RE (1991) Silicon accumulation and water uptake by wheat. Plant Soil 137:191–199. https://doi.org/10.1007/BF00011197

Mitani-Ueno N, Ma JF (2021) Linking transport system of silicon with its accumulation in different plant species. Soil Sci Plant Nutr 67:10–17. https://doi.org/10.1080/00380768.2020.1845972

Mitra D, Adhikari P, Djebaili R, Thathola P, Joshi K, Pellegrini M et al (2023) Biosynthesis and characterization of nanoparticles, its advantages, various aspects and risk assessment to maintain the sustainable agriculture: emerging technology in modern era science. Plant Physiol Biochem 196:103–120. https://doi.org/10.1016/j.plaphy.2023.01.017

Mousavi M, Fini E (2020) Silanization mechanism of silica nanoparticles in bitumen using 3-aminopropyl triethoxysilane (APTES) and 3-glycidyloxypropyl trimethoxysilane (GPTMS). ACS Sustain Chem Eng 8:3231–3240. https://doi.org/10.1021/acssuschemeng.9b06741

Mukarram M, Petrik P, Mushtaq Z, Khan MMA, Gulfishan M, Lux A (2022) Silicon nanoparticles in higher plants: uptake, action, stress tolerance, and crosstalk with phytohormones, antioxidants, and other signalling molecules. Environ Pollut 310:119855. https://doi.org/10.1016/j.envpol.2022.119855

Napierska D, Thomassen LC, Lison D, Martens JA, Hoet PH (2010) The nanosilica hazard: another variable entity. Part Fibre Toxicol 7:1–32. https://doi.org/10.1186/1743-8977-7-39

Nielsen FH (2014) Update on the possible nutritional importance of silicon. J Trace Elem Med Biol 28:379–382. https://doi.org/10.1016/j.jtemb.2014.06.024

Pantidos N, Horsfall LE (2014) Biological synthesis of metallic nanoparticles by bacteria, fungi and plants. J Nanomed Nanotechnol 5:1. https://doi.org/10.4172/2157-7439.1000233

Pineda-Vasquez TG, Casas-Botero AE, Ramírez-Carmona ME, Torres-Taborda MM, Soares CH, Hotza D (2014) Biogeneration of silica nanoparticles from rice husk ash using Fusarium oxysporum in two different growth media. Ind Eng Chem Res 53:6959–6965. https://doi.org/10.1021/ie404318w

Pommerrenig B, Diehn TA, Bienert GP (2015) Metalloidoporins: essentiality of nodulin 26-like intrinsic proteins in metalloid transport. Plant Sci 238:212–227. https://doi.org/10.1016/j.plantsci.2015.06.002

Porrang S, Rahemi N, Davaran S, Mahdavi M, Hassanzadeh B (2021) Preparation and in-vitro evaluation of mesoporous biogenic silica nanoparticles obtained from rice and wheat husk as a biocompatible carrier for anti-cancer drug delivery. Eur J Pharm Sci 163:105866. https://doi.org/10.1016/j.ejps.2021.105866

Raggi S, Demes E, Liu S, Verger S, Robert S (2020) Polar expedition: mechanisms for protein polar localization. Curr Opin Plant Biol 53:134–140. https://doi.org/10.1016/j.pbi.2019.12.001

Raleigh GJ (1939) Evidence for the essentiality of silicon for growth of the beet plant. Plant Physiol 14:823–828. https://doi.org/10.1104/pp.14.4.823

Raven JA (1983) The transport and function of silicon in plants. Biol Rev 58:179–207. https://doi.org/10.1111/j.1469-185X.1983.tb00385.x

Reffitt DM, Ogston N, Jugdaohsingh R, Cheung HFJ, Evans BAJ, Thompson RPH, Evans BA, Thompson RPH, Powell JJ, Hampson GN (2003) Orthosilicic acid stimulates collagen type 1 synthesis and osteoblastic differentiation in human osteoblast-like cells in vitro. Bone 32:127–135. https://doi.org/10.1016/S8756-3282(02)00950-X

Russell LA (2010) Osteoporosis and osteomalacia. Rheum Dis Clin 36:665–680. https://doi.org/10.1016/j.rdc.2010.09.007

Sahebi M, Hanafi MM, Akmar ASN, Rafii MY, Azizi P, Tengoua FF et al (2015) Importance of silicon and mechanisms of biosilica formation in plants. Biomed Res Int 2015:16. https://doi.org/10.1155/2015/396010

Sangster AG (1978) Silicon in the roots of higher plants. Am J Bot 65:929–935. https://doi.org/10.1002/j.1537-2197.1978.tb06157.x

Sankareswaran M, Periakaruppan R, Sasivarnam M, Danaraj J, Dhanasekaran S, Abomughaid MM (2022) Bio-fabrication of bio-inspired silica nanomaterials from Bryophyllum pinnatum leaf for agricultural applications. Appl Biochem Biotechnol 194:4266–4277. https://doi.org/10.1007/s12010-022-03996-4

Sattary M, Amini J, Hallaj R (2020) Antifungal activity of the lemongrass and clove oil encapsulated in mesoporous silica nanoparticles against wheat's take-all disease. Pestic Biochem Physiol 170:104696. https://doi.org/10.1016/j.pestbp.2020.104696

Schneider D, Kircheis R, Wassersleben S, Einicke WD, Gläser R, Enke D (2019) Low- cost microwave-assisted partial Pseudomorphic transformation of biogenic silica. Front Chem 7:575. https://doi.org/10.3389/fchem.2019.00575

Schneider D, Wassersleben S, Weiß M, Denecke R, Stark A, Enke D (2020) A generalized procedure for the production of high-grade, porous biogenic silica. Waste Biomass Valoriz 11:1–15. https://doi.org/10.1007/s12649-018-0415-6

Shamim S, Hornyak GL, Crespy D, Kyaw H, Bora T (2022) Morphology and visible photoluminescence modulation in dye-free mesoporous silica nanoparticles using a simple calcination step. Mater Res Bull 152:111842. https://doi.org/10.1016/j.materresbull.2022.111842

Sharma P, Prakash J, Palai T, Kaushal R (2022) Surface functionalization of bamboo leave mediated synthesized SiO_2 nanoparticles: study of adsorption mechanism, isotherms and enhanced adsorption capacity for removal of Cr (VI) from aqueous solution. Environ Res 214:113761. https://doi.org/10.1016/j.envres.2022.113761

Shrestha D, Nayaju T, Kandel MR, Pradhananga RR, Park CH, Kim CS (2023) Rice husk-derived mesoporous biogenic silica nanoparticles for gravity chromatography. Heliyon 9:e15142. https://doi.org/10.1016/j.heliyon.2023.e15142

Sommer AL (1926) Studies concerning the essential nature of aluminum and silicon for plant growth. Univ Calif Publ Agric Sci 5:57–81

Sommer M, Kaczorek D, Kuzyakov Y, Breuer J (2006) Silicon pools and fluxes in soils and landscapes – a review. J Plant Nutr Soil Sci 169:310–329. https://doi.org/10.1002/jpln.200521981

Stöber W, Fink A, Bohn E (1968) Controlled growth of monodisperse silica spheres in the micron size range. J Colloid Interface Sci 26:62–69. https://doi.org/10.1016/0021-9797(68)90272-5

Sun H, Duan Y, Qi X, Zhang L, Huo H, Gong H (2018) Isolation and functional characterization of CsLsi2, a cucumber silicon efflux transporter gene. Ann Bot 122:641–648. https://doi.org/10.1093/aob/mcy103

Sun H, Duan Y, Mitani-Ueno N, Che J, Jia J, Liu J, Guo J, Ma JF, Gong H (2020) Tomato roots have a functional silicon influx transporter but not a functional silicon efflux transporter. Plant Cell Environ 43:732–744. https://doi.org/10.1111/pce.13679

Takahashi E, Ma JF, Miyake Y (1990) The possibility of silicon as an essential element for higher plants. Comments Agric Food Chem 2:99–122

Tang F, Li L, Chen D (2012) Mesoporous silica nanoparticles: synthesis, biocompatibility and drug delivery. Adv Mater 24:1504–1534. https://doi.org/10.1002/adma.201104763

Tubana B, Babu T, Datnoff LE (2016) A review of silicon in soils and plants and its role in US agriculture: history and future perspective. Soil Sci 181:393–411. https://doi.org/10.1097/SS.0000000000000179

Umeda J, Kondoh K (2010) High-purification of amorphous silica originated from rice husks by combination of polysaccharide hydrolysis and metallic impurities removal. Ind Crop Prod 32:539–544. https://doi.org/10.1016/j.indcrop.2010.07.002

Van Bockhaven J, De Vleesschauwer D, Höfte M (2013) Towards establishing broad-spectrum disease resistance in plants: silicon leads the way. J Exp Bot 64:1281–1293. https://doi.org/10.1093/jxb/ers329

Venkatesan KA, Srinivasan TG, Vasudeva Rao PR (2003) Removal of complexed mercury by dithiocarbamate grafted on mesoporous silica. J Radioanal Nucl Chem 256:213–218. https://doi.org/10.1023/A:1023968813908

Waddell WH, Evans LR (2000) Silica, amorphous. Kirk-Othmer Encyclopedia of Chemical Technology. https://doi.org/10.1002/0471238961.0113151823010404.a01

Wang L, Ning C, Pan T, Cai K (2022) Role of silica nanoparticles in abiotic and biotic stress tolerance in plants: a review. Int J Mol Sci 23:1947. https://doi.org/10.3390/ijms23041947

Yaakov N, Kottakota C, Mani KA, Naftali SM, Zelinger E et al (2022) Encapsulation of Bacillus thuringiensis in an inverse Pickering emulsion for pest control applications. Colloids Surf B: Biointerfaces 213:112427. https://doi.org/10.1016/j.colsurfb.2022.112427

Yadav M, Dwibedi V, Sharma S, George N (2022) Biogenic silica nanoparticles from agro-waste: properties, mechanism of extraction and applications in environmental sustainability. J Environ Chem Eng 10:108550. https://doi.org/10.1016/j.jece.2022.108550

Zexer N, Elbaum R (2020) Unique lignin modifications pattern the nucleation of silica in sorghum endodermis. J Exp Bot 71:6818–6829. https://doi.org/10.1093/jxb/eraa127

Zielonka A, Żymańczyk-Duda E, Brzezińska-Rodak M, Duda M, Grzesiak J, Klimek-Ochab M (2018) Nanosilica synthesis mediated by aspergillus parasiticus strain. Fungal Biol 122:333–344. https://doi.org/10.1016/j.funbio.2018.02.004

Zou Y, Yang T (2019) Rice husk, rice husk ash and their applications. In: Rice bran and rice bran oil. AOCS Press, pp 207–246. https://doi.org/10.1016/B978-0-12-812828-2.00009-3

Chapter 10
New Source of Biogenic Silicon from Sugarcane Bagasse

Ntalane Sello Seroka and Lindiwe Khotseng

Abstract This paper describes the generation of biogenic silicon from sugarcane bagasse ash (SCBA). Furthermore, silica was recovered from sugarcane waste using a modified thermochemical approach, that is, tetrapropylammonium hydroxide, and then reacted with magnesium in the magnesiothermic reduction process to produce biogenic silicon. The physicochemical properties of the produced nanocrystalline silicon were examined using Powder XRD (P'XRD), Raman spectroscopy, FTIR, TEM, and SEM. X-ray diffraction spectroscopy revealed a peak at 2θ of 28 corresponding to a 30 nm crystallite size. The Raman analysis revealed a pronounced peak at 510 cm^{-1}, indicating highly ordered silicon. The surface analysis revealed two distinct bands at 445 cm^{-1} and 1046 cm^{-1}, representing the Si-O rocking and Si-O-Si stretching behavior. Nanotechnology as an enabler has proved that SCBA as a sustainable and renewable resource can be used for the production of biogenic silicon.

Keywords Biogenic silicon · Silica · Sugarcane bagasse · Sustainable sources · Biomass

N. S. Seroka
Faculty of Natural Science, University of the Western Cape, Bellville, South Africa

Council for Science and Industrial Research, Pretoria, South Africa
e-mail: 3754640@myuwc.ac.za

L. Khotseng (✉)
Faculty of Natural Science, University of the Western Cape, Bellville, South Africa
e-mail: lkhotseng@uwc.ac.za

R. de Mello Prado et al. (eds.), *Silicon Advances for Sustainable Agriculture and Human Health*, Sustainable Plant Nutrition in a Changing World,
https://doi.org/10.1007/978-3-031-69876-7_10

10.1 Introduction

To solve global climate change challenges, biomass-derived products have sparked interest in industry, academia, and government laboratories. Biosilica was synthesized in this study and used for the transformation into biogenic silicon. Production of biomaterials from biomass waste has proven to be sustainable. Therefore, sugarcane waste has been among the most renewable and sustainable resource of biogenic silicon (Wang et al. 2021).

Furthermore, nanosilicon is produced by engineering silica particles from sugarcane bagasse ash. Silica is a valuable raw resource for industrial purposes, and extensive research has been performed to extract it from agricultural wastes such as sugarcane bagasse and rice husk. Silicon is an inorganic material with superior physical and chemical properties. Because of its low cost and second abundant material in the earth's crust, the performance profile with its synergistic high surface area and small particle size makes it a good candidate for applications in a variety of applications not limited to nanoelectronics, energy production, and storage devices (Wang et al. 2021; Sarkar et al. 2021).

Another intriguing feature is the use of silicon in thermoelectric materials using magnesium alloy technology. In nanotechnology, "green" synthesis has garnered a lot of interest as a dependable, sustainable, and environmentally friendly technique for synthesizing a variety of nanomaterials for the development of nanoelectronics. In essence, green synthesis is viewed as an essential path to reducing the negative consequences associated with standard nanoparticle synthesis methods for the manufacture of functional nanomaterials, biomedical applications, energy storage, and industry (Seroka et al. 2022a, b).

It is commonly recognized that the four most critical parameters for nanoparticle synthesis utilizing the green procedure are the choice of an ecologically friendly solvent, a source of nanomaterials, a reducing agent, and a harmless stabilizing ingredient. Plants are known as natural chemical factories that are inexpensive and require no conservation. Sugarcane, also known as *Saccharum officinarum L.*, is a commercially cultivated crop that is harvested during slightly sunny and chilly weather conditions. And it is generated annually in most regions of the world, with Brazil and India producing 100 million tonnes each. South Africa produces around 19.3 million tonnes of sugarcane per year, followed by Nigeria and Uganda, which are the two largest producers of sugarcane on African soil (Seroka et al. 2022a, b; September et al. 2023).

The value of sugarcane bagasse has been intensively researched throughout the years, with niche applications in the sugar industry and bioenergy generation. Silicon dioxide (SiO_2), also known as silica, is a valuable inorganic multipurpose chemical compound and one of the fundamental materials. Silica can be found naturally as quartz, sand, or flint. Gel and amorphous forms are also possible, and the earth's crust contains both crystalline and amorphous forms. El Sayed and El-Sammi found silica in a variety of sources, including sorghum (88.8%), wheat (90.6%), corn (64.3%), bamboo (57.4%), bagasse (73.0%), lantana (23.4%), sunflower (25.3%), rice husk (93.0%), rice straw (82.0%), and bread fruit tree (81.8%) (September et al. 2023; Seroka et al. 2022a, b).

The synthesis of nanomaterials from agricultural waste is considered green since agricultural waste poses an environmental risk if not properly managed. For example, sugarcane bagasse dumps attract many insects, which impair the population's health and cause a variety of diseases. At the same time, sugarcane ash provides a sustainable source of silicon. Fortunate et al. found 71.49 wt.% silica in sugarcane bagasse ash. Several types of nanomaterials have been recovered from agricultural wastes using biogenic synthesis; these synthesis methods are an appealing alternative to traditional synthesis methods because they are green and ecologically friendly. Nanotechnology has emerged as one of the most important disciplines of research, with numerous applications in industries, such as the conversion of biomass into useful commodities in specialty niche applications (Farirai et al. 2021).

Sugarcane is one of the most promising natural resources for extracting silica and converting it into nanostructured silicon. This feedstock is historically used to produce sugar and ethanol, with the process by-product (bagasse) serving as the primary source of energy. In this paper, studies were conducted to convert silica nanoparticles to nanosilicon using magnesiothermic reduction. As a result, the produced nanoparticles were analyzed and compared in terms of physical, chemical composition, shape, and structural aspects (Negrão et al. 2021).

In this research, we report on the green synthesis of biogenic silicon from sugarcane waste that occurs in a few steps (Fig. 10.1). The physicochemical properties of the synthesized material are also discussed. Biomass waste by intuition is green technology, thus making this project cost effective and eco-friendly.

Fig. 10.1 Represents images of a typical sugarcane bagasse (**a**), sugarcane bagasse ash (**b**), extracted biosilica (**c**), and (**d**) biogenic silico

10.2 Procedures for Producing Biogenic Silica

10.2.1 Materials

The chemicals used in the production were L-cysteine hydrochloride monohydrate ≥98%, tetrapropylammonium hydroxide 1.0 M in H_2O obtained from Sigma Aldrich, and magnesium powder ≥99%. Sugarcane was acquired from the Sugar Company, South Africa. The synthesis was carried out with deionized water.

10.2.2 Extraction of Silica and Subsequent Transformation to Biogenic Silicon

Sugarcane bagasse was incompletely combusted, resulting in ash. The ash was subsequently treated with an organic acid (L-cysteine hydrochloride monohydrate). Approximately 2.00 g of sugarcane bagasse ash (SCBA) was placed in a ceramic crucible and calcined at 700 °C for 2 h in a muffle furnace at a heating rate of 10 °C/min to burn away volatile material and generate crystalline silica. This produced a pale orange-brown product(s). Following that, 1 mL of tetrapropylammonium hydroxide (TPAH) was added to the orange-brown result in a metallic reactor, as given in Eq. 1 (Fig. 10.2). The solution was then heated in a furnace at 200 °C for the first 2 h before being heated at 600 °C for an hour to burn off the surplus carbon, nitrogen, and hydrogen, as given in Eq. (2) (Fig. 10.2). Heating caused the oxidation

$$(SiO_2)_{48} + [N(C_3H_7)4]OH \rightarrow [N(C_3H_7)_4]OH(SiO_2)_{48} \ldots\ldots\ldots\ldots\ldots\ldots\ldots\ldots\ldots\ldots\ldots\ldots\ldots (1)$$

$$[N(C_3H_7)_4]OH(SiO_2)_{48} + \Delta \rightarrow (SiO_2)x(s) \downarrow +(CO)x(g) \uparrow +(H)x(g) \uparrow +(NO)x(g)$$

$$SiO_2 + 2Mg \rightarrow Si + 2MgO \ldots\ldots\ldots\ldots\ldots\ldots\ldots\ldots\ldots\ldots\ldots\ldots\ldots\ldots\ldots\ldots\ldots\ldots\ldots\ldots\ldots\ldots\ldots\ldots\ldots\ldots (3)$$

$$Si + MgO + HCl \rightarrow Si(s) \downarrow +MgCl_2(aq) \ldots\ldots\ldots\ldots\ldots\ldots\ldots\ldots\ldots\ldots\ldots\ldots\ldots\ldots\ldots\ldots\ldots\ldots (4)$$

Fig. 10.2 Represents equations of (1) extraction of biosilica using organic base tetrapropylammonium hydroxide, (2) pyrolysis of multicrystalline product to produce biosilica, (3) magnesiothermic reduction of biosilica into biogenic silicon, and (4) purification of biogenic silicon using acidic media of C, H, and N to generate the corresponding oxide(s). As a result, silica oxide(s) were produced as nanopowders labeled L cys-TPAH-Silica.

To create a homogeneous powder, 0.5 g of silica was mixed with 0.6 g of magnesium powder in a mortar and pestle. The mixture was then transferred to a crucible and burned at 650 °C in a muffle furnace to produce biogenic silicon, Eq. (3) (Fig. 10.2), which was subsequently purified in acidic media, Eq. (4) (Fig. 10.2).

10.3 Biogen Silicon: Physicochemical Properties of Nanocrystalline Silicon

10.3.1 XRD and Raman Analysis

First, the components of sugarcane bagasse were analyzed using XRD and Raman (Fig. 10.3a–d). The SiO_2 content of sugarcane bagasse was measured. According to XRD analysis, magnesium thermal reduction can take place at any time when magnesium powder (melting point of 650 °C) is in the form of magnesium vapor. The reaction temperature should be higher than magnesium vapor's melting point to ensure its presence in the reaction. At the same time, because magnesium powder explodes at high temperatures, a protective atmosphere of nitrogen gas was used during the reaction (Yadav et al. 2022).

Figure 10.3 depicts the XRD diffraction patterns of (a) amorphous carbon from sugarcane bagasse, (b) biosilica, (c) biogenic Si, and (d) silicon Raman. The illustration demonstrates that amorphous carbon has a significant bulge between 2θ

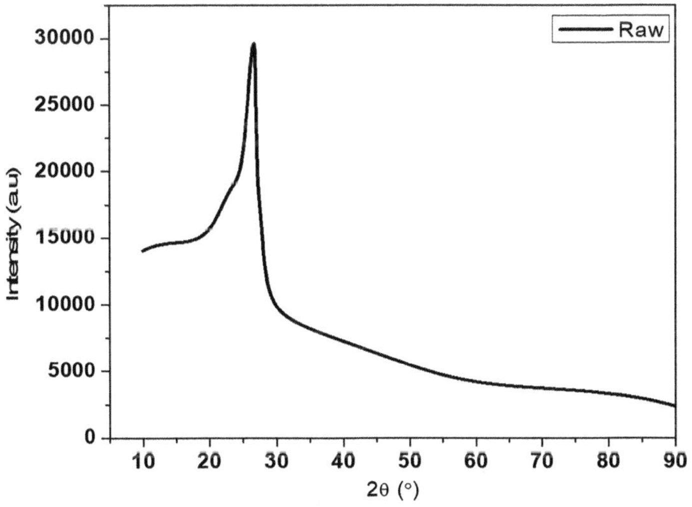

Fig. 10.3 XRD spectra of (**a**) sugarcane bagasse, (**b**) biosilica, (**c**) biogenic silicon, and (**d**) Raman spectrum of biogenic silicon

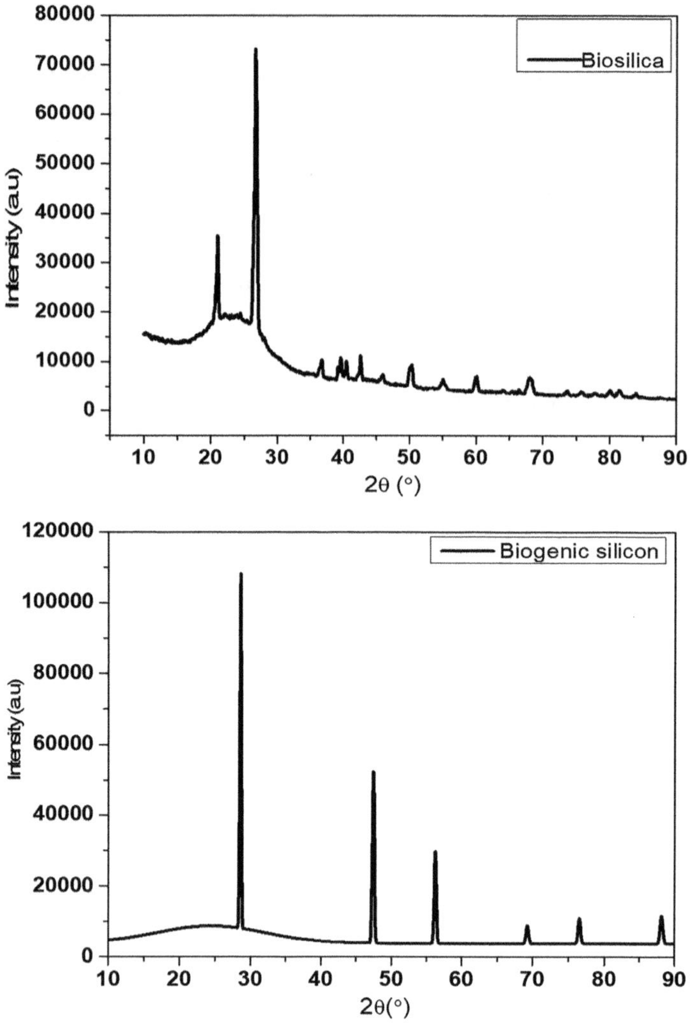

Fig. 10.3 (continued)

=20° and 30°, which corresponds to the 002 plane. Biosilica has a characteristic silica diffraction pattern, with peaks at 2θ = 20° and 26°, with the latter being the most powerful and conspicuous, all of which correlate to silica's crystalline constitution. Figure 10.3c (No. 17-0901) displays unique silicon diffraction peaks at 2θ = 28°, 47°, 56°, 69°, 76°, and 88°. These peaks correspond to the (111), (220), (311), (400), (331), and (420) crystal faces of metallic silicon, respectively (Man and Lin 2020).

Nanosilica and biogenic Si materials have visible amorphous carbon rings between 2θ =20° and 30° to the 002 plane, indicating a substantial silicon content and the presence of amorphous carbon from calcitrant bagasse. Silicon peaks at

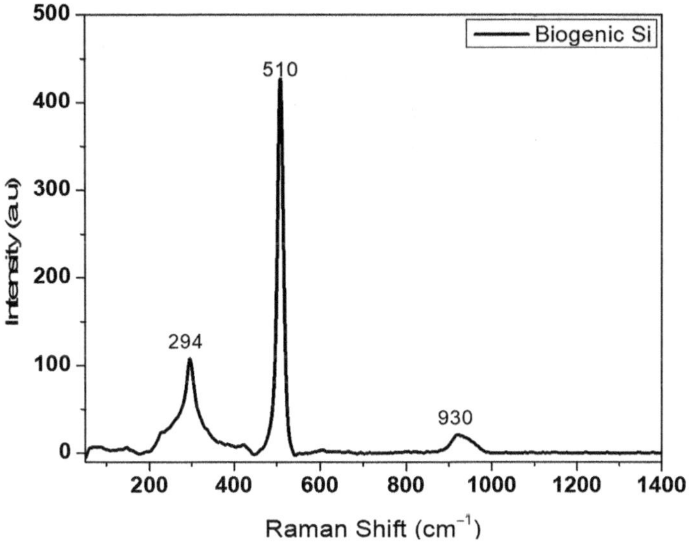

Fig. 10.3 (continued)

Table 10.1 Functional groups identified by FTIR spectroscopy

Wavenumber (cm^{-1})	Functional group
445	Si-O
788	Si-O
1046	Si-O
1590	C=O
3000	C-H
3450	-OH

$2\theta = 28.4°$ and $47.3°$ indicate a stable crystal structure (Man and Lin 2020). Raman spectra showed a peak at 510 cm^{-1}, indicating the Si-Si stretching mode, and a half-full width at half maximum (FWHM) of 6 cm^{-1}. Similar results were reported (Rotshteyn et al. 2022). The scattering of the remaining two peaks, found at 294 cm^{-1} and 930 cm^{-1}, was caused by two phonons: transverse acoustic (2TA) and optical (2TO). The peak at 935 cm^{-1} corresponds to the stretching mode of amorphous Si-Si, as described in the literature (Rotshteyn et al. 2022).

10.3.2 *FTIR Analysis*

FT-IR spectra of biogenic silicon generated by magnesiothermic reduction of biosilica recovered from sugarcane bagasse ash was done, and Table 10.1 lists the characteristic bond peaks that occurred in the spectra. The powder samples show Si-O stretching (1046 cm^{-1}) and rocking vibration (445 cm^{-1}). The aging of the Si nanoparticles contributes significantly to the increased strength of the Si-O bond. It is probable that the bonds result from the presence of O_2 in the DI water; a similar conclusion has been reached in the literature (Seroka et al. 2023; Yuvaraj et al. 2023).

FTIR spectral study at room temperature revealed important functional groups in bagasse ash, leached samples, and nanosilica. The wavelength range was 400–4000 cm^{-1}. Figure 10.4 shows the FTIR analysis of the synthesized Fig. 10.4b biosilica. The bands at 445 cm^{-1}, 788 cm^{-1}, and 1046 cm^{-1} correspond to Si~O~Si bending vibration, Si~O~Si stretch vibration, and Si~O~Si stretching vibration, respectively. Interestingly, the evolution and removal of functional groups, as well

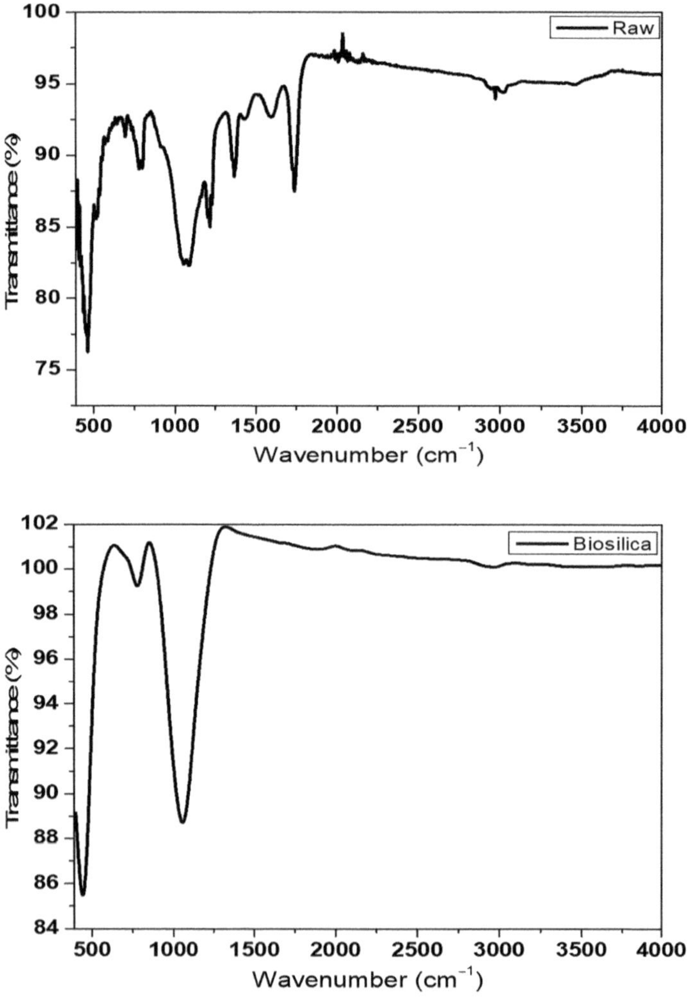

Fig. 10.4 FTIR spectra of (**a**) raw bagasse, (**b**) biosilica, (**c**) biogenic silicon

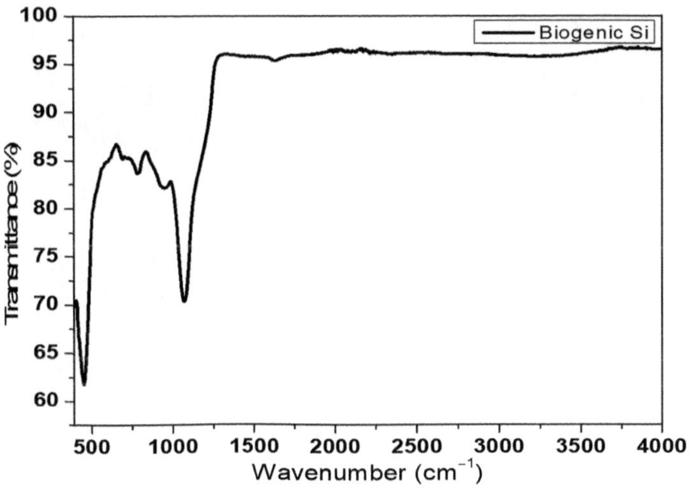

Fig. 10.4 (continued)

as the appearance of more prominent groups, indicate that new functionality have been included into our samples. The carbonyl group band is present in (a) raw samples but not in the silica spectrum (b) and (c) biogenic silicon, indicating that silica groups successfully formed to biogenic silicon, which is predominant in the material (Seroka et al. 2023; Yuvaraj et al. 2023).

10.3.3 TEM Analysis

These particles are composed of some denser particles and smaller aggregated crystalline nanoclusters, as shown in typical TEM images (Fig. 10.5a–e). The corresponding selected area electron diffraction (SAED) pattern (inset of Fig. 10.5d) reveals some bright diffraction areas associated with reflections, as expected. It is completely compatible with the cubic structure. These results are consistent with XRD and Raman data. Amorphous oxide may be present in the outermost thin layer with a thickness of around 2 nm.

Furthermore, the comparable EDX spectrum is given in Fig. 10.6d indicating the existence of a trace amount of oxygen element. This phenomenon could be produced by oxidation of the Si surface, which is widely considered to occur during the TEM specimen preparation method when Si particles are exposed to air. The findings demonstrate that biogenic silicon is predominant in considerable quantities, as shown in Table 10.2 (Croissant et al. 2020).

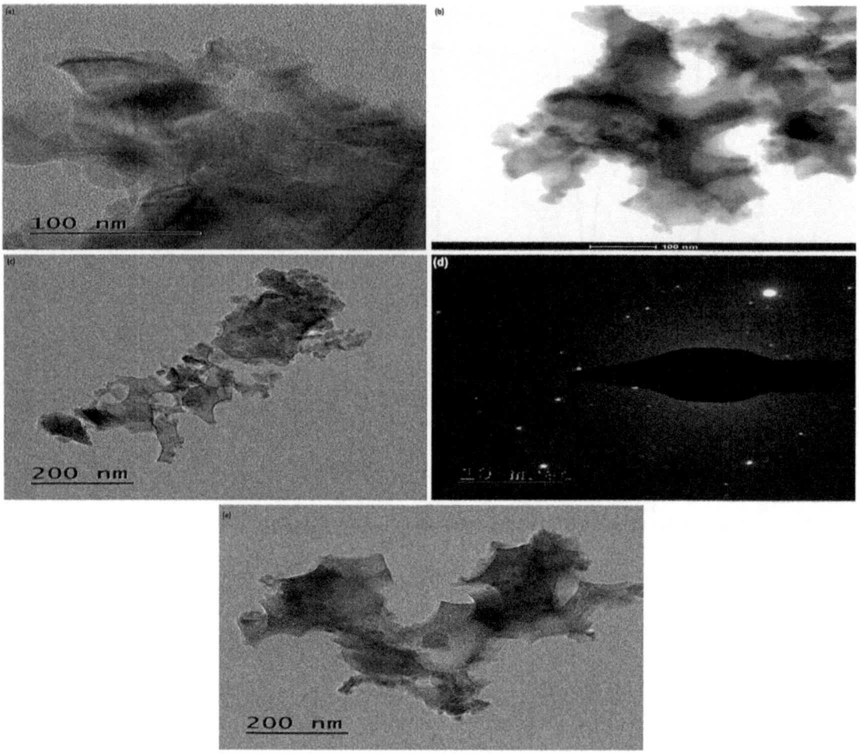

Fig. 10.5 TEM images of biogenic silicon at different magnifications (**a–d**) (Inset, SAED), and (**e**)

10.3.4 SEM Analysis

Consequently, an outline of the morphology of the as-prepared sample may be seen in the example SEM images displayed in Fig. 10.6a–d. It is evident that the product contains several particles with sizes ranging from tens to hundreds of nanometers. The elemental mapping of Si obtained from Fig. 10.6c–f then confirms the formation of pure Si phase in the specimen. The produced nanosilicon's surface morphologies show that the magnesiothermic reduction stage promotes nanoparticle aggregation. Energy dispersive X-ray spectroscopy (EDX) was used to assess the procedure efficacy, as shown in Fig. 10.6h. The presence of silica in the sample is indicated by the high oxygen and silicon intensities in the EDX spectrum. The oxygen content is a negligible contamination commonly associated with the green method. This could be due to the resulting surface of highly active silicon after exposure to an aqueous HCl solution, as well as the low presence of unreacted silica in the sample following the reduction method. The surface chemistry detected

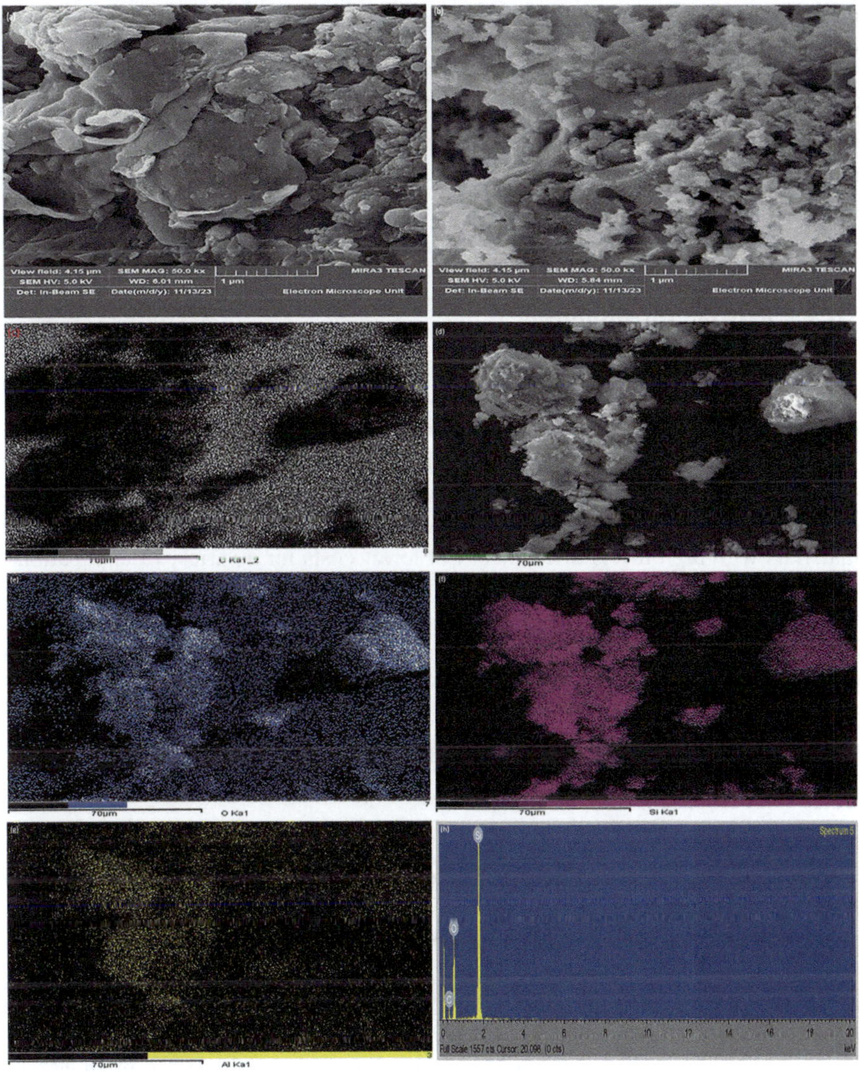

Fig. 10.6 SEM images of (**a**, **b**) biogenic silicon, (**c**) elemental mapping for carbon (**d**) (Inset), (**e**) oxygen, (**f**) silicon, and (**g**) aluminum. EDX (**h**) spectrum of biogenic silicon

Table 10.2 Quantification of elements from TEM analysis

Quantification results					
Atomic percent (%)					
Atomic % (norm.)	Area	HAADF	Si-KA	Mg-K	O-K
CA1.spx	43,472	41,629	81,74	4,19	14,07

earlier in the FTIR analysis is consistent with the observed morphological changes (Chen et al. 2020; Gadre et al. 2022).

The SEM-EDX analysis in Fig. 10.6g shows that magnesiothermic silica-to-silicon reduction is an excellent approach for turning sugarcane bagasse ash into valuable minerals like silicon nanoparticles. The sample's largely as-synthesized silicon concentration suggests that sugarcane bagasse is abundant, useful, and inexpensive for biogenic silicon synthesis (Gadre et al. 2022).

10.4 Future Perspectives

Materials derived from biomass, which range from inorganic to multifunctional, present a chance to favorably impact environmentally friendly, sustainable systems. This will speed up upscaling and industrial applications by producing better, low-cost, high-efficiency materials derived from biomass. Biogenic materials are a naturally abundant and sustainable resource, which makes it possible to produce materials with high levels of efficiency at lower costs for a variety of applications. It also helps researchers worldwide identify substitute biogenic materials that are just as effective at performing comparable tasks in a variety of systems. The green approach to produce biogenic silicon requires that the purification step utilizes hot boiling water to remove unreacted magnesium and excess silica.

10.5 Conclusion

The study explains the significance of green technology for producing nanosilicon from sugarcane bagasse. Its relatively simple (easy of operation) and inexpensive strategy to generating large silicon nanoparticles has one key limitation: the inability to scale and the chemistry of biogenic silicon (stability). Although the purity aspect has always been the most significant disadvantage of this approach due to leftover by-products. Interestingly, there is still room for development in the green way of producing extremely pure biogenic silicon nanoparticles from sugarcane bagasse ash, which will be an essential synthetic path in lowering production costs and offering low-cost biogenic silicon.

Using X-ray diffraction spectroscopy and the Scherrer formula, researchers detected a peak at 2θ of 28 with a crystallite size of 30 nm. The Raman spectra revealed a strong peak at 510 cm^{-1}, indicating crystalline silicon. The FTIR analysis revealed two distinct peaks at 445 cm^{-1} and 1046 cm^{-1}, representing the Si-O rocking mode and Si-O-Si stretching mode functional groups.

References

Chen H, Lu J, Kong Y, Li K, Yang T, Meingast A, Yang M, Lu Q, Du Y (2020) Atomic scale investigation of the crystal structure and interfaces of the B′ precipitate in Al-Mg-Si alloys. Acta Mater 185:193–203

Croissant JG, Butler KS, Zink JI, Brinker CJ (2020) Synthetic amorphous silica nanoparticles: toxicity, biomedical and environmental implications. Nat Rev Mater 5(12):886–909

Farirai F, Mupa M, Daramola MO (2021) An improved method for the production of high purity silica from sugarcane bagasse ash obtained from a bioethanol plant boiler. Part Sci Technol 39(2):252–259

Gadre CA, Yan X, Song Q, Li J, Gu L, Huyan H, Aoki T, Lee SW, Chen G, Wu R, Pan X (2022) Nanoscale imaging of phonon dynamics by electron microscopy. Nature 606(7913):292–297

Man Y, Lin RY (2020) Well-dispersed double carbon layers coated on Si nanoparticles and the enhanced electrochemical performance for lithium ion batteries. J Mater Sci Mater Electron 31:14912–14920

Negrão DR, Grandis A, Buckeridge MS, Rocha GJ, Leal MRL, Driemeier C (2021) Inorganics in sugarcane bagasse and straw and their impacts for bioenergy and biorefining: a review. Renew Sust Energ Rev 148:111268

Rotshteyn VM, Turdaliev TK, Ashurov KB (2022) Analysis of porous nanosilicon by Raman spec troscopy. J Appl Spectrosc 89(1):43–48

Sarkar J, Mridha D, Sarkar J, Orasugh JT, Gangopadhyay B, Chattopadhyay D, Roychowdhury T, Acharya K (2021) Synthesis of nanosilica from agricultural wastes and its multifaceted applications: a review. Biocatal Agric Biotechnol 37:102175

September LA, Kheswa N, Seroka NS, Khotseng L (2023) Green synthesis of silica and silicon from agricultural residue sugarcane bagasse ash–a mini review. RSC Adv 13(2):1370–1380

Seroka NS, Taziwa R, Khotseng L (2022a) Green synthesis of crystalline Silica from Sugarcane Bagasse Ash: Physico-chemical properties. Nano 12(13):2184

Seroka NS, Taziwa RT, Khotseng L (2022b) Extraction and synthesis of silicon nanoparticles (SiNPs) from sugarcane bagasse ash: a mini-review. Appl Sci 12(5):2310

Seroka NS, Taziwa R, Khotseng L (2023) Nanostructured silicon derived from an agricultural residue bagasse ash via magnesiothermic reduction method. Coatings 13(2):221

Wang Y, Zhang M, Shen X, Wang H, Wang H, Xia K, Yin Z, Zhang Y (2021) Biomass-derived carbon materials: controllable preparation and versatile applications. Small 17(40):2008079

Yadav M, Dwibedi V, Sharma S, George N (2022) Biogenic silica nanoparticles from agro-waste: properties, mechanism of extraction and applications in environmental sustainability. J Environ Chem Eng 10:108550

Yuvaraj M, Priya RS, Jagathjothi N, Saranya M, Suganthi N, Sharmila R, Cyriac J, Anitha R, Subramanian KS (2023) Silicon nanoparticles (SiNPs): challenges and perspectives for sustainable agriculture. Physiol Mol Plant Pathol 128:102161

Chapter 11
Silicon Nanoparticles and Their Benefit for Sustainable Agriculture: An Overview

Radhika Sharma, Manisha Lakhanpal, Shreya Chauhan, Tarun Verma, Sahil Mehta, Indrakant K. Singh, and Archana Singh

Abstract Recent advancements in nanotechnology have opened new frontiers in agriculture. Among various nanoparticles, silicon nanoparticles (SiNPs) offer significant advantages due to their unique physiological characteristics and structural properties. SiNPs provide a cost-effective and highly efficient method for protecting plants against a range of biotic and abiotic stresses, including insect pests, pathogenic diseases, metal toxicity, drought, and salinity. Despite considerable progress in understanding the diverse functions and mechanisms of SiNPs in enhancing agricultural sustainability over the past decade, a comprehensive summary is still needed. This chapter discusses the characteristics, synthesis, and applications of SiNPs in promoting plant growth, development, and productivity.

R. Sharma
Department of Soil Science, Punjab Agricultural University, Ludhiana, Punjab, India

M. Lakhanpal
Department of Forest Products, Dr. Yashwant Singh Parmar University of Horticulture and Forestry, Solan, Himachal Pradesh, India

S. Chauhan
Department of Forest Biology and Tree Improvement, Dr. Yashwant Singh Parmar University of Horticulture and Forestry, Solan, Himachal Pradesh, India

T. Verma
Department of Silviculture and Agroforestry, Dr. Yashwant Singh Parmar University of Horticulture and Forestry, Solan, Himachal Pradesh, India

S. Mehta
Department of Botany, Hansraj College, University of Delhi, New Delhi, India

I. K. Singh
Department of Zoology, Molecular Biology Research Lab, Deshbandhu College, University of Delhi, New Delhi, India

A. Singh (✉)
Department of Plant Molecular Biology, University of Delhi South Campus, New Delhi, India
e-mail: archanasingh@pmb.du.ac.in

R. de Mello Prado et al. (eds.), *Silicon Advances for Sustainable Agriculture and Human Health*, Sustainable Plant Nutrition in a Changing World,
https://doi.org/10.1007/978-3-031-69876-7_11

Keywords Plant growth · Stress · Sustainability · Silicon · NPs · Response

11.1 Introduction

Among global challenges such as increasing population, climate variability, and depleting natural resources, the urgency for sustainable agricultural practices has intensified. Conventional agricultural methods that heavily rely on synthetic fertilizers and pesticides, which not only contribute to environmental degradation but also pose risks to human health. Consequently, there is a growing interest in exploring innovative strategies to boost agricultural productivity while minimizing ecological footprints (Mehta et al. 2020; Yuvaraj et al. 2023; Sharma and Kumar 2024). A promising frontier in sustainable agriculture involves nanotechnology, specifically the application of silicon nanoparticles (SiNPs). Silicon (Si) is the second most abundant element in the Earth's crust and has emerged as a potential catalyst for enhancing plant growth and resilience against various stressors (Caubet et al. 2020).

Long-term intensive crop cultivation can deplete soil Si, potentially contributing to declining yields. Factors such as severe soil erosion, continuous uptake, loss through desilication, and leaching processes may lead to low Si levels in soil (Meena et al. 2014). Moreover, subtropical and tropical soils typically have low available Si and could benefit from Si fertilization. Thus, improved Si management is essential for increasing yields and sustaining crop productivity in temperate and tropical regions (Meena et al. 2014; Yuvaraj et al. 2023).

Addressing yield decline or stagnation requires assessing the Si status of agriculturally significant soils in different regions and developing a technological system that incorporates SiNPs. Characterized by a high surface area-to-volume ratio and distinct physicochemical properties, SiNPs offer several advantages over traditional silicon sources in agriculture. Plant roots can readily absorb these nanoparticles, facilitating efficient Si uptake and utilization. Furthermore, their nanostructured form improves interactions with plant tissues, facilitating cell uptake and translocation within the plant (Rastogi et al. 2019; Mehta et al. 2020).

The benefits of SiNPs in sustainable agriculture are manifold. Firstly, they enhance plant growth and development, leading to high crop yields and improved production quality (Meena et al. 2014). SiNPs also confer enhanced resistance to biotic and abiotic stresses, including pests, diseases, drought, and salinity (Rastogi et al. 2019). Silicon nanoparticles increase plant defense mechanisms and mitigate stress-induced damage, offering a sustainable alternative to chemical inputs in agriculture (Mehta et al. 2020).

Furthermore, the environmental benefits of SiNPs extend beyond direct plant effects. By promoting root development and improving nutrient uptake efficiency, these nanoparticles contribute to soil fertility and nutrient cycling, reducing reliance on synthetic fertilizers and minimizing nutrient runoff. Additionally, SiNPs have shown promise in mitigating the adverse impacts of heavy metal contamination in soils, potentially serving as a remediation strategy for polluted agricultural lands

(Mahawar et al. 2023). This chapter explores the promising applications of SiNPs in the realm of agriculture.

11.2 Major Approaches for SiNPs Synthesis

Understanding the mechanisms for synthesizing SiNPs is crucial for developing efficient production methods. SiNPs are generally formed by physical methods (top-down approach) and chemical methods (bottom-up approach) (Klabunde and Richards 2009; Sharma et al. 2019). SiNPs are synthesized physically by laser ablation, lithography, milling, sonication, sputtering, and UV irradiation. Apart from physical methods, SiNPs are also chemically formed by aerosol processes, atomic condensation, bio-reduction, reverse micro-emulsion, Stober's method, sol-gel technique, and spray pyrolysis. In addition to physical and chemical methods, there are biological techniques that are classified under chemical methods and involve the use of microbes and plants (Bhat et al. 2021; Sharma and Kumar 2024).

Among all the techniques, the most widely used for SiNP manufacturing are the sol-gel technique, Stober's technique, reverse micro-emulsion, and green and microbial synthesis (Hoffmann et al. 2006). The sol-gel method synthesizes spherical nanoparticles (7–200 nm) (Dixit et al. 2016) formed via hydrolysis from alkoxide precursors, followed by condensation with acid or base catalysts to form NPs (Gubala et al. 2020). Stober's method dates to the 1960s (Stober et al. 1968). This process requires the addition of ammonium hydroxide to a water/alcohol mix, which serves as a catalyst and initiates the disintegration of a silicon alkoxide precursor, mainly tetraethoxysilane (TEOS) (Bonacchi et al. 2011) as shown in Fig. 11.1. Contrary to Stober's and the sol-gel technique of SiNP formation, reverse micro-emulsion (Fig. 11.1) uses water droplets stabilized with surfactants distributed in a soluble oil phase to produce micelles that operate as "nanoreactors" to facilitate the restricted nucleation and development of SiNPs (Finnie et al. 2007).

In recent decades, biological techniques have become popular among researchers for synthesizing nanoparticles. These methods are cost-effective, simple, environmentally friendly, and highly effective. Plant and agricultural waste are used as precursors for producing SiNPs through calcination and acid treatment. Agricultural waste, such as rice husks, hulls, straws, and industrial remnants, such as sugarcane bagasse, are excellent sources of silicon (Karande et al. 2021).

11.3 Important Characteristics of SiNPs

According to their structural morphology, these nanoparticles can be mesoporous, non-porous, hollow mesoporous, or core-shell. The core-shell silicon nanoparticle is spherical with a solid core. Non-porous SiNPs, however, are colloidal in shape and are catalyzed-based on silica hydrolysis (Lehman et al. 2016). Alternatively,

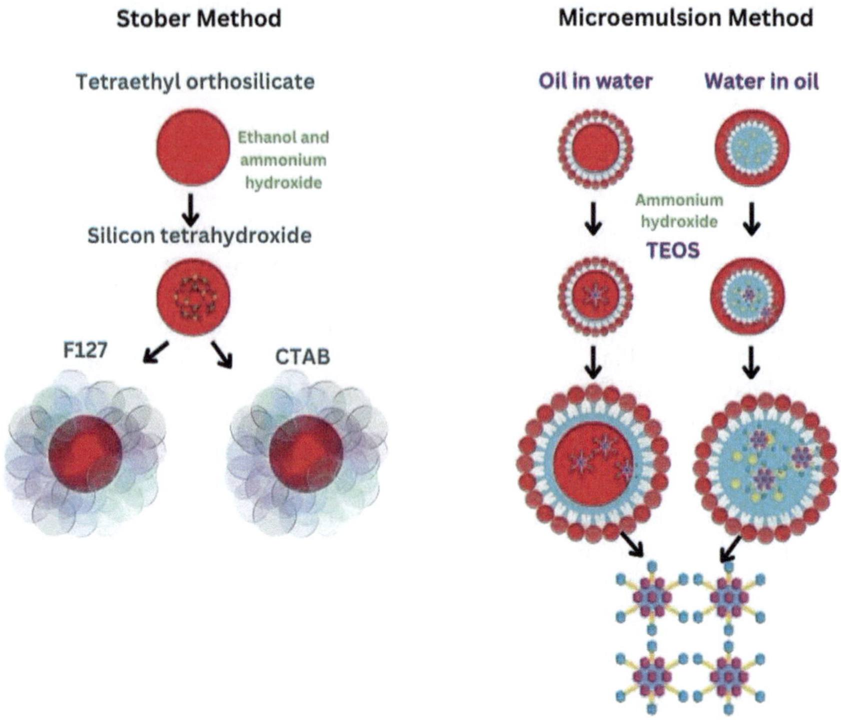

Fig. 11.1 Synthesis of nanoparticle

mesoporous SiNPs present pores ranging from 2 to 20 nm in size, making them ideal for drug delivery (Tang and Cheng 2013). Hollow mesoporous SiNPs, which are spherical, can be used to load antigens to stimulate the immune system (Lee et al. 2020). Because of their high surface area and inherent reactivity, SiNPs are highly amenable to chemical modification (Hussain and Mitra 2012). Significant properties of SiNPs include easy synthesis, high pore volume, commercial availability, small size, adjustable surface charges, and chemical and thermal stability. These characteristics make them ideal for use in drug delivery. Fig. 11.2 shows the nanoscale delivery based on different sizes (Bhat et al. 2021).

11.4 Impact of SiNPs on Plants

The application of SiNPs provides beneficial effects, assisting plants in coping with abiotic stresses such as radiation, salt, temperature (both high and low), and water availability (both high and low) (Coskun et al. 2016). SiNPs are absorbed by plants primarily as silicic acid and, in some cases, through silica transporter genes known as LSIs (1, 2, and 6) (Rao and Susmitha 2017). With their minuscule diameter and

Fig. 11.2 Diagrammatic representation for nanoscale targeted delivery

large surface area, SiNPs can easily penetrate cell wall pores and traverse stomata. Figure 11.3 illustrates the effects of SiNPs on plants.

Furthermore, SiNP application regulates the levels of sodium (Na+) and chlorine (Cl-) ions within plants, aiding in maintaining osmotic and ionic balance and controlling growth rates (Bhat et al. 2021). Exogenous application of SiNPs significantly enhances plant growth by promoting antioxidant synthesis, thereby limiting salt accumulation and the formation of reactive oxygen species (ROS) (Naguib and Abdalla 2019).

SiNPs exert direct effects on plant biology through various mechanisms. They form a binary layer on plant cell walls upon absorption, which may impart structural coloration (Strout et al. 2013; Suriyaprabha et al. 2014). Moreover, SiNPs confer disease resistance to plants by activating defense responses and safeguarding against bacterial, nematodal, and fungal infections (Rastogi et al. 2019). For instance, in wheat, SiNPs stimulate the plant's defense system and innate immune response

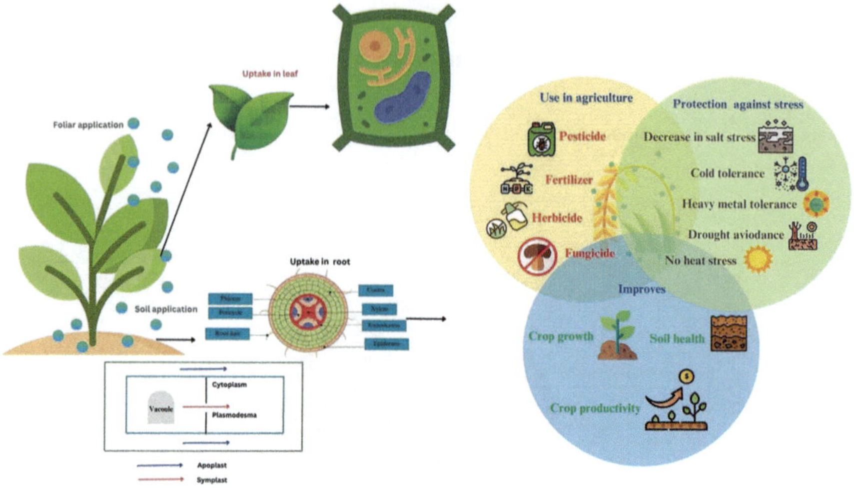

Fig. 11.3 Entry of SiNPs followed by its benefits on plants

against the fungus *Rhizoctonia solani* (Abdelrhim et al. 2021). Additionally, SiNP application induces biosilicification of the outer epidermal layer of leaves, acting as a physical barrier against phytopathogen penetration. In watermelon, deposition of a silicon layer on leaves inhibits the entry of *Fusarium oxysporum*, thus suppressing wilt disease (Kang et al. 2021).

Several studies have demonstrated the ability of SiNPs to mitigate the hazardous effects of heavy metals such as aluminum, cadmium, arsenic, and chromium. SiNPs elevate mineral content and antioxidant activities while minimizing the bioaccumulation of metal ions (Wang et al. 2022). For example, in soybean seedlings, nanosilica application reduces salt stress by increasing potassium (K^+) ion concentration and antioxidant potential while decreasing sodium (Na^+) ion concentration and ROS formation (Farhangi-Abriz and Torabian 2018). Leaf application of SiNPs in rice reduces cadmium and lead toxicity by decreasing endogenous metal ion concentrations and enhancing plant growth (Hussain et al. 2020). Similarly, in wheat seedlings, SiNP treatment reduces cadmium accumulation and enhances tissue biomass, chlorophyll content, and gaseous exchange in leaves (Ali et al. 2021). SiNPs also mitigate aluminum toxicity in maize seedlings by enhancing antioxidant defense mechanisms and reducing the effects of photorespiratory enzymes (De Sousa et al. 2019).

11.5 Use in Agriculture

The current agricultural landscape emphasizes the importance of weed control and enhancing crop yield. In this regard, SiNPs may serve as carriers for the targeted distribution of fertilizers, pesticides, and herbicides (Rastogi et al. 2019). Nanosilica has been employed in conjunction with organic fertilizers to enhance crop productivity. Mesoporous SiNPs (Janmohammadi et al. 2016) can be utilized as pesticides in two ways: firstly, by directly applying them to crops, a silica layer may form, inhibiting the growth and entry of insects and larvae; secondly, utilizing them to deliver pesticides enables target-specific delivery due to their smaller size (Naaz et al. 2023). Table 11.1 outlines the potential benefits of SiNPs on plant growth and development.

SiNPs are beneficial in agriculture for mitigating heavy metal toxicity and combating stress from UV radiation, salinity, and drought (Tripathi et al. 2017). Various types of silicate molecules categorize Si into sodium, potassium, calcium, steel slag, and mineral-based silicate fertilizers. Among these, potassium and sodium silicates are most used in agriculture. Application of potassium silicates significantly reduces the spread of powdery mildew in cucumber, zucchini, and muskmelon (Bhat et al. 2021). Nanosilica has been applied together with organic fertilizers to improve crop productivity. Mesoporous SiNPs are ideal delivery agents for nitrogen-containing fertilizers (Janmohammadi et al. 2016). Additionally, Si-based nanofertilizers have stimulated seed and root development in various crops such as cucurbits, tomatoes, and maize (Suriyaprabha et al. 2014).

SiNPs can also be utilized as pesticides in two ways: directly applying them to crops can form a silica layer that inhibits the growth and entry of insects and larvae, or they can be used to deliver pesticides for target-specific delivery due to their small size (Naaz et al. 2023). SiNPs have demonstrated fatal effects on the cowpea seed beetle (*Callosobruchus maculatus*) and major pests of stored grains (El-Helaly et al. 2016; Rouhani et al. 2013). Additionally, SiNPs exhibit antifungal activity against various pathogens affecting crops such as potato sheath blight, early blight in tomatoes, cucumber leaf spot, grain discoloration, and brown spots on rice (Debona et al. 2017).

Commercially available herbicides typically target the aboveground portions of weeds, leaving below-ground parts unaffected, allowing them to propagate. Silica-based nanoherbicidal formulations are easily absorbed through plant cuticles, releasing active constituents to targeted sites (Rastogi et al. 2019; Johnson et al. 2020). Herbicides such as anilides, benomyl, carbendazim, and chloroacetanilide, when submerged in diatom fistulae, have been effectively delivered by silicon nanocarriers in their active forms.

SiNPs also serve as nanosensors for identifying different metals in soil. Due to their core-shell structure, they offer superior accuracy, responsiveness, and sensor stability compared to other nanoparticles (Sun et al. 2016). Various studies have developed optical sensors for detecting substances like melamine and Zn ions using SiNPs, enhancing stability and accuracy (Rastogi et al. 2019).

Commercially available herbicides only kill the above-ground portion of weeds. They do not affect below-ground plant parts (rhizomes and tubers), which tends to give rise to new plants. The silica-based nanoherbicidal formulation is easily absorbed through the plant cuticle, and the active constituent is slowly released to the targeted site. Herbicides such as anilides, benomyl, carbendazim, and

Table 11.1 The potential benefit of SiNPs on plant growth and development

Crops	SiNPs used	Benefit	References
Solanum lycopersicum L.	SiNPs (11–14 nm)	Root-knot nematode resistance	Ardakani (2013)
Zea mays L.	SiNPs (10 μM), As^V (25 μM) + SiNPs (10 μM) and As^V (50 μM) + SiNPs (10 μM)	Arsenic and oxidative stress tolerance	Tripathi et al. (2016)
Triticum aestivum L.	Sodium metasilicate (Na_2SiO_3, 10 μM)	UV-B stress tolerance	Tripathi et al. (2017)
Avena sativa L.	Sodium trisilicate ($Na_2O_7Si_3$; sigma) and SiO_2 nanopowder (SiO2, +99%, 20–30 nm)	Multiple trait improvement	Asgari et al. (2018)
Triticum aestivum L.	SiNPs in deionized water (0, 300, 600, 900, 1200 mg L^{-1})	Cadmium tolerance	Hussain et al. (2019)
Cucumis sativus L.	Hydrophilic SiNPs (0, 100, 200, 300, and 400 mg kg^{-1})	Drought and salt tolerance	Alsaeedi et al. (2019)
Fragaria × ananassa Duch.	25 μM NaCl + $nSio_2$ (different concentrations) 50 μM NaCl + $nSio_2$ (different concentrations)	Salt tolerance	Avestan et al. (2019)
Triticum aestivum L.	0, 25, 50, and 100 mg kg^{-1}	Heavy metal and drought tolerance	Khan et al. (2019)
Vitis vinifera L.	Chitosan nanoparticles (CN), $Si(OC_2H_5)_4$, SiNPs (SN), and chitosan-SiNPs (CSN) at 0.40, 0.50, 0.60, and 1.00%	Protection against *Botrytis cinerea* (gray mold fungus)	Youssef et al. (2019)
Triticum aestivum L.	1.66 μM silicon dioxide nanoparticles (SiNPs)	Heat tolerance	Younis et al. (2020)
Musa acuminata Colla	Sodium metasilicate (Na_2SiO_3, 0, 200, 400, and 600 mg L^{-1})	Drought and salt tolerance	Mahmoud et al. (2020)
Cuminum cyminum L.	Different compositions of Ti, Si, and SAP (superabsorbent polymer)	Drought tolerance	Salajegheh et al. (2020)
Hordeum vulgare L.	Sodium silicate (Na_2SiO_3, sigma-Aldrich) and silicon dioxide (SiO_2 nanopowder, 99 + %, 20–30 nm, amorphous) at 125 and 250 mg L^{-1}	Drought tolerance	Ghorbanpour et al. (2020)
Tanacetum parthenium (L.) Sch.Bip.	Foliar application of glycine-nanosilicon, histidine-nanosilicon, and glutamine-nanosilicon (0 μM control, 1.5 μM, and 3.0 μM)	Drought tolerance	Esmaili et al. (2022)
Zea mays L.	15 nm SiO_2 + 12 nm for ZnO	Salinity tolerance	Shoukat et al. (2024)
Oryza sativa L.	SiNPs (2.5 mM) and FeNPs (25 mg L^{-1})	Reclamation of Pb-polluted soil	Ghouri et al. (2024)

chloroacetanilide submerged in diatom fistulae have been delivered by silicon nanocarriers in their active form (Rastogi et al. 2019; Johnson et al. 2020).

These applications underscore the diverse potential of SiNPs in revolutionizing agricultural practices and improving crop productivity.

11.6 Influence of SiNPs on Soil Microbiome

SiNPs contribute to plant growth by enhancing antioxidant enzyme activity, increasing leaf water content, and promoting the growth of beneficial microbial populations, such as plant growth-promoting rhizobacteria (PGPR). Nanosilica, a form of SiNP, positively influences soil bacterial populations and nutrient content, thereby enhancing soil fertility and crop yield. Combined application of PGPR and SiNPs has been observed to mitigate the adverse effects of saline water on barley growth and improve nutrient absorption (Gonc et al. 2023). Simultaneous application of SiNPs and nitrogen-fixing bacteria, such as *Rhizobium leguminosarum*, has resulted in higher growth rates in *Pisum sativum* plants (Epstein 1994). Leaf application not only increased bacterial communities, such as *Paenibacillus* and *Rhodobacteraceae*, but also enhanced fungal genera such as *Chaetomium* in rhizosphere soil (Tian et al. 2020). Silicate-solubilizing bacteria (SSB) convert insoluble silicates into soluble forms, thereby improving soil fertility and plant productivity (Etesami et al. 2021). Utilizing SSB enhances silicon solubilization, thereby improving phosphorus availability. Combining silicon with phosphate-solubilizing bacteria (PSB) and arbuscular mycorrhizal fungi (AMF) is an effective strategy to enhance phosphorus uptake and plant growth. Various AMF species, such as *Glomus etunicatum* and *Rhizophagus clarus*, have demonstrated increased silicon uptake in various plant species, leading to improved mycorrhizal efficiency (Etesami et al. 2021).

11.7 SiNP Application and Concerns

Three alternative methods, namely, leaf, soil, and seed priming, are employed for applying SiNPs to plants. In leaf treatment, liquid SiNPs are sprayed on leafy shoot tissues, which are then absorbed by stomata or epidermal cells of leaves (Laane 2018). A binary film develops at the epidermal cell wall after SiNPs absorption, producing a structural color (Strout et al. 2013). Soil application of SiNPs is more efficient than leaf application due to the higher uptake of SiNPs by root tissues compared to leaf tissue (Suriyaprabha et al. 2014). Additionally, SiNPs may be applied to the soil/nutrient solution. The efficacy of this technique depends on soil attributes such as pH, texture, and salt content, along with the period during which NPs release agrochemicals (Bhat et al. 2021). Root exudates combine with NPs and are transported through membrane transporters, aquaporins, and endocytosis (Rastogi et al. 2019).

SiNPs can also be used for seed priming, in addition to the approaches. Recent studies have shown that seed priming with SiNPs significantly enhances the growth and productivity of several plant species, including wheat and tomatoes (Hussain et al. 2019). This method offers significant benefits compared to the other two methods (Sundaria et al. 2019). Seed priming allows for regulated hydration and induces specific physiological conditions in plants during germination.

While numerous studies have demonstrated the positive effects of silicon nanoparticles (SiNPs) on plant growth, there is evidence of potential harm, particularly at high concentrations. For instance, certain sizes of SiNPs have been associated with delayed development in *Arabidopsis* and reduced growth in Bt cotton (Slomberg and Schoenfisch 2012; Karimi and Mohsenzadeh 2016). The adverse effects appear to depend on factors such as nanoparticle size, concentration, and plant species. Understanding these factors is crucial for maximizing the benefits of SiNPs in agriculture. Moreover, SiNPs negatively affected wheat crop seedlings when concentrations exceeded 200 mg/L, resulting in reduced growth and altered biochemical parameters (Yuvaraj et al. 2023). Integrating SiNPs with other beneficial microbes and modified insecticides could enhance their effectiveness in agricultural practices aimed at increasing crop production while reducing reliance on agrochemicals.

11.8 Conclusion

The utilization of silicon nanoparticles in agriculture holds immense potential for sustainable farming practices. SiNPs' diverse applications and proven efficacy in enhancing plant growth under stress conditions offer promising solutions to address agricultural challenges. Further research efforts are essential to fully understand their mechanisms and optimize their use for the benefit of future agriculture and global food security.

References

Abdelrhim AS, Mazrou YS, Nehela Y, Atallah OO, El-Ashmony RM, Dawood MF (2021) Silicon dioxide nanoparticles induce innate immune responses and activate antioxidant machinery in wheat against Rhizoctonia solani. Plan Theory 10:2758

Ali S, Rizwan M, Hussain A, Zia Ur Rehman M, Ali B, Yousaf B, Wijaya L, Banerjee A, Aditya, Singh A, Sudarshan M, Roychoudhury, Aryadeep (2021) Silicon nanoparticle-pulsing mitigates fluoride stress in rice by fine-tuning the economic and metabolomic balance and refining agronomic traits. Chemosphere 262:1–49

Alsaeedi A, El-Ramady H, Alshaal T, El-Garawany M, Elhawat N, Al-Otaibi A (2019) Silica nanoparticles boost the growth and productivity of cucumber under water deficit and salinity stresses by balancing nutrient uptake. Plant Physiol Biochem 139:1–10

Ardakani AS (2013) Toxicity of silver, titanium and silicon nanoparticles on the root-knot nematode, Meloidogyne incognita, and growth parameters of tomato. Nematology 15:671–677.

Asgari F, Majd A, Jonoubi P, Najafi F (2018) Effects of silicon nanoparticles on molecular, chemical, structural, and ultrastructural characteristics of oat (Avena sativa L.). Plant Physiol Biochem 127:152–160

Avestan S, Ghasemnezhad M, Esfahani M, Byrt CS (2019) Application of nano-silicon dioxide improves salt stress tolerance in strawberry plants. Agronomy 9:246

Bhat JA, Rajora N, Raturi G, Sharma S, Dhiman P, Sanand S, Shivaraj SM, Sonah H, Deshmukh R (2021) Silicon nanoparticles (SiNPs) in sustainable agriculture: major emphasis on the practicality, efficacy, and concerns. Nanoscale Adv 3:4019–4028

Bonacchi S, Genovese D, Juris R, Montalti M, Prodi L, Rampazzo E, Zaccheroni N (2011) Luminescent silica nanoparticles: extending the frontiers of brightness. Angew Chem Int Ed 50:4056–4066

Caubet M, Cornu S, Saby NP, Meunier JD (2020) Agriculture increases the bioavailability of silicon, a beneficial element for crops, in temperate soils. Sci Rep 10:19999

Coskun D, Britto DT, Huynh WQ, Kronzucker HJ (2016) The role of silicon in higher plants under salinity and drought stress. Front Plant Sci 7:1072

De Sousa A, Saleh AM, Habeeb TH, Hassan YM, Zrieq R, Wadaan MA, Hozzein WN, Selim S, Matos M, AbdElgawad H (2019) Silicon dioxide nanoparticles ameliorate the phytotoxic hazards of aluminum in maize grown on acidic soil. Sci Total Environ 693:133636

Debona D, Rodrigues FA, Datnoff LE (2017) Silicon's role in abiotic and biotic plant stresses. Annu Rev Phytopathol 55:85–107

Dixit CK, Bhakta S, Kumar A, Suib SL, Rusling JF (2016) Fast nucleation for silica nanoparticle synthesis using a sol–gel method. Nanoscale 8:19662–19667

El-Helaly AA, El-Bendary HM, Abdel-Wahab AS, El-Sheikh MAK, Elnagar S (2016) The silica-nano particles treatment of squash foliage and survival and development of Spodoptera littoralis (Bosid.) larvae. Journal of Entomology and Zoology Studies 4:175–180

Epstein E (1994) The anomaly of silicon in plant biology. Proc Natl Acad Sci 91(1):11–17

Esmaili S, Tavallali V, Amiri B, Bazrafshan F, Sharafzadeh S (2022) Foliar application of nano-silicon complexes on growth, oxidative damage and bioactive compounds of feverfew under drought stress. Silicon 14:10245–10256

Etesami H, Jeong BR, Glick BR (2021) Contribution of arbuscular mycorrhizal fungi, phosphate–solubilizing bacteria, and silicon to P uptake by plant. Front Plant Sci 12:699618

Farhangi-Abriz S, Torabian S (2018) Nano-silicon alters antioxidant activities of soybean seedlings under salt toxicity. Protoplasma 255:953–962

Finnie KS, Bartlett JR, Barbé CJ, Kong L (2007) Formation of silica nanoparticles in microemulsions. Langmuir 23:3017–3024

Ghorbanpour M, Mohammadi H, Kariman K (2020) Nanosilicon-based recovery of barley (Hordeum vulgare) plants subjected to drought stress. Environ Sci Nano 7:443–461

Ghouri F, Sarwar S, Sun L, Riaz M, Haider FU, Ashraf H, Lai M, Imran M, Liu J, Ali S, Liu X (2024) Silicon and iron nanoparticles protect rice against lead (Pb) stress by improving oxidative tolerance and minimizing Pb uptake. Sci Rep 14(1):5986.

Gonc G, Gunes A, Akca H, Taskin MB (2023) Green synthesis of biogenic nano-silicon from rice husk and its effect on combined boron and salinity stress tolerance of barley and wheat. J Soil Sci Plant Nutr. https://doi.org/10.1007/s42729-023-01577-w

Gubala V, Giovannini G, Kunc F, Monopoli MP, Moore CJ (2020) Dye-doped silica nanoparticles: synthesis, surface chemistry and bioapplications. Cancer Nanotechnol 11:1–43

Hoffmann F, Cornelius M, Morell J, Fröba M (2006) Silica-based mesoporous organic–inorganic hybrid materials. Angew Chem Int Ed 45:3216–3251

Hussain CM, Mitra S (2012) Nanomaterials for sample preparation, vol 2. In Comprehensive Sampling and Sample Preparation: Analytical Techniques for Scientists, pp 389–418

Hussain A, Rizwan M, Ali Q, Ali S (2019) Seed priming with silicon nanoparticles improved the biomass and yield while reduced the oxidative stress and cadmium concentration in wheat grains. Environ Sci Pollut Res. https://doi.org/10.1007/s11356-019-04210-5

Hussain B, Lin Q, Hamid Y, Sanaullah M, Di L, Hashmi MLUR, Khan MB, He Z, Yang X (2020) Foliage application of selenium and silicon nanoparticles alleviates Cd and Pb toxicity in rice (Oryza sativa L.). Sci Total Environ 712:136497

Janmohammadi M, Amanzadeh T, Sabaghnia N, Ion V (2016) Effect of nano-silicon foliar application on safflower growth under organic and inorganic fertilizer regimes. Botanica Lithuanica 22:53–64

Johnson SN, Rowe RC, Hall CR (2020) Silicon is an inducible and effective herbivore defence against Helicoverpa punctigera (Lepidoptera: Noctuidae) in soybean. Bull Entomol Res 110(3):417–422
Kang H, Elmer W, Shen Y, Zuverza-Mena N, Ma C, Botella P, White JC, Haynes CL (2021) Silica nanoparticle dissolution rate controls the suppression of Fusarium wilt of watermelon (Citrullus lanatus). Environ Sci Technol 55:13513–13522
Karande SD, Jadhav SA, Garud HB, Kalantre VA, Burungale SH, Patil PS (2021) Green and sustainable synthesis of silica nanoparticles. Nanotechnol Environ Eng 6:29
Karimi J, Mohsenzadeh S (2016) Effects of silicon oxide nanoparticles on growth and physiology of wheat seedlings. Russ J Plant Physiol 63:119–123
Khan ZS, Rizwan M, Hafeez M, Ali S, Adrees M, Qayyum MF, Khalid S, Rehman MZ, Sarwar MA (2019) Effects of silicon nanoparticles on growth and physiology of wheat in cadmium contaminated soil under different soil moisture levels. Environ Sci Pollut Res 27:4958–4968
Klabunde KJ, Richards RM (eds) (2009) Nanoscale materials in chemistry. John Wiley & Sons
Laane HM (2018) The effects of foliar sprays with different silicon compounds. Plan Theory 7(2):45
Lee JY, Kim MK, Nguyen TL, Kim J (2020) Hollow mesoporous silica nanoparticles with extra-large mesopores for enhanced cancer vaccine. ACS Appl Mater Interfaces 12(31):34658–34666
Lehman SE, Mudunkotuwa IA, Grassian VH, Larsen SC (2016) Nano–bio interactions of porous and nonporous silica nanoparticles of varied surface chemistry: a structural, kinetic, and thermodynamic study of protein adsorption from RPMI culture medium. Langmuir 32:731–742
Mahawar L, Ramasamy KP, Suhel M, Prasad SM, Živčák M, Brestic M, Rastogi A, Skalicky M (2023) Silicon nanoparticles: comprehensive review on biogenic synthesis and applications in agriculture. Environ Res. https://doi.org/10.1016/j.envres.2023.116292
Mahmoud LM, Dutt M, Shalan AM, El-Kady ME, El-Boray MS, Shabana YM, Grosser JW (2020) Silicon nanoparticles mitigate oxidative stress of in vitro-derived banana (Musa acuminata 'grand Nain') under simulated water deficit or salinity stress. S Afr J Bot 132:155–163
Meena VD, Dotaniya ML, Coumar V, Rajendiran S, Ajay, Kundu S, Subba Rao A (2014) A case for silicon fertilization to improve crop yields in tropical soils, vol 84. Proceedings of the National Academy of Sciences, India Section b: Biological Sciences, pp 505–518
Mehta S, Gogna M, Singh B, Patra A, Singh IK, Singh A (2020) Silicon: a plant nutritional "non-entity" for mitigating abiotic stresses. In: Giri B, Sharma MP (eds) Plant stress biology. Springer, Singapore. https://doi.org/10.1007/978-981-15-9380-2_2
Naaz H, Rawat K, Saffeullah P, Umar S (2023) Silica nanoparticles synthesis and applications in agriculture for plant fertilization and protection: a review. Environ Chem Lett 21:539–559
Naguib DM, Abdalla H (2019) Metabolic status during germination of nano silica primed Zea mays seeds under salinity stress. J Crop Sci Biotechnol 22:415–423
Rao GB, Susmitha P (2017) Silicon uptake, transportation and accumulation in Rice. J Pharm Phytochem 6:290–293
Rastogi A, Tripathi DK, Yadav S, Chauhan DK, Živčák M, Ghorbanpour M, El-Sheery NI, Brestic M (2019) Application of silicon nanoparticles in agriculture. 3 Biotech 9:1–11
Rouhani M, Samih MA, Kalantari S (2013) Insecticidal effect of silica and silver nanoparticles on the cowpea seed beetle, Callosobruchus maculatus F.(Col.: Bruchidae). J Entomol Res 4:297–305
Salajegheh M, Yavarzadeh M, Payandeh A, Akbarian MM (2020) Effects of titanium and silicon nanoparticles on antioxidant enzymes activity and some biochemical properties of Cuminum cyminum L. under drought stress. Banat J Biotechnol 1:19–25
Sharma R, Kumar V (2024) Nano enabled agriculture for sustainable soil. Waste Manag Bulletin 2:152–161
Sharma D, Kanchi S, Bisetty K (2019) Biogenic synthesis of nanoparticles: a review. Arab J Chem 12:3576–3600
Shoukat A, Pitann B, Hossain MS, Saqib ZA, Nawaz A, Mühling KH (2024) Zinc and silicon fertilizers in conventional and nano-forms: mitigating salinity effects in maize (Zea mays L.). J Plant Nutr Soil Sci. https://doi.org/10.1002/jpln.202300267

Slomberg DL, Schoenfisch MH (2012) Silica nanoparticle phytotoxicity to Arabidopsis thaliana. Environ Sci Technol 46(18):10247–10254

Stober W, Fink A, Bohn E (1968) Controlled growth of monodisperse silica spheres in the micron size range. J Colloid Interface Sci 26:62–69

Strout G, Russell SD, Pulsifer DP, Erten S, Lakhtakia A, Lee DW (2013) Silica nanoparticles aid in structural leaf coloration in the Malaysian tropical rainforest under storey herb Mapania caudata. Ann Bot 112(6):1141–1148

Sun D, Hussain HI, Yi Z, Rookes JE, Kong L, Cahill DM (2016) Mesoporous silica nanoparticles enhance seedling growth and photosynthesis in wheat and lupin. Chemosphere 152:81–91

Sundaria N, Singh M, Upreti P, Chauhan RP, Jaiswal JP, Kumar A (2019) Seed priming with iron oxide nanoparticles triggers iron acquisition and biofortification in wheat (Triticum aestivum L.) grains. J Plant Growth Regul 38:122–131

Suriyaprabha R, Karunakaran G, Yuvakkumar R, Rajendran V, Kannan N (2014) Foliar application of silica nanoparticles on the phytochemical responses of maize (Zea mays L.) and its toxicological behavior. Synth React Inorg M 44(8):1128–1131

Tang L, Cheng J (2013) Nonporous silica nanoparticles for nanomedicine application. Nano Today 8:290–312

Tian L, Shen J, Sun G, Wang B, Ji R, Zhao L (2020) Foliar application of SiO2 nanoparticles alters soil metabolite profiles and microbial community composition in the pakchoi (Brassica chinensis L.) rhizosphere grown in contaminated mine soil. Environ Sci Technol 54(20):13137–13146

Tripathi DK, Singh S, Prasad SM, Chauhan DK, Dubey NK (2016) Silicon nanoparticles more efficiently alleviate arsenate toxicity than silicon in maize cultiver and hybrid differing in arsenate tolerance. Front Environ Sci 4:46

Tripathi DK, Singh S, Singh VP, Prasad SM, Dubey NK, Chauhan DK (2017) Silicon nanoparticles more effectively alleviated UV-B stress than silicon in wheat (Triticum aestivum) seedlings. Plant Physiol Biochem 110:70–81

Wang L, Ning C, Pan T, Cai K (2022) Role of silica nanoparticles in abiotic and biotic stress tolerance in plants: a review. Int J Mol Sci 23:1947

Younis AA, Khattab H, Emam MM (2020) Impacts of silicon and silicon nanoparticles on leaf ultrastructure and TaPIP1 and TaNIP2 gene expressions in heat stressed wheat seedlings. Biol Plant 64:343–352

Youssef K, de Oliveira AG, Tischer CA, Hussain I, Roberto SR (2019) Synergistic effect of a novel chitosan/silica nanocomposites-based formulation against gray mold of table grapes and its possible mode of action. Int J Biol Macromol 141:247–258

Yuvaraj M, Priya RS, Jagathjothi N, Saranya M, Suganthi N, Sharmila R, Cyriac J, Anitha R, Subramanian KS (2023) Silicon nanoparticles (SiNPs): challenges and perspectives for sustainable agriculture. Physiol Mol Plant Pathol 102161. https://doi.org/10.1016/j.pmpp.2023.102161

Chapter 12
Exploring Silicon Nanoparticle Uptake, Transport, and Biological Functions in Plants

Shahid Ali

Abstract Nanoparticles (NPs) show a potential to greatly improve the efficacy of agrochemicals, crop yield, and soil vitality. Nevertheless, scientists have yet to understand the mechanisms by which nanoparticles enter plants, plant cells, and organelles. To improve our understanding of nanoparticle-plant interactions, our primary focus is on elucidating the mechanisms by which nanoparticles transpose barriers within plants and plant cells and identifying the key factors that impede an effective dispersion of nanoparticles. Silicon (Si) is a highly beneficial metalloid found in nature and positively mitigates environmental problems. The exceptional chemical and optoelectronic characteristics of silicon/silica nanoparticles (Si/SiO_2 NPs) have come to significant attention. Researchers have found its mesoporous structure, easy availability, and minimal biological toxicity. Chemical, physical, and biological processes can be used to create Si/SiO_2 NPs, which can then be applied to leaves, the soil, or seed priming. After Si/SiO_2 NPs are ingested and transported to target cells, they experience optimal growth, development, and resistance to environmental stressors, pest attacks, and pathogen infections. Si/SiO_2 NPs may be used as environmentally benign and economical supplements for sustainable agriculture. It facilitates the uptake of nutrients, increases plant resilience, and helps plants to manage biotic stress. This study suggests developing nanomaterials for agriculture to increase plant yields and stress tolerance.

Keywords Abiotic and biotic stress · Nanoparticles · Silver NPs · Translocation · Uptake

---

S. Ali (✉)
Guangxi Key Laboratory for Agro-Environment and Agro-Product Safety, Guangxi Colleges and Universities Key Laboratory of Crop Cultivation and Tillage, College of Agriculture, Guangxi University, Nanning, China

R. de Mello Prado et al. (eds.), *Silicon Advances for Sustainable Agriculture and Human Health*, Sustainable Plant Nutrition in a Changing World,
https://doi.org/10.1007/978-3-031-69876-7_12

12.1 Introduction

Nanotechnology is increasingly used in a variety of sectors, including agriculture. Several agricultural applications use nanoparticles, including for disease control, insect eradication, nutrient delivery, biotechnological plant modification, plant monitoring and sensing, and others (Wang et al. 2016). Nanoparticles (NPs) are available in many different forms and sizes. Nanoparticles may vary in size from 1 to 100 nm. They can be categorized as lipid-based, polymeric, metallic, ceramic, semiconductor, or carbon NPs depending on size, shape, and physicochemical properties. Silicon NPs have attracted special interest in the field of nanotechnology due to their quasi-essential effects on plants and their crucial function in reducing environmental stressors (Farooq and Dietz 2015; Mosa et al. 2022). These nanoparticles can replace various chemical fertilizers and are not harmful to the environment. Targeted distribution of SiNPs can provide effective solutions to various agricultural problems, including weeds, pathogenicity, insect infestation, abiotic stressors, and loss in crop output (Rastogi et al. 2019). Nano-Si has shown a superior performance compared to bulk Si in terms of yield, UV-B stress mitigation, and heavy metal toxicity (Afshari et al. 2021). According to Tubaña and Heckman (2015), silicon can be soluble or adsorbed with aluminum oxides or iron oxides in the soil. Si concentrations in plants range from 0.1% to 10%.

Plants absorb silicon in various forms, including silicic and monosilicic acid ($Si(OH)_4$ or H_4SiO_4). The concentration of silicon available to plants in the soil varies significantly, ranging from 10 ppm to over 100 ppm (Liang et al. 2015). Providing Si fertilizers to soils with less than 20 ppm of silicon (Si) is often recommended. Wollastonite is a highly preferred and cost-effective natural source of silicon replenishment. In a 2008 study, Maxim et al. found that wollastonite is a naturally occurring calcium metasilicate ($CaSiO_3$) consisting primarily of calcium (Ca, 34.3%) and silicon (Si, 24.3%), along with minor amounts of magnesium (Mg), aluminum (Al), manganese (Mn), iron (Fe), potassium (K), and sodium (Na). Calcite, diopside, garnet, idocrase, and quartz are considered less suitable natural silicon sources. The use of materials such as potassium silicate, sodium silicate, steel slag, and sugarcane bagasse can supplement Si for agricultural plants (Tubaña and Heckman 2015).

According to Williams and Crerar (1985), silicic acid in its monomeric form is accessible to plants, while its polymeric form, which bounds soil particles by forming silica bridges, plays a part in improving soil aggregation and water retention capacity (Norton et al. 1984). The majority of silicon found in the soil is insoluble and useless for horticulture and agronomy. As a result, soil chemical and physical weathering occurs to make Si available to plants. The physicochemical properties of the minerals and environmental variables such as pH and temperature affect silicate minerals (White and Brantley 1995; Gérard et al. 2002). Furthermore, rather than the Si concentration in the soil (Brenchley and Maskell 1927), it is the silicic acid concentration in the soil solution that determines the concentration of Si in plants (Ding et al. 2005; Henriet et al. 2008). Silicon transporters aid plants in absorbing silicon (Si) in the dissolved form of monosilicic acid (H_4SiO_4) and storing it as hydrated silica (SiO_2nH_2O) (Rao and Susmitha 2017). Si/SiO_2 nanoparticles are given to plants using one of three methods: leaf application, soil/root application, or seed priming. However, these nanoparticles are not naturally found in the soil. In the soil-plant system, there are disparities in the uptake and movement of silicon (Si) and silicon dioxide nanoparticles (Si/SiO_2 NPs). Si/SiO_2 nanoparticles show a high absorption of Si due to their mesoporous nature and tiny size (around 20 nm).

Plasmodesmata and cytoplasmic bridges facilitate the transfer of Si/SiO_2 NPs from one cell to another, ensuring cell continuity. Si/SiO_2 NPs are more accessible to seedlings than Si, and nano-Si is more efficient than bulk Si (Tripathi et al. 2016, 2017, 2021). Therefore, Si positively affects crop output and quality (Gomaa et al. 2021). Si/SiO_2 NPs increase fertilizer absorbability and the power of conventional chemical fertilizers (NPK) (El-Naggar et al. 2020). The application of mineral fertilizer to the soil, combined with a leaf spray of nanofertilizers, results in the highest values for many plant characteristics, including height, ear length, rows per ear, grains per row, grains per ear, 100 grains weight, straw yield, and grain yield (Kandil et al. 2020). Plant production increased using nanosilicon dioxide combined with organic fertilizer (Janmohammadi et al. 2016). Fertilizers based on urea, boron, and nitrogen were effectively delivered using mesoporous silica nanoparticles (MSNs), which had a pore size of 2–10 nm (Wanyika et al. 2012). Si-NPs can, therefore, transport fertilizers and herbicides to plants, and they may be a single fertilizer for some crops.

Si concentration increases and triggers physical and physiological stress defense pathways (Roohizadeh et al. 2015). Si offers supplementary benefits such as improved growth, development, and resilience against environmental stresses, insect attacks, and pathogen infection (Liu et al. 2019; Singh et al. 2020). Plants present resilient defense mechanisms to endure stress. Silicon nanoparticles (Si-NPs) have been recognized as regulators for improving plant defense against abiotic environmental stressors. Si-NPs may provide resistance to plants by creating a double layer on the epidermal cell wall and altering the antioxidant defense process. Plants improve their water efficiency under drought conditions by developing an additional layer beneath the outermost layer of leaves, which reduces water loss due to the deposition of silicon (Hattori et al. 2005). Si/SiO_2 nanoparticles are readily used as growth stimulants. They may be used in conservative and safe quantities to manage diverse pests and function as potent, non-traditional insecticides for crops throughout storage periods (Singh et al. 2021). This article provides a comprehensive overview of different methodologies for formulating Si/SiO_2 nanoparticles, their uses, and their absorption and translocation mechanisms. This study elucidates the precise role of silicon in facilitating plant growth and development, as well as its protective mechanisms against both biotic and abiotic stressors. Furthermore, it proposes future study opportunities to explore the beneficial characteristics of silicon and silicon oxide nanoparticles (Si/SiO_2 NPs) in improving global crop production.

12.2 The Dynamics and Sources of Nanoparticles in the Soil-Plant System

Both artificial and natural processes can produce NPs. Natural phenomena, such as photochemical reactions, volcanic eruptions, and forest fires, can indirectly produce NPs. NP formation is a product of soil erosion from both wind and water (Buzea et al. 2007). In addition, anthropogenic activities control NP generation. NPs are unintentionally created; they may result from combustion, food preparation, or production activities, including welding, refining, and smelting (Remedios et al. 2012). Engineered NPs, including carbon NPs, metal oxide NPs, zero valence metal NPs, and others, are NPs specifically created for industrial, agricultural, and other purposes (Barcelo et al. 2011).

There are several different chemical processes used to create NPs. Scientists might change these characteristics to best exploit NPs' chemical, physical, and surface qualities for a given function. The most common and reliable way to regularly create NPs with the desired properties is to use standardized chemical and physical methods in labs and industry (Tsuji et al. 2003; Kudu et al. 2008; Okitsu et al. 2007). However, these NPs may harm the environment and people's health directly. The processes involved in chemical synthesis are also costly and time-consuming (Narayanan et al. 2010; Gan et al. 2012). It is crucial to understand how NPs may be synthesized repeatedly while presenting appropriate properties. Recently, researchers have discovered less expensive and more environmentally friendly ways to make NPs. They may be produced from various precursors besides plants. The reductive nature of proteins and metabolites found in some species (Makarov et al. 2014) causes them to change bulk materials into their respective NPs in algae (Govindaraju et al. 2009), diatoms (Dahoumane et al. 2017; Scarano and Morelli 2002), bacteria (Lengke et al. 2007), yeast (Roychoudhury 2020), and fungus (Rautaray et al. 2003). For the synthesis of NPs, which are frequently non-toxic, plants have gained popularity as "green factories" (Salama 2012). "Green NPs" may be produced by cultivating and processing several plants; NPs made from plant extracts and microorganisms are more pharmacologically potent than NPs made from chemical synthesis. NPs derived from medicinal plants are the most pharmacologically powerful, perhaps because they contain several pharmacologically active residues (Selvakesavan and Franklin 2021; Singh and Konwar 2012).

In plants, the primary synthesis method is the buildup of the metal within the tissue, which is then released through a series of physiologic reactions, producing NPs that can be retrieved. Si is an element that occurs naturally and has a Van der Waals radius of 210 pm. Si nanoparticles, on the other hand, are intentionally produced particles that show distinct characteristics and are smaller than 100 nm. Ultra-small silicon nanoparticles have exceptional effectiveness in improving soil characteristics, in stark contrast to bulk silicon, which is relatively inferior (Nayfeh and Mitas 2008). The predominant form of Si in the soil solution was monomeric (H_4SiO_4; monosilicic acid), along with oligomeric and polysilicic acids. Plants may readily obtain the monomeric form of silicon (Frew et al. 2018). There are two different methods by which plants absorb Si: active uptake and passive uptake (Kaur and Greger 2019). During the active mode, plants show a faster rate of silicon absorption compared to that of water, resulting in a decrease in the concentration of Si in the uptaken solution. Nevertheless, when in passive mode, plants assimilate silicon at a rate comparable to the absorption of water. Consequently, there are no apparent alterations in the Si concentration in the uptake solution during passive absorption (El-Ramady et al. 2018). However, further research is needed to fully understand the processes involved in the various types of silicon absorption. Si-NPs may be transported from the roots to different aboveground regions along three pathways: cell wall pores, apoplastic route, and symplastic pathway.

Specific silicon acids, when dissolved in the soil solution, form complexes with organic and inorganic compounds. Silcretes are a form of derived soil characterized by its high silicon content. The content of Si in petrocalcic layers is significantly lower compared to that of silcrete (8%). Furthermore, the silicon content is considerably lower in specific minerals present in badly weathered Oxisols, such as bauxite and ferricrete soils (Rastogi et al. 2019). While most soils are rich in silicon, some are particularly deficient in the Si that plants can use (Coskun et al. 2019). Oxisols and Ultisols have suffered from extensive weathering, leaching, acidification, and low base saturation (Foy 1992). Conversely, the histosols show little mineral content and are highly organic (Snyder et al. 1986). Additionally, soils with prolonged agricultural productivity and a large concentration of quartz sand have low plant-available Si (Rastogi et al. 2019).

12.3 Mechanisms and Consequences of Plant Nanoparticle Uptake and Translocation

Plants are affected by nanoparticles that are present in the soil, air, and water. By quickly adhering to a variety of plant surfaces, atmospheric nanoparticles enter the plant system through wounds and stomatal pores (Rai et al. 2018). The deliberately applied water-borne nanoparticles may potentially interact with plant tissues in soil ecosystems (Mauter et al. 2018). The nanoparticles released from the soil, or a soil mixed with wastewater effluents, which are used in crop nutrition, are the first point of contact with plant roots (Cox et al. 2020; Ahmed et al. 2021). Nanoparticles can thus infiltrate plant tissues through various mechanisms, such as passive transport, ion channel transport, aquaporin-mediated transport of water molecules, endocytosis, pore formation, carrier proteins, and association with organic matter (Jha and Pudake 2016; Yanga et al. 2017).

When nanoparticles released into the soil come into touch with exudates, they can readily deposit on or stick to the surface of the root (Zhao et al. 2016; Gao et al. 2018). As a result, after targeted or random contacts with root exudates, the adsorbed nanoparticles experience a substantial physicochemical change with humic acids at the same time. Through several redox reactions, the compounds released by plants into the soil, which can oxidize and reduce, may alter nanoparticles containing metals with different valence shells (Zhang et al. 2017). Hence, the level of bioaccumulation and the ultimate fate of nanoparticles in soil or plant systems may be modified by a physicochemical change induced by plant exudates (Lv et al. 2015). Exposure to plants makes nanoparticles traverse the cell wall and cell membrane of the root epidermis, entering the xylem. Subsequently, they undergo symplastic movement inside the stem and ultimately translocate to the leaves. Nevertheless, it is necessary for NPs to transverse cell membrane holes to pass through an intact cell membrane, indicating that the absorption of nanomaterials depends on their size (Rico et al. 2011). NPs need to be passively absorbed into the apoplast of the endodermis before they can reach the stele. The xylem is the primary mechanism for the transportation and dispersion of nanoparticles (Judy 2013). The cell wall typically consists of a permeable lattice of polysaccharide fibers that allow the absorption of water molecules and other solutes (Aslani et al. 2014). The NP absorption process is often considered an active transport mechanism that also involves several other cellular functions, including recycling, signaling, and plasma membrane control (Etxeberria et al. 2009). Several studies have shown that plant cell walls function as a semipermeable barrier, allowing only certain particles to pass through holes (Miralles et al. 2012). In cell suspension media, nanoparticle internalization is altered by endocytosis occurring in the vacuole from the apoplast via vesicles produced from the plasma membrane (Chen et al. 2010).

Silicon is found in the soil, and a plant's ability to absorb and transport it depends on the soil's chemical composition and the roots. Plant roots absorb Si as monosilicic acid, which is then transferred to the shoots. Si is mostly deposited in the root endodermis, on the outer epidermal cells of inflorescence bracts, and in leaf epidermal cells. Si transporters are present in the cell membrane in rice plants because of a quicker uptake of Si (Mandlik et al. 2020). A study analyzed the molecular mechanism of silicon (Si) absorption and transport in rice using a mutagenesis approach. It showed that the transporter gene *Lsi1* plays a critical role in Si transport, in addition to *Lsi2*, a second Si transporter gene, a proton antiport, and anion transporter with 9–12 transmembrane domains. Identifying these transporters has explained the molecular basis for Si absorption and transport in rice plants (Ma et al. 2007; Siddiqui et al. 2020) (Fig. 12. 1). The *ZmLsi1* and *ZmLsi6* genes in *Z. mays* plants regulated Si absorption and deposition. *ZmLsi1* is crucial for the absorption of Si by roots. In *Z. mays* leaves, *ZmLsi6* is found in the parenchyma cells and helps Si unloading into the xylem. There are also reports of *Lsi2* and *HvLsi2* gene homologs in *H. vulgare* plants (Bokor et al. 2015).

Furthermore, the form and distinct tissue arrangement of leaves and roots exert a major impact on the absorption and transport of nanoparticles into the plant system. After entering the plant system, nanoparticles can modify physiological, biochemical, and morphological characteristics to improve resistance to biotic and abiotic stresses and control plant growth and development.

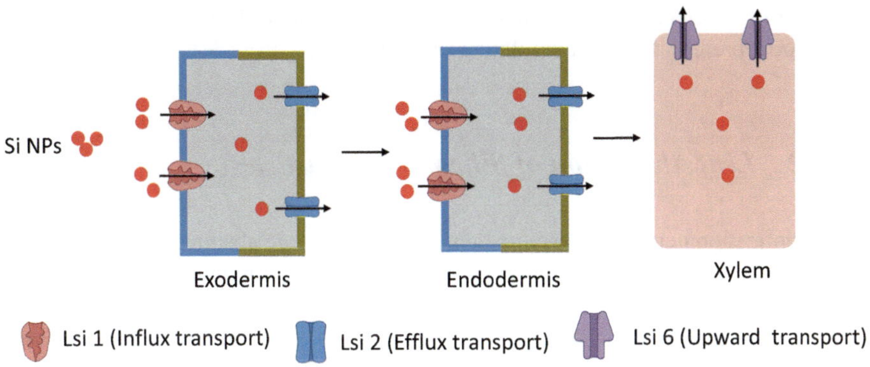

Fig. 12.1 Molecular mechanisms of silicon nanoparticle uptake and translocation within plant tissues

12.3.1 Root Absorption of Silicon Nanoparticles

The roots come into close contact with nanoparticles. They can absorb them from the soil and distribute them to different plant tissues (Galway 2006). The porous, thinner cuticle of roots and the cell wall of root hairs improve this absorption (Vithange et al. 2017; Ahmed et al. 2018). Previous studies have reported the absorption and accumulation of different nanoparticles by plant root cells. A favorable association between the rate of water absorption and the uptake of nanoparticles may help transpiration to aid in this process (Rico et al. 2013). Nanoparticle stability is a significant component in their bioavailability to plants (Von Moss et al. 2014).

In contrast, the release of enzymes and other chelating substances by soil organisms results in the modification of nanoparticles and increases their availability to plants (Schwabe et al. 2015). Size, type of coating, and homo- or hetero-aggregation of nanoparticles, all have a significant impact on bioavailability and require careful monitoring. Additionally involved in the surface adsorption and intracellular absorption of nanoparticles are proteins, humic acids, fulvic acids, and polysaccharides.

Encapsulated nanoparticles gather in the nucleus, endodermis, pericycle, walls, membranes, and vacuoles of cells, among other sub-cellular regions. They migrate to plant organs, including leaves, fruits, flowers, and stem nodes (Rajput et al. 2018). In general, the vascular bundle among plant tissues plays a significant role in the movement of nanoparticles across plant organs, even if the aggregation of nanoparticles happens in several areas (Pradas Del Real et al. 2017; Lv et al. 2019). Nanoparticles migrate quickly toward plant shoots once they enter the circulatory system or tissues such as the xylem. The phloem helps nanoparticles to accumulate in fruits as well. This implies that nanoparticles of a certain size can pass and move through phloem tissues, which is the only pathway into fruit tissues. Soybeans cultivated in a hydroponic solution containing metal and metal-oxide nanoparticles showed accumulation of these nanoparticles in their roots, nodules, stems, and pods

(Priester et al. 2017). Various nanoparticles show distinct behaviors in plant systems, most preferentially accumulating in plant roots.

12.3.2 Leaf Absorption of Silicon Nanoparticles

Leaf application helps to determine how nanoparticles are absorbed and distributed from leaves to shoots and then to below-ground parts. Engineered nanoparticles come into contact directly with exposed organs, such as open stomata, leaf hydathodes, and trichomes. Because of their small size, nanoparticles can directly infiltrate into the plant system when providing the leaf with a solution or aerosol-based spray (Wang et al. 2013). Nanoparticles with an average size of around 100 nm can readily pass through stomatal openings that are generally 100 nm in size during leaf treatments (Schwabe et al. 2015). Coating the nanoparticles with polar material increases stomatal permeability for polar substances and improves the absorption of nanoparticles (Schreiber 2005). According to an experiment, the absorption efficiency of C13 and N15 revealed that the leaf pore size in three dicot plants was greater than 100 nm (Eichert and Goldbach 2008). The absorption of individual nanoparticles or nano-sized aggregates can occasionally block stomata, slowing the generation of photosynthetic pigments by increasing the temperature of the leaf and reducing the rate of water transpiration (Hussain et al. 2013; Hirano et al. 1990).

When tiny droplets of water are hooked on a leaf, in addition to stomata, nanoparticles may be absorbed or eliminated more efficiently following guttation with leaf tip hydathodes. A study reported that cucumber plants' hydathodes released insoluble radioactive $141CeO_2$ nanoparticles. However, the interaction of nanoparticles with the plant system is a complicated process both in roots and leaves (Khan et al. 2019). The unique characteristics of nanoparticles, such as their surface area and high reactivity, facilitate their interaction with plant tissues. Moreover, size, concentration, stability, and chemical structure of nanoparticles exert a significant impact on their absorption and transportation inside plants (Lv et al. 2019). Azim et al. (2022) identified many chemical variables that affect the movement and attachment of nanoparticles in plant tissues, including Brownian motion, gravity, van der Waals forces, and double-layer forces. Nanoparticles may infiltrate the plant system through several organs using the aerial pathway, including stomata, hydathode, trichomes, or wounds produced by phytopathogens (Fincheria et al. 2020). The contact of the root with the soil also represents a crucial and intricate channel for nanoparticle absorption (Azim et al. 2022; Ali et al. 2021). As a result, the absorption and transport of nanoparticles within the plant system substantially affect their capacity to reduce biotic stress in plants. However, plant types and growth stage must also be considered when analyzing this interaction. Each plant species has unique anatomical barriers that control nanoparticle entry (i.e., the composition of cell walls controls nanoparticle transit based on the solubility and chemical form of particles; Singh et al. 2021).

12.4 Silicon and Other Nutrient Contents in Plants

Plant nutrition has a significant impact on crop quality and agricultural productivity. Approximately 40–60% of the world's food output come from crops grown using fertilizers (Roberts 2009). Fertilizers are essential for increasing agricultural output, but their nutrient use efficiency is still poor. A significant amount of fertilizer is lost by leaching, denitrification, and microbial immobilization. Since a significant portion of fertilizer nutrients are lost to the environment, plants cannot use them (Wu et al. 2008). Applying nano-based fertilizers is an innovative nanotechnology strategy that may revolutionize the fertilizer industry. These fertilizers are created by the nanoencapsulation process, which enables the regulation and gradual delivery of nutrients to plants. These methods increase the fertilizer use efficiency while reducing nutrient waste to the environment (Abobatta 2018). Nanofertilizers should ideally release nutrients in response to plant needs. DeRosa et al. (2010) stated that such a prompt release avoids the early transformation of nutrients into chemical or gaseous forms that plants cannot yet access. One example is the volatilization of NH_3 from urea. Because of its intelligent distribution technique, nanofertilizers are also known as smart fertilizers. Compared to soluble fertilizers, slow-release fertilizers have the benefit of requiring less volume since they release their nutrients gradually according to the plant's needs. Nanomaterials, which have a high surface area to volume ratio, may efficiently hold on to nutrients, acting as a longer-lasting and more reliable nutrient reservoir for plants (Navarro et al. 2008). The substances mentioned include urea, chitosan, and a fertilizer that is soluble in water in granular form. The slow-release, double-coated NPK fertilizer was created using cross-linked (poly)acrylic acid)/diatomite. NPK was applied as the outer coating, inner coating, and core of the fertilizer. This formulation resulted in a fertilizer with excellent water retention and superabsorbent capacity (Wu et al. 2008). This environmentally safe, non-toxic fertilizer has a good ability to retain water in a regulated manner for delivering nutrients, making it extremely beneficial in horticultural and agricultural applications. Chitosan has been used in combination with methacrylic acid nanoparticles to provide a controlled release of nitrogen, potassium, and phosphorus nutrients in NPK fertilizer sources such as urea, potassium chloride, and calcium phosphate (Corradini 2010). For the gradual and controlled release of urea and phosphorus fertilizers, nanocoating with sulfur (100 nm layer) offers advantages, particularly for sulfur-deficient soils (Brady and Weil 1999). The pace at which fertilizer dissolves decreases by increasing coating stability, allowing for a steady, persistent release. Synthetic apatite nanoparticles have been used as an innovative type of phosphorus (P) fertilizer for plants due to their delayed and ongoing phosphorus release (Liu and Lal 2014). This reduces the danger of water eutrophication.

Mesoporous silica nanoparticles (MSNPs) deliver nutrients to plants. The designed, synthesized, and processed components that make up MSNPs can be used in several ways to mediate plant nutrition. The physical and chemical characteristics of MSNPs vary greatly, but they are typically hydrophilic solid materials with large surfaces that have many holes or channels (Wang et al. 2016). MSNPs are highly suitable for immobilizing nutrients and toxic substances, transporting and releasing plant nutrients in a regulated manner, chemically stabilizing nutrients to improve their availability for plant absorption, and facilitating the entry of nutrients into plant cells because of their size, charged surfaces, and other chemical and physical characteristics (Luyckx et al. 2017a, b). MSNPs show specific attributes that make them highly suitable for the development of "nanofertilizers." Such fertilizers can be produced by either incorporating or enveloping nanoparticles with nutrients or by converting the nutrients themselves into nanoparticles for gradual nutrient release. This technology is comparable to other methods used for a controlled release of nutrients (Kah et al. 2018). However, untreated MSNPs behave somewhat similarly to chelating agents in terms of plant feeding because of their surface charge (Adams et al. 2018). This type of functioning depends on interactions between the MSNP and nutrient ions and the MSNP's entry into the plant. Nanotechnologies are promising in improving the delivery of micronutrients to plants, which normally show use efficiencies of less than 5% when added to soils as fertilizer (Monreal et al. 2015). Several studies have analyzed MSNP-nutrient ion interactions on plants. MSNPs interacted with nutrients in cotton when grown in a hydroponic nutrient solution. They changed the concentrations of many nutrients in both roots and shoots, depending on the cultivar (Lee et al. 2016). Slomberg and Schoenfisch (2012) observed effects on growth and leaf chlorosis in the hydroponic culture of *Arabidopsis* and treated plants with highly negatively charged silicon nanoparticles. The authors hypothesized that the dynamics of nutrient adsorption on the nanoparticle surface, which were pH-dependent, caused these effects. When MSNPs were introduced to nutrient-rich soil-less media, Sun et al. (2016) observed that *Lupinus albus* L. and *Triticum aestivum* L. showed an improved early growth and photosynthesis.

Although it is hard to collect data on natural leaf absorption testing, there is evidence that MSNPs can be taken up by plants through the roots. Sun et al. (2014) introduced MSNPs to a substrate devoid of soil and monitored the particles' root absorption and translocation throughout plant development. Measurements of both symplastic and apoplastic routes revealed that plant roots absorbed MSNPs before being transported by xylem tissues to plant shoots. Le et al. (2014) demonstrated a comparable transfer from the roots to shoots of MSNPs in cotton plants cultivated in hydroponic nutrient solutions containing Si nanoparticles. Hussain et al. (2013) also described the root absorption of MSNPs in wheat, lupin, and *Arabidopsis* using hydroponics. According to Slomberg and Schoenfisch (2012), the absorption of Si nanoparticles in hydroponic cultures is size-dependent. The only proof we could find of MSNP leaf uptake (though artificial) came from high-velocity leaf bombardment techniques.

These studies were not conducted in soils. MSNPs are more likely to interact chemically with the substrate since soil chemistry is more complicated than hydroponic solutions or artificial media. MSNPs may affect plants in various ways besides nutritional ones, some of which depend on the substance's inherent characteristics. The presence of Si benefits plants, particularly when they are under stress caused by thicker and more lignified xylem cell walls. Nazaralian et al. (2017) also reported such beneficial effects of silicon (Si) in fenugreek (*Trigonella foenum-graecum* L.). Both silicate and silica-based nanoparticles were efficient in treating UV-B stress in wheat, according to Tripathi et al. (2017). However, silica nanoparticles were more successful, showing improved plant accessibility and availability of the nanoparticle form of silicon. The study of Siddiqui et al. (2014) reported that the inclusion of Si nanoparticles in water may improve several aspects of germination and early development, such as seed germination, seedling vigor, and seeding dry mass. The characteristics were evaluated in petri dish germination assays conducted on tomato seedlings. It was unclear what led to these improvements in different parameters; however, it might have been due to supply of Si, osmotic environment optimization, or other variables.

12.5 Effects of Silicon on Plant Growth and Development

The availability of nutrients in the soil around the roots, known as the rhizospheric soil, exerts a significant impact on the growth and development of plants. Applying Si-NPs may increase agricultural growth and yields. The improved plant growth after Si-NP treatment is mostly due to the plants' increased nutrient bioavailability (Liu et al. 2015; Liu and Lal 2015). Scientists have focused on using Si-NPs in the soil to boost plant development since Si is important to plants, much as other necessary macronutrients. Si insufficiency is also connected to dietary abnormalities impairing plant growth and development (Epstein 1999).

Many studies have reported that the Si-NP priming of several crop seeds, such as wheat, pea, and *Brassica* species, increased seed germination and seedling growth (Chourasiya et al. 2021) (Fig. 12.2). Si-NPs function as a fertilizer (Verma et al. 2020; Siddiqui et al. 2014; Verma et al. 2021a, b). The application of Si-NPs led to a significant improvement in plant height, fresh and dry weight, seed quality, and overall production. In addition, it reduced the negative effects of salt stress on plant development and the volume of water in the soil. Exposure to Si-NPs caused considerable changes in the physiological and morphological characteristics of *Z. mays* plants.

Silicon NPs affect plant growth and development depending on several variables, including size, shape, application stage, and biomechanical and physical features (Rastogi et al. 2017). Si-NPs can reportedly change the morphological behavior and physiological activity of plants in various ways, according to certain findings (Siddiqui et al. 2014; Rajput et al. 2021). Several studies were on plants and Si-NPs sought to understand the underlying mechanisms better. Plants spontaneously mineralize NPs to ensure a healthy growth and development (Rajput et al. 2021). Better plant performance and production are ensured when Si-NPs are used, even in unfavorable environmental circumstances.

Previous studies showed that silicon nanoparticles significantly improved the germination of *Stevia rebaudiana* Bertoni. Also, the nano-primed seedling improves

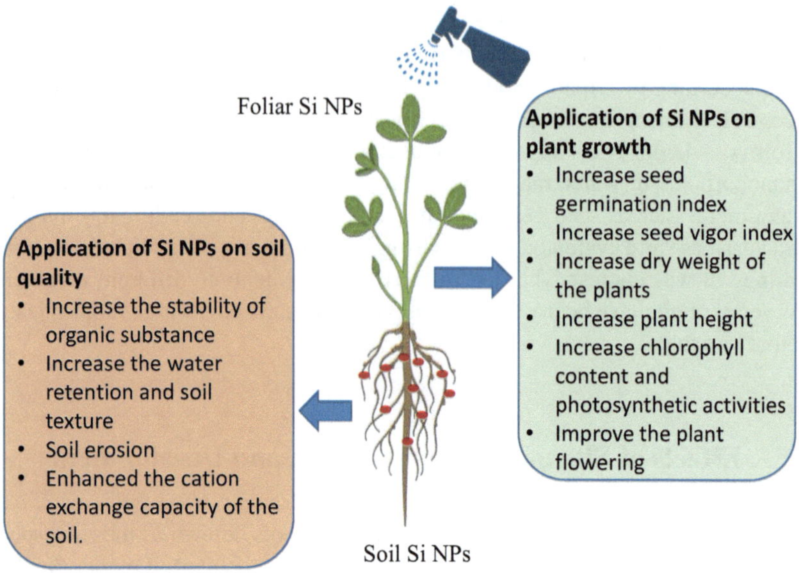

Fig. 12.2 Si-NPs improved plant growth and development by modulating different physiological and biochemical mechanisms, further improving soil conditions for growth

root, shoot, and dry weight compared to the control seedling (Hasanaklou et al. 2023). The leaf application of Si-NPs stimulates stress tolerance in lemongrass. Si-NPs amplified the photosynthetic CO_2 assimilation and stomatal conduction rate by 21% and 24% compared to salt-treated plants. The plants also significantly improved height, dry weight, and leaf area (Mukarram et al. 2023). Using SiO_2-NPs alleviated the damage of salt and low-temperature dual stress in cotton seedlings. SiO_2-NPs-treated plants improved stomatal conduction, transpiration rate, and photosynthesis (Liang et al. 2023). Si-NPs increased the growth of leaves and stem in wheat plants and increased the chlorophyll content and partly metabolomics of the plant hormones and soluble sugars. Si-NPs increase IAA (auxin) and fructose contents, directly or indirectly improving plant growth. Also, Si-NPs reduce the expression of genes related to cytokinins (IPT) while increasing the expression of genes associated with other pathways such as chlorophyll, soluble sugars, IAA, and ABA. These pathways play a role in regulating plant development (Li et al. 2023). The use of Si-NPs has the potential to improve both the development and yield of wheat plants when subjected to drought stress conditions. The findings suggested that seeds treated with varying doses of Si-NPs showed improved spike length, grains per spike, 1000 grains weight, grain yield, and plant height in comparison to the control group (Raza et al. 2023) (Table 12.1).

Table 12.1 Effects of silicon nanoparticles on plants under different environmental stress conditions

Environmental stress	Plant species	Mode of NP application	Impact of nanoparticles on plants	References
Drought stress	*Triticum aestivum*	Seed priming	To mitigate water stress, enhance plant biomass, chlorophyll levels, crop productivity, and antioxidant enzyme activity	Raza et al. (2023)
Drought stress	*Triticum aestivum*	Foliar	Lower the H_2O_2 and lipid peroxide content, and increase proline content and stress-related gene expression (*WRKY19*, *DREB2*, *MYB33*, and *SnRK2.4*)	Boora et al. (2023)
Drought stress	*Hordeum vulgare*	Soil	Enhanced plant biomass, altered osmolyte, and metabolite profile	Ghorbanpour et al. (2020)
Drought stress	*Arabidopsis thaliana*	Soil	Enhance the process of seed germination and promote the growth of plant biomass	Tran et al. (2023)
Drought stress	*Beta vulgaris*	Foliar spray	Increased leaf area index, photosynthetic rate, antioxidant activities, and glycine betaine	Namjoyan et al. (2020)
Drought stress	*Musa acuminata*	Foliar spray	Enhanced plant biomass, decreased electrolyte leakage, and MDA content	Mahmoud et al. (2020)
Salt stress	*Oryza sativa*	Soil	Enhance plant development and increase stress tolerance by manipulating The salt overly sensitive mechanisms	Gupta et al. (2021)
Salt stress	*Oryza sativa*	Hydrophonic	Lowered oxidative damage and GABA accumulation while enhancing the endogenous level of polyamine	Das et al. (2021)
Salt stress	*Medicago sativa*	Irrigation water	Lowered oxidative damage and improved total polyamine, flavonoids, and carotenoids	El Moukhtari et al. (2021)
Salt stress	*Solanum tuberosum L.*	Irrigation water	An induced antioxidant enzyme lowered the Na+/K+ ratio and increased proline and leaf-soluble carbohydrates	Kafi et al. (2021)

(continued)

Table 12.1 (continued)

Environmental stress	Plant species	Mode of NP application	Impact of nanoparticles on plants	References
Arsenic stress	*Solanum lycopersicum*	Soil	Increase enzymatic activities of APX and CAT, as well as photosynthesis	Gonzalez-Moscoso et al. (2019, 2022)
Lead (Pb) stress	*Coriandrum sativum*	Foliar spray	Enhanced POD, SOD, and CAT activities. Minimized MDA content	Fatemi et al. (2021)
Chromium (Cr) stress	*Pisum sativum*	Hydrophonic	Reduce ROS and increase antioxidant activities.	Tripathi et al. (2015)
Chromium (Cr) stress	*Brassica juncea*	Hydrophonic	Upregulated the expression of antioxidant genes such as DHAR, APX, and GR	Sharma et al. (2023)
Pseudomonas syringae	*Arabidopsis thaliana*	Foliar spray	Increased the plant's immunity by activating the salicylic acid	El-Shetehy et al. (2021)
Rhizoctonia solani	*Triticum aestivum L.*	Seed priming	Activated both non-enzymatic and enzymatic antioxidants. Induce defense-related compounds such as salicylic acid	Abdelrhim et al. (2021)
Meloidogyne incognita	*Solanum melongena*	Soil	Reduce the reproduction and population of the nematodes. Increase the growth of the plants	El-Ashry et al. (2022)
Meloidogyne incognita	*Lycopersicon esculentum*	Foliar spray	Enhanced plant growth, reduced nematode infections in the root system	Udalova et al. (2021)
Heat stress	*Triticum aestivum L.*	Foliar spray	Enhanced chlorophyll and carotenoid contents accumulate more osmoprotectants	Mustafa et al. (2021)
Heat stress	*Hordeum Vulgare*	Foliar spray	Improve the photosynthesis and decrease the MDA and H_2O_2 content	Hussain et al. (2019)
Heat stress	*Hordeum Vulgare*	Soil	Elevated the level of soluble sugars and free proline, which in turn boosted the activity of antioxidant enzymes	Naz et al. (2022)
Heat stress	*Aloe barbadensis*	Hydrophonic	Improved the concentration of soluble sugars and antioxidant enzyme activities	Azarfam et al. (2020)

12.6 Effects of Silicon on Plants Under Stress Conditions

12.6.1 The Role of Si-NPs Under Abiotic Stresses

Environmental pressures such as heavy metals, salinity, heat, and soil salinity have a negative influence on the production and quality of agriculture worldwide. Abiotic stresses alter the physiological, morphological, and biochemical characteristics of plants, resulting in reduced growth and output (Verma et al. 2021a, b; Rajput et al. 2021). As a result of reducing the effects of various abiotic stressors, Si-NPs have been found to play essential roles in plants (Wang et al. 2021; De Sousa et al. 2019; Gu et al. 2011). Adding Si-rich materials to soils improves plant growth and development and reduces metal toxicity (Verma et al. 2021a, b; Rajput et al. 2021).

Silicon nanoparticles can significantly reduce the concentration of heavy metals in plants. Data indicate that the use of Si-NPs has led to a decrease in the amount of lead present in the various tissues of *Brassica chinensis* L. in comparison to the control group (Tian et al. 2020). These findings imply that the exogenous application of Si-NPs can reduce the absorption of heavy metals (Cui et al. 2017; Cui et al. 2020). Cadmium is a dangerous poisonous metal (Shoeva and Khlestkina 2018) that inhibits photosynthesis and plant growth and development and affects plant productivity (Hayat et al. 2010). Silicon NPs boosted the plant growth and biomass in Cd-stressed plants (Hussain et al. 2020). The application of Si-NP to the soil decreases the concentration of Cd in plant tissues, particularly in grains, regardless of whether plants were under drought stress. Additionally, the Si-NP treatment stimulated plant growth and improved photosynthesis. Si-NPs also reduced oxidative stress, shown by increased superoxide dismutase and peroxidase activity, decreased H_2O_2 production, electrolyte leakage, and malondialdehyde levels. Also, Si treatment protects the cell membrane from severe damage (Merwad et al. 2018). The results indicate that plants exposed to Cd stress decreased the levels of reactive oxygen species and increased the components of antioxidative defense when treated with Si-NPs (Memari-Tabriz et al. 2021).

The leaf spray of Si-NPs significantly reduced the buildup of Pb in the leaves of *O. sativa* (Hussain et al. 2020). Another novel approach to reduce Al phytotoxicity in acidic soils was a treatment with Si-NPs in the soil. The detailed mechanism of ROS and antioxidant activities was shown at cellular and biochemical levels, which play an important role in the mitigation process (De Sousa et al 2019). Metalloid arsenic (As) harms plants and negatively affects their development (Tripathi et al. 2021; Buracherskaya et al. 2021). In *O. sativa*, Si-NP decreased stress-mediated vulnerabilities (Liu et al. 2014). Silicon NPs were used to repair the photosynthetic process and stop deterioration. Additionally, Si-NPs improved the activities of antioxidant enzymes that fight ROS production and restore cellular homeostasis (Tripathi et al. 2021). According to previous reports, chromium (Cr) accumulates in plants, changes the photosynthesis nutrient absorption, and significantly reduces plant growth (Sharma et al. 2020). Under Cr stress, Si-NPs can improve *P. sativum* seedling development. Si-NPs improved plant growth under Cr stress, such as pigment content, chlorophyll fluorescence, protein level, and nutritional status (Tripathi et al. 2021). Improvements in physiological parameters occur after reducing Cr accumulation in plant organs.

The improved stress tolerance of Si-NPs may be attributed to the upregulation of *OsNAC* genes, which in turn upregulate the expression of stress tolerance genes, including proline synthesis, soluble sugar biosynthesis, and redox homeostasis. Si-NPs play significant roles in mitigating metal toxicity in maize, leading to improved photosynthetic responses, reduced oxidative stress, and preserved antioxidative defense mechanisms. The application of Si-NPs resulted in a decrease in MDA levels during metal toxicity and in a considerable increase in cell wall width in the epidermis of roots. Also, Si-NPs reduced metal ion absorption and accumulation rates (Tripathi et al. 2017). Compared to inorganic Si, Si-NPs are more readily absorbed by plants and protect agricultural plants from toxicity by excess metal ions (Suriyaprabha et al. 2012).

Under salt stress conditions, Si-NPs dramatically increase the germination and vigor index in Cucurbita plants (Siddiqui et al. 2014). According to reports, Si can reduce abiotic stressors and maintain ion homeostasis. Under salt stress, Si-NPs created a favorable environment for seed germination (Siddiqui et al. 2014). The induction of growth may be due to Si-NPs' improved photosynthetic properties, which are necessary for nitrate absorption and photosynthetic leaf gas exchange (Rajput et al. 2021; Singhal et al. 2021).

The Si-NP deposition in various plant organs increases the strength and stiffness of plants, and it has also been related to the production of proteins and amino acids, as well as to nutrient intake (Xiao et al. 2023). Si-NPs improved the formation of organic compounds such as proteins, pigments, and phenols compared to bulk particles, improving plant morphological and photosynthetic features (Verma et al. 2021a, b; Mukarram et al. 2021). Plants synthesize suitable solutes, such as glycine betaine and proline, to sustain the osmotic potential within plant cells in unfavorable environmental circumstances. Si-NPs increase proline concentrations. Proline is a global osmoprotectant, functions as an energy source and an antioxidant (Matysik et al. 2002), and controls gene expression, which leads to osmotic adjustment (Siddiqui et al. 2014; Iyer and Caplan 1998). Si-NPs may have a significant impact on the time it takes for *S. lycopersicum* plants to germinate, on vigor index, and the amount of fresh and air-dried biomass they produce (Siddiqui et al. 2014). Janmohammadi and Sabaghnia (2015) improved the germination of *Lens culinaris* seeds and the early growth and development of seedlings in a saline environment by introducing Si-NPs into a nutrient-rich medium. When *S. Lycopersicum* plants were subjected to saline conditions and treated with Si-NPs, the expression of salt-stressed genes such as *NCED3*, *AREB*, *CRK1*, and *TAS14* increased. On the other hand, the genes *MAPK2*, *MAPK3*, *RBOH1*, *ERF5*, *APX2*, and *DDF2* downregulated (Almutairi 2016). Si caused changes in plant cell metabolism, a decrease in the absorption of heavy metals by roots, and the exudation of certain compounds, including phenols and organic acids (Wang et al. 2004; Shetty et al. 2011). Additionally, plants use Si-NPs as a slow-release supplement of Si and as a carrier of crucial macronutrients (N, P, and K) under salt stress (Mushtaq et al. 2018). These substances can, therefore, be regarded as growth-promoting agents. Si-NPs have been applied to increase plants' tolerance to salt since Si supplementation may inhibit plant salinity (Mushtaq et al. 2018). The core of controlled-release fertilizers should also contain NPK fertilizers and Si-NPs to control the release. The core can be covered with chitosan as the first semipermeable coating and sodium alginate and kaolin as the outermost superabsorbent coating (Faizan et al. 2021). These artificial coatings may gradually release nutrients, enabling plants to store large quantities of water, control salinity, and flourish under drought conditions (Mushtaq et al. 2018).

12.6.2 *The Regulatory Role of Si-NPs Under Biotic Stress*

Si-NPs are applied as alleviating agents under different biotic stress conditions (Fig. 12.3). Silicon may aid plants in avoiding pathogen penetration by creating physical barriers and preventing pathogen colonization by improving systemic acquired resistance. Si also defends plants by making silicate papillae, which stop pathogenic structures from spreading, and by thickening the cuticle and cell walls of the host plant. Si accumulating in the host plant's cell wall to produce a second layer of silica makes pathogen penetration difficult (Kim et al. 2002). Si accumulation can form a second layer under the cuticle, reducing the likelihood of diseases and blocking pathogen penetration. Si may actively function as a protector in plants. Many phenolic-like chemicals were discovered when pathogens infected plants treated with Si (Tako et al. 2020). Si increases the mechanical strength of plant cell

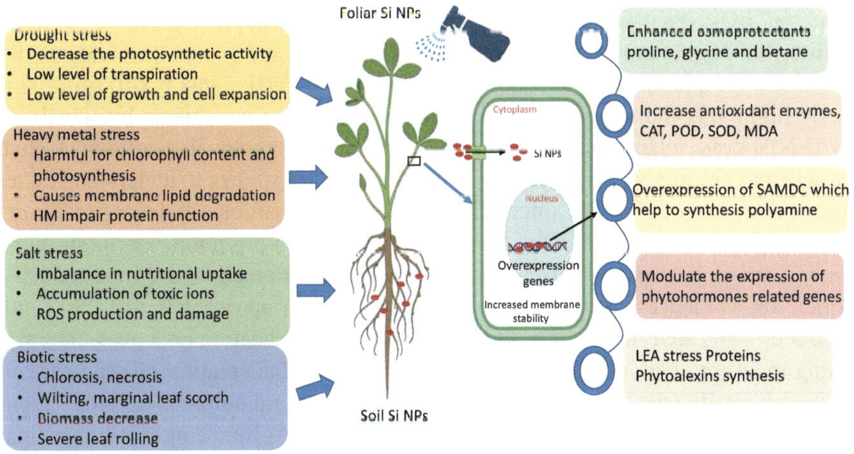

Fig. 12.3 Impacts of several abiotic and biotic stresses on plant growth: unveiling the protective mechanisms activated by silicon nanoparticles

walls by forming bonds with pectins, polyphenols, and hemicellulose (Tripathi et al. 2021; Guerriero et al. 2016). Silicon increases the cell wall hardness. It induced pathogen resistance and decreased blast illness in rice after Si treatment (Kim et al. 2002). The hyphae of *Pyricularia oryzae* could enter the *T. aestivum* effectively and then leave it, but the penetration was significantly low after Si treatment (De Sousa et al. 2019). Similarly, *Rhizoctonia solani* and *Pyricularia grisea* infection and leaf lesions were reduced in *O. sativa* after Si treatment (Rodrigues et al. 2003; Seebold et al. 2004).

Furthermore, the penetration of *P. oryzae* in *O. sativa* tissues decreased following Si treatment. The hypothesis was that the presence of Si layers prevented or delayed pathogen penetration (Hayaska et al. 2008). Si supplement decreased *O. sativa* sheath blight (Rodrigues et al. 2003). *O. sativa* blast lesion length decreased by 40–80% as a result of Si treatment (Seebold et al. 2004). In *Cucumis sativus*, the quantity of *Podosphaera fuliginea* colonies decreased by 43–94% after treatment with Si (Menzies et al. 1991).

In addition, Si-NPs are expected to aid in managing soil N levels and to lead to strong antibacterial effects against several plant pathogens (Ponmurugan et al. 2016). Si-NPs may inhibit the growth of bacteria (*Pectobacterium betavasculorum*), fungi (*Rhizoctonia solani*), and pathogenic fungi (*Fusarium oxysporum f.* sp. *niveum*) (Kang et al. 2021), as well as plant parasite nematodes (*Meloidogyne incognita*). The treatment with Si-NP boosted *P. sativum* growth, but the combined application of SiO_2-NPs and N-fixing bacteria (*Rhizobium leguminosarum*) resulted in the best growth rates (Kashyap and Siddiqui 2021).

12.7 Conclusion and Future Perspectives

Using Si/SiO_2 NPs in agriculture can improve agricultural performance and sustainability. Although nano-enabled agriculture shows a great promise, our understanding of the complex interactions between nanoparticles and plants is still little. This study highlights the significance of understanding the mechanisms by which Si/SiO_2 NPs cross plant barriers and affect crucial parameters that influence plant growth and productivity. It is crucial to conduct further studies to understand better the processes involved in the interactions between nanoparticles and plants. Gaining a thorough understanding of the role of Si/SiO_2 NPs in promoting optimum plant growth, development, and resilience will facilitate their practical use in sustainable agricultural practices. Furthermore, examining several approaches to Si/SiO_2 NP production and studying their multiple uses, such as application on leaves, incorporating them into soil, or priming seeds, may advance the creation of efficient and environmentally friendly solutions to improve agricultural output. This study highlights the potential of Si/SiO_2 NPs and establishes a basis for a continued investigation of nanoparticles that aims to mitigate problems in agriculture.

References

Abdelrhim AS, Mazrou YS, Nehela Y, Atallah OO, El-Ashmony RM, Dawood MF (2021) Silicon dioxide nanoparticles induce innate immune responses and activate antioxidant machinery in wheat against *Rhizoctonia solani*. Plan Theory 10(12):2758

Abobatta WF (2018) Nanotechnology application in agriculture, acta scientific agriculture (ISSN: 2581-365X) 2(6)

Adams CB, Thapa SB, Fan Y, Park S (2018) Agronomic and economic effects of two enhanced-efficiency urea fertilizer technologies on Southern Great Plains winter wheat. Agron J 110(3):1097–1102

Afshari F, Akhzari D, Kargar M (2021) Soil heavy metals pollution and phytoremediation potential of *Hordeum bulbosum* L. around bituminous mine, Ilam Province, Iran. ECOPERSIA 9(3):153–157

Ahmed B, Khan MS, Musarrat J (2018) Toxicity assessment of metal oxide nano-pollutants on tomato (*Solanum lycopersicon*): a study on growth dynamics and plant cell death. Environ Pollut 240:802–816. https://doi.org/10.1016/j.envpol.2018.05.015

Ahmed B, Rizvi A, Ali K, Lee J, Zaidi A, Khan MS, Musarrat J (2021) Nanoparticles in the soil–plant system: a review. Environ Chem Lett 19:1545–1609

Ali S, Mehmood A, Khan N (2021) Uptake, translocation, and consequences of nanomaterials on plant growth and stress adaptation. J Nanomater 2021:6677616

Almutairi ZM (2016) Effect of nano-silicon application on the expression of salt tolerance genes in germinating tomato (*Solanum lycopersicum* L.) seedlings under salt stress. South Cross J 9:106–114

Aslani F, Bagheri S, Muhd Julkapli N, Juraimi AS, Hashemi FSG, Baghdadi A (2014) Effects of engineered nanomaterials on plants growth: an overview. Sci World J 2014:2014

Azarfam SP, Nadian H, Moezzi A, Gholami A (2020) Effect of silicon on phytochemical and medicinal properties of aloe vera under cold stress. Appl Ecol Environ Res 18(1):561

Azim Z, Singh NB, Singh A, Amist N, Niharika, Khare S, Yadav RK, Bano C, Yadav V (2022) A review summarising uptake, translocation and accumulation of nanoparticles within the plants: current status and future prospectus. J Plant Biochem Biotechnol

Barceló D, Farré M (2011) Characterisation, analysis and risks of nanomaterials in environmental and food samples. Trends Analвt Chem 30(3):517–527

Bokor B, Bokorová S, Ondoš S, Švubová R, Lukăcová Z, Hýblová M, Szemes T, Lux A (2015) Ionome and expression level of Si transporter genes (Lsi1, Lsi2, and Lsi6) affected by Zn and Si interaction in maize. Environ Sci Pollut Res Int 22:6800–6811

Boora R, Sheoran P, Rani N, Kumari S, Thakur R, Grewal S (2023) Biosynthesized silica nanoparticles (Si NPs) helps in mitigating drought stress in wheat through physiological changes and upregulation of stress genes. Silicon 15:1–13

Brady NR, Weil RR (1999) In: Brady NR, Weil RR (eds) The nature and properties of soils. Prentice Hall, Upper Saddle River, pp 415–473

Brenchley WE, Maskell EJ (1927) The inter-relation between silicon and other elements in plant nutrition. Ann Appl Biol 14:45–82

Burachevskaya M, Minkina T, Fedorenko A, Fedorenko G, Chernikova N, Rajput VD, Mandzhieva S, Bauer T (2021) Accumulation, translocation, and toxicity of arsenic in barley grown in contaminated soil. Plant Soil 467:91

Buzea C, Pacheco II, Robbie K (2007) Nanomaterials and nanoparticles: sources and toxicity. Biointerphases 2(4):17

Chen R, Ratnikova TA, Stone MB, Lin S, Lard M, Huang G, Hudson JS, Ke PC (2010) Differential uptake of carbon nanoparticles by plant and mammalian cells. Small 6(5):612e617

Chourasiya V, Nehra A, Shukla P, Singh K, Singh P (2021) Impact of mesoporous nano-silica (SiO2) on seed germination and seedling growth of wheat, pea and mustard seed. J Nanosci Nanotechnol 21:3566

Corradini E (2010) A preliminary study of the incorporation of NPK fertilizer into chitosan nanoparticles. Express Polym Lett 4(8):509–515

Coskun D, Deshmukh R, Sonah H, Menzies JG, Reynolds O, Ma JF, Kronzucker HJ, Bélanger RR (2019) The controversies of silicon's role in plant biology. New Phytol 221:67–85

Cox AH, Wigginton SK, Amador JA (2020) Structure of greenhouse gas-consuming microbial communities in surface soils of a nitrogen-removing experimental drainfield. Sci Total Environ 739:140362

Cui J, Liu T, Li F, Yi J, Liu C, Yu H (2017) Silica nanoparticles alleviate cadmium toxicity in rice cells: mechanisms and size effects. Environ Pollut 228:363–369

Cui J, Li Y, Jin Q, Li F (2020) Silica nanoparticles inhibit arsenic uptake into rice suspension cells via improving pectin synthesis and the mechanical force of the cell wall. Environ Sci Nano 7:162–171

Dahoumane SA, Mechouet M, Wijesekera K, Filipe CD, Sicard C, Bazylinski DA, Jeffryes C (2017) Algae-mediated biosynthesis of inorganic nanomaterials as a promising route in nanobiotechnology–a review. Green Chem 19(3):552–587

Das P, Manna I, Sil P, Bandyopadhyay M, Biswas AK (2021) Silicon augments salt tolerance through modulation of polyamine and GABA metabolism in two indica rice (Oryza sativa L.) cultivars. Plant Physiol Biochem 166:41–52

De Sousa A, Saleh AM, Habeeb TH, Hassan YM, Zrieq R, Wadaan MAM, Hozzein WN, Selim S, Matos M, Abd Elgawad H (2019) Silicon dioxide nanoparticles ameliorate the phytotoxic hazards of aluminum in maize grown on acidic soil. Sci Total Environ 693:133636

DeRosa MC, Monreal C, Schnitzer M, Walsh R, Sultan Y (2010) Nanotechnology in fertilizers. Nat Nanotechnol 5:91

Ding TP, Ma GR, Shui MX, Wan DF, Li RH (2005) Silicon isotope study on rice plants from the Zhejiang province, China. Chem Geol 218:41–50

Eichert T, Goldbach HE (2008) Equivalent pore radii of hydrophilic foliar uptake routes in stomatous and astomatous leaf surfaces further evidence for a stomatal pathway. Physiol Plant 132:491–502. https://doi.org/10.1111/j.1399-3054.2007.01023.x

El Moukhtari A, Carol P, Mouradi M, Savoure A, Farissi M (2021) Silicon improves physiological, biochemical, and morphological adaptations of alfalfa (*Medicago sativa* L.) during salinity stress. Symbiosis 85:305–324. https://doi.org/10.1007/s13199-021-00814-z

El-Ashry RM, El-Saadony MT, El-Sobki AE, El-Tahan AM, Al-Otaibi S, El-Shehawi AM, Saad AM, Elshaer N (2022) Biological silicon nanoparticles maximize the efficiency of nematicides against biotic stress induced by Meloidogyne incognita in eggplant. Saudi J Biol Sci 29(2):920–932

El-Naggar ME, Abdelsalam NR, Fouda MM, Mackled MI, Al-Jaddadi MA, Ali HM, Siddiqui MH, Kandil EE (2020) Soil application of nano silica on maize yield and its insecticidal activity against some stored insects after the post-harvest. Nano 10(4):739

El-Ramady H, Alshaal T, Elhawat N, El-Nahrawy E, Omara AED, El-Nahrawy S, Elsakhawy T, Ghazi A, Abdalla N, Fári M (2018) Biological aspects of selenium and silicon nanoparticles in the terrestrial environments. Phytoremediation: Management of Environmental Contaminants 6:235–264

El-Shetehy M, Moradi A, Maceroni M, Reinhardt D, Petri-Fink A, Rothen-Rutishauser B, Mauch F, Schwab F (2021) Silica nanoparticles enhance disease resistance in Arabidopsis plants. Nat Nanotechnol 16(3):344–353

Epstein E (1999) Silicon. Annu Rev Plant Physiol Plant Mol Biol 50:641–664

Etxeberria E, Gonzalez P, Pozueta J (2009) Evidence for two endocytic transport pathways in plant cells. Plant Sci 177(4):341e348

Faizan M, Rajput VD, Al-Khuraif AA, Arshad M, Minkina T, Sushkova S, Yu F (2021) Effect of foliar fertigation of chitosan nanoparticles on cadmium accumulation and toxicity in *Solanum lycopersicum*. Biology 10:666

Farooq MA, Dietz KJ (2015) Silicon as versatile player in plant and human biology: overlooked and poorly understood. Front Plant Sci 6:994

Fatemi H, Esmaiel Pour B, Rizwan M (2021) Foliar application of silicon nanoparticles affected the growth, vitamin C, flavonoid, and antioxidant enzyme activities of coriander (*Coriandrum sativum* L.) plants grown in lead (Pb)-spiked soil. Environ Sci Pollut Res 28:1417–1425
Fincheira P, Tortella G, Duran N, Seabra AB, Rubilar O (2020) Current applications of nanotechnology to develop plant growth inducer agents as an innovation strategy. Crit Rev Biotechnol 40:15–30
Foy CD (1992) Soil chemical factors limiting plant root growth. In: Hatfield JL, Stewart BA (eds) Limitations to plant root growth. Springer, New York, pp 97–149
Frew A, Weston LA, Reynolds OL, Gurr GM (2018) The role of silicon in plant biology: a paradigm shift in research approach. Ann Bot 121:1265–1273
Galway ME (2006) Root hair cell walls: filling in the framework. Can J Bot 84:613–621. https://doi.org/10.1139/B06-006
Gan PP, Ng SH, Huang Y et al (2012) Green synthesis of gold nanoparticles using palm oil mill effluent (POME): a low-cost and eco-friendly viable approach. Bioresour Technol 113:132–135
Gao X, Avellan A, Laughton S et al (2018) CuO nanoparticle dissolution and toxicity to wheat (*Triticum aestivum*) in rhizosphere soil. Environ Sci Technol 52:2888–2897. https://doi.org/10.1021/acs.est.7b05816
Gérard F, François M, Ranger J (2002) Processes controlling silica concentration in leaching and capillary soil solutions of an acidic brown forest soil (Rhône, France). Geoderma 107:197–226
Ghorbanpour M, Mohammadi H, Kariman K (2020) Nanosilicon based recovery of barley (Hordeum vulgare) plants subjected to drought stress. Environ Sci Nano 7(2):443–461
Gomaa MA, Kandil EE, El-Dein AAZ, Abou-Donia ME, Ali HM, Abdelsalam NR (2021) Increase maize productivity and water use efficiency through application of potassium silicate under water stress. Sci Rep 11(1):224
González-Moscoso M, Martínez-Villegas NV, Cadenas-Pliego G, Benavides-Mendoza A, Rivera-Cruz MDC, González-Morales S, Juárez-Maldonado A (2019) Impact of silicon nanoparticles on the antioxidant compounds of tomato fruits stressed by arsenic. Food Secur 8(12):612
González-Moscoso M, Martínez-Villegas N, Cadenas-Pliego G, Juárez-Maldonado A (2022) Effect of silicon nanoparticles on tomato plants exposed to two forms of inorganic arsenic. Agronomy 12(10):2366
Govindaraju K, Kiruthiga V, Kumar VG et al (2009) Extracellular synthesis of silver nanoparticles by a marine alga, Sargassum Wightii Greville and their antibacterial effects. J Nanosci Nanotechnol 9(9):5497–5501
Gu HH, Qiu H, Tian T, Zhan SS, Deng TH, Chaney RL, Wang SZ, Tang YT, Morel JL, Qiu RL (2011) Mitigation effects of silicon rich amendments on heavy metal accumulation in rice (*Oryza sativa* L.) planted on multi-metal contaminated acidic soil. Chemosphere 83:1234–1240
Guerriero G, Hausman JF, Legay S (2016) Silicon and the plant extracellular matrix. Front Plant Sci 7:463
Gupta BK, Sahoo KK, Anwar K, Nongpiur RC, Deshmukh R, Pareek A, Singla-Pareek SL (2021) Silicon nutrition stimulates salt-overly sensitive (SOS) pathway to enhance salinity stress tolerance and yield in rice. Plant Physiol Biochem 166:593–604
Hasanaklou N, Mohagheghi V, Hasanaklou HT, Ma'mani L, Malekmohammadi M, Moradi F (2023) Seed nanopriming using silica nanoparticles: effects in seed germination and physiological properties of Stevia Rebaudiana Bertoni. Chem Biol Technol Agric 10
Hattori T, Inanaga S, Araki H, An P, Morita S, Luxová M, Lux A (2005) Application of silicon enhanced drought tolerance in Sorghum bicolor. Physiol Plant 123(4):459–466
Hayasaka T, Fujii H, Ishiguro K (2008) The role of silicon in preventing appressorial penetration by the rice blast fungus. Phytopathology 98:1038–1044
Hayat S, Hasan SA, Hayat Q, Ahmad A (2010) Brassinosteroids protect *Lycopersicon esculentum* from cadmium toxicity applied as shotgun approach. Protoplasma 239:3–14
Henriet C, Bodarwe L, Dorel M, Draye X, Delvaux B (2008) Leaf silicon content in banana (Musa spp.) reveals the weathering stage of volcanic ash soils in Guadeloupe. Plant Soil 313:71–82
Hirano T, Kiyota M, Kitaya Y, Aiga I (1990) The physical effects of dust on photosynthetic rate of plant leaves. J Agric Meteorol 46:1–7. https://doi.org/10.2480/agrmet.46.1

Hussain HI, Yi Z, Rookes JE et al (2013) Mesoporous silica nanoparticles as a biomolecule delivery vehicle in plants. J Nanopart Res 15(6):1676. https://doi.org/10.1007/s11051-013-1676-4

Hussain I, Parveen A, Rasheed R, Ashraf MA, Ibrahim M, Riaz S, Afzaal Z, Iqbal M (2019) Exogenous silicon modulates growth, physio-chemicals and antioxidants in barley (*Hordeum vulgare* L.) exposed to different temperature regimes. Silicon 11:2753–2762

Hussain B, Lin Q, Hamid Y, Sanaullah M, Di L, Hashmi M, Khan MB, He Z, Yang X (2020) Foliage application of selenium and silicon nanoparticles alleviates Cd and Pb toxicity in rice (*Oryza sativa* L.). Sci Total Environ 712:136497

Iyer S, Caplan A (1998) Products of proline catabolism can induce osmotically regulated genes in Rice1. Plant Physiol 116:203–211

Janmohammadi M, Sabaghnia N (2015) Effect of pre-sowing seed treatments with silicon nanoparticles on germinability of sunflower (*Helianthus annuus*). Botanica 21:13–21

Janmohammadi M, Seifi A, Pasandi M, Sabaghnia N (2016) The impact of organic manure and nano-inorganic fertilizers on the growth, yield and oil content of sunflowers under well-watered conditions. Biologija 62(4)

Jha S, Pudake RN (2016) Molecular mechanism of plant–nanoparticle interactions. Principles and Practices, Plant Nanotechnology, pp 155–181

Judy JD (2013) Bioavailability of manufactured nanomaterials in terrestrial ecosystems. Thesis and Dissertations Plant and Soil Sciences Paper 18

Kafi M, Nabati J, Ahmadi-Lahijani MJ, Oskoueian A (2021) Silicon compounds and potassium sulfate improve salinity tolerance of potato plants through instigating the defense mechanisms, cell membrane stability, and accumulation of osmolytes. Commun Soil Sci Plant Anal 52(8):843–858

Kah M, Kookana RS, Gogos A, Bucheli TD (2018) A critical evaluation of nanopesticides and nanofertilizers against their conventional analogues. Nat Nanotechnol 13:677–684

Kandil EE, Abdelsalam NR, Mansour MA, Ali HM, Siddiqui MH (2020) Potentials of organic manure and potassium forms on maize (*Zea mays* L.) growth and production. Sci Rep 10(1):8752

Kang H, Elmer W, Shen Y, Zuverza-Mena N, Ma C, Botella P, White JC, Haynes CL (2021) Silica nanoparticle dissolution rate controls the suppression of Fusarium wilt of watermelon (*Citrullus lanatus*). Environ Sci Technol 55:13513

Kashyap D, Siddiqui ZA (2021) Effect of silicon dioxide nanoparticles and rhizobium leguminosarum alone and in combination on the growth and bacterial blight disease complex of pea caused by *Meloidogyne incognita* and *Pseudomonas syringae* pv. pisi. Arch Phytopathol Plant Protect:1–17

Kaur H, Greger M (2019) A review on Si uptake and transport system. Plan Theory 8:81

Khan MR, Adam V, Rizvi TF, Zhang B, Ahamad F, Jośko I, Zhu Y, Yang M, Mao C (2019) Nanoparticle–plant interactions: twoway traffic. Small 15(37):1901794

Kim SG, Kim KW, Park EW, Choi D (2002) Silicon-induced cell wall fortification of rice leaves: a possible cellular mechanism of enhanced host resistance to blast. Phytopathology 92:1095–1103

Kundu S, Wang K, Liang H (2008) Size-controlled synthesis and self-assembly of silver nanoparticles within a minute using microwave irradiation. J Phbs Chem C 113(1):134–141

Le VN, Rui Y, Gui X, Li X, Liu S, Han Y (2014) Uptake, transport, distribution and bio-effects of SiO_2 nanoparticles in Bt transgenic cotton. J Nanobiotechnol 12. Article 50

Lee KJ, Feng YY, Choi DH, Lee BW (2016) Lead accumulation and distribution in different rice cultivars. J Crop Sci Biotechnol 19:323–328

Lengke MF, Fleet ME, Southam G (2007) Biosynthesis of silver nanoparticles by filamentous cyanobacteria from a silver (I) nitrate complex. Langmuir 23(5):2694–2699

Li Y, Xi K, Liu X, Han S, Han X, Li G, Yang L, Ma D, Fang Z, Gong S, Yin J (2023) Silica nanoparticles promote wheat growth by mediating hormones and sugar metabolism. J Nanobiotechnol 21(1):2

Liang Y, Nikolic M, Bélanger R, Gong H, Song A (2015) Silicon in agriculture: from theory to practice. Springer, Dordrecht

Liang Y, Liu H, Fu Y, Li P, Li S, Gao Y (2023) Regulatory effects of silicon nanoparticles on the growth and photosynthesis of cotton seedlings under salt and low-temperature dual stress. B.M.C. Plant Biol 23(1):504

Liu R, Lal R (2014) Synthetic apatite nanoparticles as a phosphorus fertilizer for soybean (*Glycine max*). Sci Rep 4:5686

Liu R, Lal R (2015) Potentials of engineered nanoparticles as fertilizers for increasing agronomic productions. Sci Total Environ 514:131–139

Liu C, Wei L, Zhang S, Xu X, Li F (2014) Effects of nanoscale silica sol foliar application on arsenic uptake, distribution and oxidative damage defense in rice (*Oryza sativa* L.) under arsenic stress. RSC Adv 4:57227–57234

Liu P, Yin L, Wang S, Zhang M, Deng X, Zhang S, Tanaka K (2015) Enhanced root hydraulic conductance by aquaporin regulation accounts for silicon alleviated salt-induced osmotic stress in sorghum bicolor L. Environ Exp Bot 111:42–51

Liu B, Soundararajan P, Manivannan A (2019) Mechanisms of silicon-mediated amelioration of salt stress in plants. Plants 8(9):307

Luyckx M, Hausman JF, Lutts S, Guerriero G (2017a) Impact of silicon in plant biomass production: focus on bast fibres, hypotheses, and perspectives. Plan Theory 6(3):37

Luyckx M, Hausman J-F, Lutts S, Gurriero G (2017b) Silicon and plants: current knowledge and technological perspectives. Front Plant Sci 8:411

Lv J, Zhang S, Luo L et al (2015) Accumulation, speciation and uptake pathway of ZnO nanoparticles in maize. Environ Sci Nano 2:68–77. https://doi.org/10.1039/c4en00064a

Lv J, Christie P, Zhang S (2019) Uptake, translocation, and transformation of metal-based nanoparticles in plants: recent advances and methodological challenges. Environ Sci Nano 6:41–59

Ma JF, Yamaji N, Mitani N, Tamai K, Konishi S, Fujiwara T et al (2007) An efflux transporter of silicon in rice. Nature 448:209–212. https://doi.org/10.1038/nature05964

Mahmoud LM, Dutt M, Shalan AM, El-Kady ME, El-Boray MS, Shabana YM, Grosser JW (2020) Silicon nanoparticles mitigate oxidative stress of in vitro-derived banana (*Musa acuminata* 'grand Nain') under simulated water deficit or salinity stress. S Afr J Bot 132:155–163

Makarov VV, Love AJ, Sinitsyna OV et al (2014) Green nanotechnologies: synthesis of metal nanoparticles using plants. Acta Nat 6(1):35–44

Mandlik R, Thakral V, Raturi G, Shinde S, Nikoli'c M, Tripathi DK, Sonah H, Deshmukh R (2020) Significance of silicon uptake, transport, and deposition in plants. J Exp Bot 71:6703–6718

Matysik J, Alia N, Bhalu B, Mohanty P (2002) Molecular mechanisms of quenching of reactive oxygen species by proline under stress in plants. Curr Sci 82:525–532

Mauter MS, Zucker I, Perreault F et al (2018) The role of nanotechnology in tackling global water challenges. Nat Sustain 1:166–175. https://doi.org/10.1038/s41893-018-0046-8

Memari-Tabrizi EF, Yousefpour-Dokhanieh A, Babashpour-Asl M (2021) Foliar-applied silicon nanoparticles mitigate cadmium stress through physio-chemical changes to improve growth, antioxidant capacity, and essential oil profile of summer savory (*Satureja hortensis* L.). Plant Physiol Biochem 165:71–79

Menzies JG, Ehret DL, Glass ADM, Samuels AL (1991) The influence of silicon on cytological interactions between *Sphaerotheca fuliginea* and *Cucumis sativus*. Physiol Mol Plant Pathol 39:403–414

Merwad A-RMA, Desoky E-SM, Rady MM (2018) Response of water deficit-stressed Vigna unguiculata performances to silicon, proline or methionine foliar application. Sci Hortic 228:132–144

Miralles P, Church TL, Harris AT (2012) Toxicity, uptake, and translocation of engineered nanoparticles in vascular plants. Environ Sci Technol 46(17):9224–9239

Monreal CM, DeRosa M, Mullubhotla SC, Bindraban PS, Dimkpa C (2015) Nanotechnologies for increasing the crop use efficiency of fertilizer-micronutrients. Biol Fertil Soils 52(3):423–437

Mosa MA, El-Abeid SE, Khalifa MMA, Elsharouny TH, El-Baz SM, Ahmed AY (2022) Smart pH responsive system based on hybrid mesoporous silica nanoparticles for delivery of fungicide to control Fusarium crown and root rot in tomato. J Plant Pathol 104(3):979–992

Mukarram M, Khan MMA, Corpas FJ (2021) Silicon nanoparticles elicit an increase in lemongrass (*Cymbopogon flexuosus* (Steud.) wats) agronomic parameters with a higher essential oil yield. J Hazard Mater 412:125254

Mukarram M, Khan MMA, Kurjak D, Lux A, Corpas FJ (2023) Silicon nanoparticles (SiNPs) restore photosynthesis and essential oil content by upgrading enzymatic antioxidant metabolism in lemongrass (*Cymbopogon flexuosus*) under salt stress. Front Plant Sci 14:1116769

Mushtaq A, Jamil N, Rizwan S, Mandokhel F, Riaz M, Hornyak GL, Najam Malghani M, Naeem Shahwani M (2018) Engineered silica nanoparticles and silica nanoparticles containing controlled release Fertiliser for drought and saline areas. IOP Conf Ser Mater Sci Eng 414:012029

Mustafa T, Sattar A, Sher A, Ul-Allah S, Ijaz M, Irfan M, Butt M, Cheema M (2021) Exogenous application of silicon improves the performance of wheat under terminal heat stress by triggering physio-biochemical mechanisms. Sci Rep 11(1):23170

Namjoyan S, Sorooshzadeh A, Rajabi A, Aghaalikhani M (2020) Nano-silicon protects sugar beet plants against water deficit stress by improving the antioxidant systems and compatible solutes. Acta Physiol Plant 42:1–16

Narayanan KB, Sakthivel N (2010) Biological synthesis of metal nanoparticles by microbes. Adv Colloid Interf Sci 156(1–2):1–13

Navarro E, Baun A, Behra R, Hartmann NB, Filser J, Miao AJ, Quigg A, Santschi PH, Sigg L (2008) Environmental behavior and ecotoxicity of engineered nanoparticles to algae, plants, and fungi. Ecotoxicology 17(5):372–386

Nayfeh MH, Mitas L (2008) Chapter one—silicon nanoparticles: new photonic and electronic material at the transition between solid and molecule. In: Kumar V (ed) Nanosilicon. Elsevier, Amsterdam, pp 1–78

Naz R, Gul F, Zahoor S, Nosheen A, Yasmin H, Keyani R, Shahid M, Hassan MN, Siddiqui MH, Batool S, Anwar Z (2022) Interactive effects of hydrogen sulphide and silicon enhance drought and heat tolerance by modulating hormones, antioxidant defence enzymes and redox status in barley (*Hordeum vulgare* L.). Plant Biol 24(4):684–696

Nazaralian S, Majd A, Irian S, Najafi F, Ghahremaninejad F, Lanberg T, Greger M (2017) Comparison of silicon nanoparticles and silicate treatments in fenugreek. Plant Physiol Biochem 115:25–33

Norton LD, Hall GE, Smeck NE, Bigham JM (1984) Fraginap bonding in a late-Wisconsian loss-derived soil in East-Central Ohio. Soil Sci Soc Am J 48:1360

Okitsu K, Mizukoshi Y, Yamamoto TA et al (2007) Sonochemical synthesis of gold nanoparticles on chitosan. Mater Lett 61(16):3429–3431

Ponmurugan P, Manjukarunambika K, Elango V, Gnanamangai BM (2016) Antifungal activity of biosynthesized copper nanoparticles evaluated against red root-rot disease in tea plants. J Exp Nanosci 11:1019–1031

Pradas Del Real AE, Vidal V, Carrière M et al (2017) Silver nanoparticles and wheat roots: a complex interplay. Environ Sci Technol 51:5774–5782. https://doi.org/10.1021/acs.est.7b00422

Priester JH, Moritz SC, Espinosa K et al (2017) Damage assessment for soybean cultivated in soil with either CO_2 or ZnO manufactured nanomaterials. Sci Total Environ 579:1756–1768. https://doi.org/10.1016/j.scitotenv.2016.11.149

Rai M, Ingle AP, Pandit R, Paralikar P, Shende S, Gupta I, Biswas JK, da Silva SS (2018) Copper and copper nanoparticles: role in management of insect-pests and pathogenic microbes. Nanotechnol Rev 7(4):303–315

Rajput VD, Minkina T, Sushkova S et al (2018) Effect of nanoparticles on crops and soil microbial communities. J Soils Sediments 18:2179–2187. https://doi.org/10.1007/s11368-017-1793-2

Rajput VD, Harish, Singh RK, Verma KK, Sharma L, Quiroz-Figueroa FR, Meena M, Gour VS, Minkina T, Sushkova S et al (2021) Recent developments in enzymatic antioxidant defence mechanism in plants with special reference to abiotic stress. Biology 10:267

Rao GB, Susmitha P (2017) Silicon uptake, transportation and accumulation in Rice. J Pharmacogn Phytochem 6(6):290–293

Rastogi A, Zivcak M, Sytar O, Kalaji HM, He X, Mbarki S, Brestic M (2017) Impact of metal and metal oxide nanoparticles on plant: a critical review. Front Chem 5

Rastogi A, Tripathi DK, Yadav S, Chauhan DK, Živˇcák M, Ghorbanpour M, El-Sheery NI, Brestic M (2019) Application of silicon nanoparticles in agriculture. 3 Biotech 9:90

Rautaray D, Ahmad A, Sastry M (2003) Biosynthesis of $CaCO_3$ crystals of complex morphology using a fungus and an actinomycete. J Am Chem Soc 125(48):14656–14657

Raza MAS, Zulfiqar B, Iqbal R, Muzamil MN, Aslam MU, Muhammad F, Amin J, Aslam HMU, Ibrahim MA, Uzair M, Habib-ur-Rahman M (2023) Morpho-physiological and biochemical response of wheat to various treatments of silicon nanoparticles under drought stress conditions. Sci Rep 13(1):2700

Remédios C, Rosário F, Bastos V (2012) Environmental nanoparticles interactions with plants: morphological, physiological, and genotoxic aspects. J Bot 2012:1–8

Rico CM, Majumdar S, Duarte-Gardea M, Peralta-Videa JR, GardeaTorresdey JL (2011) Interaction of nanoparticles with edible plants and their possible implications in the food chain. J Agric Food Chem 59(8):3485e3498

Rico CM, Morales MI, McCreary R et al (2013) Cerium oxide nanoparticles modify the antioxidative stress enzyme activities and macromolecule composition in rice seedlings. Environ Sci Technol 47:14110–14118. https://doi.org/10.1021/es4033887

Roberts TL (2009) The role of fertilizer in growing the world's food. Better Crops Plant Food 93:12–15

Rodrigues FA, Benhamou N, Datnoff LE, Jones JB, Bélanger RR (2003) Ultrastructural and cytochemical aspects of silicon mediated rice blast resistance. Phytopathology 93:535–546

Roohizadeh G, Majd A, Arbabian S (2015) The effect of sodium silicate and silica nanoparticles on seed germination and growth in the *Vicia faba* L. Tropic Plant Res 2(2):85–89

Roychoudhury A (2020) Yeast-mediated green synthesis of nanoparticles for biological applications. Indian J Pharm Biol Res 8(03):26–31

Salama HM (2012) Effects of silver nanoparticles in some crop plants, common bean (Phaseolus vulgaris L.) and corn (Zea mays L.). Int res. J Biotechnol 3(10):190–197

Scarano G, Morelli E (2002) Characterization of cadmium-and lead-phytochelatin complexes formed in a marine microalga in response to metal exposure. Biometals 15:145–151

Schreiber L (2005) Polar paths of diffusion across plant cuticles: new evidence for an old hypothesis. Ann Bot 95:1069–1073. https://doi.org/10.1093/aob/mci122

Schwabe F, Tanner S, Schulin R et al (2015) Dissolved cerium contributes to uptake of Ce in the presence of differently sized CeO2- nanoparticles by three crop plants. Metallomics 7:466–477. https://doi.org/10.1039/c4mt00343h

Seebold KW, Datnoff LE, Correa-Victoria FJ, Kucharek TA, Snyder GH (2004) Effects of silicon and fungicides on the control of leaf and neck blast in upland rice. Plant Dis 88:253–258

Selvakesavan RK, Franklin G (2021) Prospective application of nanoparticles green synthesized using medicinal plant extracts as novel nanomedicines. Nanotechnol Sci Appl:179–195

Sharma A, Kapoor D, Wang J, Shahzad B, Kumar V, Bali AS, Jasrotia S, Zheng B, Yuan H, Daoliang Y (2020) Chromium bioaccumulation and its impacts on plants: An overview. Plan Theory 9:100

Sharma S, Rai P, Prakash V, Tripathi S, Tiwari K, Gahlawat N, Tripathi DK, Sharma S (2023) Ameliorative effects of Si-SNP synergy to mitigate chromium induced stress in *Brassica juncea*. Environ Pollut 335:122031

Shetty R, Fretté X, Jensen B, Shetty NP, Jensen JD, Jørgensen HJL, Newman M-A, Christensen LP (2011) Silicon-induced changes in antifungal phenolic acids, flavonoids, and key phenylpropanoid pathway genes during the interaction between miniature roses and the biotrophic pathogen Podosphaera pannosa. Plant Physiol 157:2194–2205

Shoeva OY, Khlestkina EK (2018) Anthocyanins participate in the protection of wheat seedlings against cadmium stress. Cereal Res Commun 46:242–252

Siddiqui MH, Al-Whaibi MH, Faisal M, Al Sahli AA (2014) Nano-silicon dioxide mitigates the adverse effects of salt stress on *Cucurbita pepo* L. Environ Toxicol Chem 33:2429–2437

Siddiqui H, Ahmed KBM, Sami F, Hayat S (2020) Silicon nanoparticles and plants: current knowledge and future perspectives. Sustainable Agriculture Reviews 41: Nanotechnology for Plant Growth and Development, pp.129–142

Singh SP, Konwar BK (2012) Molecular docking studies of quercetin and its analogues against human inducible nitric oxide synthase. Springerplus 1(1):1–10

Singh A, Tiwari S, Pandey J, Lata C, Singh IK (2020) Role of nanoparticles in crop improvement and abiotic stress management. J Biotechnol 337:57–70

Singh RP, Handa R, Manchanda G (2021) Nanoparticles in sustainable agriculture: An emerging opportunity. J Control Release 329:1234–1248

Singhal RK, Jatav HS, Aftab T, Pandey S, Mishra UN, Chauhan J, Chand S, Indu, Saha D, Dadarwal BK et al (2021) Roles of nitric oxide in conferring multiple abiotic stress tolerance in plants and crosstalk with other plant growth regulators. J Plant Growth Regul 40:2303

Slomberg DL, Schoenfisch MH (2012) Silica nanoparticle phytotoxicity to *Arabidopsis thaliana*. Environ Sci Technol 46:10247–10254

Snyder GH, Jones DB, Gascho GJ (1986) Silicon fertilization of rice on Everglades Histosols. Soil Sci Soc Am J 50(5):1259–1263

Sun D, Hussain HI, Yi Z, Siegele R, Cresswell T, Kong L, Cahill DM (2014) Uptake and cellular distribution, in four plant species, of fluorescently labeled mesoporous silica nanoparticles. Plant Cell Rep 33:1389–1402

Sun D, Hussain HI, Yi Z, Rookes JE, Kong L, Cahill DM (2016) Mesoporous silica nanoparticles enhance seedling growth and photosynthesis in wheat and lupin. Chemosphere 152:81–91

Suriyaprabha R, Karunakaran G, Yuvakkumar R, Prabu P, Rajendran V, Kannan N (2012) Growth and physiological responses of maize (*Zea mays* L.) to porous silica nanoparticles in soil. J Nanopart Res 14:1294

Takó M, Kerekes EB, Zambrano C, Kotogán A, Papp T, Krisch J, Vágvölgyi C (2020) Plant phenolics and phenolic-enriched extracts as antimicrobial agents against food-contaminating microorganisms. Antioxidants 9(165):128

Tian L, Shen J, Sun G, Wang B, Ji R, Zhao L (2020) Foliar application of SiO2 nanoparticles alters soil metabolite profiles and microbial community composition in the Pakchoi (*Brassica chinensis* L.) rhizosphere grown in contaminated mine soil. Environ Sci Technol 54:13137–13146

Tran TLC, Guirguis A, Jeyachandran T, Wang Y, Cahill DM (2023) Mesoporous silica nanoparticle-induced drought tolerance in Arabidopsis thaliana grown under in vitro conditions. Funct Plant Biol 50:889

Tripathi DK, Singh VP, Prasad SM, Chauhan DK, Dubey NK (2015) Silicon nanoparticles (SiNp) alleviate chromium (VI) phytotoxicity in Pisum sativum (L.) seedlings. Plant Physiol Biochem 96:189–198

Tripathi DK, Singh S, Singh VP, Prasad SM, Chauhan DK, Dubey NK (2016) Silicon nanoparticles more efficiently alleviate arsenate toxicity than silicon in maize cultiver and hybrid differing in arsenate tolerance. Front Environ Sci 4:46

Tripathi DK, Singh S, Singh VP, Prasad SM, Dubey NK, Chauhan DK (2017) Silicon nanoparticles more effectively alleviated UV-B stress than silicon in wheat (*Triticum aestivum*) seedlings. Plant Physiol Biochem 110:70–81

Tripathi DK, Rai P, Guerriero G, Sharma S, Corpas FJ, Singh VP (2021) Silicon induces adventitious root formation in rice under arsenate stress with involvement of nitric oxide and indole-3-acetic acid. J Exp Bot 72:4457–4471

Tsuji H, Sugahara H, Gotoh Y, Ishikawa J (2003) Improvement of photocatalytic efficiency of rutile titania by silver negative-ion implantation. Nucl Instrum Methods Phys Res, Sect B 206:249–253

Tubaña BS, Heckman JR (2015) Silicon in soils and plants. Silicon and plant diseases:7–51

Udalova NN, Tutantsev AS, Fateev SA, Zharenova EA, Belich NA, Nemygina EM, Ryabova AV, Goodilin EA, Tarasov AB (2021) Crystallisation features of MAPbI 3 hybrid perovskite during the reaction of PbI 2 with reactive polyiodide melts. Russ J Inorg Chem 66:153–162

Verma KK, Song X-P, Li D-M, Singh M, Rajput VD, Malviya MK, Minkina T, Singh RK, Singh P, Li Y-R (2020) Interactive role of silicon and plant–rhizobacteria mitigating abiotic stresses: A new approach for sustainable agriculture and climate change. Plan Theory 9:1055

Verma KK, Song X-P, Verma CL, Chen Z-L, Rajput VD, Wu K-C, Liao F, Chen G-L, Li Y-R (2021a) Functional relationship between photosynthetic leaf gas exchange in response to silicon application and water stress mitigation in sugarcane. Biol Res 54:15

Verma KK, Song X-P, Zeng Y, Guo D-J, Singh M, Rajput VD, Malviya MK, Wei K-J, Sharma A, Li D-P et al (2021b) Foliar application of silicon boosts growth, photosynthetic leaf gas exchange, antioxidative response and resistance to limited water irrigation in sugarcane (*Saccharum officinarum* L.). Plant Physiol Biochem 166:582–592

Vithanage M, Seneviratne M, Ahmad M et al (2017) Contrasting effects of engineered carbon nanotubes on plants: a review. Environ Geochem Health 39:1421–1439. https://doi.org/10.1007/s10653-017-9957-y

Von Moos N, Bowen P, Slaveykova VI (2014) Bioavailability of inorganic nanoparticles to planktonic bacteria and aquatic microalgae in freshwater. Environ Sci Nano 1:214–232. https://doi.org/10.1039/C3EN00054K

Wang Y, Stass A, Horst WJ (2004) Apoplastic binding of aluminum is involved in silicon-induced amelioration of aluminum toxicity in maize. Plant Physiol 136:3762–3770

Wang WN, Tarafdar JC, Biswas P (2013) Nanoparticle synthesis and delivery by an aerosol route for watermelon plant foliar uptake. J Nanopart Res 15(1):1417. https://doi.org/10.1007/s11051-013-1417-8

Wang P, Lombi E, Zhoa F-J, Kopittke PM (2016) Nanotechnology: a new opportunity in plant sciences. Trends Plant Sci 21:699–712

Wang X, Jiang J, Dou F, Sun W, Ma X (2021) Simultaneous mitigation of arsenic and cadmium accumulation in rice (*Oryza sativa* L.) seedlings by silicon oxide nanoparticles under different water management schemes, vol 19. Paddy Water Environ, p 569

Wanyika H, Gatebe E, Kioni P, Tang Z, Gao Y (2012) Mesoporous silica nanoparticles carrier for urea: potential applications in agrochemical delivery systems. J Nanosci Nanotechnol 12(3):2221–2228

White AF, Brantley SL (1995) Chemical weathering rates of silicate minerals. In: Reviews in mineralogy, vol 31. Mineralogical Society of America

Williams LA, Crerar DA (1985) Silica diagenesis. II. General mechanisms. J Sediment Petrol 55:312–321

Wu L, Liu M, Liang R (2008) Preparation and properties of a double-coated slow-release NPK compound fertilizer with superabsorbent and water-retention. Bioresour Technol 99(3):547–551

Xiao Z, Fan N, Zhu W, Qian HL, Yan XP, Wang Z, Rasmann S (2023) Silicon nanodots increase plant resistance against herbivores by simultaneously activating physical and chemical defenses. ACS Nano 17(3):3107–3118

Yanga J, Cao W, Rui Y (2017) Interactions between nanoparticles and plants: phytotoxicity and defense mechanisms. J Plant Interact 12:158–169. https://doi.org/10.1080/17429145.2017.1310944

Zhang W, Dan Y, Shi H, Ma X (2017) Elucidating the mechanisms for plant uptake and in-planta speciation of cerium in radish (*Raphanus sativus* L.) treated with cerium oxide nanoparticles. J Environ Chem Eng 5:572–577. https://doi.org/10.1016/j.jece.2016.12.036

Zhao L, Ortiz C, Adeleye AS et al (2016) Metabolomics to detect response of lettuce (*Lactuca sativa*) to $Cu(OH)_2$ nanopesticides: oxidative stress response and detoxification mechanisms. Environ Sci Technol 50:9697–9707. https://doi.org/10.1021/acs.est.6b02763

Chapter 13
Silicon and Nano-Silicon in Water Use Efficiency

Gelza Carliane Marques Teixeira, Carlos Vital Gonzalez Porras, Patrícia Messias Ferreira, Juan Ricardo Rocha, and Renato de Mello Prado

Abstract This review aims to discuss and investigate the current situation of this availability and explores future scenarios that shall shape agricultural production. In general, the physiological effects that improve plant growth under water deficit stimulated by Si are the regulation of stomate opening, reduction of leaf transpiration, and increased activity of antioxidant enzymes against oxidative damage induced by water stress. However, there is a new beneficial effect that research has reported with potential accumulator plants: the modification of carbon stoichiometry, which increases the efficiency of nutritional use in plants. This review highlighted the mechanisms of Si inducing plants to increase tolerance to drought in water-restricted areas.

Keywords Silicon · Nanosilica · Water deficit · Nutritional efficiency · Water use efficiency

13.1 Introduction

Water in agricultural production is a crucial factor due to the growing demand for food in a world with a constant population growth. This chapter explores the current situation of water availability for agriculture and analyzes current challenges and existing solutions. Additionally, it examines future scenarios, considering factors such as climate change, urbanization, and technological developments.

G. C. M. Teixeira (✉)
Tocantins Institute Federal (IFTO), Tocantins, Brazil

C. V. G. Porras · P. M. Ferreira · R. de Mello Prado
São Paulo State University (UNESP), São Paulo, Brazil

J. R. Rocha
Center for Nuclear Energy in Agriculture (CENA), São Paulo, Brazil

R. de Mello Prado et al. (eds.), *Silicon Advances for Sustainable Agriculture and Human Health*, Sustainable Plant Nutrition in a Changing World,
https://doi.org/10.1007/978-3-031-69876-7_13

In this context, the use of silicon (Si) and nano-silicon (n-Si) emerges as a promising approach to optimize water use efficiency in agricultural practices. It has attracted attention of the scientific community (Teixeira and Prado 2023). This review discusses how these innovations play a key role in mitigating challenges related to water availability, promoting sustainability and food security.

This chapter analyzes the influence of Si on water absorption by plants, the impacts on tolerance to water stress, and the potential contributions of nanotechnologies to an efficient management of water resources. Furthermore, it explores prospective scenarios, considering possible transformations in the water landscape for agriculture.

In this analysis, we seek not only to understand the current situation of water availability for agricultural production but also to anticipate and prepare for the challenges and opportunities that the future holds. The importance of research on Si to boost sustainable and resilient agricultural practices is crucial.

The availability of water for agricultural production is a central concern in the context of growing global demand for food. The situation is expected to worsen with future climate changes (Richter et al. 2020; Schewe et al. 2014). Agriculture consumes a considerable amount of water resources; climate change, together with urban expansion, intensifies challenges associated with water management (Gerten et al. 2020).

This review aims to discuss and investigate the current situation of this availability and explores future scenarios that shall shape agricultural production.

13.2 Water Availability for Agricultural Production: Current Situation and Future Scenarios

13.2.1 Current Situation

The use of water in agriculture is the main global demand for this resource. Climate change, socioeconomic development, and population growth significantly affect it (Gerten et al. 2020).

Agriculture consumes approximately 70% (2800 km^3 in 4000 km^3) of the freshwater available in the world, increasing the pressure on all ecosystems (FAO 2024). Inequality in the availability and demand for water in agriculture has several consequences. Firstly, the continuous increase in the search for water for agricultural activities, driven by population growth and high individual consumption, is increasing the water deficit and putting the water security of a considerable part of the world's population at risk (Martínez-Valderrama et al. 2023).

Current perspectives on water availability for agricultural production indicate that climate change exerts an adverse influence, seriously restricting food production on a global scale (Liu et al. 2022a). The inequality between water availability and demand in agriculture has several implications, such as the continuous growth in the need for water for agricultural purposes, driven by population growth and high per capita consumption. This scenario intensifies the water deficit and compromises water security for a large part of the global population (Martínez-Valderrama et al. 2023). Also, inadequate water supplies in agriculturally dependent regions may lead to environmental and health problems (Zhu and Fang 2022).

The increase in demand for water in agriculture, together with a decrease in water availability, is expected to exert cascading effects on consumption, markets, and food security. In addition, the expansion of agricultural activities, deforestation, and the increase in irrigated areas contribute to environmental fragility and extreme water deficit events in certain regions, especially a high frequency and duration of summer droughts (Zhu and Fang 2022).

This highlights the importance of effective water management strategies and the need for sustainable agricultural practices to meet the challenges that climate change poses and ensure future food security.

13.2.2 Future Scenarios

Future climate changes could intensify water scarcity and negatively affect crop productivity in many regions of the world (Liu et al. 2022b). Climate change leads to variability in rainfall patterns, which directly affect the availability of water for agriculture (Le et al. 2023). Regions already facing water deficiency scenarios may experience a further decrease in rainfalls, while others may experience extreme events such as flooding. The implementation of climate-resilient agricultural practices, such as adapted crops and efficient water storage systems, is crucial to face these challenges.

Water use efficiency is urgent considering that population growth and urbanization increase competition for water resources between agriculture and urban areas. Urban expansion results in the conversion of agricultural land into urban areas, limiting the availability of land for food production (van Berkum 2023). Sustainable urban planning strategies and policies that balance urban and rural needs are thus essential. In addition, efficient agricultural practices and land conservation are crucial components of sustainability. Water quality issues are an additional factor considering managing water resources, as all sectors will increasingly be competing for access to uncontaminated water given the ongoing degradation of the quality of this resource (Rosegrant et al. 2009).

The development of innovative technologies, including the use of silicon (Si) and nano-silicon (n-Si), is a promising opportunity to improve water use efficiency in agriculture, especially in irrigated systems. Si increases plant tolerance to water stress, improves nutrient absorption and plant metabolism, and strengthens plant structure. Exploring these emerging technologies and integrating them into everyday agricultural practices are crucial in adapting to changing water conditions, enabling more sustainable and resilient agricultural practices.

13.3 Water Use Efficiency in Irrigated Cropping Systems

Irrigated agriculture stands out as the main contributor to individual freshwater consumption. It consumes approximately 70–75% of the global supply of water (Zhang et al. 2017). Global population growth, the growing need for food and biofuels, and global climate change are exerting increasing pressure on freshwater resources (Rosegrant et al. 2009).

However, water scarcity represents a crucial element that imposes restrictions on the growth and production of crops in agricultural areas subject to arid and semi-arid conditions. In this sense, water use efficiency in irrigated cropping systems is a critical element in facing contemporary agricultural challenges, especially in a growing scarcity of water resources. This section focuses on exploring strategies and innovations aimed at optimizing water use in irrigated cropping environments, promoting sustainability and resilient agricultural production.

Water use efficiency in irrigated cropping systems is an essential component of sustainable agricultural practices. Adopting these innovative technologies allows transforming conventional irrigation into more efficient and adaptable systems. This approach not only maximizes productivity but also promotes the conservation of water resources, contributing to creating a long-term sustainable agriculture.

Currently, many irrigated cropping systems face challenges in terms of water use efficiency. Traditional irrigation often results in excess water loss due to evaporation, leaching, and inadequate irrigation practices (Bwambale et al. 2022). Climate variability and changes in rainfall patterns exacerbate these challenges, highlighting the urgent need to improve existing practices.

Improvements in the use of water resources may occur through the introduction of technological innovations, such as incorporation of soil moisture sensors, which, combined with the use of intelligent irrigation systems, allow for a more precise water application. This ensures that plants receive the right amount of water at the right time, avoiding waste and optimizing the use of available water resources (Abioye et al. 2020; Bwambale et al. 2022).

In addition, there is concern about the use of more water-efficient irrigation systems, such as drip irrigation at a low quantity and high frequency of water supply (Kumar Jha et al. 2019). This approach is beneficial for supplying water and nutrients, especially combined with fertigation, which consists of applying fertilizers directly to the soil in the plants' root zone. This practice helps to reduce evaporation, leaching losses, and depth of percolation, resulting in a notable increase in water use efficiency compared to other irrigation methods.

Drip irrigation combined with fertigation forms an integrated management strategy, as it not only addresses the issue of water scarcity but also promotes sustainable and efficient management of resources in agricultural contexts. Fertigation, especially the supply of elements that potentially mitigate damage caused by biological water deficit, can be effective in promoting increased water productivity, producing a greater unit of food/fiber per unit of water consumed.

13.4 Benefit of Si for Improving Water Use Efficiency in Water-Deficient Plants

Cultivated plants are exposed to various environmental stress factors that play a crucial role in determining plant productivity. Such stress factors include water deficiency stress, which is the main abiotic element affecting plant development and quality of plant products (Bhardwaj and Kapoor 2021).

An important factor that must be taken into account is that the El Niño-Southern Oscillation affects greatly the occurrence of extreme rainfall events at local and global scales, influencing future global warming strongly (Sun et al. 2023). This phenomenon affects the atmospheric transport of water vapor, one of the most active components of the hydrological cycle, which has major effects on climate variability, while also being largely affected by large-scale circulation systems.

In the current climate change scenario, water scarcity is a challenge for productivity and agronomic quality. It has aggravated the occurrence of periods of drought, affecting crops in various regions of the world (Rocha et al. 2021). This phenomenon affects the growth, development, and metabolic processes of plants, resulting in a reduction in production, yield, and quality of cultivated plants. Plants respond to this stress through complex mechanisms involving a series of physiological, cellular, and molecular processes. For example, the rate of carbon assimilation, turgor pressure, photosynthesis and respiration rates, mineral nutrition, level of reactive oxygen species (ROS), and leaf gas exchange parameters may all change due to disturbances in plant function (Hussain et al. 2019). Water stress caused by water scarcity generates nutrient imbalances, ionic toxicity, and oxidative stress (Malik et al. 2021) that differ between species. In fennel plants (*Pimpinella anisum* L.), water deficit stress increases the levels of monounsaturated fatty acids and total contents of phenolic compounds and flavonoids.

Addressing such water scarcity combining it with a growing population makes it paramount to develop sustainable techniques to meet global food demand. Silicon (Si) is an ally in increasing crop productivity and ensuring food security (Bhardwaj and Kapoor 2021). Si is widely accepted as a beneficial element for multiple crop species, particularly in stressful environments (Irfan et al. 2023). Bhardwaj and Kapoor (2021) highlight Si as an "almost essential multi-talented" element due to its advantageous effects in improving plant physiology.

Among the variety of stress tolerance mechanisms, the beneficial effects of Si on plant water status are widely known (Malik et al. 2021). Several studies on different species have reported the beneficial effect of Si in preventing cell dehydration, which can lead to an increase in water use efficiency. These effects can be attributed to a decrease in the transpiration rate (Abd El-Mageed et al. 2020; Buchelt et al. 2020; Jinger et al. 2020), the regulation of multiple antioxidant defenses (Ma et al. 2016), increased content of photosynthetic pigments (Parveen et al. 2019) (Verma et al. 2020, 2021b), and adjustment in osmotic potential regulated by maintenance of ions K^+ (Umar and Moinuddin 2002). These physiological changes may act in isolation or together to increase the growth and development of plants under water deficit.

The deposition of Si on the cell walls explains the decrease in the rate of leaf transpiration, stimulating the plant's internal water storage during water shortages. In addition, the hydrophilic nature of Si compounds, such as SiO_2-nH_2O, helps to maintain the general water balance in plants, protecting them from dehydration (Agarie et al. 1998; Gong et al. 2005; Souri et al. 2020).

Studies have described the effects mentioned above in several species, such as rice, wheat, corn, tomatoes, sorghum, sugarcane, and broad beans (Cao et al. 2015; Desoky et al. 2020; Khan 2020; Ming et al. 2012; Rocha et al. 2022b; Teixeira et al. 2020b; Verma et al. 2020). It is clear that the application of Si stimulates plant resistance to water stress and promotes a better water use efficiency by maintaining the plant's water status (de Oliveira Filho et al. 2021; Rocha et al. 2022a).

In general, the physiological effects that improve plant growth under water deficit stimulated by Si are the regulation of stomate opening, reduction of leaf transpiration (Pires Barbosa et al. 2023), and increased activity of antioxidant enzymes against oxidative damage induced by water stress (Teixeira et al. 2022). The physiological effects stimulated by Si are commonly obtained in a wide variety of plant species, regardless of how Si is absorbed.

However, there is a new beneficial effect that research has reported with potential accumulator plants: the modification of carbon stoichiometry, which increases the efficiency of nutritional use in forage plants (Rocha et al. 2021). This stoichiometric change favors plant growth by stabilizing the stoichiometric homeostasis C:N and C:Si, responsible for increasing the plant's ability to convert accumulated C into dry mass, which favors the plant's water content and increases photosynthetic efficiency (Rocha et al. 2022b). Results are similar in corn plants (Teixeira et al. 2022b).

Additionally, Wang et al. (2015) reported a marked increase in the root/shoot ratio in plants supplied with Si mainly due to the regulation modulated by Si in the root configuration, resulting in increased water absorption and transport in plants subjected to Si treatment. The high absorption and translocation of water in plants after Si application during periods of water scarcity generally derive from the greater hydraulic conductivity of roots (Souri et al. 2020).

Teixeira et al. (2022) reported that the use of Si could increase water use efficiency. Gonzales-Porras et al. (2024) confirmed it. The authors observed that even with a reduced water supply, plants enriched with Si kept similar yields or showed minimal reductions compared to plants grown under the same water conditions but without the addition of Si. These results reinforce the potential of a combined use of fertigation and Si as an effective approach to ensure the stability of agricultural production in scenarios of water restriction. Noteworthy, most of these studies applied Si via fertigation (Gonzalez-Porras et al. 2024).

13.5 Benefits of Si for Improving Water Use Efficiency in Non-stressed Plants

Although studies on the use of Si in agriculture focus on mitigating the negative effects of environmental stresses, particularly water deficiency stress, non-stressed plants, i.e., those under optimum growing conditions, also benefit from the application of Si. Several authors have reported that plants in "control" treatments, i.e., those not under stressful conditions, showed better development during their growing cycle with the supply of Si (de Oliveira Filho et al. 2021; Mitani and Jian 2005; Rocha et al. 2022a).

It is important to note that Si can favor plant growth even under optimum water conditions, increasing crop response possibly by favoring other not yet studied production factors (Teixeira & Prado 2023). Recent research has revealed that the application of Si results in improvements in root architecture (Avila et al. 2020; Tripathi et al. 2021), promoting vegetative growth (Sah et al. 2022) and high resistance to pathogens (Fischer et al. 2023; Flores et al. 2023) even with adequate water availability. These findings suggest that Si plays a crucial role in improving the overall performance of plants, regardless of the presence of environmental stresses.

Other studies have corroborated the positive effects of Si on plants grown in non-stressful conditions. For example, the presence of Si promotes photosynthetic efficiency (Ramírez-Olvera et al. 2019; Rocha et al. 2022a; Gonzalez-Porras et al. 2024), increases nutrient absorption (Buchelt et al. 2020; Teixeira et al. 2020a, 2022b), and strengthens resistance to diseases (Verma et al. 2021a) even in situations of adequate water availability. These results highlight the ability of Si to substantially benefit plants, regardless of environmental pressures related to water, pointing to its potential use as a growth-promoting agent in agricultural contexts.

Si triggers positive physiological and biochemical responses in plants even in the absence of water stress (Verma et al. 2020). It also promotes greater efficiency in the use of water; Si contributes to regulating hormonal balance, strengthening resistance to pests, and improving the nutritional quality of plant products (Artyszak 2018; Tripathi et al. 2021). These findings suggest that Si exerts a positive influence on various aspects of plant development, reinforcing its relevance as a sustainable agricultural management strategy.

The implementation of fertigation in irrigated management systems has a significant impact on the economy and water use efficiency in agriculture. Wang et al. (2023) show that fertigation can result in lower water consumption compared to conventional irrigation methods due to a precise application of water and nutrients directly to the plants' root zone.

In addition, the combination of irrigation and fertigation with Si improves crop performance, as evidenced by Gonzalez-Porras et al. (2024). The authors reported gains in common bean production for the first time, highlighting the potential of Si in increasing irrigation efficiency in areas limited by low rainfall and water scarcity. These combined benefits of fertigation contribute not only to conserve water resources but also to improve the profitability of agricultural systems.

13.6 Use of Silicates to Supply Si

Fertigation consists of applying fertilizer diluted in irrigation water. It can be applied using any irrigation system, but systems with pressurized water are preferable as they allow a more uniform application of water and fertilizers and a greater water use efficiency, which is particularly advantageous in the face of increasing water scarcity (Guido et al. 2020). Supplying Si via fertigation associated with leaf spraying may be potentially viable because it may induce absorption of Si by leaf and root organs, considerably increasing the levels of Si in plants and its beneficial effects.

One of the greatest limitations to the use of Si in solution is high polymerization. It occurs in solutions with concentrations greater than 3 mmol L^{-1} (Birchall 1995). It forms dimers, trimers, and polymeric chains of Si (McKeague and Cline 1963) that plants do not absorb. In this sense, fertigation is interesting because it allows the use of low doses (<20 kg ha^{-1} of Si), with Si concentrations in solution ranging from 1.5 to 3.0 mmol L^{-1} and avoiding polymerization. In addition, it is possible to use sources with greater stability, such as sodium potassium silicate stabilized with sorbitol (Santos et al. 2020). Sorbitol is a polyol that helps to maintain the monomeric forms of Si by forming organic complexes (Kubicki and Heaney 2003) and decreases the deliquescence point of the droplet on the leaf surface (Babiker and Duncan 1974), thus delaying its drying and contributing to the absorption of the element by the leaf surface, especially during the cuticle penetration phase (Fernández and Brown 2013).

The acidification of the solution at the time of application also favors the stability of monomeric forms of Si, lowering the pH value of the solution (5.5 ± 0.2). Acidification of solutions containing Si shifts the reaction to the right ($H_2SiO_4^{-2} \rightarrow H_3SiO_4^{-1} \rightarrow H_4SiO_4$), inducing the predominance of monomeric species of Si (Kudryavtsev and Figovsky 2016). In addition, fertigation, depending on the management, allows more frequent applications. These factors together improve the quality of the Si solution applied to the plants and the absorption of Si, which can be managed using fertigation combined with leaf spraying.

Studies have reported an increase in Si absorption provided by the application of Si via fertigation combined with leaf spraying in forage plants (Rocha et al. 2021b), in sugarcane only at the pre-sprouted seedling production phase (Teixeira et al. 2021), and the first sugarcane ratoon (Oliveira Filho et al. 2021). Although the efficiency of Si supply via fertigation and leaf spraying varies depending on the species and the growing environment (Ramesh 2000), these effects occur in bean plants, which are considered non-accumulators of Si even in field conditions (Gonzalez-Porras et al. 2024).

13.7 Nano-silica as a Source of Si at Low Concentrations

Nanotechnology has played a vital role in agriculture in recent years in growth stimulants, plant protection products, additives, and nano-fertilizers (Hussain et al. 2019). Si nanoparticles (n-Si) show potential characteristics, such as greater sorption capacity, larger surface areas, and intelligent delivery system that targets sites and allows molecules to overcome the barriers of plant cell wall and cell membranes (Hussain et al. 2019).

Nanotechnology provides devices and mechanisms to synchronize the release of nutrients according to the plant's needs, preventing premature conversion into chemical/gas forms that cannot be absorbed. There are currently several commercial products available. Their use in agriculture expands rapidly. In the literature, there are several studies reporting effects of improved agronomic variables and negative effects of nanofertilizers on plant growth.

The use of Si sources in their N-Si form helps to increase the performance of beneficial roles in combating various biotic and abiotic stresses. Nanoparticulate forms of Si take the form of an amorphous silica powder synthesized with particles with nanoscale diameter between 10 and 100 nm. They come from various sources, such as tetraethyl orthosilicate ($Si(OC_2H_5)_4$), or inorganic salts, such as sodium silicate (Na_2SiO_3), or even using rice husks as an organic source (Laane 2018).

Nanoparticulate forms of Si may contribute to a greater plant absorption, avoiding losses due to high rates of polymerization in solution, which is still common in field conditions. This is because Si in solution begins to polymerize from a concentration of 3 mmol L^{-1} (Birchall 1995), forming dimers, trimers, and polymer chains (McKeague and Cline 1963), which plants do not absorb. They form, therefore, low concentrations of Si using nanoscale colloidal dispersion nanosilica (Laane 2018).

The treatment of plants with nanoparticulate sources is generally carried out by supplying them via soil, leaf, or seed treatment. Their effects have already been shown in several studies that seek to mitigate stresses of various types or in unstressed plants.

13.8 Nano-silica in Unstressed Plants

Xie et al. (2012) evaluated leaf spraying at concentrations of n-Si 0, 150, 300, and 450 mg L^{-1} in bamboo plants grown under field conditions. The results showed that the spraying at all concentrations of Si compared to the control increased gas exchange and chlorophyll fluorescence parameters. There was a 24% increase in net photosynthetic rate, a 35% increase in transpiration rate, a 20% increase in stomatal conductance, a 9% increase in PSII quantum efficiency, and a 34% increase in electron transport rate, with 300 mg L^{-1} defined as the ideal concentration.

The work of the authors showed that the mechanism of action of nanoparticles of SiO_2 initially occurs by increasing stomatal conductance, which improves leaf photosynthetic functions in absorption, transmission, and transformation of light energy. In turn, this process increases the level of opening PSII reaction centers, reduces the dissipation of antenna complexes, improves the actual photochemical efficiency and electron transport rate of PSII, and finally improves the net photosynthetic capacity of bamboo plants.

The effect of applying n-Si was compared to the use of potassium silicate (K_2SiO_3—a standard source of Si) at concentrations of 0, 5, 10, and 15 mmol L^{-1}. The results showed that Si application reduced transpiration and did not affect the chlorophyll index, photosynthetic rate, stomatal conductance, internal carbon dioxide concentration (CO_2), and mesophyll efficiency of strawberry plants. Plants that received 5 mmol L^{-1} using n-Si as a source showed significant effects of the evaluated attributes (Dehghanipoodeh et al. 2016).

The use of Si nanoparticles is a strategy to reduce Si doses when applied in sowing furrows. Alvarez et al. (2018) investigated the effects of applying Si in the form of nanosilica (<200 nm) compared to sodium silicate in rice plants grown in pots filled with Oxisol. The application of Si increased by 47% its accumulation in rice plants in relation to the control, regardless of Si source, by 1.8% in net carbon assimilation and by 48% in leaf transpiration rate. Thus, the authors concluded that the application of n-Si promotes an increase in the absorption of Si in rice plants in quantities sufficient to cause physiological changes in plants grown with no stress.

There are high densities of amorphous silica on leaf surfaces of rice plants that received n-Si application compared to plants grown under control conditions (0 g ha^{-1} of Si) (Alvarez et al. 2018). The accumulation of Si in the leaf tissue happens because after being absorbed, Si reaches the leaves due to loss of water through transpiration. It then concentrates in the leaves after the polymerization process, forming amorphous or biogenic silica. This process makes up 90% of the absorbed Si. It is then transformed into Si-cellulose structures present in the cell wall (Yoshida 1965). High silica sedimentation is the main mechanism for modifying the rate of leaf transpiration (2020), causing effects on the entire photosynthetic process.

The effects of application of n-Si on the increase of attributes related to photosynthesis, with reflexes in increased growth even in plants under suitable environmental conditions, are due to the increase in the absorption of Si due to large specific surface area and high reactivity (Hussain et al. 2019).

13.9 Nanosilica in Water-Stressed Plants

The application of SiO_2 nanoparticle forms can act as a seed priming agent to improve plant drought tolerance. Rai-Kalal et al. (2022) evaluated the effectiveness of applying 15 mg L^{-1} of n-Si to increase wheat tolerance to water stress. The authors showed that the seed treatment increased the tolerance of wheat plants to damage caused by water deficit, balancing the production of reactive oxygen species and the activity of enzymatic antioxidants, such as ascorbate peroxidase, catalase, and superoxide dismutase.

Studies on photosynthetic parameters reported that under drought conditions, nanoconditioned plants showed a higher number of active reaction centers, high absorbance, and electron transport rates compared to plants from untreated seeds. These results suggest that Si nanoconditioning is effective in increasing drought tolerance in wheat by alleviating drought-induced inhibition of the plant's photosynthetic machinery and maintaining biochemical balance, resulting in an increase in biomass production.

There is another effect of Si in the soil arising from the use of nanoparticulate amorphous silica sources. It is the increase in biogenic Si in the form of phytoliths in the soil which has a multiporous structure with pore diameters <1 μm and large specific area of 92–316 $m^2 g^{-1}$, which is capable of adsorbing organic C from the soil in its cavities and surfaces (Song et al. 2018). This effect results in an increase in micropores in the soil, with significant effects on increasing field capacity and soil available water (AlSaeedi 2022).

13.10 Doses of Si for Application in Fertigated Crops

Si is a metalloid that has been investigated in agriculture since the 1960s, demonstrating its effect on increasing yields (Tubana et al. 2016). Since 2000, it has been considered a beneficial element due to its positive effects on plant nutrition (Heckman 2012), especially in plants that were under stress conditions (Yalda Hoseinian et al. 2020). Initial reports of the beneficial effects of Si were related to a reduction in biotic damage, especially through an increase in soluble sugar content and a reduction in the rate of infection by pathogens and insects (Trejo-Téllez et al. 2020).

Si is the second most abundant element on the surface of the Earth's crust. It is found in the form of oxides and silicates in the soil. It is abundant in its non-soluble form and unavailable for plant absorption (Olle 2020). The form readily absorbed by plants, monosilicic acid (H_4SIO_4), presents a low concentration in the soil solution. This low availability is due to soil characteristics such as pH, texture, organic matter, and geological age (Tubana et al. 2016).

Crop areas with low natural Si content can receive solutions enriched by silicates or insoluble nanoparticulate sources (Vasanthi et al. 2014). The plant's form of absorption greatly influences the absorption of Si. It is greater in monocots that present an active absorption, such as Poaceae and Cyperaceae (up to 10% of dry mass), and lower in dicotyledons, such as Fabaceae, which absorb the element passively (Murali-Baskaran et al. 2020).

Several studies have reported the reduction of abiotic stresses and the increase in productivity of accumulator plants, such as sugarcane (Silva et al. 2023), rice (Dorairaj et al. 2020), wheat (Azeem et al. 2015), grasses (Rocha et al. 2022), and corn (Teixeira et al. 2022b). However, these studies have also reported positive results in plants with a low absorption rate, such as tomatoes (Reyes-Pérez et al.

2023), cucumber (Abd-Alkarim et al. 2017), pepper (Lob et al. 2023), strawberry (Liu et al. 2020), and common bean (Gonzalez-Porras et al. 2024).

The use of soluble and stable sources in applications allows a positive effect of Si on non-accumulating plants. The use of these sources increases efficiency because it allows the application of relatively low doses, considering that there is greater stability in solutions by maintaining the monomeric forms of Si, avoiding polymerization, and facilitating efficient absorption (Buchelt et al. 2023; Teixeira et al. 2021). The efficiency of application also increases through fertigation using an adequate irrigation rate to reduce polymerization and increase absorption, obtaining greater accumulation of Si (Buchelt et al. 2023).

In view of this, the Genplant (Unesp's Plant Nutrition Study Group) has invested in different research projects evaluating the benefits of Si in the soil-plant system. It has also published on the subject (Prado 2023). In addition, in recent years, the group has started pilot projects with Si fertigation, evaluating the response of some crops grown in pots filled with different soils, including eutroferric red latosol at the Unesp Experimental Farm in Jaboticabal, Brazil (Teixeira & Prado (2023). The results were promising, suggesting efficiency of fertigation with Si in optimizing this element in sugarcane crops (Teixeira et al. 2021; Oliveira Filho et al. 2021), eucalyptus (Souza Junior et al. 2021), forage (Rocha et al. 2021), and corn (Teixeira et al. 2022b). Important achievements of this management in sugarcane and energy cane cultivation were also evidenced in samples of Quartzarenic Neosol (Teixeira et al. 2022a; Silva et al. 2023).

However, the work was initially limited to controlled environments and pots. There was thus a lack of information on the best agronomic and economic dose to be applied via fertigation in field conditions in Brazil. In this sense, a study has recently reported that the optimum doses of Si for fertigation are 6.6, 7.0, and 7.1 kg ha^{-1} for the no water deficit, moderate water deficit, and severe water deficit, respectively (Gonzalez-Porras et al. 2024). This study was a pioneer in showing that the application of Si in fertigation improves plant performance, especially under severe water deficit in field conditions in Brazil. The results are of great relevance because the physiological benefits were clear, such as an increase in the relative water content in the plant, leaf water potential, and membrane resistance. This directly affected pigment content and gas exchange rates, which mitigated losses in dry mass production of common beans. Results such as these reinforce the potential of Si in increasing irrigation efficiency in areas limited by low rainfalls and water scarcity.

The application of Si using fertigation clearly mitigates the negative effects of drought on sugarcane plants (Silva et al. 2023; Teixeira et al. 2022). The same effect occurred in bean plants, improving water use efficiency (Gonzalez-Porras et al. 2024). However, the efficiency of fertigation is limited to the management of the preparation of syrups for Si application. The main factor is that the use of soluble sources requires low doses to exert an effect on different crops, varying for accumulator and non-accumulator plants (Table 13.1). Research on non-Poaceae species has advanced and suggests the benefit of Si even in species that do not accumulate Si, which calls for further research on this group of plants.

Table 13.1 Si doses and sources used in crops under fertigation

Specie	Source	Doses	References
Zea mays	Na_2SiO_3/K_2SiO_3	2.5 mmol L^{-1}	Teixeira et al. (2022)
Saccharum officinarum	Na_2SiO_3	2 mmol L^{-1}	Silva et al. (2023)
Brachiaria ruziziensis	K_2SiO_3	2.5 mmol L^{-1}	Rocha et al. (2021)
Glycine max L.	KNaSi	1 g L^{-1}	Felisberto et al. (2021)
Oryza sativa	KNaSi	1 g L^{-1}	Felisberto et al. (2021)
Phaseolus vulgaris L.	Na_2SiO_3	1.4 mmol L^{-1}	Gonzalez-Porras et al. (2024)
Capsicum annuum L.	H_4O_4Si	360 ppm	Lob et al. (2023)
Solanum lycopersicum L.	K_2SiO_3	2 mmol L^{-1}	Costan et al. (2020)
Sorghum bicolor	K_2SiO_3	2 mmol L^{-1}	Avila et al. (2021)
Lactuca sativa L.	Na_2SiO_3	2 mmol L^{-1}	Abdalla et al. (2024)

13.11 Soluble and Stable Sources

The most widely used soluble sources in agriculture are potassium silicate (K_2SiO_3), calcium silicate ($CaSio_3$), and sodium silicate (Na_2SiO_3). Calcium silicate is the result of the extraction of wollastonite (Tubaῃa and Heckman 2015). There are also alternative sources, such as sodium and potassium silicate (Felisberto et al. 2021), which are highly soluble sources and provide a high rate of absorption by plants according to several studies (Leite et al. 2024). Different sources are commonly used in agriculture (Table 13.2).

A key factor in using silicate sources to obtain greater benefits from Si is to maintain the stability of the solution to be applied in order to avoid polymerization and keep the monomeric forms available to plants (Silva et al. 2023). An example of a stabilizer with silicate sources is sorbitol, which helps to increase the absorption of Si in plants, favoring physiological and biochemical processes (Felisberto et al. 2021; Teixeira et al. 2022b).

13.12 Insoluble Nanoparticulate Sources

Nanoparticles or nanomaterials are the result of nanobiotechnology research and have gained ground over the last 5 years in different areas of knowledge (Goswami et al. 2022). In agriculture, they are regarded as an alternative to improve nutrient applications and reduce environmental impact. This is because nanoparticles are non-pollutant and generate positive effects against diseases and environmental stresses (Bhat et al. 2021).

Nano-silica is a source of Si obtained from SiO_2. It is synthesized with a size ranging from 5 to 100 nm, allowing better absorption by plants and in turn helping with the absorption of other nutrients (Mathur and Roy 2020).

Recent research has determined that nano-silica induces high root absorption via symplastic and apoplastic pathways and then through conductive tissues to other

Table 13.2 Si applied in nanoparticle of different crops

Source	Silicon content (g kg^{-1})		Chemical composition	References
	Total Si	Soluble Si		
Wollastonite	242	36	Calcium silicate	Sebastian et al. (2013)
K_2SiO_3 liquid	97	76	Potassium silicate	Sebastian et al. (2013)
Na_2SiO_3 liquid	56	–	Sodium silicate	Sebastian et al. (2013)
$MgSiO_3$ (talc)	285	1	Magnesium silicate	Sebastian et al. (2013)
CaSiO3/MgSiO3 blend	120	22	Calcium/magnesium silicate	Sebastian et al. (2013)
$KSiO_3Na$	–	115	Potassium sodium silicate	Felisberto et al. (2021)
Silica blend	121	18	Calcium silicate (mainly)	Sebastian et al. (2013)

Adapted from Tubana et al. (2016)

Table 13.3 Si applied in nanoparticle of different crops

Synthesized NPs	Material	References
SiNPs	Sugarcane bagasse	Falk et al. (2019)
SiNPs	Rice husk	Chen et al. (2013)
SiNPs	Rice straw	Sankar et al. (2016)
SiNPs	Corn cob	Okoronkwo et al. (2013)
SiNPs	Wheat straw	Surayah Osman & Sapawe (2019)
SiNPs	Bamboo leaves	Vaibhav et al. (2015)

Adapted from Goswami et al. (2022)

plant parts (Asgari et al. 2018), especially leaves. It acts by strengthening the enzymatic and non-enzymatic antioxidant defense system (Asgari et al. 2018).

The synthesis of these nanoparticles originates mainly from the SiO_2 obtained from nature in the form of crystals such as quartz (Mathur and Roy 2020). However, there are biological waste materials available for extraction of these particles, such as rice husks, sugarcane bagasse, bamboo leaves, and others (Goswami et al. 2022; Surayah Osman and Sapawe 2019). Table 13.3 shows different sources from which nanoparticles can be synthesized.

In summary, in the studies cited in this review, industrial processes have produced most of the n-Si. It is extracted from inorganic materials using different techniques to produce stable and uniform sizes (AlSaeedi 2022). However, the use of non-biogenic silicas extracted from organic sources (mainly vegetable waste, such as rice husk ash) has emerged recently. It contains 88% silica dioxide, SiO_2 (AlSaeedi 2022), or sugarcane straw. Silica extracted from organic forms is called biogenic. However, further research is needed to assess whether this process has high stability to generate stable n-SiO_2 contents between production batches and to assess its agronomic efficiency.

13.13 Conclusion and Future Prospects

Agriculture consumes a great volume of fresh water in irrigated and rainfed crops. Climate change has increased the frequency of crop droughts associated with the advance of arid or semi-arid areas. In this context, there is an urgent need for low-cost, environmentally friendly strategies to improve water use efficiency employing Si.

This review highlighted the mechanisms of Si inducing plants to increase tolerance to drought in water-restricted areas. It is also possible to reduce the water requirements of plants enriched with Si. Recent research has shown that the optimized use of Si in solution via fertigation or leaf application improves absorption by reducing the amount of Si needed per area. For this reason, the future prospect is the need for further research in field on different species that cross plant nutrition with agricultural engineering to improve the use of Si in solution with a low polymerization to strengthen agricultural sustainability without posing any risk to the environment.

It is therefore pertinent to study stable Si sources, whether biogenic or non-biogenic; to define Si doses via fertigation in different agricultural soils; to evaluate Si synergy with other nutrients, especially micronutrients; to understand the mobility of fluid Si sources in different soils better; and to understand the mechanism of Si in promoting nutritional homeostasis in plant metabolism better. In areas with low rainfalls, it can induce two simultaneous stresses: water deficit and salinity (or others) compared to field situation since most studies on Si have analyzed isolated areas with only one type of stress present.

References

Abd El-Mageed TA, Shaaban A, Abd El-Mageed SA, Semida WM, Rady MOA (2020) Silicon defensive role in maize (*Zea mays* L.) against drought stress and metals-contaminated irrigation water. Silicon:1–12. https://doi.org/10.1007/s12633-020-00690-0

Abd-Alkarim E, Bayoumi Y, Metwally E, Rakha M (2017) Silicon supplements affect yield and fruit quality of cucumber (Cucumis sativus L.) grown in net houses. Afr J Agric Res 12(31):2518–2523. https://doi.org/10.5897/AJAR2017.12484

Abdalla KA, Mousa SY, Ibrahim MF, Metwally AA (2024) Impacts of cobalt, selenium and silicon biofortification on the growth, productivity and nutritional value of lettuce. Egypt J Hortic 51(1):71–86. https://doi.org/10.21608/EJOH.2023.231176.1261

Abioye MSZ, Abidin MSA, Mahmud S, Buyamin MHI, Ishak MKIA, Rahman AO, Otuoze PO, Ramli MSA (2020) A review on monitoring and advanced control strategies for precision irrigation. Comput Electron Agric 173(2020):Article 105441. https://doi.org/10.1016/j.compag.2020.105441

Agarie S, Uchida H, Agata W, Kubota F, Kaufman PB (1998) Effects of silicon on transpiration and leaf conductance in rice plants (Oryza sativa L.). Plant Prod Sci 1:89–95. https://doi.org/10.1626/PPS.1.89

Alvarez R d CF, Prado R d M, Felisberto G, Deus ACF, de Oliveira RLL (2018) Effects of soluble silicate and Nanosilica application on Rice nutrition in an Oxisol. Pedosphere 28(4):597–606. https://doi.org/10.1016/S1002-0160(18)60035-9

AlSaeedi AH (2022) Enhancement of soil water characteristics curve (SWCC) and water use efficiency of cucumber (*Cucumis sativus* L.) in sandy soils by using silica nanoparticles. J King Saud Univ Sci 34(4):101926. https://doi.org/10.1016/j.jksus.2022.101926

Artyszak A (2018) Effect of silicon fertilization on crop yield quantity and quality—a literature review in Europe. Plan Theory 7:54. https://doi.org/10.3390/plants7030054

Asgari F, Majd A, Jonoubi P, Najafi F (2018) Effects of silicon nanoparticles on molecular, chemical, structural and ultrastructural characteristics of oat (Avena sativa L.). Plant Physiol Biochem 127:152–160. https://doi.org/10.1016/J.PLAPHY.2018.03.021

Avila RG, Magalhães PC, da Silva EM, Gomes Júnior CC, de Paula Lana UG, de Alvarenga AA, de Souza TC (2020) Silicon supplementation improves tolerance to water deficiency in sorghum plants by increasing root system growth and improving photosynthesis. Silicon 12:2545–2554. https://doi.org/10.1007/S12633-019-00349-5/METRICS

Avila RG, Magalhães PC, da Silva EM, de Souza KRD, Campos CN, de Alvarenga AA, de Souza TC (2021) Application of silicon to irrigated and water deficit sorghum plants increases yield via the regulation of primary, antioxidant, and osmoregulatory metabolism. Agric Water Manag 255:107004. https://doi.org/10.1016/J.AGWAT.2021.107004

Azeem M, Iqbal N, Kausar S, Javed MT, Akram MS, Sajid MA (2015) Efficacy of silicon priming and fertigation to modulate seedling's vigor and ion homeostasis of wheat (Triticum aestivum L.) under saline environment. Environ Sci Pollut Res 22(18):14367–14371. https://doi.org/10.1007/S11356-015-4983-8/

Babiker AGT, Duncan HJ (1974) Penetration of bracken fronds by Asulam as influenced by the addition of surfactant to the spray solution and by pH. Weed Res 14:375–377. https://doi.org/10.1111/j.1365-3180.1974.tb01078.x

Bhardwaj S, Kapoor D (2021) Fascinating regulatory mechanism of silicon for alleviating drought stress in plants. Plant Physiol Biochem 166:1044–1053. https://doi.org/10.1016/J.PLAPHY.2021.07.005

Bhat JA, Rajora N, Raturi G, Sharma S, Dhiman P, Sanand S, Shivaraj SM, Sonah H, Deshmukh R (2021) Silicon nanoparticles (SiNPs) in sustainable agriculture: major emphasis on the practicality, efficacy and concerns. Nanoscale Adv 3(14):4019–4028. https://doi.org/10.1039/D1NA00233C

Birchall JD (1995) The essentiality of silicon in biology. Chem Soc Rev 24:351–357. https://doi.org/10.1039/CS9952400351

Buchelt AC, de Mello Prado R, Caione G, de Moura Zanine A, Justina SMD, da Silva Ribeiro J, de Souza Júnior AM, de Jesus Ferreira D, da Silva de Sousa FC (2023) The use of soluble silicon via fertigation and leaf application in Panicum maximum modulates production without decreasing grass quality. Silicon 15(17):7553–7563. https://doi.org/10.1007/S12633-023-02602-4/METRICS

Buchelt AC, Teixeira GCM, Oliveira KS, Rocha AMS, de Mello PR, Caione G (2020) Silicon contribution via nutrient solution in forage plants to mitigate nitrogen, potassium, calcium, magnesium, and sulfur deficiency. J Soil Sci Plant Nutr 203(20):1532–1548. https://doi.org/10.1007/S42729-020-00245-7

Bwambale E, Abagale FK, Anornu GK (2022) Smart irrigation monitoring and control strategies for improving water use efficiency in precision agriculture: a review. Agric Water Manag 260:107324. https://doi.org/10.1016/j.agwat.2021.107324

Cao B, Ma Q, Zhao Q, Wang L (2015) Effects of silicon on absorbed light allocation, antioxidant enzymes and ultrastructure of chloroplasts in tomato leaves under simulated drought stress. Sci Hortic (Amsterdam) 194:53–62

Chen H, Wang W, Martin JC, Oliphant AJ, Doerr PA, Xu JF, DeBorn KM, Chen C, Sun L (2013) Extraction of lignocellulose and synthesis of porous silica nanoparticles from rice husks: a comprehensive utilization of rice husk biomass. ACS Sustain Chem Eng 1(2):254–259. https://doi.org/10.1021/SC300115R/ASSET/IMAGES/MEDIUM/SC-2012-00115R_0005.GIF

Costan A, Stamatakis A, Chrysargyris A, Petropoulos SA, Tzortzakis N (2020) Interactive effects of salinity and silicon application on Solanum lycopersicum growth, physiology and shelf-life of fruit produced hydroponically. J Sci Food Agric 100(2):732–743. https://doi.org/10.1002/JSFA.10076

da Silva JLF, Prado R d M, Alves TL, Lata-Tenesaca LF, Soares MB (2023) New strategy for silicon supply through fertigation in sugarcane integrating the pre-sprouted seedling phase and field cultivation. Sci Rep 13(1):1–10. https://doi.org/10.1038/s41598-022-27323-3

de Oliveira Filho ASB, de Mello Prado R, Teixeira GCM, de Cássia Piccolo M, Rocha AMS (2021) Water deficit modifies C:N:P stoichiometry affecting sugarcane and energy cane yield and its relationships with silicon supply. Sci Rep 11. https://doi.org/10.1038/s41598-021-00441-0

de Oliveira Filho ASB, de Mello Prado R, Teixeira GCM, Rocha AMS, de Souza Junior JP, de Cássia Piccolo M, Rocha JR (2021) Silicon attenuates the effects of water deficit in sugarcane by modifying physiological aspects and C:N:P stoichiometry and its use efficiency. Agric Water Manag 255:107006. https://doi.org/10.1016/j.agwat.2021.107006

de Souza Junior JP, Prado R d M, de Morais TCB, Frazão JJ, dos Santos Sarah MM, de Oliveira KR, de Paula RC (2021) Silicon fertigation and salicylic acid foliar spraying mitigate ammonium deficiency and toxicity in Eucalyptus spp. clonal seedlings. PLoS One 16:1–11. https://doi.org/10.1371/journal.pone.0250436

Dehghanipoodeh S, Ghobadi C, Baninasab B, Gheysari M, Bidabadi SS (2016) Effects of potassium silicate and nanosilica on quantitative and qualitative characteristics of a commercial strawberry (fragaria × ananassa cv. 'camarosa'). J Plant Nutr 39:502–507. https://doi.org/10.1080/01904167.2015.1086789

Desoky E-SM, Mansour E, T Yasin MA, A El-Sobky E-SE, Rady MM (2020) Improvement of drought tolerance in five different cultivars of Vicia faba with foliar application of ascorbic acid or silicon. Spanish J Agric Res 18:16. https://doi.org/10.5424/sjar/2020182-16122. ISSN 1695-971X, ISSN-e 2171-9292

Dorairaj D, Ismail MR, Sinniah UR, Tan KB (2020) Silicon mediated improvement in agronomic traits, physiological parameters and fiber content in Oryza sativa. Acta Physiol Plant 42(3):1–11. https://doi.org/10.1007/S11738-020-3024-5/TABLES/5

Falk G, Shinhe GP, Teixeira LB, Moraes EG, de Oliveira APN (2019) Synthesis of silica nanoparticles from sugarcane bagasse ash and nano-silicon via magnesiothermic reactions. Ceram Int 45(17):21618–21624. https://doi.org/10.1016/J.CERAMINT.2019.07.157

Felisberto G, de Mello Prado R, de Oliveira RLL, de Carvalho Felisberto PA (2021) Are Nanosilica, Potassium Silicate and new soluble sources of silicon effective for silicon foliar application to soybean and rice plants? SILICON 13(9):3217–3228. https://doi.org/10.1007/S12633-020-00668-Y/METRICS

Fernández V, Brown PH (2013) From plant surface to plant metabolism: the uncertain fate of foliar-applied nutrients. Front Plant Sci 4:1–5. https://doi.org/10.3389/fpls.2013.00289

Fischer IH, Morales JVP, da Silva LM, de Almeida Bertani RM, Deus ACF, de Mello PR, Pascholati SF (2023) Silicon sources on biochemical responses and corynespora cassiicola control in cucumber plants. Gesunde Pflanz 1–12. https://doi.org/10.1007/S10343-023-00956-8/FIGURES/2

Flores RA, de Lima FSR, Xavier MFN, Bueno AM, de Andrade AF, de Souza Júnior JP, Campos CNS, Cunha Júnior LC, de Oliveira Abdala K, de Mello PR (2023) Soluble silicon source via foliar application improve plant physiology and fruit quality of Solanum lycopersicum L. Silicon 1–12. https://doi.org/10.1007/s12633-023-02806-8

Food and Agriculture Organization of the United Nations (2024) Integrated agriculture water management and one health. FAO

Gerten D, Heck V, Jägermeyr J, Bodirsky BL, Fetzer I, Jalava M et al (2020) Feeding ten billion people is possible within four terrestrial planetary boundaries. Nat Sustain 3:200–208. https://doi.org/10.1038/s41893-019-0465-1

Gong H, Zhu X, Chen K, Wang S, Zhang C (2005) Silicon alleviates oxidative damage of wheat plants in pots under drought. Plant Sci 169:313–321. https://doi.org/10.1016/j.plantsci.2005.02.023

Gonzalez-Porras CV, Teixeira GCM, de Prado R, M, Ferreira PM, Palaretti LF, Oliveira KS (2024) Silicon via fertigation with and without potassium application, improve physiological aspects of common beans cultivated under three water regimes in field. Sci Reports 141(14):1–19. https://doi.org/10.1038/s41598-024-52503-8

Goswami P, Mathur J, Srivastava N (2022) Silica nanoparticles as novel sustainable approach for plant growth and crop protection. Heliyon 8(7). https://doi.org/10.1016/J.HELIYON.2022.E09908

Guido V, Finzi A, Ferrari O, Riva E, Quílez D, Herrero E et al (2020) Fertigation of maize with digestate using drip irrigation and pivot systems. Agronomy 10. https://doi.org/10.3390/agronomy10101453

Heckman JR (2012) Silicon and soil fertility. Soil Prolife 20:1–12. https://njaes.rutgers.edu/pubs/soilprofile/sp-v20.pdf

Hoseinian Y, Bahmanyar MA, Sadegh-zade F, Emadi M, Biparva P (2020) Effects of different sources of silicon and irrigation regime on rice yield components and silicon dynamics in the plant and soil. J Plant Nutr 43(15):2322–2335. https://doi.org/10.1080/01904167.2020.1771577

Hussain A, Rizwan M, Ali Q, Ali S (2019) Seed priming with silicon nanoparticles improved the biomass and yield while reduced the oxidative stress and cadmium concentration in wheat grains. Environ Sci Pollut Res 26:7579–7588. https://doi.org/10.1007/s11356-019-04210-5

Irfan M, Maqsood MA, Rehman HU, Mahboob W, Sarwar N, Hafeez OBA et al (2023) Silicon nutrition in plants under water-deficit conditions: overview and prospects. Water (Basel) 15. https://doi.org/10.3390/w15040739

Jinger D, Dhar S, Dass A, Sharma VK, Shukla L, Parihar M, Rana K, Gupta G, Jatav HS (2020) Crop productivity, grain quality, water use efficiency, and soil enzyme activity as influenced by silicon and phosphorus application in aerobic rice (*Oryza sativa*). Commun Soil Sci Plant Anal 51:2147–2162. https://doi.org/10.1080/00103624.2020.1812629

Khan A (2020) Ideal biomarkers of acute kidney injury. Kidney Biomarkers Clin Asp Lab Determ 1–18. https://doi.org/10.1016/B978-0-12-815923-1.00001-8

Kubicki JD, Heaney PJ (2003) Molecular orbital modeling of aqueous organosilicon complexes: implications for silica biomineralization. Geochim Cosmochim Acta 67:4113–4121. https://doi.org/10.1016/S0016-7037(03)00093-3

Kudryavtsev PG, Figovsky OL (2016) Nanocomposite organomineral hybrid materials. Part I. Nanotechnol Constr Sci Int J 8:16–49. https://doi.org/10.15828/2075-8545-2016-8-3-16-56

Kumar Jha S, Ramatshaba TS, Wang G, Liang Y, Liu H, Gao Y et al (2019) Response of growth, yield and water use efficiency of winter wheat to different irrigation methods and scheduling in North China Plain. Agric Water Manag 217:292–302. https://doi.org/10.1016/j.agwat.2019.03.011

Laane HM (2018) The effects of foliar sprays with different silicon compounds. Plan Theory 7:45. https://doi.org/10.3390/plants7020045

Le PVV, Randerson JT, Willett R, Wright S, Smyth P, Guilloteau C, Mamalakis A, Foufoula-Georgiou E (2023) Climate-driven changes in the predictability of seasonal precipitation. Nat Commun 14(1):1–10. https://doi.org/10.1038/s41467-023-39463-9

Leite MRL, de Alcântara Neto F, Dutra AF, Mendes LW, Miranda R d S, Melo VMM, Oliveira FAS, Sousa TK d SA, Costa RM, Rocha SMB, Pereira AP d A, Prado R d M, Araujo ASF (2024) Distinct sources of silicon shape differently the rhizospheric microbial community in sugarcane. Appl Soil Ecol 193:105131. https://doi.org/10.1016/J.APSOIL.2023.105131

Liu B, Davies K, Hall A (2020) Silicon builds resilience in strawberry plants against both strawberry powdery mildew Podosphaera aphanis and two-spotted spider mites Tetranychus urticae. PLoS One 15(12):e0241151. https://doi.org/10.1371/JOURNAL.PONE.0241151

Liu X, Liu W, Tang Q, Liu B, Wada Y, Yang H (2022) Global agricultural water scarcity assessment incorporating blue and green water availability under future climate change. Earth's. Future 10(4):e2021EF002567. https://doi.org/10.1029/2021EF002567

Lob S, Sa'ad NS, Ibrahim NF, Soh NC, Shah RM, Zaudin MSH (2023) Enhanced growth of chili (Capsicum annuum L.) by silicon nutrient application in fertigation system. Malays Appl Biol 52(2):13–19. https://doi.org/10.55230/MABJOURNAL.V52I2.2648

Ma D, Sun D, Wang C, Qin H, Ding H, Li Y, Guo T (2016) Silicon application alleviates drought stress in wheat through transcriptional regulation of multiple antioxidant defense pathways. J Plant Growth Regul 35:1–10. https://doi.org/10.1007/s00344-015-9500-2

Martínez-Valderrama J, Olcina J, Delacámara G, Guirado E, Maestre FT (2023) Complex policy mixes are needed to cope with agricultural water demands under climate change. Water Resour Manag 37(6):2805–2834. https://doi.org/10.1007/S11269-023-03481-5

Mathur P, Roy S (2020) Nanosilica facilitates silica uptake, growth and stress tolerance in plants. Plant Physiol Biochem 157:114–127. https://doi.org/10.1016/J.PLAPHY.2020.10.011

McKeague JA, Cline MG (1963) Silica in soils. Adv Agron:339–396. https://doi.org/10.1016/S0065-2113(08)60403-4

Malik MA, Wani AH, Mir SH, Rehman IU, Tahir I, Ahmad P, Rashid I (2021) Elucidating the role of silicon in drought stress tolerance in plants. Plant Physiol Biochem 165:187–195. https://doi.org/10.1016/J.PLAPHY.2021.04.021

Ming DF, Pei ZF, Naeem MS, Gong HJ, Zhou WJ (2012) Silicon alleviates PEG-induced water-deficit stress in upland rice seedlings by enhancing osmotic adjustment. J Agron Crop Sci 198:14–26. https://doi.org/10.1111/j.1439-037X.2011.00486.x

Mitani N, Jian FM (2005) Uptake system of silicon in different plant species. J Exp Bot 56:1255–1261. https://doi.org/10.1093/jxb/eri121

Murali-Baskaran RK, Senthil-Nathan S, Hunter WB (2020) Anti-herbivore activity of soluble silicon for crop protection in agriculture: a review. Environ Sci Pollut Res 28(3):2626–2637. https://doi.org/10.1007/S11356-020-11453-0

Okoronkwo EA, Imoisili PE, Olusunle SOO (2013) Extraction and characterization of a silica from corn cob ash by sol-gel method. Chem Mater Res 3(4):68–73. www.iiste.org

Olle M (2020) Silicon in a sustainable cropping system. Proc Latv Acad Sci Sect B Nat Exact Appl Sci 74(3):165–170. https://doi.org/10.2478/PROLAS-2020-0027

Parveen A, Liu W, Hussain S, Asghar J, Perveen S, Xiong Y (2019) Silicon priming regulates morpho-physiological growth and oxidative metabolism in maize under drought stress. Plan Theory 8:431. https://doi.org/10.3390/plants8100431

Pires Barbosa M, Araquém P, Cairo R, Araújo RA, Dias BL, Milton S, Sá C, Ferreira M, De Santos, Oliveira AL, Da Paz S, Fábio B, Gomes P (2023) Effects of foliar and root silicon application on mitigating water deficit stress in young Eucalyptus urophylla plants. https://doi.org/10.21203/RS.3.RS-2993031/V1

Prado R d M (2023) In: de Mello Prado R (ed) Benefits of silicon in the nutrition of plants. Springer International Publishing, Cham. https://doi.org/10.1007/978-3-031-26673-7

Ramírez-Olvera SM, Trejo-Téllez LI, Pérez Sato JA, Gómez Merino FC (2019) Silicon stimulates initial growth and chlorophyll a/b ratio in rice seedlings, and alters the concentrations of Ca, B, and Zn in plant tissues. J Plant Nutr 42:1928–1940. https://doi.org/10.1080/01904167.2019.1648678

Rai-Kalal P, Tomar RS, Jajoo A (2022) SiO2 nanopriming protects PS I and PSII complexes in wheat under drought stress. Plant Nano Biol 2:100019. https://doi.org/10.1016/j.plana.2022.100019

Ramesh P (2000) Effect of different levels of drought during the formative phase on growth parameters and its relationship with dry matter accumulation in sugarcane. J Agron Crop Sci 185:83–89. https://doi.org/10.1046/j.1439-037X.2000.00404.x

Reyes-Pérez JJ, Tipán-Torres HC, Llerena-Ramos LT, Hernandez-Montiel LG, Rivas-Garcia T (2023) Silicon increased the growth, productivity, and nutraceutical quality of tomato (Solanum lycopersicum L.). Notulae Botanicae Horti Agrobotanici Cluj-Napoca 51(2):13155–13155. https://doi.org/10.15835/NBHA51213155

Richter BD, Bartak D, Caldwell P, Davis KF, Debaere P, Hoekstra AY, Li T, Marston L, McManamay R, Mekonnen MM, Ruddell BL, Rushforth RR, Troy TJ (2020) Water scarcity and fish imperilment driven by beef production. Nat Sustain 3(4):319–328. https://doi.org/10.1038/s41893-020-0483-z

Rocha JR, de Mello Prado R, de Cássia Piccolo M (2022) New outcomes on how silicon enables the cultivation of Panicum maximum in soil with water restriction. Sci Rep 12(1):1–13. https://doi.org/10.1038/s41598-022-05927-z

Rosegrant MW, Ringler C, Zhu T (2009) Water for agriculture: maintaining food security under growing scarcity. Annu Rev Environ Resour 34:205–222. https://doi.org/10.1146/Annurev.Environ.030308.090351

Rocha JR, Mello Prado R, Teixeira GCM, Oliveira Filho ASB (2021) Si fertigation attenuates water stress in forages by modifying carbon stoichiometry, favouring physiological aspects. J Agron Crop Sci jac.12479. https://doi.org/10.1111/jac.12479

Rocha JR, de Mello PR, de Cassia PM (2022a) Mitigation of water deficit in two cultivars of panicum maximum by the application of silicon. Water Air Soil Pollut 233:1–17. https://doi.org/10.1007/s11270-022-05539-3

Rocha JR, de Mello Prado R, de Cássia Piccolo M (2022b) New outcomes on how silicon enables the cultivation of Panicum maximum in soil with water restriction. Sci Reports 2022 121 12:1–13. doi:https://doi.org/10.1038/s41598-022-05927-z

Sankar S, Sharma SK, Kaur N, Lee B, Kim DY, Lee S, Jung H (2016) Biogenerated silica nanoparticles synthesized from sticky, red, and brown rice husk ashes by a chemical method. Ceram Int 42(4):4875–4885. https://doi.org/10.1016/J.CERAMINT.2015.11.172

Santos LCN, Teixeira GCM, Prado RM, Rocha AMS, Pinto RCS (2020) Response of presprouted sugarcane seedlings to foliar spraying of potassium silicate, sodium and potassium silicate, nanosilica and monosilicic acid. Sugar Tech 22. https://doi.org/10.1007/s12355-020-00833-y

Sah SK, Reddy KR, Li J (2022) Silicon enhances plant vegetative growth and soil water retention of soybean (Glycine max) plants under water-limiting conditions. Plan Theory 11:1687. https://doi.org/10.3390/PLANTS11131687/S1

Schaller J, Cramer A, Carminati A, Zarebanadkouki M (2020) Biogenic amorphous silica as main driver for plant available water in soils. Sci Rep 10:1–7. https://doi.org/10.1038/s41598-020-59437-x

Schewe J, Heinke J, Gerten D, Haddeland I, Arnell NW, Clark DB, Dankers R, Eisner S, Fekete BM, Colón-González FJ, Gosling SN, Kim H, Liu X, Masaki Y, Portmann FT, Satoh Y, Stacke T, Tang Q, Wada Y et al (2014) Multimodel assessment of water scarcity under climate change. Proc Natl Acad Sci USA 111(9):3245–3250

Sebastian D, Rodrigues H, Kinsey C, Korndörfer G, Pereira H, Buck G, Datnoff L, Miranda S, Provance-Bowley M (2013) A 5-day method for determination of soluble silicon concentrations in nonliquid fertilizer materials using a sodium carbonate-ammonium nitrate extractant followed by visible spectroscopy with heteropoly blue analysis: single-laboratory validation. J AOAC Int 96(2):251–259. https://doi.org/10.5740/JAOACINT.12-243

Song Z, Liu C, Müller K, Yang X, Wu Y, Wang H (2018) Silicon regulation of soil organic carbon stabilization and its potential to mitigate climate change. Earth Sci Rev 185(92):463–475. https://doi.org/10.1016/j.earscirev.2018.06.020

Souri Z, Khanna K, Karimi N, Ahmad P (2020) Ssilicon and plants: current knowledge and future prospects. J Plant Growth Regul 403(40):906–925. https://doi.org/10.1007/S00344-020-10172-7

Sun Q, Zwiers F, Zhang X, Tan Y (2023) The effect of greenhouse gas–induced warming on the impact of El Niño and La Niña events on daily precipitation extremes in the boreal cold season. J Clim 36:6393–6407. https://doi.org/10.1175/JCLI-D-22-0713.1

Surayah Osman N, Sapawe N (2019) Waste material as an alternative source of silica precursor in silica nanoparticle synthesis—a review. Mater Today Proc 19:1267–1272. https://doi.org/10.1016/J.MATPR.2019.11.132

Teixeira GCM, de Mello Prado R, Rocha AMS, da Silveira Sousa Junior G, Gratão PL (2021) Beneficial effect of silicon applied through fertigation attenuates damage caused by water deficit in sugarcane. J Plant Growth Regul:1–16. https://doi.org/10.1007/S00344-021-10510-3/FIGURES/10

Teixeira GCM, de Mello Prado R, Rocha AMS, de Cássia Piccolo M (2022) Silicon as a sustainable option to increase biomass with less water by inducing Carbon:Nitrogen:Phosphorus stoichiometric homeostasis in sugarcane and energy cane. Front Plant Sci 13. https://doi.org/10.3389/fpls.2022.826512

Teixeira GCM, de Mello PR (2023) Silicon mitigates the effects of water deficit in tropical plants. In: de Mello Prado R (ed) Benefits of silicon in the nutrition of plants. Springer, Cham. https://doi.org/10.1007/978-3-031-26673-7_16

Teixeira GCM, de Mello PR, Oliveira KS, D'Amico-Damião V, da Silveira Sousa Junior G (2020a) Silicon increases leaf chlorophyll content and iron nutritional efficiency and reduces iron deficiency in sorghum plants. J Soil Sci Plant Nutr 20:1311–1320. https://doi.org/10.1007/s42729-020-00214-0

Teixeira GCM, de Mello PR, Rocha AMS, dos Santos LCN, dos Santos Sarah MM, Gratão PL, Fernandes C (2020b) Silicon in pre-sprouted sugarcane seedlings mitigates the effects of water deficit after transplanting. J Soil Sci Plant Nutr 20:849–859. https://doi.org/10.1007/s42729-019-00170-4

Teixeira GCM, de Prado RM, Rocha AMS, de Oliveira Filho ASB, da Sousa Junior GS, Gratão PL (2022a) Action of silicon on the activity of antioxidant enzymes and on physiological mechanisms mitigates water deficit in sugarcane and energy cane plants. Sci Reports 121(12):1–18. https://doi.org/10.1038/s41598-022-21680-9

Teixeira GCM, de Mello PR, de Oliveira LT et al (2022b) Silicon fertigation with appropriate source reduces water requirement of maize under water deficit. Plant Soil 477:83–97. https://doi.org/10.1007/s11104-022-05446-w

Tripathi P, Subedi S, Latif Khan A, Chung Y-S, Kim Y, Antoniou C, Maria Balestrini R, Fotopoulos V, Kolbert Z (2021) Silicon effects on the root system of diverse crop species using root phenotyping technology. Plants 10:885. https://doi.org/10.3390/PLANTS10050885

Trejo-Téllez LI, García-Jiménez A, Escobar-Sepúlveda HF, Ramírez-Olvera SM, Bello-Bello JJ, Gómez-Merino FC (2020) Silicon induces hormetic dose-response effects on growth and concentrations of chlorophylls, amino acids and sugars in pepper plants during the early developmental stage. PeerJ 2020(6):e9224. https://doi.org/10.7717/PEERJ.9224/SUPP-4

Tubana BS, Babu T, Datnoff LE (2016) A review of silicon in soils and plants and its role in us agriculture: history and future perspectives. Soil Sci 181(9–10):393–411. https://doi.org/10.1097/SS.0000000000000179

Tubaña BS, Heckman JR (2015) Silicon in soils and plants. Silicon Plant Dis:7–51. https://doi.org/10.1007/978-3-319-22930-0_2/COVER

Umar S, Moinuddin (2002) Genotypic differences in yield and quality of groundnut as affected by potassium nutrition under erratic rainfall conditions. J Plant Nutr 25:1549–1562. https://doi.org/10.1081/PLN-120005407

Vaibhav V, Vijayalakshmi U, Roopan SM (2015) Agricultural waste as a source for the production of silica nanoparticles. Spectrochim Acta A Mol Biomol Spectrosc 139:515–520. https://doi.org/10.1016/J.SAA.2014.12.083

van Berkum S (2023) How urban growth in the global south affects agricultural dynamics and food systems outcomes in rural areas: a review and research agenda. Sustainability (Switzerland) 15. https://doi.org/10.3390/su15032591

Vasanthi N, Saleena LM, Raj SA (2014) Silicon in crop production and crop protection -a review. Agric Rev 35(1):14. https://doi.org/10.5958/J.0976-0741.35.1.002

Verma KK, Liu XH, Wu KC, Singh RK, Song QQ, Malviya MK, Song XP, Singh P, Verma CL, Li YR (2020) The impact of silicon on photosynthetic and biochemical responses of sugarcane under different soil moisture levels. Silicon 12:1355–1367. https://doi.org/10.1007/s12633-019-00228-z

Verma KK, Song XP, Tian DD, Guo DJ, Chen ZL, Zhong CS, Nikpay A, Singh M, Rajput VD, Singh RK, Minkina T, Li YR (2021a) Influence of silicon on biocontrol strategies to manage biotic stress for crop protection, performance, and improvement. Plan Theory 10:2163. https://doi.org/10.3390/PLANTS10102163

Verma KK, Song XP, Tian DD, Singh M, Verma CL, Rajput VD, Singh RK, Sharma A, Singh P, Malviya MK, Li YR (2021b) Investigation of defensive role of silicon during drought stress induced by irrigation capacity in sugarcane: physiological and biochemical characteristics. ACS Omega 6:19811–19821. https://doi.org/10.1021/acsomega.1c02519

Wang W, Sardans J, Lai DYF, Wang C, Zeng C, Tong C, Liang Y, Peñuelas J (2015) Effects of steel slag application on greenhouse gas emissions and crop yield over multiple growing seasons in a subtropical paddy field in China. F Crop Res 171:146–156. https://doi.org/10.1016/J.FCR.2014.10.014

Wang N, Zhang T, Cong A, Lian J (2023) Integrated application of fertilization and reduced irrigation improved maize (Zea mays L.) yield, crop water productivity and nitrogen use efficiency in a semi-arid region. Agric Water Manag 289:108566. https://doi.org/10.1016/J.AGWAT.2023.108566

Xie Y, Li B, Tao G, Zhang Q, Zhang C (2012) Effects of nano-dióxido de silício em características de fluorescência fotossintética de Indocalamus barbatus McClure. J Nanjing Univ (Nat Sci Ed) 2:59–63

Yoshida S (1965) Chemical aspects of the role of silicon in physiology of the rice plant. Bull Nat Inst Agric Sci Ser B 15:1–58

Zhang G, Liu C, Xiao C, Xie R, Ming B, Hou P, Liu G, Xu W, Shen D, Wang K, Li S (2017) Optimizing water use efficiency and economic return of super high yield spring maize under drip irrigation and plastic mulching in arid areas of China. Field Crops Res 211:137–146. https://doi.org/10.1016/J.FCR.2017.05.026

Zhu R, Fang Y (2022) Application of a water supply-demand balance model to set priorities for improvements in water supply systems: a case study from the Koshi River Basin, Nepal. Int J Environ Res Public Health 19(3):1606. https://doi.org/10.3390/IJERPH19031606

Chapter 14
Silicon and Nano-silicon on C:N:P Stoichiometry and Nutrient Use Efficiency

Dilier Olivera-Viciedo, Kamilla Silva Oliveira, Cíntia Cármen de Faria Melo, Danilo Silva Amaral, and Renato de Mello Prado

Abstract Stoichiometric homeostasis in plants and mechanisms for resisting biotic and abiotic stresses are very active research areas. This book chapter delves into the multifaceted role of silicon, exploring how it promotes sustainability in plant production by improving C:N:P stoichiometry and nutrient use efficiency by plants under many stress conditions. It examines the interplay between silicon and the homeostasis of carbon (C), nitrogen (N), and phosphorus (P) in plants. It underscores the importance of maintaining plant C:N:P homeostasis in mitigating stress induced by climate change, emphasizing silicon's potential in bolstering plant resilience amid changing environmental conditions. Furthermore, it examines how maintaining the balance of plant C:N:P homeostasis alleviates stress caused by nutritional imbalances or toxicities, emphasizing the importance of balanced nutrient levels for optimal plant health and productivity. Lastly, the chapter outlines future perspectives and research avenues, underscoring the ongoing necessity to investigate the complex mechanisms that govern nutrient homeostasis and their implications for sustainable agriculture amid changing climatic conditions.

Keywords Stoichiometric homeostasis · Silicon · Nutrient use efficiency · Stress mitigation · Productivity · Sustainable agriculture

D. Olivera-Viciedo (✉)
Institute of Agrifood, Animals and Environmental Sciences, Universidad de O'Higgins, San Fernando, Chile
e-mail: dilier.olivera@uoh.cl

K. S. Oliveira
College of Agriculture, School of Plant, Environmental, and Soil Sciences, Louisiana State University, Baton Rouge, LA, USA

C. C. de Faria Melo · D. Silva Amaral · R. de Mello Prado
São Paulo State University, Jaboticabal, Brazil

R. de Mello Prado et al. (eds.), *Silicon Advances for Sustainable Agriculture and Human Health*, Sustainable Plant Nutrition in a Changing World,
https://doi.org/10.1007/978-3-031-69876-7_14

14.1 Introduction

There is widespread consensus on the importance of mineral nutrients essential for plant metabolism. According to the criteria of essentiality proposed by Arnon and Stout (1939), these nutrients include carbon (C), hydrogen (H), oxygen (O), nitrogen (N), phosphorus (P), potassium (K), calcium (Ca), magnesium (Mg), sulfur (S), iron (Fe), manganese (Mn), zinc (Zn), copper (Cu), boron (B), molybdenum (Mo), and chlorine (Cl). Without these nutrients, plants cannot complete their life cycle. Each essential nutrient plays multiple well-established roles in plant metabolism, acting as cellular structural constituents, enzymatic catalysts, energy transfer agents, osmotic regulators, and/or water balance regulators (Prado 2021). Based on their plant requirements, these mineral nutrients have been classified as macronutrients (N, P, K, Ca, Mg, and S) because they must be in higher concentrations ranging from 1 to 150 g per kg of plant dry matter and micronutrients (Fe, Zn, Mn, Cu, B, Mo, and Cl), which must be in lower concentrations ranging from 0.1 to 100 mg per kg of plant dry matter (Pandey 2018).

Nutritional homeostasis in plants is a central topic in plant biology, with significant implications for food production, environmental health, and agricultural sustainability. From early studies to current advances, understanding the mechanisms involved in nutrient absorption and regulation remains crucial for improving crop efficiency and resilience in a changing climate. The literature verified that the availability of nutrients and C:N:P stoichiometric ratios play a fundamental role in regulating ecosystem functions and services, such as nutrient cycling, habitat variability, litter decomposition, and crop production and quality (Huang et al. 2018; Olivera-Viciedo et al. 2021b, 2024; Habermann et al. 2022).

Studies related to plant stoichiometric relationships have increasingly focused on C, N, and P due to their crucial roles in plant growth and metabolism (Teixeira et al. 2020; Shi et al. 2021; Melo et al. 2023b; Olivera-Viciedo et al. 2024). Elemental stoichiometric ratios C:N, C:P, and N:P are crucial because the concentration of these nutrients in plant tissues can be affected by biomass production (dilution effect) and because stoichiometric patterns are species-specific (Olivera-Viciedo et al. 2019b, 2021a). Most studies on C, N, and P stoichiometric relationships in terrestrial ecosystems have primarily focused on the leaf level or on a single plant organ, overlooking interactions and differentiations among different plant organs (Olivera-Viciedo et al. 2021a, b).

Although the most metabolically active organs in plant metabolism are leaves and roots, and their ecological stoichiometry (N:P ratio) is closely interrelated, nutrient distribution in plants varies among different organs due to their specific functions and metabolic activities (Zhao et al. 2020; Zhang et al. 2020). Both N and P are closely coupled. Thus, their N:P ratio is an important indicator in plant adaptation strategies to adverse conditions (Ge and Xie 2017; Zhou et al. 2020). The N:P ratio within a plant is an index that reflects the shift between N and P limitation

(Olivera-Viciedo et al. 2019b). This ratio is often used as an indicator of changes and adaptation of plants to different habitats (Sadras 2006; Zhang et al. 2018).

Stoichiometric relationships also play a crucial role in plants' resistance to many types of stress. For example, plants can adjust their C:N, C:P, and N:P stoichiometric relationships to adapt to stress conditions such as salinity (Calero-Hurtado et al. 2019; Peña-Calzada et al. 2023b), drought, and warming (Olivera-Viciedo et al. 2021a, 2024). These adjustments in plant metabolism occur because plants need to maintain elemental homeostasis to survive adverse growth conditions.

Among the strategies under study to mitigate many types of stresses in plants is the use of silicon (Si). Although Si is not considered an essential nutrient, it is the second most abundant element (27.70% of the Earth's crust) after oxygen (Epstein 2003). Furthermore, its importance in agriculture has multiplied in the last decade, particularly due to its beneficial effects on plants under several stress conditions (Prado 2021). Despite its abundance in the Earth's crust, most of the soil Si exists in its amorphous form (i.e., not available) as silicon dioxide. Moreover, to the best of our knowledge, plants only absorb Si in the form of silicic acid or monosilicic acid [$Si(OH)_4$ or H_4SiO_4] (Epstein 1994; Olivera-Viciedo et al. 2019a; Calero-Hurtado et al. 2023).

Several studies have documented the potential impact of Si in alleviating damage caused by many plant stressors. Most of these studies have focused primarily on accumulator plants within the *Poaceae* family, including sorghum, sugar cane, rice, and forage plants (Calero-Hurtado et al. 2020a; Flores et al. 2022; Oliveira et al. 2022; Melo et al. 2023b; Olivera-Viciedo et al. 2024; Costa et al. 2024) that show active absorption mechanisms mediated by transporters. However, in the last decade, there has been a notable increase in research involving soluble sources of Si in other plant species categorized as intermediate and non-accumulator plants. Examples include *Phaseolus vulgaris* (Gonzalez-Porras et al. 2024), *Glycine max* (Peña-Calzada et al. 2023a, b), *Raphanus sativus* (Olivera-Viciedo et al. 2017, 2020), *Beta vulgaris* (Olivera-Viciedo et al. 2019a), and *Helianthus annuus* (Calero-Hurtado et al. 2020b). Therefore, these studies aim at mitigating many stress conditions, such as drought, heavy metals, and nutritional imbalances.

In recent years, one of the most prominent strategies for using Si to improve plant performance under adverse conditions is in the form of nanoparticles. Several studies have indicated that silicon nanoparticles (Si-NPs) are among the best alternatives for protecting plants from abiotic stressors (Melo et al. 2023b; Ahmadi-Nouraldinvand et al. 2024). Researchers have speculated that Si-NPs, being so small, may enter the plant cell mediated by Si transporters (Lsi1) in a simpler manner (Siddiqui et al. 2020). At the same time, other studies have indicated that Si-NPs stimulate the natural defense system in many plant species, including wheat (*Triticum aestivum* L.) (Ali et al. 2019), rice (*Oryza sativa* L.) (Hussain et al. 2020), and lemongrass (*Cymbopogon flexuosus* (Steud.) Wats) (Mukarram et al. 2021).

14.2 The Importance of Nutritional Homeostasis in Plant Life

Nutritional homeostasis is an essential mechanism for the general homeostasis of plant cells. It allows biochemical processes to function in a controlled and orderly manner for proper cell function, development, and resistance to adverse conditions and is sustained by dynamic autoregulation (Dreyer 2021).

In plants, the relevance of nutritional homeostasis is based on the fundamental premise that macronutrients and micronutrients are indispensable for plants to perform all their physiological and biochemical activities involving growth and reproduction (Marschner 2012). Therefore, a lack or excess of each negatively affects photosynthesis, protein synthesis, respiration, DNA replication, and even the response to plant stresses, negatively affecting plant growth.

Essential elements enable the life cycle of plants and are the mechanism by which these elements and their by-products enter the human and animal food chain. Studies indicate a relationship between micronutrient deficiency in the human diet and the occurrence of deficiency in plants due to cultivation in soils poor in these elements, generating plant products with low nutritional value (Assunção et al. 2022). Thus, proper plant nutrition, which in this sense is understood as balanced nutrition, is of great importance for maintaining optimal food resources to sustain the world's growing population (Rouached et al. 2010).

Plants lose nutritional homeostasis (nutritional balance) when the concentrations of these nutrients are below or above what is adequate for growth and the performance of metabolic functions. In this sense, both mineral nutrient deficiency and toxicity directly alter homeostasis due to the specific functions that each nutrient performs in plant physiology and biochemistry.

An imbalance in N concentrations induced by a deficiency of this nutrient, for example, reduces photosynthesis due to low light absorption and RuBisCo activity and content (Tantray et al. 2020). Furthermore, it reduces carbohydrate content, limiting root and shoot growth (Gao et al. 2014). On the other hand, excess N increases protein intake with a focus on vegetative growth. However, it penalizes reproductive growth, reducing productivity (Prado 2021), with a possible increase in nitrate content in leafy crops such as lettuce (Liu et al. 2014).

Insufficient P concentrations affect the production of phosphate sugars, membrane phospholipids, DNA, RNA, and nucleotide synthesis. It negatively alters all energy (ATP)-dependent processes in plant metabolism (Tariq et al. 2023). Moreover, P deficiency alters nutrient absorption in roots, leaves, and stems both negatively and positively (Meng et al. 2021). Excess P can cause an imbalance in the absorption of cationic nutrients such as Zn, Cu, Mn, and Cu (Prado 2021), altering the nutritional balance.

Potassium (K) is a nutrient that regulates many physiological processes in plants, regulating stomatal opening and closing, the osmotic potential of cells, and the activation of several plant enzymes involved in protein synthesis, sugar transport, and the metabolism of N and C (Marschner 2012). Both K deficiency and toxicity can reduce the biomass of each organ of apple plants, the root-to-shoot ratio, root activity, the rate of nitrate ion flux, photosynthesis, photochemical efficiency, the absorption and accumulation of N and C, and the distribution of these nutrients (Xu et al. 2020). Furthermore, K deficiency can increase the absorption of Na and Ca (Du et al. 2017), and its excess can trigger Mg deficiency by reducing this nutrient's absorption (Tränkner et al. 2018).

In its ionic form (Ca^{2+}), Ca provides greater rigidity to plant cells by strengthening the cell wall, neutralizes anion charges within vacuoles, and induces defense responses to many stresses (Prado 2021). Moreover, Ca^{2+} concentrations are high in plant tissues and low in the cytosol (Hepler 2005). Calcium (Ca) deficiency leads to cell wall disorganization, resulting in tissue deterioration and a decrease in calcium pectates, in the chlorophyll index, and in the growth of the aerial part and roots, with an increase in cell leakage and quality loss (de Souza Alonso et al. 2020). Calcium (Ca) toxicity can cause precipitation with P, inducing a deficiency of this nutrient and decreasing Ca cell signaling (Prado 2021).

The cell interior requires a high concentration of Mg to perform its plant functions (Shaul 2002), such as light absorption, since it is the central ion in chlorophyll and a cofactor and enzyme activator, promoting protein synthesis, photosynthesis, and nucleotide metabolism. Its deficiency accumulates sugars and amino acids, inhibiting growth and reducing photosynthesis (Verbruggen and Hermans 2013). Magnesium (Mg) toxicity induces the replacement of extracellular Ca^{2+}, affecting cell wall stability and plasma membrane permeability (Marschner 2012) and inhibiting plant growth (Guo et al. 2014).

Sulfur (S) is an essential macronutrient for synthesizing amino acids, proteins, chloroplasts, vitamins, and coenzymes. It helps in maintaining photosynthesis, respiration, and the formation of structures in cell membranes (Sung et al. 2018; Li et al. 2020). Furthermore, S and molecules containing this nutrient act as signaling agents in stress control in plants, aiding in the maintenance of metabolic processes (Narayan et al. 2023). The imbalance in this nutrient through deficiency negatively affects crop yield and quality by reducing protein and vitamin content in grains (Yu et al. 2021). Moreover, it can increase the amide and nitrate contents in plants (Carciochi et al. 2017).

It is worth noting that the adequate nutritional status of each nutrient must be prioritized. However, the balance between nutrients in the soil-plant system should be highlighted because interactions between the elements, from the soil solution to the inside of the plant after absorption, influence plant development. These interactions alter the processes of absorption, transport, redistribution, and metabolism. They can induce deficiencies and/or toxicity of other nutrients, altering the relationships between these nutrients and, consequently, the physiological and biochemical responses, influencing plant production (Prado 2021).

Each nutrient has a network of mechanisms to manage its homeostasis in plant cells. It includes transporters, chelating molecules, assimilation routes, gene transcriptions, and even plant signaling (Hanikenne et al. 2020).

The homeostatic regulation of N includes N acquisition, transport, and assimilation in plants involving cytokinin and sugar signaling pathways (Asad et al. 2023).

In order to maintain P homeostasis, plants use long-distance communication between the aerial part and the roots, and vice versa. They use regulatory proteins and small RNAs through a pathway that contains transcription factors in response to phosphate deprivation (PHRs), phosphate transporters (PHTs), long-distance signaling molecules (miR399) or PTGS, and plant hormones, such as auxins, which stimulate root growth and help increase P uptake (Prathap et al. 2022).

Potassium (K) homeostasis in plant cells is regulated by mechanisms that address the sensitivity of roots to inefficient K supply, triggering physiological responses such as hyperpolarization of the root cell membrane and extrusion of H^+, maximizing the action of H^+-ATPases that pump H^+ (Wang and Wu 2010; Srivastava et al. 2019). The same happens in the leaf epidermis, where homeostasis depends on the exchange of K^+ for H^+ in the plasma membrane and tonoplast and an interconnection between cell pH and Ca^{2+} (Li et al. 2024).

In order to maintain Ca homeostasis, the Ca influx into the cytosol is controlled by the Ca-dependent signaling responses (Kanwar 2021). In order to avoid toxicity, Ca is exported out of the cytoplasm or stored in intracellular organelles (Luans and Wang 2021), as well as proteins, to buffer cytosolic Ca by direct binding (Schwaller 2009). Magnesium (Mg) homeostasis can be achieved through transporters in the chloroplast, such as MGT10, MGR8, and MGR9 (Dukic et al. 2023). In order to avoid damage from S deficiency or excess, plants can regulate S assimilation through genes involved in this process, regulating it positively in cases of deficiency or negatively in cases of toxicity (Maruyama-Nakashita 2008). Furthermore, the H_2S molecule itself participates in S homeostasis in plants, undergoing conversion reactions in other molecules to maintain optimal concentrations of this nutrient (Narayan et al. 2023).

Overall, plant nutritional homeostasis involves a complex network of mechanisms that regulate nutrient acquisition, transport, redistribution, disorder signaling, and storage. These mechanisms have specific actions for each nutrient in disarray. However, they also have joint responses with many other nutrients that help with signaling and/or nutritional balance, which have yet to be fully elucidated (Dreyer 2021).

Bouain et al. (2019) indicated that there is a homeostatic adjustment at the molecular level and actions in different ways in plant homeostasis: (i) one element depends on the other, one depends on the other for proper absorption, and deficiency or toxicity positively or negatively affects the absorption of the other because they share internal mechanisms or regulation processes; (ii) hidden responses greater than a single stress. Therefore, plants can identify the levels of these nutrients in the soil and in their cells and adjust the mechanisms that lead to nutrient accumulation.

Some well-known interactions include K, Ca and Mg, P and Zn, N and S, and N and K. Studies involving the combined effect of nutritional deficiency of N, P, K, and S regarding the individual deficiency of each nutrient on the root parameters of *Arabidopsis* have indicated that synergistic, antagonistic, and null effects occur in the plants (Kellermeier et al. 2014). This study showed that a combination of transporters, receptors, and kinases acts as signalers, especially in N and K interactions.

The interaction between P and S homeostasis in plants is well known because, in cases of S deficiency, plants can replace sulfolipids with phospholipids, and the opposite also happens in P deficiency (Yu et al. 2002). Paul et al. (2015) showed how N, P, and S homeostasis processes are interconnected through miRNAs that encode transporter genes for each nutrient.

The antagonistic interaction between Ca and Mg in plant metabolism has been studied for several decades. Tang and Luan (2017) demonstrated the connection between the homeostasis of Ca^{2+} and Mg^{2+}. According to them, the deficiency of one of these nutrients stimulates responses that induce the absorption of the other, favoring plant development. Similarly, an excess of one of these nutrients reduces the absorption of the other. In this interaction, a Ca^{2+} signaling network occurs in the tonoplast in case of excess Mg^{2+}. This excess is perceived by calcineurin B (CBL) proteins that activate a set of protein kinases (CIPKs), inducing the sequestration of Mg^{2+} in the vacuole and maintaining homeostasis (Tang et al. 2015).

In addition to nutritional interactions, plant homeostasis studies widely discuss elementary or ecological stoichiometry. Besides providing information on the physiological and biochemical functioning of crops, it also adds information regarding the ecosystem's functioning as a whole (Ding et al. 2022). Luo et al. (2020) evidenced that several species of shrubs growing in the desert in China have lower P concentrations but higher K concentrations. In other words, they have a low P:K ratio compared to other plants growing in other desert conditions worldwide (Luo et al. 2020). They pointed out how the soil evaluated can alter the stoichiometric response conditions of plants. Depending on the environmental and stress conditions to which the plants are subjected, stoichiometric adjustment plays a role in adaptation.

Ågren and Weih (2020) tested scaling relationships between other nutrients (K, Ca, Mg, S, Cu, Zn, Fe, and Mn). They concluded that the scaling relationships reported for N and P drive plant stoichiometric relationships, and these scales can be extended to the other plant nutrients. These findings are in line with Tian et al. (2018), who evaluated a set of nutritional data from several plant species. They found that increasing N and P in the soil improved the N:K, N:Ca, N:Mg, P:K, P:Ca, and P:Mg ratios. They concluded that soil concentrations of N and P affect stoichiometric relationships, that is, the balance and homeostasis between nutrients. Moreover, these relationships are also modified by the plants' demand for nutrients to maintain several functions under climate change.

It is worth noting that the C:N:P ratio is the stoichiometric ratio widely used in plant and environmental studies. It reflects a concept that evaluates the interaction between essential elements C, N, and P by assessing the balance in ecological systems (Ren et al. (2016) that involve nutrient cycling in the plant-soil-microorganism system (Xu et al. 2024). It is due to the influence of many adverse conditions on this relationship in plants, including stresses such as drought water stress and high temperature (Olivera-Viciedo et al. 2021a), high CO_2 (Habermann et al. 2024) and nutritional deficiency (Costa et al. 2023b). It has been reported that water stress decreases nutrient uptake and yield in forage plants (Barreto et al. 2020; Olivera-Viciedo et al. 2021b), resulting in decreased N and P content in plant tissues. This decrease leads to increased C:N and C:P ratios and lower biomass production (Habermann et al. 2019; Olivera-Viciedo et al. 2021b, 2024), and the adjustments of nutrient concentrations and their C:N:P ratios were adjusted due to stress.

Therefore, the C:N:P ratio provides broader answers to the homeostatic adjustments plants make to several growth conditions, whether soil, environmental, or nutritional. The changes in the C:N:P ratios indicate internal adjustments for plant adaptation. However, recent evidence shows that these responses are modified in different ecosystems and stress scenarios, such as warming (Yu et al. 2024). Thus, there may be different responses depending on the species' growing conditions and locations.

Therefore, nutritional homeostasis in plants is essential for coordinating metabolic and physiological responses during their development. Homeostatic adjustment is necessary for the species' defense, adaptation, and survival reactions to be achieved, especially in crops subject to adversity, whether in the soil or the environment.

14.3 The Importance of Silicon in the Uptake and Transportation of N and P in Plants

As discussed above, plants have incredible mechanisms for absorbing, transporting, redistributing, and maintaining cellular homeostasis of nutrients. However, elements beneficial to plants can interact with these processes, influencing nutrient concentrations and, consequently, the stoichiometric relationships between nutrients in plant tissues. Silicon is an element capable of promoting beneficial effects on plant growth. It interacts with the absorption and utilization of macronutrients, such as N and P. Silicon fertilization can improve the N nutrition of crops by influencing the nutrient's absorption, assimilation, or remobilization (Pavlovic et al. 2021). In a study on canola, cotton, and wheat crops, Mehrabanjoubani et al. (2015) elucidated that Si especially favors the symplastic route of nutrient absorption.

Silicon accumulating species (content >10 mg kg^{-1}) that have specific transporters in the roots (Ma and Takahashi 2002), such as those of the Poaceae family, are very responsive to Si supply. Wu et al. (2017) observed in rice plants (*Oryza sativa* L.) that Si can regulate the expression of genes involved in the absorption (OsAMT1;1) and assimilation of N (OsGS1;1), under high or low N fertilization. Maize seedlings (*Zea mays* L.) that received Si showed increases in N content, in the activity of the nitrate reductase enzyme, in the dark green color index, and in dry matter production (da Silva et al. 2021a, b). Galindo et al. (2021) describe a high positive correlation between Si content and N accumulation in wheat (*Triticum aestivum* L.) and maize plants, with gains in the agronomic efficiency of N use at low and high fertilizations.

Silicon is also beneficial in the assimilation of N and other nutrients in non-accumulating crops such as vegetables and fruit trees (Liang et al. 2015). In strawberry plants, Si application improved the activity of glutamate synthase (GOGAT) and glutamate dehydrogenase (GDH), with an increase in leaf chlorophyll content (Xu et al. 2023). In cucumber grown under water deficit, the application of Si nanoparticles promoted an increase in leaf N content (Alsaeedi et al. 2019).

In an effort to elucidate the mechanisms of Si in P uptake, Hu et al. (2018) compared P uptake in the presence and absence of Si in wild rice and in a mutant deficient in Si uptake and quantified the expression of Pi transporter genes. The authors observed that the expression of the OsPT6 transporter in the roots was negatively regulated in the presence of Si. Ma and Takahashi (1990) already indicated that Si benefits the nutrition of rice grown at low or high P levels.

Silicon can also interact with the acquisition of P by plants, directly influencing its absorption and transport (Greger et al. 2018), as observed in the soybean crop (Shen et al. 2009). In wheat grown under low P, the supply of Si (Na_2SiO_3) positively regulates the expression of P transporters (TaPHT1.1 and TaPHT1), allowing the plant to reach adequate P content in the aerial part (Kostic et al. 2017). Silicon also improved P absorption in sorghum plants grown under salinity or P deficiency (Calero-Hurtado et al. 2020a; da Silva et al. 2021a, b).

Silicon application also optimizes P nutrition in rice, leading to greater root growth and improved use of phosphate fertilizer (Pati et al. 2016). However, Si can also minimize P absorption in rice plants if the nutrient is in excess in the growing medium (Hu et al. 2020).

There is also the indirect effect of Si on the P nutrition of crops, as it can increase the availability of phosphate in soils (Greger et al. 2018). It is worth noting that this benefit depends on the source applied because silicates in particular can compete with phosphate for adsorption sites in the soil (Etesami and Schaller 2023). Furthermore, silicates can increase the soil's pH (corrective effect) by favoring the solubilization of Fe and Al phosphates, contributing in both ways to greater P availability in the soil (Crusciol et al. 2014).

Silicon directly and indirectly influences the absorption of N and P by plants. Thus, it can act to regulate the nutritional balance of crops (homeostasis) when the growing environment imposes stress and even boost the productivity of non-stressed crops.

14.4 The Role of Silicon in the Homeostasis of C, N, and P and in the Efficiency of Nutrient Use in Plants

Nutritional homeostasis analyzed based on stoichiometric relationships makes it possible to infer the response of crops to environmental conditions and to measure the effects of anthropogenic actions on a management system (i.e., when fertilizing) (Prado and Silva 2017; Xiao et al. 2024). The C:N:P ratio is noteworthy because it relates nutrients that have a major influence on plant composition and metabolism

(Ågren and Weih 2020). It acts as an indicator of the plant's photosynthetic capacity to fix C as a function of the accumulation of P and N (Costa et al. 2024).

The literature commonly reports gains in the biological and agronomic efficiency of P and N use after Si fertilization for several crops, such as coffee, wheat, forage plants, and sugar beet (Parecido et al. 2022; Araújo et al. 2022; Carvalho et al. 2022; Taskin et al. 2023). The scientific community has consistently elucidated that the benefits of Si in plants may be closely related to its influence on elemental stoichiometric relationships, given its interaction with the absorption and assimilation of N, P, and C in plants.

Silicon must accumulate in the tissues before it can benefit plants, and only once it has accumulated can there be an interaction with the C:N:P stoichiometry. Therefore, the variation in the capacity of plant species to absorb nutrients and Si (due to genetic factors such as the expression of genes that encode transporters) directly influences the magnitude of the stoichiometric changes promoted by Si. Thus, in accumulator species, the action of this element on nutritional homeostasis is expected to be more significant than in intermediate accumulator or non-accumulator species.

In addition to the absorption of mineral nutrients, Si also interacts with the incorporation of C into plants. As a result of the incorporation of Si into the plant structure, there is generally a decrease in the C content of the tissues, especially in the leaves (Neu et al. 2017; Lata-Tenesaca et al. 2021). Silicon plays a structural role in plant tissues through the deposition of amorphous silica and the formation of covalent bonds with some cell wall components, such as hemicellulose, pectin, and lignin (Sheng and Chen 2020). The incorporation of Si has a reduced energy cost compared to the synthesis of structural C compounds (Raven 1983). Thus, in addition to the stoichiometric modification, the plant can direct more energy toward its growth and/or adaptation to adverse conditions.

Even though it is a recent approach, studies have already shown that Si (applied exogenously or present in natural soil reserves such as phytoliths), when absorbed, triggers a promising action by improving nutritional homeostasis in many species and growing conditions, with or without stress. The supply of root or foliar Si altered the C:N:P stoichiometry in sugarcane (*Saccharum officinarum* L.) seedlings and ratoons under water deficit, restoring stoichiometric relationships similar to those of plants grown under full irrigation (Teixeira et al. 2020; Costa et al. 2023a). By adjusting these relationships between nutrients in the tissues as a result of the greater absorption of N and P, Si improves the photosynthetic rate and increases C use efficiency (Frazão et al. 2020).

In commercial cultivation, sugarcane crop productivity was increased by up to 11 Mg ha^{-1} in stalk yield and by 20% in total recoverable sugar in the industry due to greater C:P and N:P homeostasis under Si fertilization (Dutra et al. 2023). Similarly, Souza Júnior et al. (2022) observed greater accumulation and use efficiency of N, P, and C in sugarcane cultivars fertigated with Si due to changes in the C:N, C:P, and C:Si ratios in the leaves. Increases in height, stalk diameter, number of tillers, leaf area, and dry mass of leaves and stalks benefited the plants.

The increase in Si supply (hydrophilic pyrogenic SiO_2) in wheat cultivation increased the Si content in leaf tissues, with a decrease in C content and an increase in P and N content and use efficiency, balancing the stoichiometry of these nutrients with C and increasing grain yield (Neu et al. 2017). A similar effect of Si was observed in maize, with a decrease in the C:P and N:P ratios in stressed and non-stressed plants by improving absorption and increasing the N and P content in the roots and aerial part of the plants (Costa et al. 2023b).

In sorghum (*Sorghum bicolor* L.) and sunflower (*Helianthus annuus* L.) crops, the application of Si recovered the C:N ratio that had been reduced by salt stress. It improved biomass production by 39 and 24% for sorghum and sunflower, respectively (Calero-Hurtado et al. 2020b). The beneficial effects of Si also extend to forage species. In *Megathyrsus maximus* cv. Zuri, fertigation with nanosilica improved the C:N, C:P, and N:P balance, increasing N use efficiency, tillering, and dry matter production in unstressed plants grown in Oxisol (Melo et al. 2023b).

Olivera-Viciedo et al. (2024) reported that plants of *M. maximus* cv. Mombaça absorbed Si even without exogenous Si application, possibly from biogenic deposits. This absorption of Si promoted more balanced stoichiometric relationships, leading to greater dry matter production by the pasture under double stress (water deficit and temperature rise). Grasses of the genus *Urochloa* also responded to Si fertigation by decreasing the C content, which was reflected in greater efficiencies in the use of C and photosystem II, improving dry matter production in plants with or without water restriction (Rocha et al. 2021).

Lata-Tenesaca et al. (2021) observed the best nutritional performance of quinoa plants (*Chenopodium quinoa* Willd.) fertilized with Si (sodium potassium silicate) in a hydroponic medium. The study revealed that even though quinoa is an intermediate accumulator of Si with contents of 2–9 g kg^{-1}, it benefited from changes in the C:N and C:P ratios, associated with greater absorption of N and P, with an increase in the use efficiency of C, N, and P, and optimized productivity. The use of Si stimulates the synthesis of antioxidant compounds (Silva et al. 2021a, b) and enriches food with Si, an element that is required for human health, thus providing food biofortification.

In the plantain crop (*Plantago lanceolata* L.) subjected to water deficit, the increase in leaf Si caused the recovery of the C:N and C:P ratios, contributing to maintaining a balance of nutrients during stress and also in periods without stress (Long et al. 2018). Tropical tree species can also benefit from the effects of Si on elemental stoichiometry. According to Klotz et al. (2023), the availability of Si in the substrate promoted its absorption to a greater or lesser extent depending on the species, with an increase in foliar P concentrations in *Sorecea* even without a change in P availability in the soil, modifying N:P and C:P ratios in *Sorecea* and *Theobroma* and increasing aerial biomass in several tree species. These results indicate the potential use of Si in nutritional homeostasis in tree species, pointing to the need for further research in commercially cultivated species such as the cacao tree (*Theobroma cacao*) and other fruit trees.

14.5 Involvement of Silicon in C:N:P Homeostasis and Mitigation of Stress Induced by Climate Change

Climate change has exacerbated the incidence and severity of many abiotic stresses worldwide. Unexpected rainfall, prolonged droughts, extreme temperatures, salinity, and heavy metals are among the primary limitations affecting plant growth and development and impacting crop yield and quality. Plant growth and development under stress result from plant-environment interactions in which several climatic and edaphic stressors drive biochemical, physiological, and molecular responses (Ahire et al. 2021).

Studies related to the mitigating effects of Si on many abiotic stresses have been framed within the understanding of the beneficial role of this element in regulating the C:N:P homeostasis of plants. These abiotic stresses include salinity, drought, and increased temperature, which increased in recent years due to climate change. This regulatory effect of Si on C:N:P homeostasis has allowed plants under different stresses to show greater resistance to each of these stressors in a changing climate (Costa et al. 2022; Melo et al. 2023b; Olivera-Viciedo et al. 2024).

At the biochemical level, plants alter metabolism in many ways to adapt to environmental stress, including antioxidant defense, ion homeostasis, and osmotic adjustment. Many plant species respond to environmental stress with an effective ROS elimination system involving antioxidant enzymes such as superoxide dismutase (SOD), peroxidase (POX), and catalase (CAT) (Peña-Calzada et al. 2023a). When plants are subjected to several environmental stresses, they alter their C, N, and P balance and thus their elemental stoichiometric ratios, leading to structural, nutritional, biochemical, and physiological impacts and reduced productivity (Olivera-Viciedo et al. 2019b, 2021a, 2024; Costa et al. 2024).

Several studies have indicated that Si can increase salt stress tolerance in several plant species (Calero-Hurtado et al. 2020a; Peña-Calzada et al. 2023a, b). In sorghum and sunflower plants, Calero-Hurtado et al. (2021) reported that Si applied both via root and foliar application alleviates NaCl toxicity in sorghum and sunflower plants by improving ion homeostasis in roots and shoots and increasing dry matter accumulation. Furthermore, an additional study reported that the impact of Si on C:N, C:P, and N:P stoichiometric ratios varies with the species' ability to accumulate this element (i.e., accumulator, intermediate, or non-accumulator plants) and with the method of Si application used (Calero-Hurtado et al. 2020b). In soybean plants, Peña-Calzada et al. (2023a) indicated that the use of exogenous Si can be an efficient strategy to mitigate the harmful effects of salt stress, as Si incorporation increased potassium (K) accumulation, plant water status, photosynthetic pigment content, chlorophyll fluorescence parameters, and gas exchange attributes, decreasing Na+ uptake by the plants.

The regulatory role of silicon in nutrient absorption, growth, and plant production under stress conditions has been documented in many crops. In bean plants with potassium (K) nutritional deficiency, Si application provided through the nutrient solution or foliar application mitigated K deficiency stress. Furthermore, it increased chlorophyll and carotenoid content, photosynthetic activity, water use efficiency, and vegetative growth (dos Santos Sarah et al. 2021). In sugarcane, Si application promoted the mitigation of aluminum (Al) stress by regulating nutrient stoichiometric homeostasis, improving dry biomass production (Sousa Junior et al. 2022). Similarly, Si mitigated the effects of manganese (Mn) deficiency in sugarcane, even during prolonged stress periods, by reducing H2O2 content by 18% and MDA by 32%. It also increased Mn absorption efficiency by 53%, SOD activity by 23%, and GPOX activity by 76%, along with phenol content, thereby enhancing growth (Oliveira et al. 2022).

In the forage grass *M. maximus*, Si nanoparticle application triggered modifications in C:N:P stoichiometry and alleviated stress caused by phosphorus (P) toxicity, improving grass biomass production (Melo et al. 2023b). In another recent study, Olivera-Viciedo et al. (2024) revealed for the first time that *Panicum maxim* tolerance to warming (2 °C above ambient temperature) and soil water restriction may be associated with Si absorption and changes in plant C:N:P stoichiometry. This study found that under soil water restriction conditions, C, N, and P stoichiometric ratios increase, while under warming and no irrigation, C:Si, N:Si, and P:Si ratios decrease, improving biomass production only under warming and irrigation conditions.

14.6 Implication of Silicon in C:N:P Homeostasis and in the Mitigation of Stress Caused by Nutritional Deficiency or Toxicity

The nutrients N and P are required by many forms of life on earth and have fundamental biochemical, physiological, and structural functions in plants. However, low natural availability in the soil and limitations in the nutrient absorption process can compromise crop production (Prado 2021; Melo et al. 2023a). For example, N and P deficiencies are a limiting factor for the production of many crops and pastures worldwide (Euclides et al. 2022; Kebede et al. 2024).

Fertilization is used to overcome nutritional deficiencies. However, in many regions worldwide, fertilization is excessive, reducing nutrient use efficiency (Pavinato et al. 2020; Feyisa et al. 2024). Furthermore, the accumulation of N and P in soils increases the risk of toxicity to plants and alters the N:P stoichiometry, unbalancing terrestrial and aquatic ecosystems (Oliveira Filho et al. 2020; Penuelas et al. 2023).

Silicon use can promote balanced crop nutrition and mitigate the biological damage of plant deficiency or toxicity. In tomato plants (*Solanum lycopersicum* L.), Si mitigated the damage caused by oxidative stress in N-deficient plants under water restriction, with an increase in root length and, consequently, the plants' dry weight (Machado et al. 2024). According to Gou et al. (2020), the root application of Si in cucumber plants (*Cucumis sativus* L.) under excess nitrate promoted less oxidative damage and better assimilation of N, increasing chlorophyll and its precursors and optimizing the photosynthetic rate. These studies reinforce that Si can help mitigate the damage caused by plant nutritional imbalances.

Research results have strengthened the hypothesis that Si has a modulating action (positive or negative) on the absorption of N and P in many crops, depending on the level of the nutrient in the growing medium (Wu et al. 2017; Hu et al. 2018; Gou et al. 2020; da Silva et al. 2021a, b; Carvalho et al. 2022). This interaction with nutrient absorption is the basis of the changes in elemental stoichiometry in plants fertilized with Si. Many studies have placed C:N:P homeostasis as a basic mechanism of Si that results in greater crop resilience against nutritional imbalances, such as deficiency or toxicity.

In rice under N deficiency, there were changes in the C:N:P:Si ratio responsible for reducing cell leakage and increasing production by 18% under silicate fertilization (Deus et al. 2020). Araújo et al. (2022) reported that Si added to the nutrient solution increased the use efficiency of N, P, and Ca in plants of *U. brizantha* and *M. maximum* deficient in these nutrients, increasing the aerial part's dry matter production. The authors did not analyze stoichiometric changes. However, it may have occurred since the adjustment in homeostasis in the presence of Si is related to the better nutritional efficiency of forage plants (Rocha et al. 2021; Melo et al. 2023b).

Some studies report the mitigation of P toxicity via nutritional homeostasis mediated by Si. In eggplant (*Solanum aethiopicum*), P toxicity occurred after an improvement in C:N homeostasis based on 13% lower P absorption compared to plants that did not receive Si, reducing oxidative stress damage to membranes and improving the efficiency of micronutrient use (Alves et al. 2023).

Furthermore, nanosilica fertigation mitigated P toxicity in forage Poaceae grown in Oxisol and Entisol (Melo et al. 2023b). According to the authors, there was a decrease in the P content of *M. maximus* plants grown under an excess of the nutrient when they received Si. It allowed the C:N, C:P, and N:P ratios to recover their balance, with an increase in the efficiency of N and P use, less leaf senescence, and recovery of dry matter production. Figure 14.1 shows some visual aspects of P toxicity and its mitigation via Si-mediated C:N:P homeostasis in the crops mentioned.

The adjustment in elemental stoichiometry was also the basis for the mitigation of potassium (K) deficiency mediated by Si in the bean plant (*Phaseolus vulgaris* L. BRS Estilo). The supply of sodium potassium silicate (SiNaK) increased the Si content in the aerial part, with changes in the C:N, C:Si, N:P, N:Si, and P:Si ratios in the aerial part and C:N, C:P, C:Si, N:Si, N:P, and P:Si ratios in the roots, increasing the K content and efficiency and minimizing biomass loss (Costa et al. 2023b).

The benefits of Si in mitigating nitrate toxicity in cucumber (Gou et al. 2020), ammonium toxicity in passion fruit (Silva Júnior et al. 2019), sugar beet (Olivera-Viciedo et al. 2019a), and other crops have also been described, as reviewed by Campos et al. (2020). Moreover, Si can also mediate the mitigation of P deficiency in wheat plants (Kostic et al. 2017) and several other crops, as reviewed by Hu et al. (2020). However, the relationship between the mitigation of these stresses and nutritional homeostasis has yet to be elucidated, indicating the need for further research into this subject.

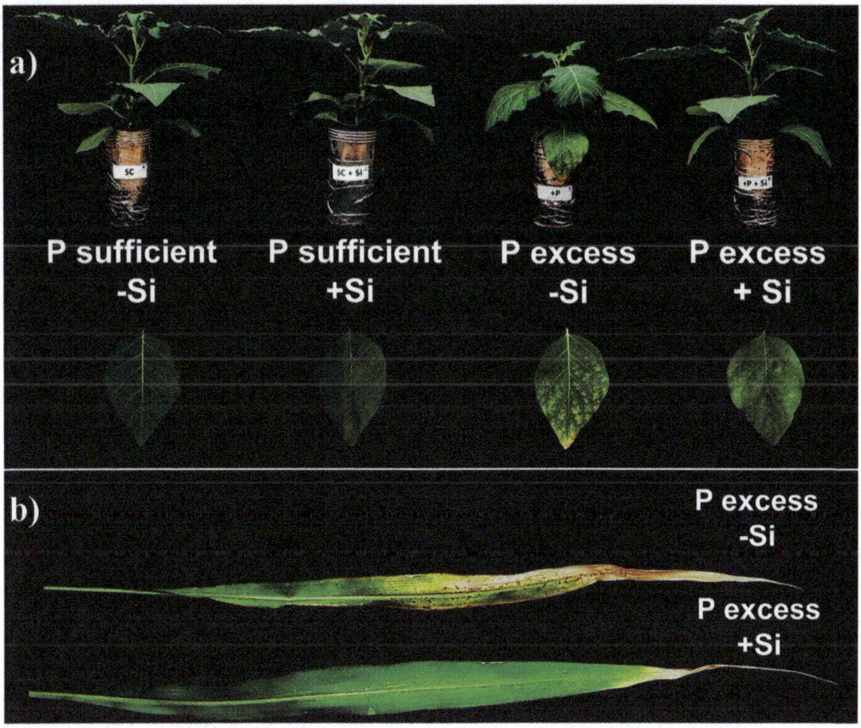

Fig. 14.1 Visual aspect of plants and leaves of eggplant (*Solanum aethiopicum*) and leaves of grass *M. maximus* cv. Zuri under phosphorus toxicity in the absence (-Si) and presence of Si (+Si) supplied as nanosilica. (Adapted from Alves et al. (2023) and Melo et al. (2023a, b))

Therefore, homeostasis in C:N:P ratios contributes to mitigating nutritional imbalances due to both deficiency and toxicity. Currently, this homeostasis in C:N:P ratios is most explored regarding the nutrients N and P. Studies to date indicate that nutritional homeostasis may be an essential mechanism of Si in mitigating stresses. They also indicate that it can increase the nutritional efficiency of crops and production levels (Souza Júnior et al. 2022; Carvalho et al. 2022; Dutra et al. 2023).

14.7 Conclusions and Future Perspectives

Studies involving nutritional homeostasis are essential for understanding how plants adapt to stress conditions in their life cycle. Studies in this area of nutrition are wide-ranging and require in-depth knowledge at the proteomic and metabolomic levels to investigate the relationships between nutrients. Much remains unclear, especially regarding the signaling pathways and the interdependence between nutrients and responses to environmental growth conditions. Identifying the dependence between nutrients for most relationships is still difficult. However, the homeostasis between C:N:P is still the focus of major research because of its direct connection with biogeochemical cycles, climate change, and other abiotic stresses, which explains well the adaptation and growth responses of plants. Although Si is not a plant nutrient, it has been prevalent in research involving stress and stoichiometry due to its role in inducing homeostatic regulation of the C:N:P ratio and improving plants' nutrient use efficiency. However, further research is needed into the many plant groups, sources of Si (especially nanosilica), forms of application, and growing conditions, emphasizing stress-free crops. Further research into the mechanisms by which nanoparticulate Si can influence adaptive responses by regulating nutrient homeostasis through nutrient absorption and assimilation in different crops and the interaction between Si and other nutrients will be of utmost importance since this knowledge is still incipient. If these gaps are prioritized in future studies, they could be the key to achieving new, higher levels of crop production while maintaining the quality of the harvested product and the safety of ecosystems.

References

Ågren GI, Weih M (2020) Multi-dimensional plant element stoichiometry: looking beyond carbon, nitrogen, and phosphorus. Front Plant Sci 11:23. https://doi.org/10.3389/fpls.2020.00023

Ahire ML, Mundada PS, Nikam TD, Bapat VA, Penna S (2021) Multifaceted roles of silicon in mitigating environmental stresses in plants. Plant Physiol Biochem 169:291–310. https://doi.org/10.1016/J.PLAPHY.2021.11.010

Ahmadi-Nouraldinvand F, Sharifi RS, Siadat SA, Khalilzadeh R (2024) Fascinating role of silicon dioxide nanoparticles and co-inoculation of mycorrhiza and rhizobacteria to combat NaCl stress: changes in physiological characteristics, uptake of nutrient elements, and enhancing photosystem II activities in wheat. J Soil Sci Plant Nutr 24:1–18. https://doi.org/10.1007/S42729-024-01640-0/TABLES/5

Ali S, Rizwan M, Hussain A, Zia ur Rehman M, Ali B, Yousaf B, Wijaya L, Alyemeni MN, Ahmad P (2019) Silicon nanoparticles enhanced the growth and reduced the cadmium accumulation in grains of wheat (Triticum aestivum L.). Plant Physiol Biochem 140:1–8. https://doi.org/10.1016/j.plaphy.2019.04.041

Alonso TAS, Ferreira Barreto R, de Mello PR, Pereira de Souza J, Falleiros CR (2020) Silicon spraying alleviates calcium deficiency in tomato plants, but Ca-EDTA is toxic. J Plant Nutr Soil Sci 183:659–664. https://doi.org/10.1002/jpln.202000055

Alsaeedi A, El-Ramady H, Alshaal T et al (2019) Silica nanoparticles boost growth and productivity of cucumber under water deficit and salinity stresses by balancing nutrients uptake. Plant Physiol Biochem 139. https://doi.org/10.1016/j.plaphy.2019.03.0

Alves DMR, Oliveira JN, Prado RM et al (2023) O silício na forma de nanosílica mitiga a toxicidade do P em berinjela vermelha. Sci Rep 13:9190. https://doi.org/10.1038/s41598-023-36412-w

Araújo WBS, Teixeira GCM, Prado RM et al (2022) Silicon mitigates nutritional stress of nitrogen, phosphorus, and calcium deficiency in two forages plants. Sci Rep 12:6611. https://doi.org/10.1038/s41598-022-10615-z

Arnon DI, Stout PR (1939) The essentiality of certain elements in minute quantity for plants with special reference to copper. Plant Physiol 14:371. https://doi.org/10.1104/PP.14.2.371

Asad MAU, Guan X, Zhou L, Qian Z, Yan Z, Cheng F (2023) Involvement of plant signaling network and cell metabolic homeostasis in nitrogen deficiency induced early leaf senescence. Plant Sci 336:111855. https://doi.org/10.1016/j.plantsci.2023.111855

Assunção AGL, Cakmak I, Clemens S, González-Guerrero M, Nawrocki A, Thomine S (2022) Micronutrient homeostasis in plants for more sustainable agriculture and healthier human nutrition. J Exp Bot 73:1789–1799. https://doi.org/10.1093/jxb/erac014

Barreto RF, de Prado RM, Habermann E et al (2020) Warming change nutritional status and improve Stylosanthes capitata Vogel growth only under well-watered conditions. J Soil Sci Plant Nutr. https://doi.org/10.1007/s42729-020-00255-5

Bouain N, Krouk G, Lacombe B, Rouached H (2019) Getting to the root of plant mineral nutrition: combinatorial nutrient stresses reveal emergent properties. Trends Plant Sci 24:6. https://doi.org/10.1016/j.tplants.2019.03.008

Calero Hurtado A, Olivera Viciedo D, de Mello Prado R (2023) Beneficial role of silicon in plant nutrition under salinity conditions. Benefits Silicon Nutr Plants 253–273. https://doi.org/10.1007/978-3-031-26673-7_15/COVER

Calero-Hurtado A, Aparecida Chiconato D, de Mello Prado R, da Silveira Sousa Junior G, Felisberto G (2019) Silicon attenuates sodium toxicity by improving nutritional efficiency in sorghum and sunflower plants. Plant Physiol Biochem 142:224–233. https://doi.org/10.1016/j.plaphy.2019.07.010

Calero-Hurtado A, Chiconato DA, Prado R d M, Sousa Junior G d S, Gratão PL, Felisberto G, Olivera Viciedo D, Mathias dos Santos DM (2020a) Different methods of silicon application attenuate salt stress in sorghum and sunflower by modifying the antioxidative defense mechanism. Ecotoxicol Environ Saf 203:110964. https://doi.org/10.1016/j.ecoenv.2020.110964

Calero-Hurtado A, Chiconato DA, Prado R d M, Sousa Junior G d S, Olivera Viciedo D, Piccolo M d C (2020b) Silicon application induces changes C:N:P stoichiometry and enhances stoichiometric homeostasis of sorghum and sunflower plants under salt stress. Saudi J Biol Sci 27:3711–3719. https://doi.org/10.1016/j.sjbs.2020.08.017

Calero-Hurtado AC, Chiconato DA, de Mello Prado R, da Silveira Sousa Junior G, Viciedo DO, Díaz YP, Calzada KP, Gratão PL (2021) Silicon alleviates sodium toxicity in sorghum and sunflower plants by enhancing ionic homeostasis in roots and shoots and increasing dry matter accumulation. SILICON. https://doi.org/10.1007/s12633-020-00449-7

Campos CNS, Silva Júnior da GB, de Prado RM et al (2020) Silicon mitigates ammonium toxicity in plants. Agron J 112:635–647. https://doi.org/10.1002/agj2.20069

Carciochi WD, Divito GA, Fernández LA, Echeverría HE (2017) Sulfur affects root growth and improves nitrogen recovery and internal efficiency in wheat. J Plant Nutr 40:1231–1242. https://doi.org/10.1080/01904167.2016.1187740

Carvalho LTS, Prado RM, Silva JLF et al (2022) Impact of nanotechnology from nanosilica to mitigate N and P deficiencies favoring the sustainable cultivation of sugar beet. Nano materials 12:22. https://doi.org/10.3390/nano12224038

Costa MG, dos Santos Sarah MM, de Mello Prado R, Palaretti LF, de Cássia Piccolo M, de Souza Júnior JP (2022) Impact of Si on C, N, and P stoichiometric homeostasis favors nutrition and stem dry mass accumulation in sugarcane cultivated in tropical soils with different water regimes. Front Plant Sci 0:2739. https://doi.org/10.3389/FPLS.2022.949909

Costa MG, Prado RM, Sarah MMS et al (2023a) New approaches to the effects of Si on sugarcane ratoon under irrigation in quartzipsamments, eutrophic red oxisol, and dystrophic red oxisol. BMC Plant Biol 23:51. https://doi.org/10.1186/s12870-023-04077-2

Costa MG, Prado RM, Sarah MMS et al (2023b) Silicon mitigates K deficiency in maize by modifying C, N, and P stoichiometry and nutritional efficiency. Sci Rep 13:16929. https://doi.org/10.1038/s41598-023-44301-5

Costa MG, Prado RM, Palaretti LF et al (2024) The effect of abiotic stresses on plant C: N: P homeostasis and their mitigation by silicon. Crop J. https://doi.org/10.1016/j.cj.2023.11.012

Crusciol CAC, Foltran R, Rossato OB et al (2014) Effects of surface application of calcium-magnesium silicate and gypsum on soil fertility and sugarcane yield. Rev Bras Cienc Solo 38. https://doi.org/10.1590/S0100-06832014000600019

Deus ACF, Prado RM, Alvarez RCF et al (2020) Role of silicon and salicylic acid in the mitigation of nitrogen deficiency stress in rice plants. SILICON 12. https://doi.org/10.1007/s12633-019-00195-5

Ding DD, Arif M, Liu M, Li J, Hu X, Geng Q, Yin V, Li C (2022) Plant-soil interactions and C:N:P stoichiometric homeostasis of plant organs in riparian plantation. Front Plant Sci 13:2022. https://doi.org/10.3389/fpls.2022.979023

Dreyer I (2021) Nutrient cycling is an important mechanism for homeostasis in plant cells. Plant Physiol 4:2246–2261. https://doi.org/10.1093/plphys/kiab217

Du Q, Zhao XH, Jiang CJ, Wang XG, Han Y, Wang J et al (2017) Effect of potassium deficiency on root growth and nutrient uptake in maize (Zea mays L.). Agric Sci 8:1263–1277. https://doi.org/10.4236/as.2017.811091

Dukic E, van Maldegen KA, Shaikh KM et al (2023) Chloroplast magnesium transporters play essential but differential roles in maintaining magnesium homeostasis. Front Plant Sci 14:2023. https://doi.org/10.3389/fpls.2023.1221436

Dutra AF, Leite MRL, Melo CCF et al (2023) Soil and foliar Si fertilization alters elemental stoichiometry and increases yield of sugarcane cultivars. Sci Rep 13:16040. https://doi.org/10.1038/s41598-023-43351-z

Epstein E (1994) The anomaly of silicon in plant biology. Proc Natl Acad Sci USA. https://doi.org/10.1073/pnas.91.1.11

Epstein E (2003) SILICON. Annu Rev Plant Biol 50:641–664. https://doi.org/10.1146/annurev.arplant.50.1.641

Etesami H, Schaller J (2023) Improving phosphorus availability to rice through silicon management in paddy soils: a review of the role of silicate-solubilizing bacteria. Rhizosphere:100749. https://doi.org/10.1016/j.rhisph.2023.100749

Euclides VPB, Montagner DB, Araújo AR (2022) Biological and economic responses to increasing nitrogen rates in Mombaça guinea grass pastures. Sci Rep 12:1937. https://doi.org/10.1038/s41598-022-05796-6

Feyisa DS, Jiao X, Mojo D (2024) Wheat yield response to chemical nitrogen fertilizer application in Africa and China: a meta-analysis. J Soil Sci Plant Nutr. https://doi.org/10.1007/s42729-023-01609-5

Flores RA, Pessoa-de-Souza MA, de Andrade AF, Bueno AM, de Oliveira Abdala K, de Souza Júnior JP, de Mello Prado R, Santos GG, Mesquita M (2022) Does foliar application of silicon under natural water stress conditions increase rice yield in subtropical dry regions? SILICON 14:3591–3600. https://doi.org/10.1007/S12633-021-01109-0/METRICS

Frazão JJ, Prado RM, Souza Júnior JP et al (2020) Silicon changes C:N:P stoichiometry of sugarcane and its consequences for photosynthesis, biomass partitioning and plant growth. Sci Rep 10:12492. https://doi.org/10.1038/s41598-020-69310-6

Galindo FS, Pagliari PH, Rodrigues WL et al (2021) Silicon amendment enhances agronomic efficiency of nitrogen fertilization in maize and wheat crops under tropical conditions. Plan Theory 10:7. https://doi.org/10.3390/plants10071329

Gao K, Chen F, Yuan L, Zhang F, Mi G (2014) A comprehensive analysis of root morphological changes and nitrogen allocation in maize in response to low nitrogen stress. Plant Cell Environ 38:740–750. https://doi.org/10.1111/pce.12439

Ge J, Xie Z (2017) Leaf litter carbon, nitrogen, and phosphorus stoichiometric patterns as related to climatic factors and leaf habits across Chinese broad-leaved tree species. Plant Ecol 218:1063–1076. https://doi.org/10.1007/s11258-017-0752-8

Gonzalez-Porras CV, Teixeira GCM, Prado R d M, Ferreira PM, Palaretti LF, Oliveira KS (2024) Silicon via fertigation with and without potassium application, improve physiological aspects of common beans cultivated under three water regimes in field. Sci Rep 14:1–19. https://doi.org/10.1038/s41598-024-52503-8

Gou T, Yang L, Hu W et al (2020) Silicon improves the growth of cucumber under excess nitrate stress by enhancing nitrogen assimilation and chlorophyll synthesis. Plant Physiol Biochem 152:53–61. https://doi.org/10.1016/j.plaphy.2020.04.031

Greger M, Landberg T, Vaculík M (2018) Silicon influences soil availability and accumulation of mineral nutrients in various plant species. Plan Theory 7(2):41. https://doi.org/10.3390/plants7020041

Guo W, Cong Y, Hussain N, Wang Y, Liu Z, Jiang L, Liang Z, Chen K (2014) The remodeling of seedling development in response to long-term magnesium toxicity and regulation by ABA–DELLA signaling in Arabidopsis. Plant Cell Physiol 55:1713–1726. https://doi.org/10.1093/pcp/pcu102

Habermann E, Dias de Oliveira EA, Contin DR, Delvecchio G, Viciedo DO, de Moraes MA, de Mello Prado R, de Pinho Costa KA, Braga MR, Martinez CA (2019) Warming and water deficit impact leaf photosynthesis and decrease forage quality and digestibility of a C4 tropical grass. Physiol Plant 165:383–402. https://doi.org/10.1111/ppl.12891

Habermann E, Contin DR, Afonso LF, Barosela JR, de Pinho Costa KA, Viciedo DO, Groppo M, Martinez CA (2022) Future warming will change the chemical composition and leaf blade structure of tropical C3 and C4 forage species depending on soil moisture levels. Sci Total Environ 821:153342. https://doi.org/10.1016/J.SCITOTENV.2022.153342

Habermann E, Dias de Oliveira EA, Bianconi ME, Contin DR, Lemos MTO, Costa JVCP, Oliveira KS, Riul BN, Bonifácio-Anacleto F, Viciedo DO, Approbato AU, Alzate-Marin AL, Prado R d M, Costa KA d P, Martinez CA (2024) Balancing trade-offs: enhanced carbon assimilation and productivity with reduced nutritional value in a well-watered C4 pasture under a warmer CO2 enriched atmosphere. Plant Physiol Biochem 207:108408. https://doi.org/10.1016/J.PLAPHY.2024.108408

Hanikenne M, Esteves SM, Fanara S, Rouached H (2020) Coordinated homeostasis of essential mineral nutrients: a focus on iron. J Exp Bot 72:2136–2153. https://doi.org/10.1093/jxb/eraa483

Hepler HP (2005) Calcium: a central regulator of plant growth and development. Plant Cell 17:2142–2155. https://doi.org/10.1105/tpc.105.032508

Hu AY, Che J, Shao JF et al (2018) Silicon accumulated in the shoots results in down-regulation of phosphorus transporter gene expression and decrease of phosphorus uptake in rice. Plant Soil 423. https://doi.org/10.1007/s11104-017-3512-6

Hu AY, Xu SN, Qin DN et al (2020) Role of silicon in mediating phosphorus imbalance in plants. Plan Theory 10(1):51. https://doi.org/10.3390/plants10010051

Huang J, Wang P, Niu Y, Yu H, Ma F, Xiao G, Xu X (2018) Changes in C:N:P stoichiometry modify N and P conservation strategies of a desert steppe species Glycyrrhiza uralensis. Sci Rep 8:12668. https://doi.org/10.1038/s41598-018-30324-w

Hussain B, Lin Q, Hamid Y, Sanaullah M, Di L, Hashmi ML u R, Khan MB, He Z, Yang X (2020) Foliage application of selenium and silicon nanoparticles alleviates Cd and Pb toxicity in rice (Oryza sativa L.). Sci Total Environ 712:136497. https://doi.org/10.1016/J.SCITOTENV.2020.136497

Kanwar P (2021) Mechanism of Ca2+ homeostasis across the plant membranes. In: Upadhyay SK (ed) Calcium transport elements in plants. Academic, pp 39–54. https://doi.org/10.1016/B978-0-12-821792-4.00006-0

Kebede G, Worku W, Jifar H et al (2024) Grain yield productivity, nutrient uptake and use efficiency and profitability of oat (Avena sativa L.) as influenced by variety and levels of nitrogen and phosphorus fertilizers in the central highlands of Ethiopia. J Crop Sci Biotechnol. https://doi.org/10.1007/s12892-023-00230-3

Kellermeier F, Armengaud P, Seditas TJ, Danku J, Salt DE, Amtmann A (2014) Analysis of the root system architecture of Arabidopsis provides a quantitative readout of crosstalk between nutritional signals. Plant Cell 26:1480–1496

Klotz M, Schaller J, Engelbrecht BM (2023) Effects of plant-available soil silicon on seedling growth and foliar nutrient status across tropical tree species. Oikos:10030. https://doi.org/10.1111/oik.10030

Kostic L, Nikolic N, Bosnic D et al (2017) Silicon increases phosphorus (P) uptake by wheat under low P acid soil conditions. Plant Soil 419:447–455. https://doi.org/10.1007/s11104-017-3364-0

Lata-Tenesaca LF, Prado RM, Piccolo MC et al (2021) Silicon modifies C: N: P stoichiometry, and increases nutrient use efficiency and productivity of quinoa. Sci Rep 11:1. https://doi.org/10.1038/s41598-021-89416-9

Li Q, Gao Y, Yang A (2020) Sulfur homeostasis in plants. Int J Mol Sci 21:8926. https://doi.org/10.3390/ijms21238926

Li K, Grauschopf C, Rainer H, Dreyer I, Konrad KR (2024) K+ and pH homeostasis in plant cells is controlled by a synchronized K+/H+ antiport at the plasma and vacuolar membrane. New Phytol 241:1525–1542. https://doi.org/10.1111/nph.19436

Liang Y, Nikolic M, Bélanger R et al (2015) Effect of silicon on crop growth, yield and quality. In: Silicon in agriculture. Springer, Dordrecht, pp 209–223

Liu C-W, Sung Y, Chen B-C, Lai H-Y (2014) Effects of nitrogen fertilizers on the growth and nitrate content of lettuce (Lactuca sativa L.). Int J Environ Res Public Health 1:4427–4440. https://doi.org/10.3390/ijerph110404427

Long M, Guo L, Li J et al (2018) Effects of water and exogenous Si on element concentrations and ecological stoichiometry of plantain (Plantago lanceolata L.). J Plant Nutr 41:10. https://doi.org/10.1080/01904167.2018.1443128

Luans S, Wang C (2021) Calcium signaling mechanisms across kingdoms. Annu Rev Cell Dev Biol 37:311–340. https://doi.org/10.1146/annurev-cellbio-120219-035210

Luo Y, Peng Q, He M, Zhang M, Liu Y, Gong Y et al (2020) N, P and K stoichiometry and resorption efficiency of nine dominant shrub species in the deserts of Xinjiang, China. Ecol Res. https://doi.org/10.1111/1440-1703.12111

Ma J, Takahashi E (1990) Effect of silicon on the growth and phosphorus uptake of rice. Plant Soil 126. https://doi.org/10.1007/BF00041376

Ma JF, Takahashi E (2002) Soil, fertilizer, and plant silicon research in Japan. Elsevier

Machado J, Fernandes APG, Bokor B et al (2024) The effect of silicon on the antioxidant system of tomato seedlings exposed to individual and combined nitrogen and water deficit. Ann Appl Biol 184:1. https://doi.org/10.1111/aab.12849

Marschner P (ed) (2012) Marschner's mineral nutrition of higher plants, 3rd edn. Elsevier/Academic, London

Maruyama-Nakashita A (2008) Transcriptional regulation of genes involved in sulfur assimilation in plants: understanding from the analysis of high-affinity sulfate transporters. Plant Biotechnol 25:323–328. https://doi.org/10.5511/plantbiotechnology.25.323

Mehrabanjoubani P, Abdolzadeh A, Sadeghipour HR et al (2015) Silicon affects transcellular and apoplastic uptake of some nutrients in plants. Pedosphere 25:2. https://doi.org/10.1016/S1002-0160(15)60004-2

Melo CCF, Amaral DS, Prado RM (2023a) Review on the impact of nitrogen and phosphorus nutritional disorders in Poaceae and future perspectives. Científica 51. https://doi.org/10.5016/1984-5529.2023.v51.1436

Melo CCF, Amaral DS, Prado RM et al (2023b) Nanosilica modulates C:N:P stoichiometry attenuating phosphorus toxicity more than deficiency in Megathyrsus maximus cultivated in an oxisol and entisol. Sci Rep 13:10284. https://doi.org/10.1038/s41598-023-37504-3

Meng X, Chen W-W, Wang Y-Y, Huang Z-R, Ye X, Chen L-S et al (2021) Effects of phosphorus deficiency on the absorption of mineral nutrients, photosynthetic system performance and antioxidant metabolism in Citrus grandis. PLoS One 16(2):e0246944. https://doi.org/10.1371/journal.pone.0246944

Mukarram M, Khan MMA, Corpas FJ (2021) Silicon nanoparticles elicit an increase in lemongrass (Cymbopogon flexuosus (Steud.) Wats) agronomic parameters with a higher essential oil yield. J Hazard Mater 412:125254. https://doi.org/10.1016/J.JHAZMAT.2021.125254

Narayan OP, Kumar P, Yadav B, Dua B, Johri AK (2023) Sulfur nutrition and its role in plant growth and development. Plant Signal Behav 18:1. https://doi.org/10.1080/15592324.2022.2030082

Neu S, Schaller J, Dudel E (2017) Silicon availability modifies nutrient use efficiency and content, C:N:P stoichiometry, and productivity of winter wheat (Triticum aestivum L.). Sci Rep 7:40829. https://doi.org/10.1038/srep40829

Olivera-Viciedo D, de Mello PR, Lizcano Toledo R et al (2017) Response of radish seedlings (Raphanus sativus L.) to different concentrations of ammoniacal nitrogen in absence and presence of silicon. Agron Colomb 35:198–204. https://doi.org/10.15446/agron.colomb.v35n2.62772

Oliveira Filho JS, Ferrari AC, Pereira MG et al (2020) Phosphorus accumulation in soil after successive applications of swine manure: a long-term study in Brazil. Environ Earth Sci 79:62. https://doi.org/10.1007/s12665-019-8805-z

Oliveira KS, de Mello Prado R, Checchio MV, Gratão PL (2022) Interaction of silicon and manganese in nutritional and physiological aspects of energy cane with high fiber content. BMC Plant Biol 22:1–13. https://doi.org/10.1186/S12870-022-03766-8/FIGURES/5

Olivera-Viciedo D, de Mello Prado R, Lizcano Toledo R, dos Santos L, Calero Hurtado A, Nedd L, Castellanos L (2019a) Silicon supplementation alleviates ammonium toxicity in sugar beet (Beta vulgaris L.). J Soil Sci Plant Nutr 19:413–419. https://doi.org/10.1007/s42729-019-00043-w

Olivera-Viciedo D, de Mello Prado R, Martínez CA, Habermann E, de Cássia Piccolo M (2019b) Short-term warming and water stress affect Panicum maximum Jacq. stoichiometric homeostasis and biomass production. Sci Total Environ 681:267–274. https://doi.org/10.1016/j.scitotenv.2019.05.108

Olivera-Viciedo D, Mello Prado R, Lizcano Toledo R, Salas Aguilar D, Nascimento C, dos Santos L, Calero Hurtado A, Peña Calzada K, Betancourt Aguilar C (2020) Physiological role of silicon in radish seedlings under ammonium toxicity. J Sci Food Agric jsfa.10587. https://doi.org/10.1002/jsfa.10587

Olivera-Viciedo D, de Mello Prado R, Martinez CA, Habermann E, de Cássia Piccolo M, Calero Hurtado A, Barreto RF, Peña Calzada K (2021a) Changes in soil water availability and air-temperature impact biomass allocation and C:N:P stoichiometry in different organs of Stylosanthes capitata Vogel. J Environ Manag 278. https://doi.org/10.1016/j.jenvman.2020.111540

Olivera-Viciedo D, Mello Prado R, Martinez CA, Habermann E, Branco RBF, Cássia Piccolo M, Calero Hurtado A, Peña Calzada K, Lata Tenesaca LF (2021b) Water stress and warming impact nutrient use efficiency of Mombasa grass (Megathyrsus maximus) in tropical conditions. J Agron Crop Sci 207:128–138. https://doi.org/10.1111/jac.12452

Olivera-Viciedo D, Oliveira KS, de Mello Prado R, Habermann E, Martínez CA, de Moura Zanine A (2024) Silicon uptake and utilization on Panicum maximum grass modifies C:N:P stoichiometry under warming and soil water deficit. Soil Tillage Res 235:105884. https://doi.org/10.1016/J.STILL.2023.105884

Pandey N (2018) Role of plant nutrients in plant growth and physiology. In: Plant nutrients and abiotic stress tolerance, pp 51–93. https://doi.org/10.1007/978-981-10-9044-8_2/COVER

Parecido RJ, Soratto RP, Guidorizzi FV et al (2022) Soil application of silicon enhances initial growth and nitrogen use efficiency of Arabica coffee plants. J Plant Nutr 45:7. https://doi.org/10.1080/01904167.2021.2006707

Pati S, Pal B, Badole S et al (2016) Effect of silicon fertilization on growth, yield, and nutrient uptake of rice. Commun Soil Sci Plant Anal 47:3. https://doi.org/10.1080/00103624.2015.1122797

Paul S, Datta S, Datta K (2015) miRNA regulation of nutrient homeostasis in plants. Front Plant Sci 6:2015. https://doi.org/10.3389/fpls.2015.00232

Pavinato PS, Cherubin MR, Soltangheisi A et al (2020) Revealing soil legacy phosphorus to promote sustainable agriculture in Brazil. Sci Rep 10:15615. https://doi.org/10.1038/s41598-020-72302-1

Pavlovic J, Kostic L, Bosnic P et al (2021) Interactions of silicon with essential and beneficial elements in plants. Front Plant Sci 12:697592. https://doi.org/10.3389/fpls.2021.697592

Peña-Calzada K, Calero Hurtado A, Olivera Viciedo D, Habermann E, de Mello Prado R, de Oliveira R, Ajila G, Tenesaca LFL, Rodríguez JC, Gratão PL (2023a) Regulatory role of silicon on growth, potassium uptake, ionic homeostasis, proline accumulation, and antioxidant capacity of soybean plants under Salt stress. J Plant Growth Regul 1–13. https://doi.org/10.1007/S00344-023-10921-4/FIGURES/6

Peña-Calzada K, Olivera-Viciedo D, Calero-Hurtado A, De R, Prado M, Habermann E, Felipe L, Tenesaca L, Ajila G, De Oliveira R, Carlos Rodríguez J, Gratão PL (2023b) Silicon mitigates the negative impacts of salt stress in soybean plants. J Sci Food Agric. https://doi.org/10.1002/JSFA.12503

Penuelas J, Coello F, Sardans J (2023) A better use of fertilizers is needed for global food security and environmental sustainability. Agric Food Secur 12:5. https://doi.org/10.1186/s40066-023-00409-5

Prado RM (2021) Mineral nutrition of tropical plants. Springer, Cham, 339p

Prado RM, Silva GP (2017) Ecological response to global change: changes in C: N: P stoichiometry in environmental adaptations of plants. In: Yousaf Z (ed) Plant ecology-traditional approaches to recent trends. IntechOpen, London, pp 147–163

Prathap V, Kumar A, Maheshwari C et al (2022) Phosphorus homeostasis: acquisition, sensing, and long-distance signaling in plants. Mol Biol Rep 49:8071–8086. https://doi.org/10.1007/s11033-022-07354-9

Raven JA (1983) The transport and function of silicon in plants. Biol Rev 58:2. https://doi.org/10.1111/j.1469-185X.1983.tb00385.x

Ren CJ, Zhao FZ, Kang D, Yang GH, Han XH, Ren GX et al (2016) Linkages of C:N:P stoichiometry and bacterial community in soil following afforestation of former farmland. For Ecol Manag 376:59–66. https://doi.org/10.1016/j.foreco.2016.06.004

Rocha JR, Prado RM, Teixeira GCM et al (2021) Si fertigation attenuates water stress in forages by modifying carbon stoichiometry, favoring physiological aspects. Agron Crop Sci 207:4. https://doi.org/10.1111/jac.12479

Rouached H, Secco D, Arpat BA (2010) Regulation of ion homeostasis in plants: current approaches and future challenges. Plant Signal Behav 5:501–502. https://doi.org/10.4161/psb.11027

Sadras VO (2006) The N:P stoichiometry of cereal, grain legume and oilseed crops. F Crop Res 95:13–29. https://doi.org/10.1016/j.fcr.2005.01.020

Santos Sarah MM, de Mello Prado R, de Souza Júnior JP, Teixeira GCM, dos Santos Duarte JC, de Medeiros RLS (2021) Silicon supplied via foliar application and root to attenuate potassium deficiency in common bean plants. Sci Rep 11:1–13. https://doi.org/10.1038/S41598-021-99194-Z/TABLES/1

Schwaller B (2009) The continuing disappearance of "pure" Ca2+ buffers. Cell Mol Life Sci 66:275–300. https://doi.org/10.1007/s00018-008-8564-6

Shaul O (2002) Magnesium transport and function in plants: the tip of the iceberg. Biometals 15:309–323. https://doi.org/10.1023/A:1016091118585

Shen X, Li J, Duan L et al (2009) Nutrient acquisition by soybean treated with and without silicon under ultraviolet-B radiation. J Plant Nutr 32:10

Sheng H, Chen S (2020) Plant silicon-cell wall complexes: identification, model of covalent bond formation and biofunction. Plant Physiol Biochem 155. https://doi.org/10.1016/j.plaphy.2020.07.020

Shi L, Li Q, Fu X, Kou L, Dai X, Wang H (2021) Foliar, root and rhizospheric soil C:N:P stoichiometries of overstory and understory species in subtropical plantations. Catena 198:105020. https://doi.org/10.1016/j.catena.2020.105020

Siddiqui H, Ahmed KBM, Sami F, Hayat S (2020) Silicon nanoparticles and plants: current knowledge and future perspectives. In: Sustainable agriculture reviews 41. Springer, Cham, pp 129–142. https://doi.org/10.1007/978-3-030-33996-8_7

Silva DL, Prado RM, Lata-Tenesaca LF et al (2021a) Silicon attenuates calcium deficiency in rocket plants by increasing the production of non-enzymatic antioxidants compounds. Sci Hortic 285:110169. https://doi.org/10.1016/j.scienta.2021.110169

Silva ES, Prado RM, Soares AAVL et al (2021b) Response of corn seedlings (Zea mays L.) to different concentrations of nitrogen in absence and presence of silicon. SILICON 13. https://doi.org/10.1007/s12633-020-00480-8

Silva Júnior GB, Prado RM, Campos CNS et al (2019) Silicon mitigates ammonium toxicity in yellow passionfruit seedlings. Chil J Agric Res 79:425–434. https://doi.org/10.4067/S0718-58392019000300425

Sousa Junior G, Hurtado AC, de Souza Junior JP, de Mello Prado R, de Cássia Piccolo M, Dos Santos DMM (2022) Beneficial role of silicon on regulating C, N, and P stoichiometric homeostasis and the growth of sugarcane seedlings under aluminum toxicity. J Soil Sci Plant Nutr 22:4138–4152. https://doi.org/10.1007/S42729-022-01013-5/FIGURES/6

Souza Júnior JP, Oliveira TL, Prado RM et al (2022) Analyzing the role of silicon in leaf C:N:P stoichiometry and its effects on nutritional efficiency and dry weight production in two sugarcane cultivars. J Soil Sci Plant Nutr 22. https://doi.org/10.1007/s42729-022-00836-6

Srivastava AK, Shankar A, Nalini Chandran AK, Sharma M, Jung K-H, Suprasanna P, Pandey GK (2019) Emerging concepts of potassium homeostasis in plants. J Exp Bot. https://doi.org/10.1093/jxb/erz458

Sung J, Baek S, Kim J, Kim Y, Lee Y, Lee S, Lee D, Jung H-I (2018) Responses of primary metabolites and glucosinolates in sulfur deficient-cabbage (Brassica rapa L. ssp. Pekinensis). J Plant Biochem Physiol 6:223. https://doi.org/10.4172/2329-9029.1000223

Tang R-J, Luan S (2017) Regulation of calcium and magnesium homeostasis in plants: from transporters to signaling network. Curr Opin Plant Biol 39:97–105. https://doi.org/10.1016/j.pbi.2017.06.009

Tang R-J, Zhao F-G, Garcia VJ, Kleist TJ, Yang L, Zhang H-X, Luan S (2015) Tonoplast CBL–CIPK calcium signaling network regulates magnesium homeostasis in Arabidopsis. Proc Natl Acad Sci 112(10):3134–3139. https://doi.org/10.1073/pnas.1420944112

Tantray AY, Bashir SS, Ahmad A (2020) Low nitrogen stress regulates chlorophyll fluorescence in coordination with photosynthesis and Rubisco efficiency of rice. Physiol Mol Biol Plants 26:83–94. https://doi.org/10.1007/s12298-019-00721-0

Tariq A, Zeng F, Graciano C, Ullah A, Sadia S, Ahmed Z, Zhang Z (2023) Regulation of metabolites by nutrients in plants. In: Singh P, Siddiqui MH (eds) Plant ionomics: sensing, signaling, and regulation. Wiley, pp 1–18. https://doi.org/10.1002/9781119803041.ch1

Taskin MB, Akca H, Babar SK et al (2023) Evaluating the comparative effects of acid modified rice husk and nano-silicon derived from rice husk on phosphorus use efficiency in wheat and lettuce plants with differing silicon contents. J Plant Nutr 46:10. https://doi.org/10.1080/01904167.2022.2155535

Teixeira GCM, Prado RM, Rocha MAS et al (2020) Root- and foliar-applied silicon modifies C:N:P ratio and increases the nutritional efficiency of pre-sprouted sugarcane seedlings under water deficit. PLoS One 15:10. https://doi.org/10.1371/journal.pone.0240847

Tian D, Reich PB, Chen HYH, Xiang Y, Luo Y, Shen Y et al (2018) Global changes alter plant multi-element stoichiometric coupling. New Phytol. https://doi.org/10.1111/nph.15428

Tränkner M, Tavakol E, Jákli B (2018) Functioning of potassium and magnesium in photosynthesis, photosynthate translocation and photoprotection. Physiol Plant 163:414–431. https://doi.org/10.1111/ppl.12747

Verbruggen N, Hermans C (2013) Physiological and molecular responses to magnesium nutritional imbalance in plants. Plant Soil 368:87–99. https://doi.org/10.1007/s11104-013-1589-0

Wang Y, Wu W-H (2010) Plant sensing and signaling in response to K+-deficiency. Mol Plant 3(2):280–287. https://doi.org/10.1093/mp/ssq006

Wu X, Yu Y, Baerson SR et al (2017) Interactions between nitrogen and silicon in rice and their effects on resistance toward the brown planthopper Nilaparvata lugens. Front Plant Sci 8:28. https://doi.org/10.3389/fpls.2017.00028

Xiao H, Li P, Monaco TA et al (2024) Nitrogen and phosphorus additions alter foliar nutrient concentrations of dominant grass species and regulate primary productivity in an Inner Mongolian meadow steppe. Sci Total Environ 912:168791. https://doi.org/10.1016/j.scitotenv.2023.16879

Xu XX, Du X, Wang F, Sha JC, Chen Q, Tian G, Zhu Z, Ge S, Jiang Y (2020) Effects of potassium levels on plant growth, accumulation and distribution of carbon, and nitrate metabolism in apple dwarf rootstock seedlings. Front Plant Sci 11:2020. https://doi.org/10.3389/fpls.2020.00904

Xu X, Zou G, Li Y et al (2023) Silicon application improves strawberry plant antioxidation ability and fruit nutrition under both full and deficit irrigation. Sci Hortic 309:111684. https://doi.org/10.1016/j.scienta.2022.111684

Xu H, Wang M, You C et al (2024) Warming effects on C:N:P stoichiometry and nutrient limitation in terrestrial ecosystems. Soil Tillage Res 235:105896. https://doi.org/10.1016/j.still.2023.105896

Yu B, Xu C, Benning C (2002) Arabidopsis disrupted in SQD2 encoding sulfolipid synthase is impaired in phosphate-limited growth. Proc Natl Acad Sci USA 99:5732–5737. https://doi.org/10.1073/pnas.082696499

Yu Z, She M, Zheng T, Diepeveen D, Islam S, Zhao Y, Zhang Y, Tang G, Zhang Y, Zhang J et al (2021) Impact and mechanism of sulphur-deficiency on modern wheat farming nitrogen-related sustainability and gliadin content. Commun Biol 4:1–6. https://doi.org/10.1038/s42003-021-02458-7

Yu Z, Zhang C, Liu X, Lei J, Zhang Q, Yuan Z, Peng C, Koerner SE, Xu J, Guo L (2024) Responses of C:N:P stoichiometric correlations among plants, soils and microorganisms to warming: a meta-analysis. Sci Total Environ 912:168827. https://doi.org/10.1016/j.scitotenv.2023.168827

Zhang J, Yan X, Su F, Li Z, Wang Y, Wei Y, Ji Y, Yang Y, Zhou X, Guo H, Hu S (2018) Long-term N and P additions alter the scaling of plant nitrogen to phosphorus in a Tibetan alpine meadow. Sci Total Environ 625:440–448. https://doi.org/10.1016/J.SCITOTENV.2017.12.292

Zhang J, He N, Liu C, Xu L, Chen Z, Li Y, Wang R, Yu G, Sun W, Xiao C, Chen HYH, Reich PB (2020) Variation and evolution of C:N ratio among different organs enable plants to adapt to N-limited environments. Glob Chang Biol 26:2534–2543. https://doi.org/10.1111/gcb.14973

Zhao N, Yu G, Wang Q, Wang R, Zhang J, Liu C, He N (2020) Conservative allocation strategy of multiple nutrients among major plant organs: from species to community. J Ecol 108:267–278. https://doi.org/10.1111/1365-2745.13256

Zhou TC, Sun J, Liu M, Shi PL, Zhang XB, Sun W, Yang G, Tsunekawa A (2020) Coupling between plant nitrogen and phosphorus along water and heat gradients in alpine grassland. Sci Total Environ 701:134660. https://doi.org/10.1016/j.scitotenv.2019.134660

Chapter 15
Silicon Stimulates Flowering and Improves Crop Quality

Jonas Pereira de Souza Junior, Milton Garcia Costa, Cid Naudi Silva Campos, and Rilner Alves Flores

Abstract Silicon (Si) has garnered increasing interest in agriculture due to its beneficial impacts on crop development. This study reviews the influence of Si on flowering and fruit quality in various plants, including cotton, tomato, sunflower, and sorghum. Si plays a critical role in floral development, influencing the formation and morphology of pollen grains and contributing to fertility and crop yield. Furthermore, the application of Si during the reproductive period has been shown to significantly enhance both yield and fruit quality. The research highlights Si's positive impact on flower quality, resistance against biotic and abiotic stresses, and stimulation of antioxidant and metabolic mechanisms. Another observed benefit is mineral biofortification, with Si application resulting in a substantial increase in essential elements in fruits, such as iron, zinc, and selenium, thereby helping to meet the nutritional needs of the population. These findings underscore the importance of Si in optimizing agricultural production and promoting food security, providing valuable insights for sustainable agricultural practices and biofortification strategies. This study emphasizes the potential of Si as a crucial tool for enhancing crop productivity and quality in the face of global environmental and nutritional challenges.

Keywords Beneficial element · Yield · Reproductive stage

J. P. de Souza Junior (✉)
Institute of Food and Agriculture Sciences, Citrus Research and Education Center, University of Florida, Lake Alfred, FL, USA
e-mail: jpereiradesouza@ufl.edu

M. G. Costa
School of Veterinarian and Agricultural Science, Sao Paulo State University, Jaboticabal, Brazil

C. N. S. Campos
Federal University of Mato Grosso do Sul, Chapadao do Sul, Brazil

R. A. Flores
Federal University of Goias, Goiania, Brazil

R. de Mello Prado et al. (eds.), *Silicon Advances for Sustainable Agriculture and Human Health*, Sustainable Plant Nutrition in a Changing World,
https://doi.org/10.1007/978-3-031-69876-7_15

15.1 Introduction

The relationship between silicon (Si) and the development of agricultural crops has been an area of increasing interest and research over the past few decades. From studies highlighting the presence of Si in floral structures (Schoelynck et al. 2023) to investigations into the effects of Si on fruit quality and yield (Santos et al. 2022), a deeper understanding of this element and its role in plants has proven crucial. The application of Si, whether through foliar spray, soil amendment, or fertigation, has shown significant impacts on various aspects of plant development, from vegetative growth to the reproductive phase, across various crops such as *Gossypium hirsutum* (Lopes et al. 2023; Souza Júnior et al. 2022, 2023), *Brassica oleracea* (Wenneck et al. 2021), *Solanum lycopersicum* L. (Flores et al. 2023), *Helianthus annuus* (Peixoto et al. 2020), and *Sorghum bicolor* (Peixoto et al. 2020).

Studies highlight the importance of Si in the floral development of crops, revealing its fundamental presence in reproductive structures and its ability to influence the metabolism of phytohormones that play crucial roles in flowering (Mir et al. 2022). Additionally, they demonstrate Si's effects on the formation and morphology of pollen grains, contributing to crop fertility and productivity (Crang and May 1974). However, despite significant advances, gaps remain in understanding the underlying mechanisms of Si effects on plants. Deepening knowledge about how Si influences the biochemical and physiological processes of plants may open new opportunities to optimize its use in agriculture.

Therefore, this text aims to provide a comprehensive overview of recent studies on the relationship between Si and the development of agricultural crops, from its influence on flowering to its effects on fruit quality. By reviewing and synthesizing the latest findings, the goal of this work is to provide valuable insights for researchers, farmers, and professionals in the agricultural sector, as well as to stimulate further research and sustainable agricultural practices.

15.2 The Role of Silicon in the Floral Development of Crops

A classic study on the reproductive structures of *Lychnis alba* L. highlighted that Si is an elementary component present in the structure of pollen grain sporoderm (Crang and May 1974). More recently, research involving 56 plant species grouped into basal angiosperms, magnoliids, chloranthales, monocots, and eudicots revealed that flowers are silicified (Schoelynck et al. 2023). The authors detected Si in all floral structures without significant differences among them, except for the sepals, which showed a significantly higher Si content compared to other floral structures.

The concentration of Si in reproductive structures varies considerably among different plant species, with some species showing high concentrations, like *Poales cympogogon* N. (51.65 mg g^{-1}), while others exhibit Si exclusion behavior, like *Fuchsia* sp. (<0.04 mg g^{-1}). This variation reflects the Si concentration in the leaves, suggesting that the processes of Si absorption and precipitation in the leaves are also responsible for its presence in flowers, aligning with the notion that flowers are indeed modified leaves (Schoelynck et al. 2023).

Silicon accumulation in leaves is regulated by both passive and active processes. Passive mechanisms, mainly regulated by transpiration, play a fundamental role in the absorption and distribution of Si (McLarnon et al. 2017). Variations in stomatal density and conductance contribute to the differences in Si concentration among species (Manivannan and Ahn 2017). Flowers have a cuticle to prevent water loss (Riederer and Müller 2006), and stomata are present on the epidermis of petals and sepals, though generally in fewer numbers than on leaves, with some even being non-functional (Zhang et al. 2018). Additionally, the longevity of flowers varies among species and depends on several factors, but generally, it is much shorter than that of leaves, meaning less time for Si accumulation (Schoelynck et al. 2023). Considering these factors together, the lower transpiration and shorter lifespan of flowers compared to leaves may explain why the Si concentration in flowers is generally about half that observed in the leaves of the same plant (Schoelynck et al. 2023).

The effect of Si on the activation of reproductive development is not yet fully clarified in the literature. However, it is known that Si regulates the metabolism of phytohormones, including auxin, cytokinins, ethylene, gibberellin, salicylic acid, abscisic acid, brassinosteroids, and jasmonic acids, which play crucial roles in the floral development phase (Mir et al. 2022). Moreover, studies suggest that Si can increase the nutritional efficiency of flowers, resulting in an enhancement of floral development, with plants of *Zinnia elegans* L. and *Gerbera jamesonii* L. showing increased diameter and height (Kamenidou et al. 2010).

The process of pollen grain formation may also be favored by the presence of Si, as this element is recognized as one of the elementary components in the structure of pollen grain sporoderm (Crang and May 1974). During the metabolism of Si, complexes can be formed with pectins, calloses, tannins, and starches (Boylston et al. 1990), which are essential compounds for the formation of pollen grain sporoderm (Golaszweska and Bednarska 1999). Additionally, mixed bindings of Si with (1;3,1;4)-β-D-glucan can increase the physical resistance of the pollen grain sporoderm (Kido et al. 2015).

The influence of Si on the morphology of pollen grains was studied in *Oryza sativa* L. plants grown under conditions of water scarcity (Das et al. 2022). The findings indicate that Si application not only increases the diameter of pollen grains but also helps in maintaining their structural integrity during drought conditions.

Pollen grain deformation is closely associated with infertility, primarily due to the grains' reduced ability to expand (Kumar et al. 2019; Singh et al. 2013). This results in decreased dehiscence and smaller grain size, which can lead to reduced fertility and, consequently, lower crop yields (Kumar et al. 2019; Singh et al. 2013). Proper development of pollen grains and successful fertilization are critical processes influenced by Si fertilization, which has been shown to boost crop yield (Das et al. 2022).

15.3 The Relationship Between Silicon Application during Flowering and Crop Productivity

Silicon can be supplied to plants through soil application, fertigation, or foliar spraying. Regardless of the chosen method, it is crucial to select the appropriate sources and doses for application. The benefits of Si in plant biochemistry and physiology become more pronounced with proper application. Recent research has highlighted the significant impact of the timing of application on the effects of Si on crop productivity. Consequently, foliar application of Si to plants during the reproductive period has become the focus of numerous studies, yielding promising results in enhancing reproductive growth across various crops.

Currently, numerous studies demonstrate that the application of Si during the reproductive period enhances crop yield. Examples include *Gossypium hirsutum* (Lopes et al. 2023; Souza Júnior et al. 2022, 2023), *Brassica oleracea* (Wenneck et al. 2021), *Solanum lycopersicum* L. (Flores et al. 2023), *Helianthus annuus* (Peixoto et al. 2020), and *Sorghum bicolor* (Peixoto et al. 2020).

In this context, foliar application of Si during the reproductive period can contribute to increased yield and even improve the quality of agricultural products. In a study conducted on cotton crops, Lopes et al. (2023) experimented with doses of boron in the soil and four foliar applications of Si (920 g ha^{-1} of Si), using potassium silicate stabilized with sorbitol (SiKE—115 g L^{-1} Si, 113.85 g L^{-1} K_2O, 100 mL L^{-1} sorbitol, pH 12.0) as the Si source. The applications began at the reproductive stage B1 (a flower bud on the first reproductive branch), with intervals of 7–11 days, varying according to weather conditions. The authors observed a synergistic effect of foliar Si use with borate fertilization in the soil. The results of this study demonstrated that foliar supplementation with Si increases the yield of cotton fibers (Fig. 15.1).

In another study conducted with two experiments on cotton crops, the combination of Si and boron for foliar spraying in both experiments resulted in significant increases in cotton production and fiber quality, enhancing tensile strength and reducing the content of short fiber (Souza Júnior et al. 2022). In this study, doses of boron and six foliar applications of Si (1.15 g L^{-1} of Si) were used, with SiKE as the source of Si. Applications began at the reproductive stage B1 with intervals of 7–11 days depending on weather conditions.

In another experiment conducted on cotton flowers, with doses of B, application of water, and Si (1.0 g L^{-1} of Si), with ten replicates and using SiKE as the Si source, starting when the plants were at reproductive stage B1, it was reported that the application of Si contributed to an increase in dry matter production in both petals and anthers (Souza Junior et al. 2023). However, the combination of Si and B demonstrated greater efficacy in enhancing these variables (Fig. 15.1).

In a study carried out on tomato crops (*Solanum lycopersicum* L.), Flores et al. (2023) investigated the effects of foliar application of Si. Five concentrations of Si (0, control, 1, 2, 3, and 4 g L^{-1} of Si) were evaluated, using SiKCuE (stabilized potassium copper silicate) as the Si source. Applications were made at the

Fig. 15.1 Summary of the role of foliar-applied silicon during flowering in different plant species

beginning of flowering, mid-flowering, and full flowering. The results indicated that the use of Si in tomatoes demonstrated economic viability at all evaluated concentrations, in addition to favoring photosynthetic parameters, production, and fruit quality (Fig. 15.1).

In a study conducted on *Sorghum bicolor* L., Flores et al. (2018) investigated the effects of applying five doses of foliar Si: T1 (0, control); T2 (0.84 g L^{-1}); T3 (1.68 g L^{-1}); T4 (2.52 g L^{-1}); and T5 (3.36 g L^{-1} of Si), applied as potassium and sodium silicate (Si, 1.4 g L^{-1}; 42 g K_2O L^{-1}; 31 g Na_2O L^{-1}; d, 1.15 g L^{-1}). Three applications were made, with 1/3 of the Si rate applied at 15 days after emergence (DAE), 1/3 at 30 DAE, and 1/3 at the end of the vegetative stage in pre-flowering (45 DAE). The results showed that foliar application of Si resulted in significant increases in sorghum biomass production.

15.4 The Relationship between Silicon and Flowering for Enhancing the Quality of Flowers and Fruits

The relationship between silicon (Si) and the flowering process has become a key area of study in the pursuit of innovative solutions for producing biofortified foods. Silicon emerges as a crucial agent in promoting flowering across various agricultural crops, significantly influencing the growth, quality, and longevity of flowers (Kumari et al. 2023). This positive effect on flowering is directly related to Si's role in mitigating plant stress caused by both abiotic and biotic factors (Debona et al. 2017).

15.4.1 Silicon in Enhancing Flower Quality and Stress Mitigation Mechanisms

Silicon regulates water balance and transpiration, protects plant cells from biotic and abiotic stresses, and activates protection mechanisms at the levels of organs, tissues, and cells (Nedukha 2022). It also enhances plants' capacity to detoxify reactive oxygen species, modulates antioxidant defense systems, and improves plant tolerance to abiotic constraints (Souri et al. 2020). The role of Si in flowers is linked to its ability to trigger growth and development, regulate physiological and metabolic characteristics, and mitigate conditions of biotic and abiotic stress (Mir et al. 2022; Mostofa et al. 2021).

Silicon plays a crucial role in mitigating the adverse effects of water deficit during the flowering of corn, resulting in tangible benefits such as increased carbohydrate content and oil yield (Atta et al. 2022). This underscores not only the importance of Si in flowering but also its potential in overcoming environmental challenges.

Moreover, in *Gerbera* hybrid L. "Acapella," it was found that the addition of Si led to the formation of more robust floral peduncles, a significant increase in flower diameter, greater plant height, and earlier flowering compared to control plants that were not supplemented (Kamenidou et al. 2010). These findings highlight the benefits of Si supplementation, demonstrating notable improvements in the structural and temporal characteristics of floral development in this specific variety of *Gerbera*. In conditions of water deficiency for *Gerbera jamesonii* Adlam plants, the addition of Si not only improved visual characteristics but also had metabolic impacts, increasing the levels of proteins, free amino acids, and proline, as well as enhancing the activity of antioxidant enzymes such as catalase, peroxidase, superoxide dismutase, and ascorbate peroxidase (Ahsan et al. 2023).

In experiments conducted with *Polianthes tuberosa* L. plants under saline stress conditions, it was observed that the application of Si had a significant positive impact on biometric growth characteristics. Increases in plant height, leaf number, flower number, flower length, fresh flower weight, and vase life were notable after Si application (Shahzad et al. 2022). This improvement can be attributed to the enhanced activity of antioxidant enzymes, including superoxide dismutase, peroxidase, catalase, glutathione reductase, and ascorbate peroxidase. These enzymes play a crucial role in protecting plants against saline stress, contributing to a more robust and adaptive response to adverse environments.

Lilium sp. flowers are notable for their vibrant colors and distinct shapes; however, their quality and durability often fall short. A recent study revealed that the application of nanosilica led to a significant increase in flower longevity, with a notable increment of 21.6% (Sánchez-Navarro et al. 2021). This benefit was accompanied by marked improvements in photosynthetic pigments, with an increase of 65%, and a remarkable rise in glutathione peroxidase activity, reaching up to 69.9%. Additionally, phenol levels in the flowers also saw a substantial increase, up to 25.93%. These promising results suggest that the use of nanosilica could be an

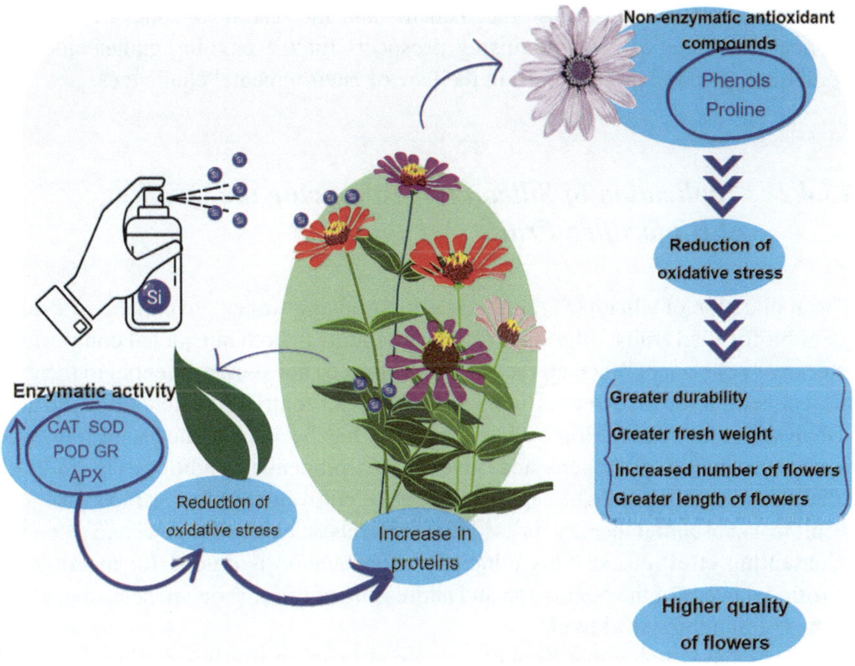

Fig. 15.2 Summary of the role of silicon in reducing oxidative stress and enhancing glower quality

effective strategy to enhance both the longevity and biochemical characteristics of *Lilium* sp. flowers (Fig. 15.2).

In cotton plants subjected to boron deficiency conditions, the direct application of Si to the flowers showed a significant reduction in the levels of hydrogen peroxide and malondialdehyde (Souza Junior et al. 2023). Simultaneously, there was an increase in phenol and proline contents, indicating that Si acts as an antioxidant mechanism. In the absence of Si, carotenoids emerge as the primary antioxidant defense mechanism in the plant. These results suggest that Si modifies the antioxidant defense mechanism in boron-deficient plants, resulting in better preservation of the flower dry mass and reduction of biological damage caused by boron deficiency in cotton flowers.

In summary, the presence of Si emerges as a promising agent in optimizing flower quality, triggering mechanisms that mitigate both biotic and abiotic stress. Silicon's regulatory properties in water balance, transpiration, and detoxification of reactive oxygen species are crucial for enhancing structural and biochemical characteristics in flowering plants. Whether mitigating water deficits during corn flowering, strengthening floral development in specific varieties such as *Gerbera hybrid* L. "Acapella," or providing resistance against saline stress in *Polianthes tuberosa* L., Si stands out as a crucial ally. Furthermore, future perspectives suggest potential enhancements in visual and metabolic characteristics associated with the addition of Si, as evidenced in recent studies with *Gerbera jamesonii* Adlam and *Lilium* sp.

Continued exploration of these mechanisms and the search for innovative silicon application strategies offer promising prospects for the ongoing enhancement of floral quality and plant resilience in the face of environmental challenges.

15.4.2 Application of Silicon for Enhancing the Quality of Biofortified Fruits

The application of silicon (Si) emerges as a promising strategy to enhance the quality of biofortified fruits, offering significant benefits in both nutritional composition and stress resistance. When applied appropriately, Si has shown potential to increase the concentration of essential elements in fruits, contributing to biofortification (Dhaliwal et al. 2022). Moreover, its presence has been associated with improved plant resistance to pathogens and adverse environmental conditions (Islam et al. 2020). Studies suggest that Si application in the cultivation of biofortified fruits can result in a substantial increase in essential minerals, such as iron, zinc, and selenium (Consentino et al. 2023). This mineral biofortification is crucial for meeting the nutritional needs of the population and addressing nutritional deficiencies that affect many communities worldwide.

Studies have shown that Si application can increase fruit yield, cluster weight, berry count, berry crispness, total soluble solids content, titratable acidity, macro- and micronutrients, and photosynthetic efficiency (Carbajal-Vázquez et al. 2023; Nascimento et al. 2022). The use of Si enhances the content of bioactive compounds, such as total soluble solids, titratable acidity, vitamin C, phenols, flavonoids, and antioxidant activity in fruits (Grankina et al. 2022).

In quinoa crops, grain productivity was significantly increased by the application of Si both foliarly and through nutrient solution (Lata-Tenesaca et al. 2022). This practice not only resulted in an increase in the quantity of grains produced but also contributed to a notable increase in the vitamin C content in the grains.

The combined application of Si and zinc has shown promising results in enhancing productivity and the weight of 100 grains in corn plants subjected to water deficit, as well as a significant increase in the protein content of the grains (Kandil et al. 2023). Additionally, the application of Si was associated with an increase in proline content. The study's results suggest that combining a 15-day irrigation interval with the application of Si and zinc has notable potential to enhance both the productivity and quality of corn under water-deficit stress conditions.

Foliar supplementation with nanosilica has been shown to significantly increase both the above-ground and root biomass of chervil (*Anthriscus cerefolium* L. Hoffm.), both individually and in combination with selenium and iodine (Golubkina et al. 2021). Nano-Si also promoted the growth of chervil shoots and roots, highlighting its stimulative role in plant development. Furthermore, the joint application of selenium, iodine, and nano-Si resulted in enhanced levels of biofortification, particularly for iodine, indicating significant potential in the production of functional foods. These results suggest promising prospects for the use of nanosilica in improving growth and simultaneous iodine/selenium biofortification in chervil plants, offering an innovative approach to the production of biofortified foods.

In tomato plants of the Micro-Tom variety, foliar spraying of Si during the reproductive phase led to a significant increase in ascorbic acid content and fruit firmness (Santos et al. 2022). Notably, a concentration of 0.4 g L^{-1} of Si from monosilicic acid resulted in the greatest increase in ascorbic acid, titratable acidity, and fruit firmness. These results highlight the potential of foliar Si application as a valuable approach to enhance both the nutritional and technological quality of tomatoes, providing relevant insights for agricultural practices aimed at Si biofortification.

The application of Si proves to be indispensable for the cultivation of chard and kale, especially in leafy vegetables, resulting in a significant increment in the content and accumulation of this element, as well as in the growth and production of these plants. The benefits extend to a notable improvement in development, yield, and crop quality, increasing the ascorbate content and reducing water loss after harvest (Souza et al. 2019). It is noteworthy that a concentration of 2.52 g L^{-1} of Si, in the form of potassium silicate, emerged as the most effective among those tested, underscoring its potential as a relevant strategy to optimize both the production and quality of chard and kale through foliar fertilization with Si. These results highlight the practical and economic importance of careful Si use in the cultivation of leafy vegetables.

The incorporation of cobalt, selenium, and Si in the biofortification of lettuce cv. Limor resulted in significant improvements in vegetative growth parameters, mineral content, chlorophyll levels, sugars, and free amino acids, without affecting the levels of nitrate and nitrite in lettuce heads (Abdalla et al. 2024). The increase in the fresh weight of the head ranged from 11.35% to 14.54% in the first harvest and from 13.07% to 16.95% in the second harvest. Foliar application of cobalt sulfate at 100 μM, sodium selenite at 20 μM, or sodium silicate at 2 mM not only optimized the vegetative growth and yield of lettuce plants but also significantly enhanced the nutritional quality and strengthened the content of these elements in lettuce heads. These results provide valuable insights for agricultural practices aimed at biofortification, contributing to a more comprehensive understanding of the role of these beneficial elements in optimizing vegetable production and quality.

The application of Si stands out as a promising strategy to enhance the quality of biofortified fruits, offering substantial benefits in both nutritional composition and stress resistance. Reviewed studies highlight the potential of Si to increase the concentration of essential elements in fruits, significantly contributing to biofortification and, consequently, meeting the nutritional needs of the population. Furthermore, the presence of Si is associated with improved plant resistance to pathogens and environmental adversities, solidifying its relevance in the cultivation of biofortified fruits.

In summary, the reviewed studies suggest that the application of Si and its combination with other beneficial elements represent promising strategies to enhance the quality of biofortified fruits (Fig. 15.3). Prospects should focus on optimizing the doses, sources, and methods of application of these elements, as well as

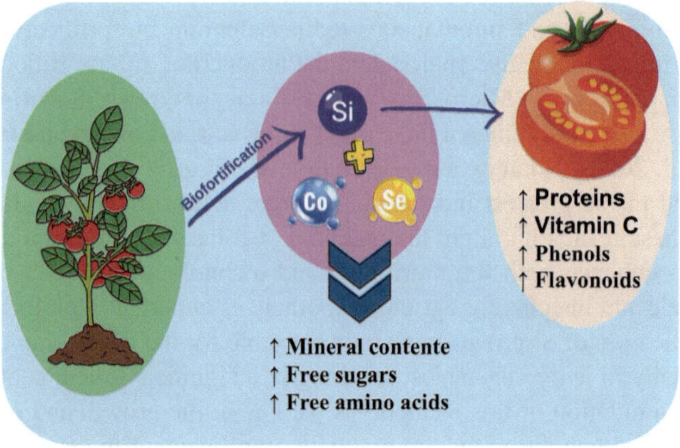

Fig. 15.3 Role of silicon in the biofortification of fruits and its association with cobalt and selenium in enhancing fruit quality

exploring their interaction in different crops and environmental conditions. Moreover, it is crucial to expand research to understand the molecular mechanisms underlying the observed benefits, providing a solid foundation for the practical implementation of these innovative approaches in sustainable agriculture.

15.5 Final Remarks and Prospects

The potential of strategic silicon uses to enhance the reproductive development of plants has significant implications for global food security and environmental sustainability. By strengthening plant reproductive structures and increasing seed and fruit production, silicon can play a crucial role in ensuring more abundant and resilient crops, even under adverse conditions.

Looking forward, it is imperative that research on the use of silicon in agriculture continues to advance to unlock its full potential. This includes more in-depth investigations into the molecular mechanisms underlying silicon's effects on plants, as well as its direct and indirect impacts on plant reproductive development, to support the pursuit of more efficient, safe, and sustainable food systems.

References

Abdalla K, Mousa S, Ibrahim M, Metwally A (2024) Impacts of cobalt, selenium and silicon biofortification on the growth, productivity and nutritional value of lettuce. Egypt J Hortic 51:71–86. https://doi.org/10.21608/ejoh.2023.231176.1261

Ahsan M, Valipour M, Nawaz F, Raheel M, Abbas HT, Sajid M, Manan A, Kanwal S, Mahmoud EA, Casini R, Elansary HO, Radicetti E, Zulfiqar H (2023) Evaluation of silicon supplementation for drought stress under water-deficit conditions: an application of sustainable agriculture. Agronomy 13:599. https://doi.org/10.3390/agronomy13020599

Atta MMM, Abd-El-Salam RM, Abdel-Lattif HM, Garang MA (2022) Effect of silicon on maize under water deficit conditions at flowering stage. J Breed Genet 54:948–962. https://doi.org/10.54910/sabrao2022.54.4.25

Boylston EK, Hebert JJ, Hensarling TP, Bradow JM, Thibodeaux DP (1990) Role of silicon in developing cotton fibers. J Plant Nutr 13:131–148. https://doi.org/10.1080/01904169009364063

Carbajal-Vázquez VH, Gómez-Merino FC, Alcántar-González G, Herrera-Corredor JA, Contreras-Oliva A, Trejo-Téllez LI (2023) Sodium chloride and silicon affect yield and quality of tomato fruits. Agrociencia 57:1–14. https://doi.org/10.47163/agrociencia.v57i3.2967

Consentino BB, Ciriello M, Sabatino L, Vultaggio L, Baldassano S, Vasto S, Rouphael Y, La Bella S, De Pascale S (2023) Current acquaintance on agronomic biofortification to modulate the yield and functional value of vegetable crops: a review. Horticulturae 9:219. https://doi.org/10.3390/horticulturae9020219

Crang RE, May G (1974) Evidence for silicon as a prevalent elemental component in pollen wall structure. Can J Bot 52:2171–2174. https://doi.org/10.1139/b74-280

Das S, Panwar GS, Shankhdhar D, Shankhdhar SC (2022) Silicon-mediated modulation of physiological attributes, and pollen morphology under normal and water-deficit conditions in rice (Oryza sativa L.). Cereal Res Commun 50:929–939. https://doi.org/10.1007/s42976-022-00246-5

Debona D, Rodrigues FA, Datnoff LE (2017) Silicon's role in abiotic and biotic plant stresses. Annu Rev Phytopathol 55:85–107. https://doi.org/10.1146/annurev-phyto-080516-035312

Dhaliwal SS, Sharma V, Shukla AK, Verma V, Kaur M, Shivay YS, Nisar S, Gaber A, Brestic M, Barek V, Skalicky M, Ondrisik P, Hossain A (2022) Biofortification—A frontier novel approach to enrich micronutrients in field crops to encounter the nutritional security. Molecules 27:1340. https://doi.org/10.3390/molecules27041340

Flores RA, Martins Arruda E, Damin V, Pereira Souza Junior J, Diego Carvalho Maranhão D, André Ribeiro Correia M, de Mello Prado R (2018) Physiological quality and dry mass production of Sorghum bicolor following silicon (Si) foliar application. Aust J Crop Sci 12:631–638. https://doi.org/10.21475/ajcs.18.12.04.pne967

Flores RA, de Lima FSR, Xavier MFN, Bueno AM, de Andrade AF, de Souza Júnior JP, Campos CNS, Cunha Júnior LC, de Oliveira Abdala K, de Mello Prado R (2023) Soluble silicon source via foliar application improve plant physiology and fruit quality of Solanum lycopersicum L. SILICON 16:1943. https://doi.org/10.1007/s12633-023-02806-8

Golaszweska B, Bednarska E (1999) Immunocytochemical localization of pectins in the maturing anther of Allium cepa L. Folia Histochem Cytobiol 37:199

Golubkina N, Moldovan A, Fedotov M, Kekina H, Kharchenko V, Folmanis G, Alpatov A, Caruso G (2021) Iodine and selenium biofortification of chervil plants treated with silicon nanoparticles. Plan Theory 10:2528. https://doi.org/10.3390/plants10112528

Grankina A, Bocharnikova E, Matichenkov V (2022) Silicon-based biostimulators for sustainable agriculture, in: biostimulants for crop production and sustainable agriculture. CABI, GB, pp 85–94. https://doi.org/10.1079/9781789248098.0006

Islam W, Tayyab M, Khalil F, Hua Z, Huang Z, Chen HYH (2020) Silicon-mediated plant defense against pathogens and insect pests. Pestic Biochem Physiol 168:104641. https://doi.org/10.1016/j.pestbp.2020.104641

Kamenidou S, Cavins TJ, Marek S (2010) Silicon supplements affect floricultural quality traits and elemental nutrient concentrations of greenhouse produced gerbera. Sci Hortic 123:390–394. https://doi.org/10.1016/j.scienta.2009.09.008

Kandil EE, Lamlom SF, Gheith E-SMS, Javed T, Ghareeb RY, Abdelsalam NR, Hussain S (2023) Biofortification of maize growth, productivity and quality using nano-silver, silicon and zinc particles with different irrigation intervals. J Agric Sci 161:339–355. https://doi.org/10.1017/S0021859623000345

Kido N, Yokoyama R, Yamamoto T, Furukawa J, Iwai H, Satoh S, Nishitani K (2015) The matrix polysaccharide (1;3,1;4)-2- d -glucan is involved in silicon-dependent strengthening of rice cell wall. Plant Cell Physiol 56:268–276. https://doi.org/10.1093/pcp/pcu162

Kumar N, Jeena N, Singh H (2019) Elevated temperature modulates rice pollen structure: a study from foothill of Himalayan agro-ecosystem in India. 3 Biotech 9:175. https://doi.org/10.1007/s13205-019-1700-1

Kumari P, Sharma R, Panwar S, Paul S, Banyal N (2023) Silicon as vital element in flower crop production. J Plant Nutr 46:2747–2762. https://doi.org/10.1080/01904167.2022.2160739

Lata-Tenesaca LF, de Mello Prado R, de Cássia Piccolo M, da Silva DL, da Silva JLF, Ajila-Celi GE (2022) Forms of application of silicon in quinoa and benefits involved in the association between productivity with grain biofortification. Sci Rep 12:12732. https://doi.org/10.1038/s41598-022-17181-4

Lopes PM d SL, Campos CNS, Teodoro PE, Costa MG, de Souza Júnior JP, de Mello Prado R (2023) The synergistic effects of soil-applied boron and foliar-applied silicon on cotton fiber quality and yield. BMC Plant Biol 23:520. https://doi.org/10.1186/s12870-023-04542-y

Manivannan A, Ahn Y-K (2017) Silicon regulates potential genes involved in major physiological processes in plants to combat stress. Front Plant Sci 8. https://doi.org/10.3389/fpls.2017.01346

McLarnon E, McQueen-Mason S, Lenk I, Hartley SE (2017) Evidence for active uptake and deposition of Si-based defenses in tall fescue. Front Plant Sci 8. https://doi.org/10.3389/fpls.2017.01199

Mir RA, Bhat BA, Yousuf H, Islam ST, Raza A, Rizvi MA, Charagh S, Albaqami M, Sofi PA, Zargar SM (2022) Multidimensional role of silicon to activate resilient plant growth and to mitigate abiotic stress. Front Plant Sci 13. https://doi.org/10.3389/fpls.2022.819658

Mostofa MG, Rahman MM, Ansary MMU, Keya SS, Abdelrahman M, Miah MG, Phan Tran L-S (2021) Silicon in mitigation of abiotic stress-induced oxidative damage in plants. Crit Rev Biotechnol 41:918–934. https://doi.org/10.1080/07388551.2021.1892582

Nascimento CWA, Silva FBV, Lima LHV, Silva JR, Veloso V d L, Silva FL, Freitas ST, Santos LF, Santos MA (2022) Silicon application to soil increases the yield and quality of table grapes (Vitis vinifera L.) grown in a semiarid climate of Brazil. Silicon. https://doi.org/10.1007/s12633-022-02129-0

Nedukha OM (2022) The role of silicon in plant under normal conditions and stress. Not Sci Biol 14:10973. https://doi.org/10.15835/nsb14110973

Peixoto M d M, Flores RA, do Couto CA, Pacheco HDN, de Mello Prado R, Souza-Junior JP, Castro-Netto JA, Graciano-Ribeiro D (2020) Silicon application increases biomass yield in sunflower by improving the photosynthesizing leaf area. SILICON 14:275. https://doi.org/10.1007/s12633-020-00818-2

Riederer M, Müller C (2006) Biology of the plant cuticle. Blacwell Publishing Ltd, Oxford

Sánchez-Navarro JF, González-García Y, Benavides-Mendoza A, Morales-Díaz AB, González-Morales S, Cadenas-Pliego G, García-Guillermo M d S, Juárez-Maldonado A (2021) Silicon nanoparticles improve the shelf life and antioxidant status of Lilium. Plan Theory 10:2338. https://doi.org/10.3390/plants10112338

Santos MMM d, Silva GP d, Prado R d M, Pinsetta Junior JS, Mattiuz B-H, Braun H (2022) Biofortification of tomato with stabilized alkaline silicate and silicic acid, nanosilica, and potassium silicate via leaf increased ascorbic acid content and fruit firmness. J Plant Nutr 45:896–903. https://doi.org/10.1080/01904167.2021.1994599

Schoelynck J, De Block P, Van Dyck E, Cooke J (2023) Is there silicon in flowers and what does it tell us? Ecol Evol 13:e10630. https://doi.org/10.1002/ece3.10630

Shahzad S, Ali S, Ahmad R, Ercisli S, Anjum MA (2022) Foliar application of silicon enhances growth, flower yield, quality and postharvest life of tuberose (Polianthes tuberosa L.) under saline conditions by improving antioxidant defense mechanism. SILICON 14:1511–1518. https://doi.org/10.1007/s12633-021-00974-z

Singh DK, Agnihotri RK, Ganie SA, Singh G, Ambedkar BR (2013) Pollen fertility and scanning electron microscopic studies of Sida cordifolia L. Indian Res J Genet Biotechnol 5:98–104

Souri Z, Khanna K, Karimi N, Ahmad P (2020) Silicon and plants: current knowledge and future prospects. J Plant Growth Regul 40:906–925. https://doi.org/10.1007/S00344-020-10172-7

Souza Júnior JP, de Mello Prado R, Campos CNS, Oliveira DF, Cazetta JO, Detoni JA (2022) Silicon foliar spraying in the reproductive stage of cotton plays an equivalent role to boron in increasing yield, and combined boron-silicon application, without polymerization, increases fiber quality. Ind Crop Prod 182:114888. https://doi.org/10.1016/j.indcrop.2022.114888

Souza Junior JP d, Prado R d M, Silva Campos CN, Sousa Junior G d S, Costa MG, de Pádua Teixeira S, Gratão PL (2023) Silicon modulate the non-enzymatic antioxidant defence system and oxidative stress in a similar way as boron in boron-deficient cotton flowers. Plant Physiol Biochem 197:107594. https://doi.org/10.1016/j.plaphy.2023.02.024

Souza JZ, De Mello Prado R, Silva SL d O, Farias TP, Neto JG, Souza Junior JP d (2019) Silicon leaf fertilization promotes biofortification and increases dry matter, ascorbate content, and decreases post-harvest leaf water loss of chard and kale. Commun Soil Sci Plant Anal 50:164–172. https://doi.org/10.1080/00103624.2018.1556288

Wenneck GS, Saath R, Rezende R, Andrean AFBA, Santi DC (2021) Agronomic response of cauliflower to the addition of silicon to the soil under water deficit1. Pesqui Agropecu Trop 51. https://doi.org/10.1590/1983-40632021v5166908

Zhang F, Carins Murphy MR, Cardoso AA, Jordan GJ, Brodribb TJ (2018) Similar geometric rules govern the distribution of veins and stomata in petals, sepals and leaves. New Phytol 219:1224–1234. https://doi.org/10.1111/nph.15210

Chapter 16
Mechanisms of Action of Nanosilica to Reduce Oxidative Stress in Plants

Carlos Alberto Garza-Alonso, Yolanda González-García, Fernando de Jesús Carballo-Méndez, and Antonio Juárez-Maldonado

Abstract This chapter describes oxidative stress in plants, its generation, and the adverse impacts it produces. It also describes the defense system plants have to deal with oxidative stress. Furthermore, it shows how nanosilica induces tolerance to oxidative stress in plants through changes at different levels, such as genetic and biochemical. Nanosilica can trigger a signaling cascade at the cellular level, which results in genetic modifications and the overexpression of genes related to the antioxidant defense system. In turn, it translates into greater production of enzymatic and non-enzymatic antioxidants.

Keywords Silicon · Nanosilica · Oxidative stress · Antioxidant defense

16.1 Introduction

Plants are exposed to many environmental conditions because they remain fixed during their growth and development. At certain times, these conditions can be adverse and cause stress. The generation of reactive oxygen species (ROS) is one of the most significant effects of stress on plants, regardless of the origin (biotic or abiotic). Reactive oxygen species inhibit cellular metabolism, negatively affecting plant growth and development. Thus, ROS affect productivity and quality (Gao et al. 2023; Verma et al. 2022a).

C. A. Garza-Alonso · F. de Jesús Carballo-Méndez
Facultad de Agronomía, Universidad Autónoma de Nuevo León, General Escobedo, Mexico

Y. González-García
Instituto Nacional de Investigaciones Forestales, Agrícolas y Pecuarias, Centro de Investigación Regional del Noroeste, La Paz, Mexico

A. Juárez-Maldonado (✉)
Departamento de Botánica, Universidad Autónoma Agraria Antonio Narro, Saltillo, Mexico
e-mail: antonio.juarez@uaaan.edu.mx

R. de Mello Prado et al. (eds.), *Silicon Advances for Sustainable Agriculture and Human Health*, Sustainable Plant Nutrition in a Changing World,
https://doi.org/10.1007/978-3-031-69876-7_16

Under normal conditions, there is a balance between producing and eliminating ROS in plants. However, this balance changes under stress. This change induces overproduction of these molecules, generating oxidative stress in cellular components and organelles (Mansoor et al. 2022). Oxidative stress causes cellular damage and harmful and irreversible effects on the development and growth of tissues and organs (Faisal et al. 2023). Oxidative stress refers to the alteration of the balance between generating and accumulating reactive molecular species (RMS) within plant cells. Their imbalance can generate some oxidative modification of biomolecules, leading to cell death due to necrosis or apoptosis (Kumar et al. 2023; Li et al. 2024; Yalcinkaya et al. 2019).

Regarding oxidative stress, plants have evolved and developed a complex antioxidant defense system to deal with ROS (Bela et al. 2022; Sun et al. 2024). In this system, an antioxidant molecule reacts with some reactive species to produce a less reactive product or, through other actions, to neutralize the reactivity of ROS (Spiegel et al. 2022). Antioxidants can be enzymatic or non-enzymatic based on their catalytic capacity. However, they can also be primary or secondary according to their mechanism of action (Stoia and Oancea 2022).

Nanomaterials (NMs) are natural or artificial materials comprising particles or aggregates, independently or mixed, ranging in size from 1 to 100 nm. They have unique properties and can be used in different areas of science, such as agriculture (Behl et al. 2022). Silicon nanoparticles (Si NPs) represent one of the nanomaterials with a favorable influence on agriculture, especially in plants subjected to stress (Bhat et al. 2021). Unlike natural silica, nano-SiO_2 is amorphous and ranges in size from 5 to 100 nm. It comprises 47% Si and 53% oxygen. Moreover, it is inert to pH variations, less toxic, and optically transparent, and some forms are degradable. Thus, it has several advantages, such as a relatively high surface-to-volume ratio, unique thermal and electrical properties, and higher permeability in plant cells (Van Hoang et al. 2022).

Silicon nanoparticles are also more effective than bulk Si as a plant fertilizer. Unlike bulk Si, Si NPs positively impact Si-accumulating and non-accumulating plants (Bhat et al. 2021). Silicon nanoparticles offer alternatives to synthetic chemical fertilizers. They are sustainable and environmentally acceptable because they do not interfere with nature's biological processes (Okeke et al. 2023).

This chapter describes oxidative stress in plants, its generation, and the adverse impacts it produces. It also describes the defense system plants have to deal with oxidative stress. Furthermore, it shows how nanosilica induces tolerance to oxidative stress in plants through changes at different levels, such as genetic and biochemical.

16.2 Oxidative Stress in Plants

Oxidative stress refers to the alteration of the balance between generating and accumulating reactive molecular species (RMS) within plant cells and their ability to neutralize and/or eliminate them. Their imbalance can generate some oxidative modification of the biomolecules, leading to cell death by necrosis or apoptosis (Kumar et al. 2023; Li et al. 2024; Yalcinkaya et al. 2019).

Reactive molecular species are atoms or molecules derived from cellular metabolism that contain one or more unpaired electrons in the valence shell or outer orbit and can exist independently. Having an odd number of electrons makes them unstable, short-lived, and highly reactive (Khudyakov 2023). Given the high reactivity, they can obtain electrons from other molecules or compounds to achieve their stability so that the attacked molecule loses its electron and becomes a free radical, starting a cascade of chain reactions that cause multiple damages to the cell (Phaniendra et al. 2015).

Reactive molecular species are a broad group, mainly including ROS, reactive nitrogen species (RNS), and reactive sulfur species (RSS) (Sandalio et al. 2023). These molecules are inevitable by-products of metabolic processes in plants. However, some molecules, such as ROS, have beneficial effects at moderate or low levels. Furthermore, they are involved in many physiological functions in plants, such as the induction of tolerance to abiotic stress, cell signaling, mitogenic response, and redox regulation (Medrano-Macías et al. 2022; Phaniendra et al. 2015). In contrast, they generate oxidative stress at high concentrations, causing potential damage to several biomolecules (Lee et al. 2023; Xu et al. 2023a).

ROS are the most studied group of reactive molecular species because they generate the greatest oxidative stress in plant cells if there is little or no ability to neutralize and/or eliminate them (Sandalio et al. 2023). This group of RMS includes free radicals, such as the hydroxyl radical ($^{\bullet}OH$), the superoxide anion ($O_2^{\bullet-}$), the alkoxyl radical ($RO^{\bullet}$), and the peroxyl radical ($ROO^{\bullet}$). Moreover, it includes non-radical forms that do not have unpaired electrons but are also highly reactive, such as hydrogen peroxide (H_2O_2), singlet molecular oxygen ($^1O_2$), hypochlorous acid (HClO), ozone (O_3), hydroperoxides (ROOH), and excited carbonyls (RO*) (Dumanović et al. 2021; Hasanuzzaman et al. 2020).

The ability of the plant cell to neutralize and eliminate ROS to avoid oxidative stress depends on the antioxidant defense system, which can be enzymatic and/or non-enzymatic. The imbalance between high levels of reactive species and an insufficient antioxidant defense system damages cellular structures and deteriorates molecules, such as lipids, fatty acids, proteins, and nucleic acids (DNA and RNA), due to stress (Kumar et al. 2023).

In the plant cell, the chloroplast is the main site of ROS production, specifically in the thylakoids, where the light reactions of photosynthesis occur. However, they can also be produced in the mitochondria, peroxisomes, cytosol, plasma membrane, and the cell wall (Singh 2022). For example, $O_2^{\bullet-}$ is produced in Photosystem I (PSI) of the thylakoids of the chloroplast derived from the electron transport chain of photosynthesis (Hasanuzzaman et al. 2019). Meanwhile, $^{\bullet}OH$ is produced by the Fenton reaction in chloroplasts (Rahman et al. 2020). In turn, H_2O_2 is produced mainly in peroxisomes since these organelles house a complex enzymatic system, such as glycolate oxidase (GOX), acyl-CoA oxidase (AOX), urate oxidase (UO), polyamine oxidase, copper amine oxidase (CuAO), sulfite oxidase (SO), sarcosine oxidase (SOX), or superoxide dismutase (SOD), that generates this reactive species

(Corpas et al. 2020; Sousa et al. 2019). However, it can also be produced in the mitochondria and the chloroplast, where it is generated from the dismutation of the $O_2^{\bullet-}$ by SOD or under conditions of low cellular pH (Zhao et al. 2020). Finally, $^1O_2$ is produced in the cell wall and membrane, as well as in the tonoplast and cytosol (Zhang et al. 2023b).

The unwanted side effects of oxidative stress mainly include alteration of oxidative lipid metabolism, such as lipid peroxidation. In lipid peroxidation, lipid molecules, mainly phospholipids containing polyunsaturated fatty acids (PUFAs), undergo oxidation by ROS (mainly $^{\bullet}OH$) and become a source of lipid peroxide radicals, which react with molecular oxygen to form lipid peroxyl and lipid alkoxyl radicals (de Alché 2019; Ben et al. 2023). In turn, these are further oxidized to form lipid hydroperoxides (LOOH), which, when fragmented, form compounds of unsaturated α,β-aldehydes, such as 4-hydroxynonenal (4-HNE) or malondialdehyde (MDA) (Sadowska-Bartosz and Bartosz 2024).

Damage to nucleic acids also occurs through different mechanisms, such as modification of the nucleotide base, because some ROS, such as $^{\bullet}OH$ and $^1O_2$, modify the guanine (G) bases to 8-hydroxydeoguanosine (8-OH-dG) in DNA and 8-hydroxyguanosine (8-OH-G) in RNA. Thus, they cause base pairing mutations, erroneous copies during replication, and genetic mutations (Huang et al. 2021; Martí-Guillén et al. 2022). Furthermore, oxidation of the deoxyribose sugar residue, single-strand breaks, cross-linking of DNA with associated proteins, obstruction of cytosine (C) methylation, and ribosomal blockage of translation can be induced (Akter et al. 2021; Martí-Guillén et al. 2022; Tripathi et al. 2023).

Reactive molecular species can degrade proteins either directly due to chemical modifications, such as nitrosylation, carboxylation, disulfide bond formation, and glutathionylation, or indirectly due to the interaction of cellular proteins with lipid peroxidation products. In this case, amino acids undergo site-specific modifications that lead to proteolytic degradation (Tripathi et al. 2023). All of these changes affect the formation of adducts with different cellular components that significantly modify their structure, thus affecting biological activity and causing alterations in cellular metabolism. These can lead to cellular dysfunction, inhibition of plant growth and performance, and apoptosis or programmed cell death (Sadowska-Bartosz and Bartosz 2024; Singh 2022).

Oxidation of carbohydrates can also be induced, especially free carbohydrates, such as sugars, polyols, and structural polysaccharides of the cell wall. It causes the formation of reactive molecules such as dicarbonyls, especially glyoxal, methylglyoxal, and 3-deoxyglucosone (Martí-Guillén et al. 2022). These cause the glycation of protein residues, such as the glycation of lysine (Lys) and arginine (Arg) by glucose to form fructosyl-lysine (FL) and hydroimidazolone (MG-H1) residues, respectively. Thus, they affect the electrostatic interaction between proteins, enzyme substrates, and nucleic acids, generating protein degradation, enzymatic inactivation, and gene mutations (Rabbani et al. 2020).

As mentioned above, RMS are inevitable products of plant metabolism under suitable conditions. However, when plants undergo stress conditions, their production rates increase significantly, resulting in oxidative stress (Yalcinkaya et al.

2019). For example, *Glycine max* L. (Merr) plants subjected to hypoxia stress increased the content of H_2O_2 and $O_2^{\bullet-}$ in leaves and roots, in addition to lipid peroxidation, substantially decreasing biomass accumulation compared to non-stressed plants (Posso et al. 2023). Under salinity stress conditions, ×*Triticale* plants increased lipid peroxidation and H_2O_2 content, decreasing crop yield (Rasouli et al. 2024). Synthetic organic pollutants such as aniline ($C_6H_5NH_2$) induce oxidative stress in *Oryza sativa* L. plants by increasing H_2O_2 and $O_2^{\bullet-}$, causing lipid peroxidation and damage to the DNA of the cells (Tao et al. 2022). Contamination by passed metals such as arsenic (As) increased the H_2O_2 content and lipid peroxidation, generating severe damage to the cell membrane, causing electrolyte leakage (EL) and cell death in *Lens culinaris* Medik plants (Hossain et al. 2023).

16.3 Antioxidant Defense System in Plants

Plant cells carry out three vital processes of plant metabolism, which involve molecular oxygen (O_2), photosynthesis, photorespiration, and oxidative phosphorylation (Rodrigues de Queiroz et al. 2023). In its ground state, O_2 has two unpaired electrons with parallel spins in two separate π antibonding orbitals, making it a triplet diradical that subsequently transitions to the singlet state ($^1O_2$) by absorbing energy (Khorobrykh et al. 2020). All other forms of active oxygen are produced by a $^1O_2$ transfer mechanism, and the resulting partially reduced forms have greater reactivity toward several organic molecules and are, therefore, collectively called ROS (Khorobrykh et al. 2020; Zeppilli et al. 2023).

Under normal conditions, there is a balance between producing and eliminating ROS in plants. However, this balance changes under stress. This change induces overproduction of these molecules, generating oxidative stress in the components and cellular organelles (Mansoor et al. 2022). Oxidative stress causes cellular damage and harmful and irreversible effects on the development and growth of tissues and organs (Faisal et al. 2023). However, plants have evolved and developed a complex antioxidant defense system to deal with ROS (Bela et al. 2022; Sun et al. 2024). In this system, an antioxidant molecule reacts with some reactive species to produce a less reactive product (whose unpaired electron is no longer located in a single atom but is distributed throughout the structure) or, through other actions, to neutralize the reactivity of the ROS (Spiegel et al. 2022). Hence, the antioxidant defense system can be classified according to catalytic issues (enzymatic and non-enzymatic antioxidants) and the mechanism of antioxidant action (primary and secondary antioxidants) (Stoia and Oancea 2022).

Two different processes intervene in the elimination of free radicals to prevent oxidative stress according to the mechanism of action: (1) direct reactions with ROS through primary antioxidants that neutralize and/or eliminate free radicals and (2) reactions with molecules that generate free radicals, preventing the production of ROS through secondary antioxidants that neutralize and/or eliminate the processes that generate free radicals (Martínez and Reina 2017).

Primary antioxidants generally scavenge free radicals directly or delay or interrupt the propagation reaction by donating a hydrogen atom or transferring a single electron to a free radical. Hydrogen is released from the active OH or NH groups of antioxidants and is subsequently donated to free radicals (Couttolenc et al. 2020; Varatharajan and Pushparani 2018). Enzymatic and non-enzymatic antioxidants are the most common primary antioxidants. The main enzymes are superoxide dismutase (SOD), catalase (CAT), ascorbate peroxidase (APX), glutathione reductase (GR), monodehydroascorbate reductase (MDHAR), dehydroascorbate reductase (DHAR), glutathione peroxidase (GPX), guaiacol peroxidase (GOPX), glutathione S-transferase (GST), ferritin nicotinamide adenine dinucleotide phosphate (NADPH), alternative oxidase-like oxidase (AOX), peroxiredoxins (PRX), thioredoxins (TRX), and glutaredoxin (GRX) (Hasanuzzaman et al. 2020). Meanwhile, the main non-enzymatic compounds include ascorbic acid (AsA), glutathione (GSH), phenolic acids, alkaloids, flavonoids, carotenoids, α-tocopherol, and non-protein amino acids (Siddeeg et al. 2021).

Secondary antioxidants can chelate copper (Cu II) or iron (Fe III) ions and reduce the rate constant of their reduction process, hindering the Fenton reaction (the second step of the Haber-Weiss cycle, which produces $^{\bullet}OH$ radicals) (García-Díez and Mora-Diez 2020). Polyphenols are the most common secondary antioxidants. They are secondary metabolites present in many fruits, especially berries, as well as in some vegetables and cereals, which are divided according to their structural difference into flavonoids, phenolic acids, lignans, tannins, aromatic amines, and stilbenes (Lang et al. 2024; Zhang et al. 2023a).

The reaction mechanisms by which antioxidants can exert their activity depend on their intrinsic capacities in the plant cell through several mechanisms. Hydrogen atom transfer (HAT) ($HA + R^{\bullet} \rightarrow HA^{\bullet +} RH$) is the concerted movement of a proton (H^+) and an electron (e^-) between two functional groups in a single kinetic step, which can be inter- and intramolecular (Bonciolini et al. 2022; Subramanian and Sibi 2023). Single-electron transfer (SET) ($HA + R^{\bullet} \rightarrow HA^{+\bullet} + R^-$) is the reaction in which an electron is transferred from one molecule called the electron donor to another molecule called the electron acceptor in a single step (Bericat-Vadell et al. 2023). Radical adduct formation (RAF) ($HA + R^{\bullet} \rightarrow [HA\text{-}RH]^{\bullet}$) is the reaction in which the presence of multiple bonds interferes. However, it is more likely to occur in neutral and/or positive charge molecules deficient in electrons (electrophilic radicals) (Galano et al. 2016). Finally, proton-coupled electron transfer (PCET) is a one-step reaction like HAT but, however, differs in that the target substrate undergoes a redox event and a deprotonation with two different molecular entities (an oxidant and a base, respectively) (Bonciolini et al. 2022; Spiegel et al. 2022).

Each plant cell or subcellular compartment has a specialized set of antioxidants that, upon accumulation of ROS, induce transcriptional and metabolic changes to prevent the toxicity of these ROS through efficient scavenging (Sun et al. 2024). It can be through the combination and concentration of appropriate antioxidants and also by allowing controlled transport of ROS from organelle to organelle, either through channels such as aquaporins or a direct association, such as chloroplasts and nuclei via stromules (Kerchev and Van Breusegem 2022).

16.4 Impact of Nanomaterials on Plants

Nanomaterials (NMs) are natural or artificial materials comprising particles or aggregates, independently or mixed, ranging in size from 1 to 100 nm (Behl et al. 2022). The most commonly used nanoparticles (NPs) in agriculture can be of organic and inorganic nature. In organic NPs, based on carbon, we can find graphene, fullerenes, and carbon nanotubes (single-walled and multi-walled). In inorganic NPs, we can find metal or metalloid-based, such as silver (Ag), gold (Au), copper (Cu), copper oxide (CuO_2), iron (Fe), iron oxide (Fe_3O_4), zinc (Zn), zinc oxide (ZnO_2), manganese (Mn), titanium dioxide (TiO_2), selenium (Se), silicon (Si), silicon dioxide (SiO_2), and cerium dioxide (CeO_2), among others. Nanoparticles are characterized by physical and chemical characteristics different from their conventional forms (Almutairi et al. 2023). These materials have been widely used to promote and improve the exchange between plants, the environment, and the stress tolerance of crops (Azameti and Imoro 2023; Faizan et al. 2023; Fu et al. 2023).

Research is currently being conducted worldwide on using nanoparticles to improve plant growth and productivity to increase production and improve plant quality. Nanoparticles used as fertilizers have a high impact on plants, particularly because of their small size and large surface area, which allows them to pass through the cells more efficiently than their non-nanometric counterparts, thus increasing nutrient availability (Almutairi et al. 2023). These particles can enter the plant tissue through different structures, such as stomata, root hairs, and cracks in the leaf surface. Once inside the plant, the nanoparticles move through the vascular bundle and are distributed throughout the organism (Wang et al. 2023b).

Plants face unfavorable environmental conditions that negatively impact their growth, productivity, and quality. Drought, salinity, heavy metals, and nutritional imbalance are some environmental stresses they face. These adverse conditions cause alterations in genetic, physiological, biochemical, and morphological processes in plants (El-Ramady et al. 2023). The generation of ROS is one of the most significant effects of stress on plants, regardless of the origin (biotic or abiotic). ROS inhibit cellular metabolism, negatively affecting plant growth and development. Thus, ROS affect productivity and quality. It has been demonstrated that NPs can change the level of ROS and improve the internal conditions of plants (Gao et al. 2023; Verma et al. 2022a).

Silicon nanoparticles represent one of the nanomaterials with a favorable influence on agriculture, especially in plants subjected to stress (Bhat et al. 2021).

16.4.1 Nanosilica

Silicon dioxide (SiO_2) exists mainly in crystalline form in nature. It comprises SiO_4 units arranged in a tetrahedral geometry, such as sand, quartz, and others. Despite the abundance of silicon at the earth's surface, the limited amount of the plant-available form of Si in most cultivated soils and the limited effectiveness of silicate fertilizers are the main concerns for the exploration of new forms of Si (Bhat et al. 2021). Unlike natural silica, nano-SiO_2 is amorphous and ranges in size from 5 to 100 nm. It comprises 47% Si and 53% oxygen. Moreover, it is inert to pH variations, less toxic, and optically transparent, and some forms are degradable. Thus, it has several advantages, such as a relatively high surface-to-volume ratio, unique thermal and electrical properties, and higher permeability in plant cells (Van Hoang et al. 2022). Silicon nanoparticles are also more effective than bulk Si as a plant fertilizer. Unlike bulk Si, Si NPs positively impact Si-accumulating and non-accumulating plants (Bhat et al. 2021).

There is a wealth of evidence demonstrating the positive impact of nanosilicon in improving stressed plants, benefiting growth and productivity (Verma et al. 2022b). The most important effects include the increase in the concentration and protection of pigments that maintain photosynthetic activity, the accumulation of biochemical antioxidant compounds, the increase of compounds that maintain osmotic balance, and the maintenance and increase of the activity of antioxidant enzymes (Fig. 16.1) (Alharbi et al. 2022; Badawy et al. 2021).

Silicon nanoparticles offer alternatives to synthetic chemical fertilizers. They are sustainable and environmentally acceptable because they do not interfere with nature's biological processes (Okeke et al. 2023).

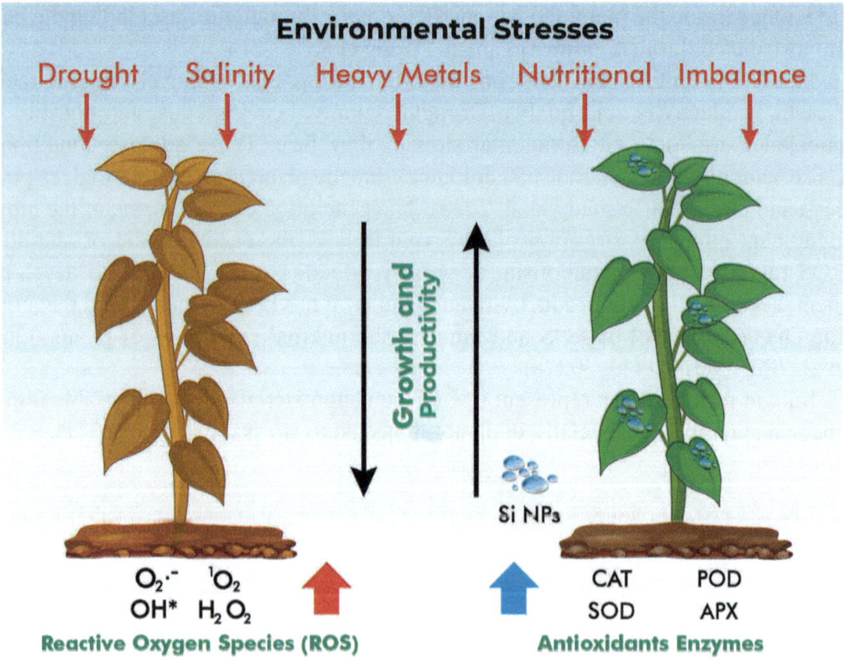

Fig. 16.1 Representation of the impact of environmental stress on plants without and with Si NPs

16.4.2 Impact of Nanosilica on Plants

The application of Si NPs has shown positive effects in plants under optimal growth conditions, such as in *Zea mays* L., where it increased growth, showing greater plant height, fresh and dry weight of foliage and root, leaf area, and relative water content. Furthermore, it benefited biochemical avidity with increased photosynthetic enzyme activity (Hao et al. 2023). Similarly, in *Glycine max* L., the number of pods per plant and the number of grains per pod increased. Meanwhile, in *Oryza sativa* L., it reduced the percentage of electrolyte leakage and increased the dry weight of the plant as well as the number of panicles and grain production (Felisberto et al. 2021). Regarding *Megathyrsus maximus* Jacq., the application of SiO_2 NPs increased the plant's yield (de Faria Melo et al. 2023). Exposure of *Raphanus sativus* L. roots to Si NPs increased plant growth and yield due to the increase in phospho-synthetic pigments, such as chlorophyll and carotenoids; nutritional compounds, such as sugars and proteins; and essential elements, such as manganese (Mn) (Xu et al. 2023b).

When plants do not grow under optimal conditions but are subjected to different types of stress, they increase the generation of reactive oxygen species (ROS), which include singlet oxygen ($^1O_2$), hydrogen peroxide (H_2O_2), superoxide ($O_2^{\bullet-}$), and hydroxyl radicals ($^{\bullet}OH$). These negatively affect several cellular components, mainly by damaging the integrity of the membrane, leading to high lipid peroxidation levels. Thus, several cellular processes are compromised, which does not allow the cell to perform its normal functions, affecting growth, development, and productivity (Mansoor et al. 2022).

Several studies have demonstrated the beneficial effect of Si NPs in plants subjected to stress. There are reports on the impact of Si NPs on plants subjected to drought stress, as in *Mangifera indica* L., where the application of Si NPs showed increased growth by increasing the length, thickness, and number of shoots, as well as leaf area, chlorophyll and carotene content, and concentrations of nitrogen (N), phosphorus (P), and potassium (K). Moreover, fruit set percentages increased, and fruit drop percentages were reduced, which increased fruit size, firmness, and quality, improving yield (Almutairi et al. 2023). *Fragaria × ananassa* Duch. is another plant affected by drought but benefited from Si NPs. When fertilized with Si NPs, it showed a reduction in malondialdehyde (MDA) and hydrogen peroxide H_2O_2 content. Furthermore, it showed an increase in the number of leaves, leaf area, root dry weight, and the concentration of phenolic compounds, anthocyanins, chlorophyll, carotenoids, and vitamin C. Moreover, it increased water use efficiency, phenolic compounds, anthocyanins, chlorophyll, carotenoids, vitamin C, and water use efficiency (WUE) (Zahedi et al. 2020).

The excessive presence of salts in the growing medium and irrigation water generates, among other problems, an imbalance in the ionic homeostasis of the plants. The application of Si NPs has reduced this problem by impacting other parameters of tolerance to this stress. The application of Si NPs to *Pisum sativum* L. grown under saline conditions showed a reduction of EL%, MDA, and H_2O_2 content while increasing height and biomass (number of leaves, flowers, pods, and seeds), as well as relative water content (RWC) and chlorophyll a and b concentration (Al-Shammari et al. 2023). Alharbi et al. (2022) reported that the presence of Si NPs in plants of *Oryza sativa* L. subjected to salinity improved their growth due to the increase in the concentration of N, P, K, chlorophylls, and carotenoids. The reduction of sodium (Na), which allowed an increased activity of the enzymes catalase (CAT), superoxide dismutase (SOD), and peroxidase (POD), as well as the stomatal conductance and the RWC and consequently a reduction of MDA, H_2O_2, and EL%, was presented.

The harmful effects of heavy metals on plants generate ROS, which impairs plant biomass growth and imbalances nutrient uptake, resulting in low production. Jalil et al. (2023) reported that *Oryza sativa* L. plants subjected to cadmium (Cd) toxicity increased their biomass when fertilized with Si NPs. The main positive impacts of these nanoparticles included the increase in pigments (chlorophyll a, chlorophyll b, total chlorophyll contents, and carotenoids) and essential elements (magnesium (Mg), calcium (Ca), and potassium (K)), as well as photosynthetic activity (net photosynthetic rate (Pn), transpiration rate (E), stomatal conductance (gs), and intercellular CO_2 concentrations (Ci)). In contrast, the concentration of Cd, H_2O_2, and MDA decreased. Riaz et al. (2022) showed that excessive copper (Cu^{+2}) inhibited the root length and plant height of *Triticum aestivum* L., in addition to generating membrane peroxidation (MDA) and high H_2O_2 contents in root tissues. However, the application of Si NPs benefited growth. It reduced the accumulation of MDA and H_2O_2 by increasing the concentration of chlorophylls and carotenoids, in addition to increasing the antioxidants CAT, SOD, POD, and ascorbic peroxidase (APX) and photosynthetic activity (Pn, E, and gs).

Nutrient imbalance is considered a type of stress since the deficiency or excess of any of the essential elements negatively impacts plants. Plants of *Solanum aethiopicum* L. subjected to P excess showed an increase in EL% caused by oxidative stress. Contrary to the above, the application of Si NPs decreased the absorption of P, improved C:N homeostasis, and increased the efficiency of Fe, Cu, and Mn use, as well as the concentration of phenolic compounds, chlorophylls, and carotenoids. At the same time, it favored photosynthetic efficiency and plant growth (Alves et al. 2023). Carvalho et al. (2022) reported that in N-deficient *Beta vulgaris* L. plants, EL% increased, while the application of Si NPs increased biomass, N concentration, total phenols, chlorophyll, carotenoids, and PSII quantum efficiency. Table 16.1 shows other studies on the effect of Si NPs on plants subjected to stress.

16.5 Nanosilica and Antioxidant Defense System

Nanomaterials can produce many responses when they come into contact with plant cells, and these responses can be positive or negative for plant metabolism (Juárez-Maldonado et al. 2021). In the first contact with cell membranes, several physical, chemical, and biochemical alterations occur, and the cell itself can recognize and trigger a signaling cascade (Juárez-Maldonado et al. 2019). The response mechanism to the application of NMs is similar to other stimuli (physical, chemical, and

Table 16.1 Recent studies on the impact of Si NPs on plants subjected to some type of stress

Stress	Specie	Effect	Author
Drought	*Crocus sativus* L.	Increased activity of SOD and APX enzymes	Tavakoli et al. (2021)
	Solidago canadensis L.	Increased K content, flavonoids, and stomatal conductance (gs)	Othman et al. (2023)
	Solanum tuberosum L.	Increased RWC of leaves, proline content, chlorophylls, and carotenoid concentration and photosynthetic activity (Pn, E, gs, and Ci)	Al-Selwey et al. (2023b)
Salinity	*Pisum sativum* L.	Improved height and biomass production, increased K content and CAT enzyme activity	Barceló et al. (2022)
	Solanum lycopersicum L.	Improved height of plant and fruit production, increased N, P, Ca, Mg, Fe and Zn content and gibberellic acid (GA3), abscisic acid (ABA), and proline levels in leaves	Sayed et al. (2022)
	Oryza sativa L.	Improved root biomass production, leaf area and yield, increased RWC and K content Reduced Na content	Badawy et al. (2021)
Cd toxicity	*Phaseolus vulgaris* L.	Improved biomass production, increased RWC, Pn, and CAT enzyme activity Reduced MDA and EL%	Koleva et al. (2022)
Pb toxicity	*Coriandrum sativum* L.	Increased flavonoid concentration and CAT enzyme activity Reduced Pb shoot and root content	Fatemi et al. (2020)
P excess	*Megathyrsus maximus* Jacq.	Reduced tiller mortality and improved biomass production	De Faria Melo et al. (2023)
P deficiency	*Beta vulgaris* L.	Increased chlorophyll, carotenoid, and total phenol concentration and P content Improved PSII quantum efficiency	Carvalho et al. (2022)

biological), where the objective is always to maintain a balance regarding oxidation-reduction (redox) to avoid damage to the genetic material, macromolecules, and other plant structures (Garza-Alonso et al. 2022).

16.5.1 Nanosilica Impact at the Genetic Level

One of the main effects of nanocomposites includes the alteration in the expression of many genes involved in the antioxidant response of plants. Examples include some transcription factors or directly related genes involved in metabolic pathways that synthesize non-enzymatic antioxidant compounds, as well as genes responsible for the assembly of certain antioxidant enzymes (Arif et al. 2021).

The application of Si NPs (2 mM) in *Fragaria* × *ananassa* plants resulted in the overexpression of the *DREB*, *cAPX*, *Mn-SOD*, and *GST* genes, all of which are related to the plants' antioxidant defense system (Moradi et al. 2022). On the other hand, Kiany et al. (2022) reported that using Si NPs (50 and 100 mg L^{-1}) showed the overexpression of the *GSH1*, *PCS*, and *ABC1* genes related to the non-enzymatic antioxidant system. Furthermore, the overexpression of the *Lsi1*, *Lsi2*, and *Lsi3* genes occurred, favoring the reduction of oxidative stress produced by As in *Oryza sativa* L. plants. In another study, the use of Si NPs at 50, 100, and 150 mg L^{-1} in plants of *Zingiber officinale* Roscoe produced alterations in some transcription factors of the *MYB* family, specifically for *MYB_c49329_g1*, *MYB_c66024_g5*, *MYB_c56909_g1*, *MYB_c60640*, and *MYB:c59591_g1*, in addition to overexpressing the LysM genes*: Marker00002932* and *LysM:Marker00003969* (Peng et al. 2022). In the same study, an increase in the activity of the enzymes POD, CAT, SOD, and PPO also occurred, as well as an increase in total flavonoid levels. On the other hand, Ismail et al. (2022) found that the application of Si NPs (3 mM) promoted the overexpression of *SOD*, *POD*, *CAT*, and *PslecRLK* genes in *Pisum sativum* L. plants. Similarly, other reports indicate that the application of Si NPs at 50 and 100 mg L^{-1} in *Oryza sativa* L. plants resulted in the overexpression of the genes *OsSOD-Cu/Zn*, *OsCATA*, *OsCATB*, and *OsAPX1*, all involved in the enzymatic defense system (Jalil et al. 2023).

Boora et al. (2023) applied Si NPs at different levels (30, 60, 90, and 120 mg L^{-1}) in *Triticum aestivum* L. plants subjected to water stress. The results demonstrated the overexpression of some transcription factors involved in the antioxidant response of plants; specifically, these TFs were DREB2, MYB33, MYB33R, WRKY19, and *SnRK2*.4. Similarly, Guerriero et al. (2023) showed that the applications of Si NPs fused with quercetin in *Solanum lycopersicum* L. plants produced the overexpression of several genes involved in the enzymatic and non-enzymatic antioxidant system, including *4CL*, *APX1*, *PAL2*, *EXPA4*, *CHI1*, *XETB2*, *EXPA5*, *CHI2*, *SOS1*, *4CL3*, *AREB2*, *GLUR*, *APX6*, *SOD2*, *PAL6*, *SOD1*, *XETB1*, *HSF30*, *SOD3*, and *TAS14*.

On the other hand, Biju et al. (2023) showed that the application of Si NPs in *Lens culinaris* Medik plants produced the overexpression of genes involved in secondary metabolism pathways (*Lc15103.1*, *Lc29018.1*), alkaloid synthesis (*Lc07235.1*), and flavonoid pathways (*Lc15103.1*). Another study verified the overexpression of the *PDH2*, *FLS*, *PAL*, and *NCED1* genes in *Fragaria* × *ananassa* plants subjected to the application of SiO_2 NPs at 50 mg L^{-1} (Zahedi et al. 2023). The same study also reported increases in the concentration of phenolic compounds and increased activity of antioxidant enzymes, such as *POD*, *CAT*, *SOD*, and *APX*.

On the other hand, Ulhassan et al. (2023) showed that using Si NPs promoted the overexpression of the genes *BnCAT*, *BnPAL*, *BnPPO*, *BnPAO*, *BnSOD*, Bn *APX*, and *BnGR* in leaves and roots of *Brassica napus* L. plants. Similarly, the application of Si NPs at doses of 2000–4000 mg L^{-1} produced the overexpression of the *POX* and *PPO* genes in *Vicia faba* L. plants subjected to salt stress (El-Flaah et al. 2021).

Rashad et al. (2021) found that applications of Si NPs at 150 mg L^{-1} in *Vitis vinifera* L. plants produced the overexpression of the peroxidase enzyme (*POX*) encoding gene, as well as other genes involved in defense against pathogens, such as β-1,3-glucanase (*GLU*), pathogenesis-related-protein 1 (*PR1*), and chitinase class II (*CHI II*), encoding them, in addition to overexpressing the transcription factor jasmonate and ethylene-responsive factor 3 (*JERF3*).

Although most reports have found an overexpression of genes involved in the antioxidant defense system, an opposite effect occurs in some cases. For example, Ahmed et al. (2023) found that applying Si NPs at 100 and 250 mg L^{-1} decreased the expression of *BnSOD*, *BnCAT*, *BnPOD*, and *BnAPX* genes in *Brassica napus* L. plants.

On the other hand, reports indicate genetic modifications in processes that are indispensable in the antioxidant system of plants, such as cell signaling by phytohormones. For example, Li et al. (2023) showed that applying Si NPs at a dose of 200 mg L^{-1} in *Triticum aestivum* L. plants produced the overexpression of the genes *PYR*, *PYL*, *PP2C*, *SnRK2*, *ABF*, and *TIR1*, all of which are involved in signaling by IAA and ABA. Similarly, some reports indicate genetic alterations in hormonal signaling by JA and SA, as is the case of Wang et al. (2023a), who carried out applications of Si NPs (50 mg L^{-1}) in *Oryza sativa* L. plants. They found overexpression of the genes *PLD, LOX, AOS, OPR, JAR, JAZ, MYC2, ICS, NPR1, WRKY70*, and *WRKY45*, as well as increased expression of the *PAL* gene.

All the genetic changes mentioned above will produce enzymatic and non-enzymatic antioxidant compounds in plant cells to cope with the different reactive species that produce oxidative stress (Fig. 16.2).

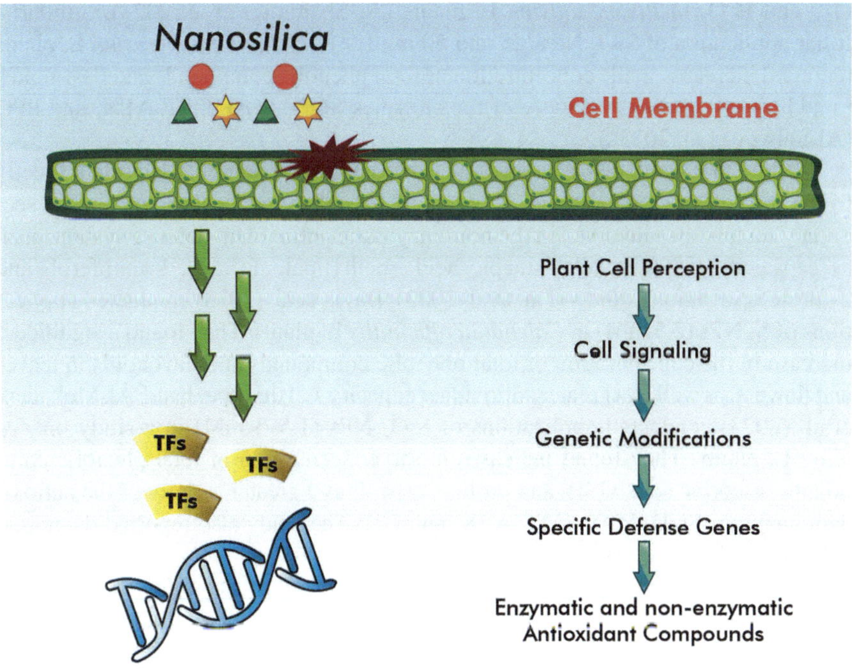

Fig. 16.2 Graphical representation of the impact of nanosilica on the antioxidant defense system of cells

16.5.2 Nanosilica Impact at the Biochemical Level

As mentioned in the previous section, plant cells seek to maintain a redox balance in the environment, so Si NPs are a promising tool for producing changes in the production of antioxidant compounds to counteract the effects of several types of stress. An example of this is Mukarram et al. (2023), where foliar applications of Si NPs (150 mg L^{-1}) were carried out on *Cymbopogon flexuosus* (Nees ex Steud.) Watson plants subjected to salt stress, finding a greater activity of the enzymes SOD, CAT, POD, and PRO, which decreased the oxidative stress caused by high levels of NaCl. Another report showed that applications of SiO_2 NPs deposited directly in the soil at doses of 1–4 g pot^{-1} in *Cunninghamia lanceolata* (Lamb.) Hook plants resulted in increased activity of SOD, CAT, and POD enzymes, as well as increased concentration of vitamin C and GSH (Liu et al. 2023).

Emamverdian et al. (2023) found that applications of SiO_2 NPs (150 μM) in *Pleioblastus pygmaeus* (Mitford) Nakai plants subjected to As stress increased the concentration of GSH, AsA, and phytochelatin compounds, in addition to promoting the activity of the enzymes SOD, POX, GPX, APX, and CAT. The same study also reported a decrease in the concentration of the free radicals H_2O_2 and $O_2^{\bullet-}$, as well as lower levels of MDA. Other reports have shown that drench applications of Si NPs (50 mg L^{-1}) increased the levels of total phenolic compounds and the activity of the enzymes CAT, SOD, and POX, in addition to reducing the concentration of $O_2^{\bullet-}$ and H_2O_2 in *Pisum sativum* L. plants (Al-Shammari et al. 2023). Similarly, foliar application of SiO_2 NPs (25 and 50 mg L^{-1}) in *Solanum tuberosum* L. plants recorded increases in the concentration of total phenolic compounds and flavonoids, in addition to increased activity of the enzymes SOD, POD, CAT, APX, and PPO (Al-Selwey et al. 2023a).

On the other hand, using Si quantum dots at a dose of 50 mg L^{-1} applied directly to the soil in *Raphanus sativum* L. plants increased the concentration of some secondary metabolites involved in the non-enzymatic antioxidant defense system, such as capsaicin, GSH, trifolin, sinapic acid, methylmalonic acid, kaempferol, and 2-foroic acid, among others (Fan et al. 2023). Bayat et al. (2023) conducted applications of Si NPs (2.5 mM) in *Calendula officinalis* L. plants. They found a significant increase in the concentration of total phenolic compounds and flavonoids in leaves and flowers, as well as a greater antioxidant capacity. On the other hand, Al-Mokadem et al. (2023) conducted applications of SiO_2 NPs (1.5–3 mM) in a study on *Zea mays* L. plants. They found increases in the concentration of total phenolic compounds, ascorbic acid, GSH, and proline, as well as a greater activity of the antioxidant enzymes SOD, POX, CAT, APX, and GR. The study also reported decreased MDA and ROS levels (H_2O_2, $^{\bullet}OH$, and $O_2^{\bullet-}$).

Chen et al. (2023) applied Si NPs (3, 30, and 300 mg L^{-1}) to *Solanum tuberosum* L. plants infected with the pathogen *Phytophthora infestans*. They found an increase in the activity of the POX and PPO enzymes, as well as a decrease in the activity of the enzyme ascorbic acid oxidase (AAO). On the other hand, applying Si NPs at

300, 600, and 900 mg L^{-1} in *Triticum aestivum* L. plants showed a significant reduction in H_2O_2 and MDA levels, in addition to increases in the activity of SOD and POX enzymes (Raza et al. 2023). Sharma et al. (2023) applied SiO_2 NPs (5–125 mg L^{-1}) in *Stevia rebaudiana* Bertoni plants. They reported an increase in the total concentration of carotenoids in the leaves, in addition to greater activity of the antioxidant enzymes PPO, CAT, POX, and SOD.

Another study investigated the effect of SiO_2 NPs and a hybrid nanomaterial of graphene oxide + SiO_2 NPs, which were applied under in vitro conditions at doses of 5, 15, and 25 mg L^{-1} in the tissue culture of *Ammi visnaga Gaertn*. Thus, a considerable increase in H_2O_2 levels was reported. However, increases in the concentration of total phenolic compounds, total flavonoids, and increased FRAP antioxidant capacity also occurred, in addition to increases in the activity of PPO, SOD, CAT, GPX, and APX enzymes (Golkar et al. 2023).

On the other hand, Ahmed et al. (2023) showed that foliar application of SiO_2 NPs at levels of 100, 200, 300, and 400 mg L^{-1} increased the concentration of total carotenoids in *Zea mays* L. plants, in addition to a higher concentration of total proteins, amino acids, and proline. The same study also reported an increase in the activity of the enzymes POX, CAT, SOD, and APX.

Other reports indicate that using SiO_2 NPs (10 mg L^{-1}) can reduce the levels of $O_2^{\bullet-}$ and H_2O_2 radicals, which is mainly attributed to increased activity of the enzymes SOD and APX (Anand and Pandey 2023). It was demonstrated in *Vigna radiata* (L.) R. Wilczek plants. Similarly, in another study that conducted foliar applications of SiO_2 NPs (50 mg L^{-1}) to *Fragaria* × *ananassa* plants, there were decreases in the concentration of MDA and H_2O_2, as well as increases in the activity of the enzymes POD, CAT, SOD, APX, and PAL (Zahedi et al. 2023). Furthermore, the concentration of total carotenoids and proline also increased.

Ulhassan et al. (2023) applied Si NPs (400 mg L^{-1}) on *Brassica napus* L. seeds subjected to stress by high levels of Cr. They found that NPs reduced the oxidative stress caused by this heavy metal by reducing the concentration of MDA, H_2O_2, and $O_2^{\bullet-}$, which could be explained thanks to a higher concentration of GSH and total carotenoids, as well as increased activity of the enzymes SOD, CAT, APX, and GR. Continuing with the topic of stress caused by heavy metals, another study of the effect of Si NPs on *Raphanus sativus* L. plants subjected to high concentrations of As found that NPs (100 mM) decreased the levels of MDA and H_2O_2, in addition to increasing the total concentration of phenolic compounds, flavonoids, anthocyanins, AsA, GSH, and total protein, as well as increasing the activity of CAT, APX, POD, and PPO enzymes (Bhardwaj et al. 2023). On the other hand, Riaz et al. (2023) conducted foliar applications of Si NPs (2.5 mM) in *Oryza sativa* L. plants subjected to these by Cd. They found a decrease in the levels of MDA and H_2O_2, as well as increases in the concentration of GSH and total carotenoids, in addition to increases in the activity of the APX enzyme. Table 16.2 presents other studies demonstrating the effects of nano Si at the biochemical level on several plant species.

Table 16.2 Recent studies on the biochemical impact of nano Si on the antioxidant system of plants

Material	Concentration applied	Plant species	Effect	References
Si NPs	50, 100, and 150 mg L^{-1}	*Mangifera indica* L.	Increased vitamin C and β-carotene	Almutairi et al. (2023)
SiO_2 NPs	25, 50, and 100 mg L^{-1}	*Triticum aestivum* L.	SOD, POD, CAT, carotenoids	Manzoor et al. (2023)
Si NPs	35 and 60 mg L^{-1}	*Triticum aestivum* L.	Increase in CAT, POX, PPO and decrease in H_2O_2, MDA	Ahmadi-Nouraldinvand et al. (2023)
$NaSiO_2$ NPs	50 and 100 mg L^{-1}	*Fragaria* × *ananassa* Duch.	SOD, CAT	Şener et al. (2023)
SiO_2 NPs	50–2000 mg L^{-1}	*Arachis hypogaea* L.	Increased total carotenoids	Prasad et al. (2023)
SiO_2 NPs	500 mg L^{-1}	*Ocimum basilicum* L.	POD, CAT	Ullah et al. (2023)
Si NPs + biochar	0.1% mix with soil	*Pisum sativum* L.	Increased the concentration of total phenolic compounds	Salman et al. (2023)
SiO_2 NPs	10 and 30 mg L^{-1}	*Momordica charantia* L.	SOD, POD, CAT, APX, and MDA reduction	Sun et al. 2023
Si NPs	500, 1000, and 1500 mg L^{-1}	*Malus domestica* Borkh.	Increased total phenolic compounds, CAT, POD	Jelodarian et al. (2023)
SiO_2 NPs	20 mg L^{-1}	*Vigna radiata* L.	SOD, POD, APX, CAT	Zaheer et al. (2023)

16.6 Conclusions

Plants undergo different conditions of both biotic and abiotic stress. As a consequence, biochemical reactions are triggered at the cellular level, resulting in the production of reactive molecules and, therefore, oxidative stress. Plants have an antioxidant defense system comprising different enzymatic or non-enzymatic antioxidants to deal with this stress. This antioxidant defense system can directly neutralize reactive molecules (primary antioxidants) or indirectly prevent their production (secondary antioxidants).

Nanomaterials have unique characteristics that can be very useful in inducing the activation of the antioxidant defense system. In particular, silicon is a beneficial element associated with many positive plant responses. This nanoscale element has even more potential to induce positive responses in plants. Particularly, nanosilica can efficiently activate the antioxidant defense system and make it more efficient in reducing oxidative stress in plants. Nanosilica can trigger a signaling cascade at the cellular level, resulting in genetic modifications and the overexpression of genes related to the antioxidant defense system, which translates into greater production of enzymatic and non-enzymatic antioxidants.

References

Ahmadi-Nouraldinvand F, Sharifi RS, Siadat SA, Khalilzadeh R (2023) Reduction of salinity stress in wheat through seed bio-priming with mycorrhiza and growth-promoting bacteria and its effect on physiological traits and plant antioxidant activity with silicon nanoparticles application. Silicon. https://doi.org/10.1007/s12633-023-02552-x

Ahmed S, Iqbal M, Ahmad Z, Iqbal MA, Artyszak A, Sabagh AEL, Alharby HF, Hossain A (2023) Foliar application of silicon-based nanoparticles improve the adaptability of maize (Zea mays L.) in cadmium contaminated soils. Environ Sci Pollut Res 30:41002–41013. https://doi.org/10.1007/s11356-023-25189-0

Akter S, Khan MS, Smith EN, Flashman E (2021) Measuring ROS and redox markers in plant cells. RSC Chem Biol 2:1384–1401. https://doi.org/10.1039/D1CB00071C

Alharbi K, Osman HS, Rashwan E, Hafez EM, Omara AED (2022) Stimulating the growth, anabolism, antioxidants, and yield of rice plants grown under salt stress by combined application of bacterial inoculants and Nano-silicon. Plan Theory 11. https://doi.org/10.3390/plants11243431

Al-Mokadem AZ, Sheta MH, Mancy AG, Hussein HAA, Kenawy SKM, Sofy AR, Abu-Shahba MS, Mahdy HM, Sofy MR, Al Bakry AF, Agha MS (2023) Synergistic effects of kaolin and silicon nanoparticles for ameliorating deficit irrigation stress in maize plants by upregulating antioxidant defense systems. Plan Theory 12:1–24. https://doi.org/10.3390/plants12112221

Almutairi KF, Górnik K, Awad RM, Ayoub A, Abada HS, Mosa WFA (2023) Influence of selenium, titanium, and silicon nanoparticles on the growth, yield, and fruit quality of mango under drought conditions. Horticulturae 9:1231. https://doi.org/10.3390/horticulturae9111231

Al-Selwey WA, Alsadon AA, Alenazi MM, Tarroum M, Ibrahim AA, Ahmad A, Osman M, Seleiman MF (2023a) Morphological and biochemical response of potatoes to exogenous application of ZnO and SiO_2 nanoparticles in a water deficit environment. Horticulturae 9. https://doi.org/10.3390/horticulturae9080883

Al-Selwey WA, Alsadon AA, Ibrahim AA, Labis JP, Seleiman MF (2023b) Effects of zinc oxide and silicon dioxide nanoparticles on physiological, yield, and water use efficiency traits of potato grown under water deficit. Plan Theory 12. https://doi.org/10.3390/plants12010218

Al-Shammari WB, Altamimi HR, Abdelaal K (2023) Improvement in physiobiochemical and yield characteristics of pea plants with nano silica and melatonin under salinity stress conditions. Horticulturae 9. https://doi.org/10.3390/horticulturae9060711

Alves DMR, de Oliveira JN, de Mello Prado R, Ferreira PM (2023) Silicon in the form of nanosilica mitigates P toxicity in scarlet eggplant. Sci Rep 13. https://doi.org/10.1038/s41598-023-36412-w

Anand V, Pandey A (2023) Synthesis and characterization of CeO_2 and SiO_2 nanoparticles and their effect on growth parameters and the antioxidant defense system in Vigna mungo L. Hepper. Environ Sci Pollut Res 30:100814–100827. https://doi.org/10.1007/s11356-023-29415-7

Arif Y, Singh P, Bajguz A, Alam P, Hayat S (2021) Silicon mediated abiotic stress tolerance in plants using physio-biochemical, omic approach and cross-talk with phytohormones. Plant Physiol Biochem 166:278–289. https://doi.org/10.1016/j.plaphy.2021.06.002

Azameti MK, Imoro A-WM (2023) Nanotechnology: a promising field in enhancing abiotic stress tolerance in plants. Crop Design 2:100037. https://doi.org/10.1016/j.cropd.2023.100037

Badawy SA, Zayed BA, Bassiouni SMA, Mahdi AHA, Majrashi A, Ali EF, Seleiman MF (2021) Influence of nano silicon and nano selenium on root characters, growth, ion selectivity, yield, and yield components of rice (Oryza sativa l.) under salinity conditions. Plan Theory 10. https://doi.org/10.3390/plants10081657

Barceló J, Ismail LM, Soliman MI, Abd El-Aziz MH, Abdel-Aziz HMM (2022) Impact of silica ions and nano silica on growth and productivity of pea plants under salinity stress. Plants (Basel). https://doi.org/10.3390/plants

Bayat H, Shahraki B, Aminifard MH, Azarmi-Atajan F (2023) Alleviating adverse effects of salt stress in pot marigold (Calendula officinalis L.) by foliar spray of silicon and nano-silicon under greenhouse and field conditions. J Horticult Postharvest Res 6:287–298. https://doi.org/10.22077/JHPR.2023.6307.1315

Behl T, Kaur I, Sehgal A, Singh S, Sharma N, Bhatia S, Al-Harrasi A, Bungau S (2022) The dichotomy of nanotechnology as the cutting edge of agriculture: Nano-farming as an asset versus nanotoxicity. Chemosphere 288:132533. https://doi.org/10.1016/J.CHEMOSPHERE.2021.132533

Bela K, Riyazuddin R, Csiszár J (2022) Plant glutathione peroxidases: non-Heme peroxidases with large functional flexibility as a core component of ROS-processing mechanisms and signalling. Antioxidants 11:1624. https://doi.org/10.3390/antiox11081624

Ben Y, Cheng M, Liu Y, Wang X, Wang L, Yang Q, Huang X, Zhou Q (2023) Biomarker changes and oxidative damage in living plant cells as new biomonitoring indicators for combined heavy metal stress assessment. Ecol Indic 154:110784. https://doi.org/10.1016/j.ecolind.2023.110784

Bericat-Vadell R, Sekar P, Patehebieke Y, Zou X, Kaul N, Broqvist P, Lindblad R, Lindblad A, Arkhypchuk A, Walletin C-J, Sa J (2023) Single-electron transfer reactions on surface-modified gold plasmons. Mater Today Chem 34:101783. https://doi.org/10.1016/j.mtchem.2023.101783

Bhardwaj S, Verma T, Raza A, Kapoor D (2023) Silicon and nitric oxide-mediated regulation of growth attributes, metabolites and antioxidant defense system of radish (Raphanus sativus L.) under arsenic stress. Phyton-Int J Exp Bot 92:763–782. https://doi.org/10.32604/phyton.2023.025672

Bhat JA, Rajora N, Raturi G, Sharma S, Dhiman P, Sanand S, Shivaraj SM, Sonah H, Deshmukh R (2021) Silicon nanoparticles (SiNPs) in sustainable agriculture: major emphasis on the practicality, efficacy and concerns. Nanoscale Adv. https://doi.org/10.1039/d1na00233c

Biju S, Fuentes S, Gupta D (2023) Novel insights into the mechanism(s) of silicon-induced drought stress tolerance in lentil plants revealed by RNA sequencing analysis. BMC Plant Biol 23:1–29. https://doi.org/10.1186/s12870-023-04492-5

Bonciolini S, Noël T, Capaldo L (2022) Synthetic applications of photocatalyzed halogen-radical mediated hydrogen atom transfer for C-H bond functionalization. Eur J Org Chem 202200417:24–44. https://doi.org/10.1002/ejoc.202200417

Boora R, Sheoran P, Rani N, Kumari S, Thakur R, Grewal S (2023) Biosynthesized silica nanoparticles (Si NPs) helps in mitigating drought stress in wheat through physiological changes and upregulation of stress genes. Silicon 15:5565–5577. https://doi.org/10.1007/s12633-023-02439-x

Carvalho LT d S, de Prado RM, Da Silva JLF, Ferreira PM, Antonio RI (2022) Impact of nanotechnology from Nanosilica to mitigate N and P deficiencies favoring the sustainable cultivation of sugar beet. Nanomaterials 12. https://doi.org/10.3390/nano12224038

Chen S, Guo X, Zhang B, Nie D, Rao W, Zhang D, Lü J, Guan X, Chen Z, Pan X (2023) Mesoporous silica nanoparticles induce intracellular peroxidation damage of phytophthora infestans: a new type of green fungicide for late blight control. Environ Sci Technol 57:3980–3989. https://doi.org/10.1021/ACS.EST.2C07182/SUPPL_FILE/ES2C07182_SI_001.PDF

Corpas FJ, González-gordo S, Palma JM (2020) Plant peroxisomes: a factory of reactive species. Front Plant Sci 11:1–12. https://doi.org/10.3389/fpls.2020.00853

Couttolenc A, Díaz-Porras Á, Espinoza C, Medina ME, Trigos Á (2020) On the primary and secondary antioxidant activity from hydroxy-methylcoumarins: experimental and theoretical studies. J Phys Org Chem 33:1–10. https://doi.org/10.1002/poc.4025

de Alché JD (2019) A concise appraisal of lipid oxidation and lipoxidation in higher plants. Redox Biol 23:101136. https://doi.org/10.1016/j.redox.2019.101136

de Faria Melo CC, Amaral DS, de Moura Zanine A, de Jesus Ferreira D, de Mello Prado R, de Cássia Piccolo M (2023) Nanosilica enhances morphogenic and chemical parameters of

Megathyrsus maximus grass under conditions of phosphorus deficiency and excess stress in different soils. BMC Plant Biol 23. https://doi.org/10.1186/s12870-023-04521-3

Dumanović J, Nepovimova E, Natić M, Kuča K, Jaćević V (2021) The significance of reactive oxygen species and antioxidant defense system in plants: a concise overview. Front Plant Sci 11. https://doi.org/10.3389/fpls.2020.552969

El-Flaah RF, El-Said RAR, Nassar MA, Hassan M, Abdelaal K (2021) Effect of Rhizobium, nano silica and ascorbic acid on morpho-physiological characters and gene expression of POX and PPO in faba bean (Vicia faba L.) under salinity stress conditions. Fresenius Environ Bull 30:5751–5764

El-Ramady H, Abdalla N, Sári D, Ferroudj A, Muthu A, Prokisch J, Fawzy ZF, Brevik EC, Solberg S (2023) Nanofarming: promising solutions for the future of the global agricultural industry. Agronomy. https://doi.org/10.3390/agronomy13061600

Emamverdian A, Ding Y, Hasanuzzaman M, Barker J, Liu G, Li Y, Mokhberdoran F (2023) Insight into the biochemical and physiological mechanisms of nanoparticles-induced arsenic tolerance in bamboo. Front Plant Sci 14:1–19. https://doi.org/10.3389/fpls.2023.1121886

Faisal M, Shahbaz M, Khalid F, Rasheed Y (2023) Plant stress biogenic nanoparticles application in agriculture for ROS mitigation and abiotic stress tolerance: a review ☆. Plant Stress 10:100281. https://doi.org/10.1016/j.stress.2023.100281

Faizan M, Alam P, Rajput VD, Faraz A, Afzal S, Ahmed SM, Yu F-Y, Minkina T, Hayat S (2023) Nanoparticle mediated plant tolerance to heavy metal stress: what we know? Sustain For 15:1446. https://doi.org/10.3390/su15021446

Fan N, Zhao C, Chang Z, Yue L, He F, Xiao Z, Wang Z (2023) Silicon quantum dots promote radish resistance to root herbivores without impairing rhizosphere microenvironment health. Environ Sci Nano 10:2232–2244. https://doi.org/10.1039/d3en00368j

Fatemi H, Behrooz Pour E, Rizwan M (2020) Foliar application of silicon nanoparticles affected the growth, vitamin C, flavonoid, and antioxidant enzyme activities of coriander (Coriandrum sativum L.) plants grown in lead (Pb)-spiked soil. Environ Sci Pollut Res Int. https://doi.org/10.1007/s11356-020-10549-x/Published

Felisberto G, De Mello Prado R, Raimundo Lima De Oliveira L, Aparecida P, Felisberto C (2021) Are nanosilica, potassium silicate and new soluble sources of silicon effective for silicon foliar application to soybean and rice plants? Silicon. https://doi.org/10.1007/s12633-020-00668-y/Published

Fu C, Khan MN, Yan J, Hong X, Zhao F, Chen L, Ma H, Li Y, Li J, Wu H (2023) Mechanisms of nanomaterials for improving plant salt tolerance. Crop Environ. https://doi.org/10.1016/j.crope.2023.03.002

Galano A, Mazzone G, Alvarez-Diduk R, Marino T, Alvarez-Idaboy JR, Russo N (2016) Food antioxidants: chemical insights at the molecular level. Annu Rev Food Sci Technol 7:335–352. https://doi.org/10.1146/annurev-food-041715-033206

Gao M, Chang J, Wang Z, Zhang H, Wang T (2023) Advances in transport and toxicity of nanoparticles in plants. J Nanobiotechnol. https://doi.org/10.1186/s12951-023-01830-5

García-Díez G, Mora-Diez N (2020) Theoretical study of the iron complexes with Aminoguanidine: investigating secondary antioxidant activity. Antioxidants 9:756. https://doi.org/10.3390/antiox9080756

Garza-Alonso CA, Olivares-Sáenz E, González-Morales S, Cabrera-De la Fuente M, Juárez-Maldonado A, González-Fuentes JA, Tortella G, Valdés-Caballero MV, Benavides-Mendoza A (2022) Strawberry biostimulation: from mechanisms of action to plant growth and fruit quality. Plan Theory 11:1–34. https://doi.org/10.3390/plants11243463

Golkar P, Akbari R, Bazarganipour M, Javed R (2023) Biochemical and phytochemical responses of Ammi visnaga L. (Apiaceae) callus culture elicited by SiO2 and graphene Oxide-SiO2 nanoparticles. Plant Physiol Biochem 200. https://doi.org/10.1016/j.plaphy.2023.107741

Guerriero G, Sutera FM, Hoffmann J, Leclercq CC, Planchon S, Berni R, Hausman JF, Renaut J, Torabi-Pour N, Pennington HC, Ahmed M, Welsh M, Dehsorkhi A, Saffie-Siebert S (2023)

Nanoporous quercetin-loaded silicon-stabilized hybrid lipid nanoparticles alleviate salt stress in tomato plants. ACS Appl Nano Mater 6:3647–3660. https://doi.org/10.1021/acsanm.2c05384

Hao Y, Yu Y, Sun G, Gong X, Jiang Y, Lv G, Zhang Y, Li L, Zhao Y, Sun D, Gu W, Qian C (2023) Effects of multi-walled carbon nanotubes and Nano-silica on root development, leaf photosynthesis, active oxygen and nitrogen metabolism in maize. Plan Theory 12. https://doi.org/10.3390/plants12081604

Hasanuzzaman M, Bhuyan MHMB, Anee TI, Parvin K, Nahar K, Al Mahmud J, Fujita M (2019) Regulation of ascorbate-glutathione pathway in mitigating oxidative damage in plants under abiotic stress. Antioxidants 8:384. https://doi.org/10.3390/antiox8090384

Hasanuzzaman M, Bhuyan MHM, Zulfiqar F, Raza A, Mohsin S, Mahmud J, Fujita M, Fotopoulos V (2020) Reactive oxygen species and antioxidant defense in plants under abiotic stress: revisiting the crucial role of a universal defense regulator. Antioxidants 9:1–52. https://doi.org/10.3390/antiox9080681

Hossain MS, Sohag MMH, Fardus J, Rob MM, Fujita M (2023) Acetate alleviates As toxicity via improving ROS metabolism and antioxidant defense system in lentil seedlings. Plant Stress 9:100190. https://doi.org/10.1016/j.stress.2023.100190

Huang D, Jing G, Zhang L, Chen C, Zhu S (2021) Interplay among hydrogen sulfide, nitric oxide, reactive oxygen species, and mitochondrial DNA oxidative damage. Front Plant Sci 12:1–10. https://doi.org/10.3389/fpls.2021.701681

Ismail LM, Soliman MI, Abd El-Aziz MH, Abdel-Aziz HMM (2022) Impact of silica ions and Nano silica on growth and productivity of pea plants under salinity stress. Plan Theory 11:1–21. https://doi.org/10.3390/plants11040494

Jalil S, Nazir MM, AL-Huqail AA, Ali B, Al-Qthanin RN, Asad MAU, Eweda MA, Zulfiqar F, Onursal N, Masood HA, Yong JWH, Jin X (2023) Silicon nanoparticles alleviate cadmium toxicity in rice (Oryza sativa L.) by modulating the nutritional profile and triggering stress-responsive genetic mechanisms. Ecotoxicol Environ Saf 268:115699. https://doi.org/10.1016/j.ecoenv.2023.115699

Jelodarian S, Abdossi V, Zarrinnia V, Talaie A, Larijani K (2023) Application of Si and Ag green nanoparticles, Epibrassinolide, and methyl Jasmonate causes delay in decay of Malus Domestica fruits via improving postharvest physiology at ambient conditions. Agric Res. https://doi.org/10.1007/s40003-023-00675-6

Juárez-Maldonado A, Ortega-Ortiz H, González-Morales S, Morelos-Moreno Á, Cabrera-de la Fuente M, Sandoval-Rangel A, Cadenas-Pliego G, Benavides-Mendoza A (2019) Nanoparticles and nanomaterials as plant biostimulants. Int J Mol Sci 20:1–19. https://doi.org/10.3390/ijms20010162

Juárez-Maldonado A, Tortella G, Rubilar O, Fincheira P, Benavides-Mendoza A (2021) Biostimulation and toxicity: the magnitude of the impact of nanomaterials in microorganisms and plants. J Adv Res. https://doi.org/10.1016/j.jare.2020.12.011

Kerchev PI, Van Breusegem F (2022) Improving oxidative stress resilience in plants. Plant J 109:359–372. https://doi.org/10.1111/tpj.15493

Khorobrykh S, Havurinne V, Mattila H (2020) Oxygen and ROS in Photosynthesis. Plants 9:91. https://doi.org/10.3390/plants9010091

Khudyakov IV (2023) Electronically excited states of free radicals. Physchem 3:332–341. https://doi.org/10.3390/physchem3030023

Kiany T, Pishkar L, Sartipnia N, Iranbakhsh A, Barzin G (2022) Effects of silicon and titanium dioxide nanoparticles on arsenic accumulation, phytochelatin metabolism, and antioxidant system by rice under arsenic toxicity. Environ Sci Pollut Res 29:34725–34737. https://doi.org/10.1007/s11356-021-17927-z

Koleva L, Umar A, Yasin NA, Shah AA, Siddiqui MH, Alamri S, Riaz L, Raza A, Javed T, Shabbir Z (2022) Iron oxide and silicon nanoparticles modulate mineral nutrient homeostasis and metabolism in cadmium-stressed Phaseolus vulgaris. Front Plant Sci 13. https://doi.org/10.3389/fpls.2022.806781

Kumar H, Dhalaria R, Guleria S, Cimler R, Sharma R, Siddiqui SA, Valko M, Nepovimova E, Dhanjal DS, Singh R, Kumar V, Pathera AK, Verma N, Kaur T, Manickam S, Alomar SY, Kuča K (2023) Anti-oxidant potential of plants and probiotic spp. in alleviating oxidative stress induced by H_2O_2. Biomed Pharmacother 165:115022. https://doi.org/10.1016/j.biopha.2023.115022

Lang Y, Gao N, Zang Z, Meng X, Lin Y, Yang S, Yang Y, Jin Z, Li B (2024) Classification and antioxidant assays of polyphenols: a review. J Future Foods 4:193–204. https://doi.org/10.1016/j.jfutfo.2023.07.002

Lee J, Han M, Shin Y, Lee J-M, Heo G, Lee Y (2023) How extracellular reactive oxygen species reach their intracellular targets in plants. Mol Cells 46:329–336. https://doi.org/10.14348/molcells.2023.2158

Li Y, Xi K, Liu X, Han S, Han X, Li G, Yang L, Ma D, Fang Z, Gong S, Yin J, Zhu Y (2023) Silica nanoparticles promote wheat growth by mediating hormones and sugar metabolism. J Nanobiotechnol 21:1–12. https://doi.org/10.1186/s12951-022-01753-7

Li X, Zhang J, Liu G, Wu G, Wang R, Zhang J (2024) High altitude hypoxia and oxidative stress: the new hope brought by free radical scavengers. Life Sci 336:122319. https://doi.org/10.1016/j.lfs.2023.122319

Liu C, Sun H, Xu Y, Wu C (2023) Effects of SiO2 nanoparticles on root structures, gas exchange, and antioxidant activities of Cunninghamia lanceolata seedlings under drought stress. J Plant Nutr 46:3771–3793. https://doi.org/10.1080/01904167.2023.2211617

Mansoor S, Ali Wani O, Lone JK, Manhas S, Kour N, Alam P, Ahmad A, Ahmad P (2022) Reactive oxygen species in plants: from source to sink. Antioxidants 11:225. https://doi.org/10.3390/antiox11020225

Manzoor N, Ali L, AL-Huqail AA, Alghanem SMS, Al-Haithloul HAS, Abbas T, Chen G, Huan L, Liu Y, Wang G (2023) Comparative efficacy of silicon and iron oxide nanoparticles towards improving the plant growth and mitigating arsenic toxicity in wheat (Triticum aestivum L.). Ecotoxicol Environ Saf 264. https://doi.org/10.1016/j.ecoenv.2023.115382

Martí-Guillén JM, Pardo-Hernández M, Martínez-Lorente SE, Almagro L, Rivero RM (2022) Redox post-translational modifications and their interplay in plant abiotic stress tolerance. Front Plant Sci 13:1–19. https://doi.org/10.3389/fpls.2022.1027730

Martínez A, Reina M (2017) Copper or free radical scavenger? Comput Theor Chem 1104:1–11. https://doi.org/10.1016/j.comptc.2017.01.033

Medrano-Macías J, Flores-Gallegos AC, Nava-Reyna E, Morales I, Tortella G, Solís-Gaona S, Benavides-Mendoza A (2022) Reactive oxygen, nitrogen, and sulfur species (RONSS) as a metabolic cluster for signaling and biostimulation of plants: an overview. Plan Theory 11:3203. https://doi.org/10.3390/plants11233203

Moradi P, Vafaee Y, Mozafari AA, Tahir NA-r (2022) Silicon nanoparticles and methyl Jasmonate improve physiological response and increase expression of stress-related genes in strawberry cv. Paros under salinity stress. Silicon 14:10559–10569. https://doi.org/10.1007/s12633-022-01791-8

Mukarram M, Khan MMA, Kurjak D, Lux A, Corpas FJ (2023) Silicon nanoparticles (SiNPs) restore photosynthesis and essential oil content by upgrading enzymatic antioxidant metabolism in lemongrass (Cymbopogon flexuosus) under salt stress. Front Plant Sci 14:1–18. https://doi.org/10.3389/fpls.2023.1116769

Okeke ES, Nweze EJ, Ezike TC, Nwuche CO, Ezeorba TPC, Nwankwo CEI (2023) Silicon-based nanoparticles for mitigating the effect of potentially toxic elements and plant stress in agroecosystems: a sustainable pathway towards food security. Sci Total Environ. https://doi.org/10.1016/j.scitotenv.2023.165446

Othman EZ, El-Attar AB, El-Bahbohy RM, Abd El-Khalek SN, Morgan SH, Mahmoud AWM (2023) Exogenous appliance of nano-zeolite and Nano-silicon elevate Solidago canadensis invasive plant tolerance to water deficiency. Horticulturae 9. https://doi.org/10.3390/horticulturae9020172

Peng H, Hu H, Xi K, Zhu X, Zhou J, Yin J, Guo F, Liu Y, Zhu Y (2022) Silicon nanoparticles enhance ginger rhizomes tolerance to postharvest deterioration and resistance to Fusarium solani. Front Plant Sci 13. https://doi.org/10.3389/fpls.2022.816143

Phaniendra A, Jestadi DB, Periyasamy L (2015) Free radicals: properties, sources, targets, and their implication in various diseases. Indian J Clin Biochem 30:11–26. https://doi.org/10.1007/s12291-014-0446-0

Posso DA, Da-Silva CJ, Shimoia EP, da Martins TS, Reissig GN, de Oliveira ACB, Borella J, van Dongen JT, do Amarante L (2023) Root-hypoxia tolerance in soybean sister-lines plants indicates a better balance in energy use/dissipation and oxidative stress control. Plant Stress 10:100225. https://doi.org/10.1016/j.stress.2023.100225

Prasad TNVKV, Swethasree M, Satisha GC, Nirmal Kumar AR, Sudhakar P, Ravindra Reddy B, Saritha M, Sabitha N, Bhaskar Reddy BV, Rajasekhar P, Prasanthi L, Girish BP, Roy Choudhury S (2023) Nanoparticulate silica internalization and its effect on the growth and yield of groundnut (Arachis hypogaea L.). Environ Sci Technol 57:5881–5890. https://doi.org/10.1021/acs.est.3c00327

Rabbani N, Al-Motawa M, Thornalley PJ (2020) Protein glycation in plants—an under-researched field with much still to discover. Int J Mol Sci 21:3942. https://doi.org/10.3390/ijms21113942

Rahman MM, Byanju B, Grewell D, Lamsal BP (2020) High-power sonication of soy proteins: hydroxyl radicals and their effects on protein structure. Ultrason Sonochem 64:105019. https://doi.org/10.1016/j.ultsonch.2020.105019

Rashad YM, El-Sharkawy HHA, Belal BEA, Abdel Razik ES, Galilah DA (2021) Silica nanoparticles as a probable anti-oomycete compound against downy mildew, and yield and quality enhancer in grapevines: field evaluation, molecular, physiological, ultrastructural, and toxicity investigations. Front Plant Sci 12. https://doi.org/10.3389/fpls.2021.763365

Rasouli F, Yun P, Kiani-Pouya A, Movahedi A, Rasouli M, Salehi M, Shabala S (2024) One size does not fit all: different strategies employed by triticale and barley plants to deal with soil salinity. Environ Exp Bot 218:105585. https://doi.org/10.1016/j.envexpbot.2023.105585

Raza MAS, Zulfiqar B, Iqbal R, Muzamil MN, Aslam MU, Muhammad F, Amin J, Aslam HMU, Ibrahim MA, Uzair M, Habib-ur-Rahman M (2023) Morpho-physiological and biochemical response of wheat to various treatments of silicon nano-particles under drought stress conditions. Sci Rep 13:1–13. https://doi.org/10.1038/s41598-023-29784-6

Riaz M, Zhao S, Kamran M, Ur Rehman N, Mora-Poblete F, Maldonado C, Hamzah Saleem M, Parveen A, Ahmed Al-Ghamdi A, Al-Hemaid FM, Ali S, Elshikh MS (2022) Effect of nano-silicon on the regulation of ascorbate-glutathione contents, antioxidant defense system and growth of copper stressed wheat (Triticum aestivum L.) seedlings. Front Plant Sci 13. https://doi.org/10.3389/fpls.2022.986991

Riaz M, Kamran M, Fahad S, Wang X (2023) Nano-silicon mediated alleviation of Cd toxicity by cell wall adsorption and antioxidant defense system in rice seedlings. Plant Soil 486:103–117. https://doi.org/10.1007/s11104-022-05588-x

Rodrigues de Queiroz A, Hines C, Brown J, Sahay S, Vijayan J, Stone JM, Bickford N, Wuellner M, Glowacka K, Buan NR, Roston RL (2023) The effects of exogenously applied antioxidants on plant growth and resilience. Phytochem Rev 22:407–447. https://doi.org/10.1007/s11101-023-09862-3

Sadowska-Bartosz I, Bartosz G (2024) The cellular and organismal effects of Nitroxides and Nitroxide-containing nanoparticles. Int J Mol Sci 25:1446. https://doi.org/10.3390/ijms25031446

Salman S, Shahzad L, Khan WUD, Riaz U, Sharif F, Tahir A, Hayyat MU, Hussain A, Tariq Saeed M (2023) Influence of silicon doped biochar on germination and defense mechanisms of pea (Pisum sativum L.) under copper and salinity stresses. J Plant Nutr 46:3933–3953. https://doi.org/10.1080/01904167.2023.2220720

Sandalio LM, Collado-Arenal AM, Romero-Puertas MC (2023) Deciphering peroxisomal reactive species interactome and redox signalling networks. Free Radic Biol Med 197:58–70. https://doi.org/10.1016/j.freeradbiomed.2023.01.014

Sayed EG, Mahmoud AWM, El-Mogy MM, Ali MAA, Fahmy MAM, Tawfic GA (2022) The effective role of nano-silicon application in improving the productivity and quality of grafted tomato grown under salinity stress. Horticulturae 8. https://doi.org/10.3390/horticulturae8040293

Şener S, Sayğı H, Duran CN (2023) Responses of in vitro strawberry plants to drought stress under the influence of Nano-silicon dioxide. Sustain For 15:15569. https://doi.org/10.3390/su152115569

Sharma S, Gupta S, Jain R, Kothari SL, Kachhwaha S (2023) SiO2 nanoparticles as elicitor for increased rebaudioside-A in Stevia rebaudiana micropropagated in solid and liquid cultures: a comparative study. Plant Cell Tissue Organ Cult:541–552. https://doi.org/10.1007/s11240-023-02604-5

Siddeeg A, AlKehayez NM, Abu-Hiamed HA, Al-Sanea EA, AL-Farga AM (2021) Mode of action and determination of antioxidant activity in the dietary sources: an overview. Saudi J Biol Sci 28:1633–1644. https://doi.org/10.1016/j.sjbs.2020.11.064

Singh D (2022) Juggling with reactive oxygen species and antioxidant defense system – a coping mechanism under salt stress. Plant Stress 5:100093. https://doi.org/10.1016/j.stress.2022.100093

Sousa RHV, Carvalho FEL, Lima-melo Y, Alencar VTCB (2019) Impairment of peroxisomal APX and CAT activities increases protection of photosynthesis under oxidative stress. J Exp Bot 70:627–639. https://doi.org/10.1093/jxb/ery354

Spiegel M, Marino T, Prejanò M, Russo N (2022) Primary and secondary antioxidant properties of scutellarin and scutellarein in water and lipid-like environments : a theoretical investigation. J Mol Liq 366:120343. https://doi.org/10.1016/j.molliq.2022.120343

Stoia M, Oancea S (2022) Low-molecular-weight synthetic antioxidants: classification, pharmacological profile, effectiveness and trends. Antioxidants 11:638. https://doi.org/10.3390/antiox11040638

Subramanian H, Sibi MP (2023) Stereoselective hydrogen atom transfer in free radical reactions. Asian Chem Editor Soc 12:1–22. https://doi.org/10.1002/ajoc.202300175

Sun H, He S, Liu T, Zhang Q, Yu J, Gao Y, Wang X (2023) Alleviation of cadmium toxicity by Nano-silicon dioxide in Momordica charantia L. Seedlings. J Soil Sci Plant Nutr 23:1060–1069. https://doi.org/10.1007/S42729-022-01103-4/METRICS

Sun Y, Xie X, Jiang C-J (2024) Antioxidant agriculture for stress-resilient crop production: field practice. Antioxidants 13:164. https://doi.org/10.3390/antiox13020164

Tao N, Zhang W, Si L, Zhang R, Wang D, Guo C (2022) Effects of aniline on growth, oxidative and DNA damage of rice (Oryza sativa L.) seedlings. Environ Technol Innov 28:102583. https://doi.org/10.1016/j.eti.2022.102583

Tavakoli F, Rafieiolhossaini M, Ravash R (2021) Effects of PEG and Nano-Silica elicitors on secondary metabolites production in Crocus sativus L. Russ J Plant Physiol 68:931–940. https://doi.org/10.1134/S1021443721050216

Tripathi D, Oldenburg DJ, Bendich AJ (2023) Ribonucleotide and R-loop damage in plastid DNA and mitochondrial DNA during maize development. Plan Theory 12:3161. https://doi.org/10.3390/plants12173161

Ulhassan Z, Yang S, He D, Khan AR, Salam A, Azhar W, Muhammad S, Ali S, Hamid Y, Khan I, Sheteiwy MS, Zhou W (2023) Seed priming with nano-silica effectively ameliorates chromium toxicity in Brassica napus. J Hazard Mater 458. https://doi.org/10.1016/j.jhazmat.2023.131906

Ullah H, Khan I, Mustafa G, Shehzad J, Hasan M, Saeed U, Chaudhari SK (2023) Molecular characterization and Mitigative role of silicon dioxide nanoparticles in Ocimum Basilicum under Lead (Pb) stress. Silicon 15:2551–2565. https://doi.org/10.1007/s12633-022-02178-5

Van Hoang C, Thoai DN, Cam NTD, Phuong TTT, Lieu NT, Hien TTT, Nhiem DN, Pham TD, Tung MHT, Tran NTT, Mechler A, Vo QV (2022) Large-scale synthesis of Nanosilica from silica sand for plant stimulant applications. ACS Omega 7:41687–41695. https://doi.org/10.1021/acsomega.2c05760

Varatharajan K, Pushparani DS (2018) Screening of antioxidant additives for biodiesel fuels. Renew Sust Energ Rev 82:2017–2028. https://doi.org/10.1016/j.rser.2017.07.020

Verma KK, Song XP, Singh M, Huang HR, Bhatt R, Xu L, Kumar V, Li YR (2022a) Influence of nanosilicon on drought tolerance in plants: an overview. Front Plant Sci. https://doi.org/10.3389/fpls.2022.1014816

Verma KK, Zeng Y, Song XP, Singh M, Wu KC, Rajput VD, Li YR (2022b) Nanosilicon: an approach for abiotic stress mitigation and sustainable agriculture. Front Plant Sci. https://doi.org/10.3389/fpls.2022.1025974

Wang J, Cao X, Wang C, Xiao Z, Chen F, Feng Y, Yue L, Wang Z, Xing B (2023a) Transcriptomics and metabolomics reveal the mechanisms of enhanced constitutive resistance in rice (Oryza Sativa L.) by silica nanomaterials. Environ Sci Nano 10:2831–2848. https://doi.org/10.1039/d3en00504f

Wang X, Xie H, Wang P, Yin H (2023b) Nanoparticles in plants: uptake, transport and physiological activity in leaf and root. Materials. https://doi.org/10.3390/ma16083097

Xu W, Miao Y, Kong J, Lindsey K, Zhang X, Min L (2023a) ROS signaling and its involvement in abiotic stress with emphasis on heat stress-driven anther sterility in plants. Crop Environ. https://doi.org/10.1016/j.crope.2023.12.002

Xu X, Guo Y, Hao Y, Cai Z, Cao Y, Fang W, Zhao B, Haynes CL, White JC, Ma C (2023b) Nanosilicon fertiliser increases the yield and quality of cherry radish. Mod Agric. https://doi.org/10.1002/moda.19

Yalcinkaya T, Uzilday B, Ozgur R, Turkan I (2019) The roles of reactive carbonyl species in induction of antioxidant defence and ROS signalling in extreme halophytic model Eutrema parvulum and glycophytic model Arabidopsis thaliana. Environ Exp Bot 160:81–91. https://doi.org/10.1016/j.envexpbot.2019.01.009

Zahedi SM, Moharrami F, Sarikhani S, Padervand M (2020) Selenium and silica nanostructure-based recovery of strawberry plants subjected to drought stress. Sci Rep 10. https://doi.org/10.1038/s41598-020-74273-9

Zahedi SM, Hosseini MS, Fahadi Hoveizeh N, Kadkhodaei S, Vaculík M (2023) Comparative morphological, physiological and molecular analyses of drought-stressed strawberry plants affected by SiO2 and SiO2-NPs foliar spray. Sci Hortic 309. https://doi.org/10.1016/j.scienta.2022.111686

Zaheer S, Shehzad J, Chaudhari SK, Hasan M, Mustafa G (2023) Morphological and biochemical responses of Vigna radiata L. seedlings towards green synthesized SiO2 NPs. Silicon 15:5925–5936. https://doi.org/10.1007/s12633-023-02470-y

Zeppilli D, Aldinio-colbachini A, Ribaudo G, Tubaro C, Tiezza MD, Bortoli M, Zagotto G, Orian L (2023) Antioxidant chimeric molecules: are chemical motifs additive? The case of a selenium-based ligand. Int J Mol Sci 24. https://doi.org/10.3390/ijms241411797

Zhang Y, Shang C, Sun C, Wang L (2023a) Understanding prominent effects of the intramolecular hydrogen bond on the photophysical properties and antiradical abilities of six flavonoids. J Mol Liq 386:122534. https://doi.org/10.1016/j.molliq.2023.122534

Zhang Z-W, Fu Y-F, Yang X-Y, Yuan M, Zheng X-J, Luo X-F, Zhang M-Y, Xie L-B, Shu K, Reinbothe S, Reinbothe C, Wu F, Feng L-Y, Du J-B, Wang C-Q, Gao X-S, Chen Y-E, Zhang Y-Y, Li Y, Tao Q, Lan T, Tang X-Y, Zeng J, Chen G-D, Yuan S (2023b) Singlet oxygen induces cell wall thickening and stomatal density reducing by transcriptome reprogramming. J Biol Chem 299:105481. https://doi.org/10.1016/j.jbc.2023.105481

Zhao Y, Yu H, Zhou J-M, Smith SM, Li J (2020) Malate circulation: linking chloroplast metabolism to mitochondrial ROS. Trends Plant Sci 25:446–454. https://doi.org/10.1016/j.tplants.2020.01.010

Chapter 17
Benefits of Silicon on Plant Tolerance to Stress from Signaling with Phytohormones

Rafael Ferreira Barreto

Abstract Silicon (Si) is a non-essential element. However, it plays a crucial role in plant growth and productivity. Its application activates defense mechanisms, involving the modulation of hormones under biotic and abiotic stresses. In turn, plant hormones (auxins, gibberellins, cytokinins, ethylene, abscisic acid, brassinosteroids, jasmonates, salicylic acid, and strigolactones) contribute to coordinate several processes, including adaptation to biotic and abiotic stresses. The objective of this chapter is to present information on the interaction between the use of Si to alleviate abiotic and biotic stresses and modulation of plant hormones. Except of strigolactones, the interaction with Si is already established for other hormones under abiotic stresses (drought; salinity; cadmium, copper, ammonium, and arsenic toxicity; potassium, sulfur, and iron deficiency; chilling; heat; high pH; and submergence) and biotic stresses (mildew; brown spot fungus; bacterial wilt; pink stem borer; chewing herbivores; late blight; and anthracnose and clubroot disease). This chapter provides a comprehensive overview of hormonal regulation in relation to stress tolerance induced by Si application.

Keywords Abiotic stress · Biotic stress · Beneficial element · Plant hormones

17.1 Introduction

Published in 1994, the revision article "The anomaly of silicon in plant biology" (Epstein 1994) was a key turning point in research involving silicon (Si). Thereafter, there has been numerous publications seeking to understand the benefits of Si for plants (Coskun et al. 2019). In 2006, a group of authors published the first research on the importance of Si-mediated biotic and abiotic stresses involving hormonal regulation. In total, there are 30 publications on Si and abiotic stresses and eight on

R. F. Barreto (✉)
Federal University of Mato Grosso do Sul, Chapadão do Sul, MS, Brazil
e-mail: rafael.barreto@ufms.br

R. de Mello Prado et al. (eds.), *Silicon Advances for Sustainable Agriculture and Human Health*, Sustainable Plant Nutrition in a Changing World,
https://doi.org/10.1007/978-3-031-69876-7_17

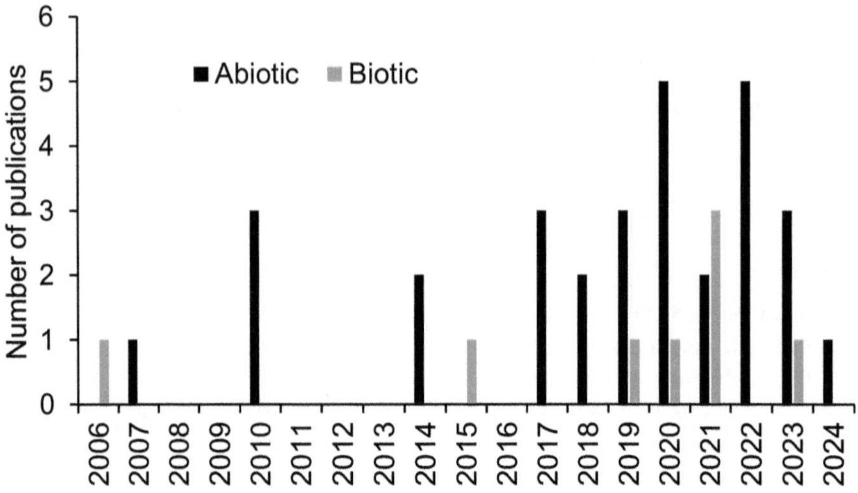

Fig. 17.1 Number of publications per year studying silicon to mitigate biotic and abiotic stresses evaluating plant hormones. The search was conducted on Web of Science using topics and the terms "silicon" and "hormone" or "phytohormone" or "phytohormonal". The document type was articles

Si and biotic stresses evaluating plant hormones (Fig. 17.1). Additionally, there are literature reviews on the topic (Arif et al. 2021; Khan et al. 2023a; Thakur et al. 2023).

Nine main hormones regulate plant development: auxins, gibberellins, cytokinins, ethylene, abscisic acid, brassinosteroids, jasmonates, salicylic acid, and strigolactones (Monshausen and Peer 2017). As signaling molecules, they act at very low concentrations. Through signaling pathways, they contribute to coordinate various processes, including adaptation to biotic and abiotic stresses (Semeradova et al. 2020). Thus, the objective of the chapter is to present information on the interaction between the use of Si to alleviate abiotic (Table 17.1) and biotic (Table 17.2) stresses and each plant hormone. This chapter, however, does not address the use of Si in normal growth conditions (no stress) and the application of plant regulators.

17.2 Role of Si in Hormonal Regulation Under Stress Conditions

17.2.1 Auxin

Auxins are a class of plant hormones. The most common auxin in plants is indoleacetic acid (IAA). It is produced mainly in the apical meristems, both stems and roots, and transported to other locations in the plant. In the target cells, auxins participate in cell growth and act on cell elongation (Monshausen and Peer 2017).

Table 17.1 Hormonal regulation by Si when alleviating abiotic stress

Plant	Stress	Hormonal regulation by Si	Reference
Oryza sativa	Salinity	↑ GA1 and GA20	Hwang et al. (2007)
Glycine max	Drought	↓ GA1, JA, and AS; ↑GA4	Hamayun et al. (2010)
Glycine max	Salinity	↑ GA1, GA4, and SA; ↓ JA	Hamayun et al. (2010)
Glycine max	Salinity	↑ GA1, GA12, and GA20; ↓ ABA	Lee et al. (2010)
Oryza sativa	Cadmium and copper toxicity	↓ ABA, SA, and JA	Kim et al. (2014a)
Oryza sativa	Salinity	↑ SA and ABA; ↓ JA	Kim et al. (2014b)
Mangifera indica	Drought	↑ IAA, ↑ GA; ↑ CK; ↓ ABA	Helaly et al. (2017)
Hordeum vulgare	K deficiency	= iP and ABA	Hosseini et al. (2017)
Hordeum vulgare	Drought	↑ iP and ABA	Hosseini et al. (2017)
Zea mays	Chilling	↑ IAA, GA, ABA, SA, and JA	Moradtalab et al. (2018)
Hordeum vulgare	S deficiency	Root: ↑ IAA; ↓ ABA Shoot: = IAA and ABA	Maillard et al. (2018)
Glycyrrhiza uralensis	Salinity	↑ IAA, GA3, and ABA	Lang et al. (2019)
Solanum lycopersicum	High pH	= SA and ↓ ABA	Khan et al. (2019)
Solanum tuberosum	Salinity	↑ GA3 and ↓ ABA	Mahmoud et al. (2019)
Oryza sativa	S deficiency	↓ SA and ABA; = JA	Réthoré et al. (2020)
Phoenix dactylifera	Heat	↑ SA; ↓ ABA	Khan et al. (2020a)
Solanum lycopersicum	Heat	↓ SA and ABA	Khan et al. (2020b)
Cucumis sativus	Salinity	↓↑ IAA, ABA, and CK (changes with duration of stress); ↑ SA	Zhu et al. (2020)
Cucumis sativus	Fe deficiency	↑ IAA, CK; ↓ JA	Hernández-Apaolaza et al. (2020)
Triticum aestivum	Salinity	↓ ABA; ↑ JA	Zia et al. (2021)
Solanum lycopersicum	Ammonium toxicity	Shoot: ↑ tZ; = iP Root: ↓ tZ; = iP	Barreto et al. (2021)
Hordeum vulgare	Heat and drought	↑ IAA; ↓ ABA	Naz et al. (2022)
Solanum lycopersicum	Salinity	↑ GA3, ↓ ABA	Sayed et al. (2022)

(continued)

Table 17.1 (continued)

Plant	Stress	Hormonal regulation by Si	Reference
Oryza sativa	Submergence	↓ ET and GA; = ABA and ↑ JA	Pan et al. (2022)
Phoenix dactylifera	Arsenic toxicity	↓ SA and JA; ↑ ABA	Khan et al. (2022)
Prunus persica	Drought	↑ tZ, GA and IAA; ↓ ABA	Gao et al. (2022)
Elymus sibiricus	Drought	↑ IAA, CK, GA, BR e ET; ↓ ABA	Khan et al. (2023b)
Lens culinaris	Drought	↑ AUX; ↓ ET and JA	Biju et al. (2023)
Zea mays	Cadmium	↓ AS; ↑ JA and ABA	Shah et al. (2023)
Zea mays	Cadmium	↓ ABA, GA3 e AS; ↑IAA, JA e ZR	An et al. (2024)

↑: upregulation; ↓ downregulation; = no regulation

Table 17.2 Hormonal regulation by Si when alleviating biotic stress

Plant	Stress	Hormonal regulation by Si	Reference
Arabidopsis thaliana	Mildew (*Erysiphe cichoracearum*)	↑AS; ↑JA; ↑ET	Fauteux et al. (2006)
Oryza sativa	Brown spot fungus (*Cochliobolus miyabeanus*)	↓ ET	Van Bockhaven et al. (2015)
Solanum lycopersicum	Bacterial wilt (*Ralstonia solanacearum*)	Delay in the increase of ET and SA	Jiang et al. (2019)
Eleusine coracana	Pink stem borer (*Sesamia inferens*)	Genes encoding ↑ ET, SA and JA	Jadhao et al. (2020)
Brachypodium distachyon	Chewing herbivores	↓ JA	Johnson et al. (2021)
Solanum tuberosum	Late blight (*Phytophthora infestans*)	↑ ET, JA; ↓ SA	Xue et al. (2021)
Sorghum bicolor	Anthracnose (*Colletotrichum sublineolum*)	↑ expression of ET, JA, and SA marker genes	Resende et al. (2021)
Brassica napus	Clubroot disease (*Plasmodiophora brassicae*)	↓ AS, JA, ABA	Sarkar et al. (2023)

↑: upregulation; ↓ downregulation; = no regulation

Lack of water causes many physiological changes that limit plant growth. Some of the strategies to survive and maintain growth are to enlarge the root system attempting to obtain water, decrease water loss by closing stomata, and adjust the osmotic potential in the tissues. Many hormones play different roles in tolerance to water stress. Abscisic acid (ABA) is the best described in the literature (Jing et al. 2023). However, auxins are also modulated in response to drought stress, generally decreasing their levels (Farooq et al. 2009).

Researchers conducted a study with three mango cultivars in three consecutive harvests and reported that water stress resulted in a decrease in vegetative growth, productivity, fruit quality, and early flowering. These effects of water stress followed a decrease in leaf IAA concentration. However, the supply of Si via fertigation led to an increase in vegetative growth, productivity, fruit quality, and a normal flowering period. In addition, IAA levels were high (Helaly et al. 2017).

In the forage grass *Elymus sibiricus*, water deficiency resulted in a decrease in leaf IAA content. However, researchers supplied doses of Si via seed treatment and observed that the differences between the levels of the hormone in plants with and without stress decreased. In other words, Si onsets a tendency for hormonal levels to be closer to those observed in the control (Khan et al. 2023b).

In barley plants, drought and high temperatures, combined or isolated, decreased the IAA content. However, plants treated with Si showed IAA levels similar to those of the control (Naz et al. 2022). In peach seedlings, Si mitigated the symptoms of lack of water, and there was an increase in root IAA (Gao et al. 2022).

A study on differentially expressed genes (DEG) in lentil leaves analyzing the mechanisms of tolerance to drought stress induced by Si identified positively regulated DEG for auxins (Biju et al. 2023). Although the authors did not quantify auxins in roots, they suggested that Si may have triggered the upregulation of DEG related to the metabolism of this hormone by increasing the development of the plant's root system, which is vital for tolerance to drought.

Plants are frequently exposed to cold stress, including chilling (0–12 °C) and freezing (<0 °C). Cold inhibits plant growth, but freezing temperatures destroy cell membranes, causing cell death (Jing et al. 2023). In this context, during a 7-day cold stress period, IAA concentrations in maize plants decreased. However, after seed treatment with Si, hormonal concentrations were restored, comparable to non-stressed plants (Moradtalab et al. 2018).

Salt stress affects plants throughout their life cycle. It inhibits seed germination and modulates growth and development. There are three main effects of salt on plants: water deficit, ionic toxicity, and ionic imbalance. Plants employ a series of mechanisms to deal with salt stress, including the modulation of some plant hormone levels (Jing et al. 2023). In *Glycyrrhiza uralensis*, a plant in which roots and rhizomes are used as natural sweeteners and herbal medicines, saline stress resulted in a decrease in IAA, but there was an increase in the content of this hormone depending on the dose of Si and the season of evaluation of seedlings (Lang et al. 2019). In cucumber, saline stress changed the root IAA content, that is, it led to an increase and a decrease depending on the time of evaluation. However, Si made changes less extreme, and levels were closer to those of the control treatment (Zhu et al. 2020). In maize under cadmium toxicity, hormonal regulation was similar as that of saline stress, that is, there was a decrease in IAA content due to toxicity and an increase in the content of this hormone in plants that received Si (An et al. 2024).

Finally, in general, mitigating abiotic stresses mediated by Si follows an increase in auxin levels at levels similar as those of control conditions in different plant species.

17.2.2 Gibberellins

Gibberellins (GA) control several aspects of plant growth and development, including seed germination, stem elongation, leaf expansion, and flower and seed development. Among the more than 100 GA identified in plants, a small number of them, such as GA1 (shoot elongation), GA3 (seed germination), GA4 (converts into GA1), and GA7 (converts into GA3), act as active hormones (Yamaguchi 2008).

As observed for auxins, GA levels generally decrease due to lack of water (Farooq et al. 2009). Similarly, Si, which alleviates water deficiency, also leads to an increase in GA contents to levels similar as those of normal growth conditions, as reported for mango (Helaly et al. 2017), the forage grass *Elymus sibiricus* (Khan et al. 2023b), and peach seedlings (Gao et al. 2022). However, in soybeans, PEG-induced drought stress did not change leaf GA1 contents. When Si was used to mitigate the effects of drought stress, there was a decrease in GA1. Water stress resulted in GA4 decrease, but using Si resulted in GA4 increase (Hamayun et al. 2010). The authors also noted that salt stress caused a decrease in GA1 and GA4; however, Si led to an increase in the contents of both GAs.

Cold stress decreased GA concentration in maize, but Si seed treatment restored GA concentrations to levels similar as those of the control (Moradtalab et al. 2018). The results were similar for saline stress; for example, in *Glycyrrhiza uralensis*, saline stress resulted in a decrease in GA3, but there was an increase in the content of this hormone depending on Si dose and time of evaluation of seedlings (Lang et al. 2019). In potato and tomato, salt stress resulted in a decrease in GA3, but there was an increase in this hormone using Si (Mahmoud et al. 2019; Sayed et al. 2022). Furthermore, in maize plants, cadmium toxicity also led to a decrease in GA3, but Si mitigated this damage.

In submersion-stressed rice, Si treatment decreased leaf GA. This decrease was related to a high productivity because increased leaf GA accelerates chlorophyll degradation and senescence in submerged leaves (Pan et al. 2022).

17.2.3 Cytokinins

Cytokinins (CK) are a family of plant hormones that regulate multiple responses, such as cell division and nutrient transport. The search for natural plant CK led to the identification of *trans*-zeatin (*t*Z), isopentenyladenine (iP), and dihydrozeatin. There are unfavorable scenarios that affect CK content in plants, changing CK levels, which alter tolerance to stress (Ha et al. 2012).

Similar as IAA and GA contents, water stress also results in low CK contents, as reported for mango (Helaly et al. 2017), the forage grass *Elymus sibiricus* (Khan et al. 2023b), and peach seedlings (Gao et al. 2022). However, these authors also noted that Si attenuated this adverse effect by increasing CK levels and mitigating water stress. Regarding this, Si may have increased the expression of the IPT7 and ARR5 genes, which act in the biosynthesis of CK (Markovich et al. 2017).

Furthermore, researchers evaluated the interaction of Si with CK under other types of stress. For example, in barley leaves, K deficiency resulted in an increase in iP, and there was no difference in the concentration of this hormone with the supply of Si (Hosseini et al. 2017). In cucumber, saline stress caused changes in root CK content, that is, there were an increase and a decrease in levels depending on the time of evaluation. However, Si made the changes less extreme; the levels were closer as those of the control treatment (Zhu et al. 2020). In maize, cadmium toxicity caused a decrease in CK, but Si mitigated this adverse effect (An et al. 2024). Finally, in Micro-tom tomato, Si mitigated ammonium toxicity, followed by a decrease in *tZ* in roots and an increase in shoots. This suggests that Si may be involved in the transport of *tZ* (Barreto et al. 2021). However, the role of Si on tZ transporters is unknown.

17.2.4 Ethylene

Ethylene (ET) is not a plant hormone like the others described above, as it is a gaseous compound. It is produced in all plants and acts on the maturation of climacteric fruits, leaf abscission, and several other important physiological aspects (Monshausen and Peer 2017).

The first report on the hormonal regulation by Si under biotic stress conditions was powdery mildew inoculation. That study conducted a complete transcriptome analysis on a control and powdery mildew-stressed *Arabidopsis thaliana* plants with or without Si application. In general, genes related to ET biosynthesis upregulated in the presence of Si (Fauteux et al. 2006).

Whenever attacked by necrotrophic pathogens, that is, pathogens causing death of tissue and then feeding on it, plants trigger defense mechanisms, including increases in ET content (Van Bockhaven et al. 2015). In inoculated rice leaves, ET production by fungi was the main source of this hormone and an important determinant of the pathogen's virulence. Furthermore, the pathogen can interrupt ET signaling in the plant to hinder the signaling of defense mechanisms. In this context, Si alleviates damage caused by fungi and acts to deactivate the production of ET by fungi, compromising the virulence of the pathogen and favoring the signaling of plant defense mechanisms, resulting in turn in plant immunity induced by Si (Van Bockhaven et al. 2015).

In tomato plants inoculated with the bacterium *Ralstonia solanacearum*, which causes bacterial wilt, there was an increase in ET production on the first day after inoculation. The first symptoms of wilt occurred after 2 days, and all plants were completely wilted at 7 days. However, Si delayed the increase in ET production 7 days after inoculation; the first symptoms of wilting occurred at 4 days, and only 20–30% of plants were slightly wilted at 7 days (Jiang et al. 2019).

Regarding the hormonal pathways involved in Si-mediated resistance to potato late blight compared to a control, there was an approximately threefold increase in ET 15 minutes after Si application. It subsequently fell to a level similar to that of the control. This increase triggered other hormonal changes, that is, there was an increase in JA and a decrease in SA, which resulted in tolerance to potato late blight (Xue et al. 2021). In sorghum, the greatest resistance to anthracnose using Si occurred with an increased expression of ET marker genes (Resende et al. 2021). In millet attacked by stalk borer, the expression of genes encoding ET increased in the presence of Si 72 hours after infestation (Jadhao et al. 2020).

Researchers have also studied the interaction between Si and ET under abiotic stresses. For example, in the forage grass *Elymus sibiricus*, drought stress (45% of water retention capacity) resulted in a decrease in leaf ET content. However, Si mitigated this change (Khan et al. 2023b). However, Si can also act to reduce ethylene production, as reported by a study on DEG in lentil leaves (Biju et al. 2023). The authors suggested that the downregulation by Si of genes involved in ET production may have contributed to delayed senescence following upregulated genes for chlorophyll biosynthesis. Another study by the same group of researchers evidenced this fact, that is, Si supplementation effectively mitigated drought stress in lentil plants by regulating reactive oxygen species and maintaining chlorophyll pigments under drought (Biju et al. 2017).

Research on rice plants subjected to stress by submersion and mitigation of stress by Si at the hormonal level revealed that submersion caused an increase in ethylene in roots. However, in plants treated with Si, the increase was less pronounced. Thus, since Si negatively regulated ET synthesis, there were gains in productivity because ET production accelerates chlorophyll degradation and leaf senescence under submersion (Pan et al. 2022).

17.2.5 Abscisic Acid

Abscisic acid (ABA) is a growth inhibitor. It is produced under a wide variety of environmental stresses, including drought. All plants respond to drought and many other stresses by accumulating ABA (Farooq et al. 2009). Typically, ABA is responsible for defending plants against abiotic stresses because environmental conditions, such as drought, salinity, cold, heat stress, and wounding, trigger increases in ABA levels (Verma et al. 2016).

The research results for the interaction between Si and ABA revealed that the mitigation of stress by Si caused a decrease in the levels of this hormone; this was observed in mango crops (Helaly et al. 2017), the forage *Elymus sibiricus* (Khan et al. 2023b), peach seedlings (Gao et al. 2022), and barley (Naz et al. 2022).

Researchers have studied the interaction between Si and ABA under nutrient deficiency stress. In barley leaves, there was no difference in ABA concentration depending on K deficiency in the presence or absence of Si (Hosseini et al. 2017).

In rice, S deficiency resulted in increased leaf ABA content, but Si-treated plants did not show such an increase (Réthoré et al. 2020).

On the other hand, studies on the interaction between Si and toxicity of chemical elements revealed different regulation patterns. In date palm under arsenic toxicity, Si increased ABA (Khan et al. 2022). In maize under cadmium toxicity, Si increased ABA in both shoots and roots (Shah et al. 2023). However, other research on maize under cadmium toxicity reported that Si resulted in a decrease in ABA (An et al. 2024).

Regarding salt stress, its attenuation by Si involves the regulation of ABA, as in *Glycyrrhiza uralensis*. Salt stress resulted in a decrease in ABA, but there was an increase in the content of this hormone depending on the dose of Si and the time of evaluation of seedlings (Lang et al. 2019). In potatoes under salt stress, there was a decrease in ABA with the use of Si (Mahmoud et al. 2019). In cucumber, salt stress changed root ABA contents, that is, there were an increase and a decrease depending on the time of evaluation. However, Si made the changes less extreme; the levels were closer to those of the control treatment (Zhu et al. 2020). In rice, alleviation of salt stress by Si resulted in increased ABA content (Kim et al. 2014b). Finally, in wheat and tomato, salt stress increased the ABA content, but there was a decrease using Si (Sayed et al. 2022; Zia et al. 2021). Although it is not possible to identify a well-defined pattern of increase or decrease in ABA with salt stress, its mitigation by Si results in the opposite of stress and contributes to a less limited plant development.

Researchers have studied the modulation of ABA by Si under other stresses. In the case of cold-stressed maize, there was a decrease in ABA concentration. However, seed treatment with Si restored ABA concentrations to levels similar as those of the control (Moradtalab et al. 2018). In tomato grown at pH 9, there was an increase in ABA in shoots, but Si resulted in ABA decrease (Xue et al. 2021). Other studies reported similar results also in tomatoes, that is, an increase in ABA caused by heat and a decrease in the levels of this hormone using Si (Khan et al. 2020b).

Research on the modulation of ABA by Si under conditions of biotic stress reported that canola plants with clubroot disease reduced gall formation and slowed infection progression in the presence of Si, accompanied by lower levels of ABA (Sarkar et al. 2023). The authors suggested that the reduction in gall formation in Si-treated plants led to mitigated drought stress in infected plants and, consequently, lower ABA.

17.2.6 Brassinosteroids

Brassinosteroids (BR) are a group of plant hormones crucial to many aspects of the plant's life. BR were originally characterized as for their function in cell elongation. However, it is becoming clear that they play important roles in plant growth, development, and response to various stresses such as extreme temperatures and drought (Nolan et al. 2020).

The only study assessing Si and BR on the forage *Elymus sibiricus* under water deficit (45% of water retention capacity) reported a decrease in leaf BR content. However, using different doses of Si via seed treatment, the differences between BR levels in plants with and without stress were smaller (Khan et al. 2023b).

17.2.7 Jasmonates

Jasmonates (JA) are plant hormones whose main bioactive form, a conjugate with the amino acid isoleucine (JA-Ile), is widely considered as one of the main signaling molecules that regulate plants' response and defense to stress. High levels of JA are related to low plant growth, presumably as a result of reallocation of carbon sources from primary growth to the biosynthesis of defense compounds (Nguyen et al. 2022).

In lentil, as expected, water deficiency positively regulated DEG. However, Si negatively regulated DEG (Biju et al. 2023). In this regard, its production is generally low in Si treatments. The decrease in JA may thus be correlated to low stress levels in Si treatments (Mir et al. 2022). Similarly, in soybeans, lack of water led to an increase in JA. Si attenuated this increase (Hamayun et al. 2010).

In addition to the increase in JA under water stress and its decrease using Si, this response pattern occurred under other stresses, as is the case of date palm under arsenic toxicity (Khan et al. 2022) and rice under salt stress (Kim et al. 2014b). However, studies have reported other responses. For example, in maize plants stressed by cold, there was a decrease in JA, but Si attenuated this damage (Moradtalab et al. 2018); in rice, S deficiency, in the presence or absence of Si, did not change the concentration of JA (Réthoré et al. 2020); in wheat, salt stress caused an increase in JA, and this increase was marked in the presence of Si (Zia et al. 2021); finally, in maize under cadmium toxicity, Si increased JA in both shoots and roots (An et al. 2024; Shah et al. 2023).

Research on rice under submersion stress revealed an important role of Si regulating JA, that is, JA increased in the presence of Si. The authors proposed that JA inhibited submersion-induced internode growth. Furthermore, the application of Si can inhibit GA, increasing the JA content in rice leaves, thereby reducing the elongation and length of internodes. Inhibiting the stretching of submerged rice (quiescent strategy) is important to reduce carbohydrate consumption. If elongation occurs (escape strategy), it is promoted by GA, but carbohydrates may deplete before the rice emerges, which then leads to death (Pan et al. 2022).

Regarding biotic stresses, Si caused a reduction in JA concentrations in *Brachypodium distachyon* in response to masticatory herbivory. This occurred because Si plants invest in physical defenses. Although chewing herbivores trigger the JA pathway associated with greater Si absorption, it does not appear to be necessary to trigger this pathway to the same extent as in plants without Si in order to maintain effective defenses (Johnson et al. 2021). However, there are other biotic stresses in which attenuation by Si coincides with the increase in JA, as observed for sugarcane with infestation by sugarcane borer (Jadhao et al. 2020), in soybeans with powdery mildew (Fauteux et al. 2006), and sorghum with anthracnose (Resende et al. 2021).

17.2.8 Salicylic Acid

Salicylic acid (SA) is a plant hormone belonging to a diverse group of substances known as phenolic compounds. Several studies have analyzed the functions of SA involving plant-pathogen relationships. It acts as a signaling molecule that modulates plant responses to biotic stresses, mainly in the translation signals of the main pathways for acquired systemic resistance. However, research has shown that SA, in addition to participating in several physiological processes in plants, also promotes protective effects in plants subjected to abiotic stress (Rivas-San Vicente and Plasencia 2011).

In soybeans, water stress caused an increase in SA concentration and Si attenuated this increase. That same study reported that salt stress in soybeans did not cause changes in SA contents. However, salt stress combined with Si resulted in increased SA (Hamayun et al. 2010). In cucumber, salt stress resulted in a decrease in SA; however, with Si, the levels increased and were equal to those of the control (Zhu et al. 2020). In rice, alleviation of salt stress by Si resulted in increased SA (Kim et al. 2014b).

In different abiotic stresses and plant species, as well as under saline stress, there is no pattern of response to Si. For example, in maize plants stressed by cold, there was a decrease in SA, but Si attenuated this damage (Moradtalab et al. 2018). In tomato grown at pH 9, SA increased, but the supply of Si resulted in SA levels equal to those of the control (Xue et al. 2021). In rice, S deficiency resulted in increased leaf SA content, but Si-treated plants did not show such an increase (Réthoré et al. 2020). In date palm under arsenic toxicity, Si decreased SA (Khan et al. 2022). In maize under cadmium toxicity, Si decreased SA in both shoots and roots (An et al. 2024; Shah et al. 2023). In date palm, alleviating heat stress with Si increased SA (Khan et al. 2020a). In tomato, the result was the opposite, that is, the mitigation of heat stress with Si decreased SA (Khan et al. 2020b).

Studies involving biotic stresses on soybeans inoculated with powdery mildew reported that the application of Si resulted in the positive regulation of genes related to SA biosynthesis (Fauteux et al. 2006). In tomato plants inoculated with the bacterium *Ralstonia solanacearum*, there was an increase in SA production on the first day after inoculation, the first symptoms of wilt occurred at 2 days, and all plants were completely wilted at 7 days. However, Si resulted in a delay in the increase in SA production at 7 days after inoculation, the first symptoms of wilting occurred at 4 days, and only 20–30% of the plants were slightly wilted at 7 days (Jiang et al. 2019). In millet, 72 h after sugarcane borer infestation, there was an increase in SA with the use of Si (Jadhao et al. 2020). In sorghum, the greatest resistance to anthracnose with the use of Si occurred with the increased expression of SA marker genes (Resende et al. 2021). In canola, Si alleviated the symptoms of clubroot disease, and there was a decrease in SA (Sarkar et al. 2023).

17.2.9 Strigolactones

Strigolactones (SL) comprise another hormonal class that negatively controls the development of axillary buds. Discovered in 2008, they are terpenoid lactones derived from carotenoids, produced in stems and roots, and transported by the xylem (Zwanenburg and Blanco-Ania 2018). As far as it is currently known, the modulation of SL by Si under stress relief conditions has not yet been explored.

References

An T, Gao Y, Kuang Q et al (2024) Unveiling Si's shield: a holistic examination of cd stress alleviation in maize through physiological and transcriptomic insights. Environ Exp Bot 219:105626. https://doi.org/10.1016/j.envexpbot.2023.105626

Arif Y, Singh P, Bajguz A et al (2021) Silicon mediated abiotic stress tolerance in plants using physio-biochemical, omic approach and cross-talk with phytohormones. Plant Physiol Biochem 166:278–289. https://doi.org/10.1016/j.plaphy.2021.06.002

Barreto RF, de Mello PR, Lúcio JCB et al (2021) Ammonium toxicity alleviation by silicon is dependent on cytokinins in tomato cv. Micro-Tom. J Plant Growth Regul. https://doi.org/10.1007/s00344-021-10314-5

Biju S, Fuentes S, Gupta D (2017) Silicon improves seed germination and alleviates drought stress in lentil crops by regulating osmolytes, hydrolytic enzymes and antioxidant defense system. Plant Physiol Biochem 119:250–264. https://doi.org/10.1016/j.plaphy.2017.09.001

Biju S, Fuentes S, Gupta D (2023) Novel insights into the mechanism(s) of silicon-induced drought stress tolerance in lentil plants revealed by RNA sequencing analysis. BMC Plant Biol 23:498. https://doi.org/10.1186/s12870-023-04492-5

Coskun D, Deshmukh R, Sonah H et al (2019) The controversies of silicon's role in plant biology. New Phytol 221:67–85. https://doi.org/10.1111/nph.15343

Epstein E (1994) The anomaly of silicon in plant biology. Proc Natl Acad Sci 91:11–17. https://doi.org/10.1073/pnas.91.1.11

Farooq M, Wahid A, Kobayashi N et al (2009) Plant drought stress: effects, mechanisms and management. In: Sustainable Agriculture. Springer Netherlands, Dordrecht, pp 153–188

Fauteux F, Chain F, Belzile F et al (2006) The protective role of silicon in the *Arabidopsis* – powdery mildew pathosystem. Proc Natl Acad Sci 103:17554–17559. https://doi.org/10.1073/pnas.0606330103

Gao H, Yu W, Yang X et al (2022) Silicon enhances the drought resistance of peach seedlings by regulating hormone, amino acid, and sugar metabolism. BMC Plant Biol 22:422. https://doi.org/10.1186/s12870-022-03785-5

Ha S, Vankova R, Yamaguchi-Shinozaki K et al (2012) Cytokinins: metabolism and function in plant adaptation to environmental stresses. Trends Plant Sci 17:172–179. https://doi.org/10.1016/j.tplants.2011.12.005

Hamayun M, Sohn EY, Khan SA et al (2010) Silicon alleviates the adverse effects of salinity and drought stress on growth and endogenous plant growth hormones of soybean (*Glycine max* L.). Pakistan J Bot 42:1713–1722

Helaly MN, El-Hoseiny H, El-Sheery NI et al (2017) Regulation and physiological role of silicon in alleviating drought stress of mango. Plant Physiol Biochem 118:31–44. https://doi.org/10.1016/j.plaphy.2017.05.021

Hernández-Apaolaza L, Escribano L, Zamarreño ÁM et al (2020) Root silicon addition induces Fe deficiency in cucumber plants, but facilitates their recovery after Fe resupply. A comparison with Si foliar sprays. Front Plant Sci 11. https://doi.org/10.3389/fpls.2020.580552

Hosseini SA, Maillard A, Hajirezaei MR et al (2017) Induction of barley silicon transporter HvLsi1 and HvLsi2, increased silicon concentration in the shoot and regulated starch and ABA homeostasis under osmotic stress and concomitant potassium deficiency. Front Plant Sci 8. https://doi.org/10.3389/fpls.2017.01359

Hwang S, Hamayun M, Kim H et al (2007) Effect of nitrogen and silicon nutrition on bioactive gibberellin and growth of rice under field conditions. J Crop Sci Biotechnol 10:281–286

Jadhao KR, Bansal A, Rout GR (2020) Silicon amendment induces synergistic plant defense mechanism against pink stem borer (*Sesamia inferens* Walker.) in finger millet (*Eleusine coracana* Gaertn.). Sci Rep 10:4229. https://doi.org/10.1038/s41598-020-61182-0

Jiang N, Fan X, Lin W et al (2019) Transcriptome analysis reveals new insights into the bacterial wilt resistance mechanism mediated by silicon in tomato. Int J Mol Sci 20:761. https://doi.org/10.3390/ijms20030761

Jing H, Wilkinson EG, Sageman-Furnas K, Strader LC (2023) Auxin and abiotic stress responses. J Exp Bot 74:7000–7014. https://doi.org/10.1093/jxb/erad325

Johnson SN, Hartley SE, Ryalls JMW et al (2021) Targeted plant defense: silicon conserves hormonal defense signaling impacting chewing but not fluid-feeding herbivores. Ecology 102:e03250. https://doi.org/10.1002/ecy.3250

Khan A, Kamran M, Imran M et al (2019) Silicon and salicylic acid confer high-pH stress tolerance in tomato seedlings. Sci Rep 9:19788. https://doi.org/10.1038/s41598-019-55651-4

Khan A, Bilal S, Khan AL et al (2020a) Silicon and gibberellins: synergistic function in harnessing ABA signaling and heat stress tolerance in date palm (*Phoenix dactylifera* L.). Plan Theory 9:620. https://doi.org/10.3390/plants9050620

Khan A, Khan AL, Imran M et al (2020b) Silicon-induced thermotolerance in *Solanum lycopersicum* L. via activation of antioxidant system, heat shock proteins, and endogenous phytohormones. BMC Plant Biol 20:248. https://doi.org/10.1186/s12870-020-02456-7

Khan T, Bilal S, Asaf S et al (2022) Silicon-induced tolerance against arsenic toxicity by activating physiological, anatomical and biochemical regulation in *Phoenix dactylifera* (Date Palm). Plan Theory 11:2263. https://doi.org/10.3390/plants11172263

Khan I, Awan SA, Rizwan M et al (2023a) Silicon: an essential element for plant nutrition and phytohormones signaling mechanism under stressful conditions. Plant Growth Regul 100:301–319. https://doi.org/10.1007/s10725-022-00872-3

Khan I, Awan SA, Rizwan M et al (2023b) Silicon-induced modulation of photosynthetic pigments, osmolytes, and phytohormonal regulation boosted the drought tolerance in *Elymus sibiricus* L. J Plant Growth Regul 43:998. https://doi.org/10.1007/s00344-023-11155-0

Kim Y-H, Khan AL, Kim D-H et al (2014a) Silicon mitigates heavy metal stress by regulating P-type heavy metal ATPases, Oryza sativa low silicon genes, and endogenous phytohormones. BMC Plant Biol 14:13. https://doi.org/10.1186/1471-2229-14-13

Kim YH, Khan AL, Waqas M et al (2014b) Silicon application to rice root zone influenced the phytohormonal and antioxidant responses under salinity stress. J Plant Growth Regul 33:137–149. https://doi.org/10.1007/s00344-013-9356-2

Lang DY, Fei PX, Cao GY et al (2019) Silicon promotes seedling growth and alters endogenous IAA, GA 3 and ABA concentrations in *Glycyrrhiza uralensis* under 100 mM NaCl stress. J Hortic Sci Biotechnol 94:87–93. https://doi.org/10.1080/14620316.2018.1450097

Lee SK, Sohn EY, Hamayun M et al (2010) Effect of silicon on growth and salinity stress of soybean plant grown under hydroponic system. Agrofor Syst 80:333–340. https://doi.org/10.1007/s10457-010-9299-6

Mahmoud AWM, Abdeldaym EA, Abdelaziz SM et al (2019) Synergetic effects of zinc, boron, silicon, and zeolite nanoparticles on confer tolerance in potato plants subjected to salinity. Agronomy 10:19. https://doi.org/10.3390/agronomy10010019

Maillard A, Ali N, Schwarzenberg A et al (2018) Silicon transcriptionally regulates sulfur and ABA metabolism and delays leaf senescence in barley under combined sulfur deficiency and osmotic stress. Environ Exp Bot 155:394–410. https://doi.org/10.1016/j.envexpbot.2018.07.026

Markovich O, Steiner E, Kouřil Š et al (2017) Silicon promotes cytokinin biosynthesis and delays senescence in Arabidopsis and Sorghum. Plant Cell Environ 40:1189–1196. https://doi.org/10.1111/pce.12913

Mir RA, Bhat BA, Yousuf H et al (2022) Multidimensional role of silicon to activate resilient plant growth and to mitigate abiotic stress. Front Plant Sci 13. https://doi.org/10.3389/fpls.2022.819658

Monshausen GB, Peer W (2017) Sinais e transdução de sinal. In: Taiz L, Zeiger E, Moller IM, Murphy A (eds) Fisiologia e desenvolvimento vegetal, 6th edn. Artmed, Porto Alegre, pp 407–441

Moradtalab N, Weinmann M, Walker F et al (2018) Silicon improves chilling tolerance during early growth of maize by effects on micronutrient homeostasis and hormonal balances. Front Plant Sci 9. https://doi.org/10.3389/fpls.2018.00420

Naz R, Gul F, Zahoor S et al (2022) Interactive effects of hydrogen sulphide and silicon enhance drought and heat tolerance by modulating hormones, antioxidant defence enzymes and redox status in barley (*Hordeum vulgare* L.). Plant Biol 24:684–696. https://doi.org/10.1111/plb.13374

Nguyen TH, Goossens A, Lacchini E (2022) Jasmonate: a hormone of primary importance for plant metabolism. Curr Opin Plant Biol 67:102197. https://doi.org/10.1016/j.pbi.2022.102197

Nolan TM, Vukašinović N, Liu D et al (2020) Brassinosteroids: multidimensional regulators of plant growth, development, and stress responses. Plant Cell 32:295–318. https://doi.org/10.1105/tpc.19.00335

Pan T, Wang L, Peng Z et al (2022) Silicon enhances the submergence tolerance of rice by regulating quiescence strategy and alleviating oxidative damage. Plant Physiol Biochem 182:124–132. https://doi.org/10.1016/j.plaphy.2022.04.018

Resende RS, De VD, Höfte M, Ávila Rodrigues F (2021) New insights into the hormonal regulation of silicon-supplied sorghum plants challenged with *Colletotrichum sublineolum*. Physiol Mol Plant Pathol 115:101682. https://doi.org/10.1016/j.pmpp.2021.101682

Réthoré E, Ali N, Yvin J-C, Hosseini SA (2020) Silicon regulates source to sink metabolic homeostasis and promotes growth of rice plants under sulfur deficiency. Int J Mol Sci 21:3677. https://doi.org/10.3390/ijms21103677

Rivas-San Vicente M, Plasencia J (2011) Salicylic acid beyond defence: its role in plant growth and development. J Exp Bot 62:3321–3338. https://doi.org/10.1093/jxb/err031

Sarkar A, Kisiala A, Adhikary D et al (2023) Silicon ameliorates clubroot responses in canola (*Brassica napus*): a "multi-omics"-based investigation into possible mechanisms. Physiol Plant 175:e13900. https://doi.org/10.1111/ppl.13900

Sayed EG, Mahmoud AWM, El-Mogy MM et al (2022) The effective role of nano-silicon application in improving the productivity and quality of grafted tomato grown under salinity stress. Horticulturae 8:293. https://doi.org/10.3390/horticulturae8040293

Semeradova H, Montesinos JC, Benkova E (2020) All roads lead to auxin: post-translational regulation of auxin transport by multiple hormonal pathways. Plant Commun 1:100048. https://doi.org/10.1016/j.xplc.2020.100048

Shah T, Khan Z, Khan SR et al (2023) Silicon inhibits cadmium uptake by regulating the genes associated with the lignin biosynthetic pathway and plant hormone signal transduction in maize plants. Environ Sci Pollut Res 30:123996–124009. https://doi.org/10.1007/s11356-023-31044-z

Thakur A, Singh A, Tandon A, Sharma V (2023) Insights into the molecular mechanisms of uptake, phytohormone interactions and stress alleviation by silicon: a beneficial but non-essential nutrient for plants. Plant Growth Regul 101:1–13. https://doi.org/10.1007/s10725-023-01002-3

Van Bockhaven J, Spíchal L, Novák O et al (2015) Silicon induces resistance to the brown spot fungus Cochliobolus miyabeanus by preventing the pathogen from hijacking the rice ethylene pathway. New Phytol 206:761–773. https://doi.org/10.1111/nph.13270

Verma V, Ravindran P, Kumar PP (2016) Plant hormone-mediated regulation of stress responses. BMC Plant Biol 16:86. https://doi.org/10.1186/s12870-016-0771-y

Xue X, Geng T, Liu H et al (2021) Foliar application of silicon enhances resistance against phytophthora infestans through the ET/JA- and NPR1- dependent signaling pathways in potato. Front Plant Sci 12. https://doi.org/10.3389/fpls.2021.609870

Yamaguchi S (2008) Gibberellin metabolism and its regulation. Annu Rev Plant Biol 59:225–251. https://doi.org/10.1146/annurev.arplant.59.032607.092804

Zhu Y, Jiang X, Zhang J et al (2020) Silicon confers cucumber resistance to salinity stress through regulation of proline and cytokinins. Plant Physiol Biochem 156:209–220. https://doi.org/10.1016/j.plaphy.2020.09.014

Zia A, Hegazy HS, Hassan NS et al (2021) Biochemical responses of wheat to silicon application under salinity. J Plant Nutr Soil Sci 184:255–262. https://doi.org/10.1002/jpln.202000181

Zwanenburg B, Blanco-Ania D (2018) Strigolactones: new plant hormones in the spotlight. J Exp Bot 69:2205–2218. https://doi.org/10.1093/jxb/erx487

Chapter 18
Enhancing Resilience in Hydroponic Crops with Silicon: Insights into Growth Enhancement and Stress Mitigation

Rohit Dilip Kambale, Qiong Su, R. Karthikeyan, Jeffrey Adelberg, and Byoung Ryong Jeong

Abstract Hydroponics is a soil-independent cultivation method using nutrient solutions and a promising alternative to other methods by offering benefits such as enhanced yield, lower water consumption, and minimized risks of diseases and pests. In this context, silicon (Si) has gained attention due to its various roles in enhancing plant growth and mitigating stress in hydroponic systems. This book chapter delves into the significant roles of Si in hydroponics and emphasizes its contributions to enhancing plant growth, resilience, and yield. Although it is not considered essential, Si markedly improves plant health, aids in the creation of physical barriers against pathogens, bolsters cell wall integrity, and improves nutrient uptake and photosynthesis. Furthermore, Si supplementation in hydroponics mitigates a range of biotic and abiotic stresses, including drought, salinity, and high temperatures, by reinforcing plant physiological and biochemical mechanisms. This chapter synthesizes the current understanding of Si's multifaceted benefits for hydroponic cultivation from its impact on disease resistance and stress mitigation to its influence on nutrient efficiency and plant structural strength. Given the critical need for further in-depth research to optimize Si application in various crops, this review underscores the potential of Si as a sustainable approach to enhancing agricultural productivity and sustainability in the face of environmental challenges.

Keywords Plant growth and development · Abiotic stress · Biotic stress · Stress reduction · Sustainability · Inorganic nutrition

R. D. Kambale · Q. Su · R. Karthikeyan
Department of Agricultural Sciences, Clemson University, Clemson, SC, USA

J. Adelberg
Department of Plant and Environmental Sciences, Clemson University, Clemson, SC, USA

B. R. Jeong (✉)
Division of Horticultural Science, College of Agriculture and Life Sciences, Gyeongsang National University, Jinju, South Korea

R. de Mello Prado et al. (eds.), *Silicon Advances for Sustainable Agriculture and Human Health*, Sustainable Plant Nutrition in a Changing World,
https://doi.org/10.1007/978-3-031-69876-7_18

18.1 Introduction

The sustainability and resilience of global food security are increasingly challenged due to climate change, rapid population growth, limited arable land, and scarcity of freshwater resources (Singh and Su 2022). These challenges are further exacerbated by escalating biotic stressors, including pathogen invasions and pest infestations, which pose significant threats to crop growth and development, phenotypic stability, and yield (Jana and Jeong 2014; Soundararajan et al. 2014; Etesami and Jeong 2018, 2022; Shilpha et al. 2023). As conventional soil-based cultivation increasingly faces limitations due to such environmental and demographic shifts, the exploration and adoption of innovative agricultural methodologies have become critical.

Hydroponics, a soil-independent cultivation technique that uses nutrient solutions for plant growth, offers several advantages, including higher yield, lower water consumption, and reduced risks of pest attacks and diseases (Jana and Jeong 2014). Furthermore, hydroponics facilitates year-round cultivation independent of climatic conditions, thus greatly improving food security (Park et al. 2013; Etesami and Jeong 2018; Xiao et al. 2022; Yang et al. 2022). The success of hydroponic systems is highly dependent on nutrient solution management. Achieving an optimal nutrient formula involves carefully balancing the composition of essential macronutrients and micronutrients to meet the plant's specific physiological needs. The assimilation of dissolved nutrient ions by the plant is facilitated through both passive diffusion and active transport mechanisms, highlighting the importance of a deep understanding of plant nutritional physiology to maximize yield and ensure a robust plant health (Muneer et al. 2017; Etesami and Jeong 2020; Noh and Jeong 2022; Song et al. 2022a, b).

Several studies have explained the role of various elements, such as nitrogen, phosphorus, potassium, and silicon, in maintaining a healthy plant growth and improving tolerance to environmental stresses (Mir et al. 2022; Xuebin et al. 2020; Ma 2004). Among these elements, Si stands out for its extensive use in various fields, particularly in agriculture (Muneer et al. 2017; Etesami and Jeong 2020; Noh and Jeong 2022; Song et al. 2022b). Research has highlighted that silicon fertilization markedly improves crop health and productivity, though it is not strictly essential for most plant species (Asimov 1954; Petkowski et al. 2020). Although Si is considered a non-essential element, it provides plants with a wide range of benefits, particularly under environmental stress, making it a powerful element in enhancing plant growth and resilience (Ma 2004; Zia et al. 2017; Zargar et al. 2019).

Silicon is abundant in the soil, but its bioavailable form significantly hinders plant uptake. Plants primarily absorb Si in the form of monosilicic acid (H_4SiO_4), and the degree of accumulation varies across species. Monocots, particularly those in the Poaceae (grasses) and Cyperaceae (sedges) families, tend to accumulate more Si than dicots (broadleaf plants) (Epstein 1994; Savvas and Ntatsi 2015). Cornelis et al. (2011) have explained three main mechanisms of Si uptake: active, passive, and rejective. Plants take up monosilicic acid, which is then transported via xylem and ultimately deposited on the surfaces of the leaf epidermis as polymerized silica gel, forming structures known as phytolith (Jones and Handreck 1965; Yoshida et al. 1962). The incorporation of silicon into the plant physiology notably fortifies cell wall properties, improving resistance against biotic threats such as bacteria, fungi, and insects (Lux et al. 2020, 2003; Iwasaki et al. 2002; Schuerger et al. 2021). This defense is crucial in controlled environments, where stress factors such as temperature and humidity fluctuations may detrimentally affect plant health. Moreover, Si strengthens plant structure, forming a more robust outer layer that improves the plant's ability to withstand wind damage or lodging (Rafi et al. 1997; Bélanger et al. 2003). This is a significant advantage for fruit-bearing plants such as tomatoes (McGinnity 2015). Furthermore, new research has pointed to silicon's role in enhancing nutrient absorption and use, potentially leading to a more efficient plant growth (Souri et al. 2021). Several studies (Muneer et al. 2014, 2017; Muneer and Jeong 2015; Park et al. 2018a, b) have suggested that silicon may contribute to increase chlorophyll content in leaves, which is vital for photosynthesis and photosynthesis-related physiology, thereby potentially increasing plant growth and yield.

Existing research highlights the benefits of using Si in hydroponic systems, yet there is a lack of detailed studies seeking to understand its long-term effects and refine application techniques for different crops. The aim of this chapter is to illustrate the benefits of Si for crop growth and development. It also explores how Si application can enhance plant defenses, potentially leading to a higher crop yield productivity. Section 18.2 discusses the availability and absorption of Si and its supplementation strategies to optimize plant growth in hydroponic cultures. Section 18.3 presents the impacts of Si on plant biomass, yield, health, nutrient uptake efficiency, and structural integrity, underscoring its vital roles in agricultural success. Section 18.4 delves into the critical function of Si in improving the activities of antioxidant enzymes, highlighting its capacity to mitigate oxidative stress through the upregulation of enzymes that are critical for plant defense. Section 18.5 explores the capacity of Si in mitigating biotic and abiotic stresses. Finally, Sect. 18.6 summarizes the benefits of Si application in hydroponic systems and emphasizes the need for further comprehensive research aiming to elucidate its long-term impacts and refine application methods in various crops.

18.2 Silicon in Hydroponic Systems

18.2.1 Silicon Availability and Absorption

In hydroponic cultivation, 17 essential elements play a crucial role in plant growth and development. Traditionally, silicon (Si) is considered a non-essential element. Unlike soil-based agriculture, where Si is naturally available, hydroponic systems often lack this element since these systems do not use natural soils. The introduction of Si in hydroponics through supplementation with calcium silicate ($CaSiO_3$), potassium silicate (K_2SiO_3), or sodium silicate (Na_2SiO_3) has proven effective in making silicon available to plants (Son et al. 2012, 2013; Li et al. 2020). Silicon uptake by plants occurs through both passive and active mechanisms, as shown in studies on rice (*Oryza sativa*), corn (*Zea mays*), sunflower (*Helianthus annuus*), and wax gourd (*Benincasa hispida*), which showed Si absorption in hydroponic conditions (Liang et al. 2006). Plants primarily absorb Si in the form of silicic acid [$Si(OH)_4$], with Si ions moving through the root apoplast and influx carriers to the plant's vascular system for distribution.

Notably, rice absorbs silicic acid faster than water, highlighting the efficiency of Si uptake in certain plants (Ma et al. 2001). Recent research on the uptake kinetics of Si by Tamai and Ma (2003) has shed light on the role of transporter proteins in this process, suggesting that the rate of Si uptake varies according to its concentration. In hydroponic systems, the availability of Si depends on factors such as Si supplementation and the composition of the nutrient solution, in contrast to soil-based agriculture, in which characteristics such as soil type, pH, and organic matter content exert the greatest influence. Also, the presence of elements such as calcium (Ca) and phosphorus (P) can compete with Si uptake, thus highlighting the need for a balanced nutrient solution (Pontigo et al. 2015; Pavlovic et al. 2021).

18.2.2 Silicon Supplementation Strategies

Unlike traditional soil-based agriculture, hydroponic solutions often lack silicon. Therefore, growers increasingly turn to a strategic supplementation with calcium silicate ($CaSiO_3$), potassium silicate (K_2SiO_3), or sodium silicate (Na_2SiO_3), which dissolve easily in hydroponic solution, making Si readily available for plant uptake (Guntzer et al. 2012). Additionally, alternatives such as crushed rock dust or specialized silicate fertilizers are explored for efficacy in supplementing Si (Benevides Filho et al. 2023). The optimal concentration of Si in a hydroponic solution varies depending on the type of crop grown and the specific desired outcome (Savvas et al. 2002; Sambo et al. 2019). Different plant species have different Si requirements. For instance, crops such as rice, corn, and sugarcane are known as Si accumulators, and rice has a notably high Si uptake, showing a mean relative shoot Si concentration of 4.167 (Ma et al. 2001; Ma 2004; Mitani and Ma 2005). A beneficial Si concentration range for hydroponic solutions is generally between 60 and 220 mg L^{-1} (Meena et al. 2014).

Monitoring the pH and electrical conductivity (EC) of the nutrient solution is essential, as these factors affect the availability of Si to crops (Gocke et al. 2013). This approach to Si supplementation underscores the importance of nutrient management in hydroponic systems, leading to the optimization of plant growth and development through precise Si provision.

18.3 Silicon Impact on Crop Growth

18.3.1 Growth Enhancement

Silicon plays a significant role in enhancing plant growth, yield, and overall plant health. The incorporation of Si into plant cell walls is known to substantially improve their mechanical strength (Currie and Perry 2007; Epstein 1994, 2009; He et al. 2013), allowing plants to better support their own structure. This structural reinforcement often translates into increased biomass production, as plants more effectively sustain their weight and optimize their growth (Savant et al. 1997).

Furthermore, the application of Si has been associated with improvements in fruit size and other quality traits, as evidenced in crops such as apple and cantaloupe (Karagiannis et al. 2021; Alam et al. 2021). In addition to these benefits, Si enhances the plant's resilience to several environmental stresses, including drought (Bhardwaj and Kapoor 2021) and salinity (Zhu and Gong 2014), which are critical for maintaining optimal productivity under challenging conditions.

Silicon supplementation may notably improve biomass production and leaf area. It may also improve the nutritional value of several hydroponically grown crops such as corn salad [*Valerianella locusta* (L.) Laterr] (Gottardi et al. 2012), bamboo (Collin et al. 2013), wheat (Rizwan et al. 2015; Rizwan et al. 2016); tomato (Yilmaz and Korkmaz 2023), sweet pepper (Altuntas et al. 2018; Trejo-Téllez et al. 2020; López-Pérez et al. 2018), and cucumber (Mohaghegh et al. 2010). Savvas et al. (2002) analyzed the effects of Si on gerbera plants grown in a closed soilless culture system and highlighted the impact of Si concentrations in nutrient solutions, revealing that a higher Si concentration resulted in a greater proportion of high-quality flowers.

18.3.2 Nutrient Uptake and Efficiency: How Silicon Affects the Absorption and Use of Essential Nutrients

Silicon, while not classified as a macronutrient, plays a significant role in improving the absorption and use of essential nutrients in plants. Si improves the functionality of roots by depositing on the cell walls of root hairs, thereby enhancing their rigidity and operational efficiency (Lux et al. 2020; Mandlik et al. 2020; Hu et al. 2020). Additionally, Si increases membrane permeability, which may regulate the uptake of essential nutrients and reduce the passive absorption of non-beneficial elements (Chaiwong and Prom-U-Thai 2022).

The contribution of Si to nutrient availability also causes indirect effects, such as enhancing plant resilience against certain stress, thereby facilitating a more efficient nutrient acquisition (Broadley et al. 2012; Irfan et al. 2023). Studies have suggested that Si may work synergistically with key nutrients such as phosphorus (P) and calcium (Ca) to improve their uptake and use, which in turn contributes to improved plant growth and development (Etesami 2018; Ma et al. 2022). However, silicon might also compete with other cations for uptake sites on plant roots, potentially leading to a slight reduction in the uptake of potassium (Pontigo et al. 2015; Pavlovic et al. 2021).

Hydroponics allow an exact nutrient composition of the solution compared to natural soil. This advantage potentially mitigates any competitive interactions caused by Si while maximizes its benefits. The exact mechanisms by which Si affects nutrient uptake and consumption are still unknown. Further research is needed to fully understand the complex interplay between Si and nutrient ions, thereby improving plant nutrition and growth in hydroponic systems.

18.3.3 Structural and Mechanical Strength

The contribution of Si in enhancing the structural and mechanical strength of plants, particularly by fortifying cell walls and improving structural integrity, is a significant aspect of plant defense mechanisms. The initial hypothesis was its role in developing a physical barrier against pathogens, but the polymerization of Si within cell walls has been identified as a key factor in bolstering plants' resistance to diseases, such as increased resilience of rice against rice blast fungus (*P. oryzae*) as a result of a delayed infection incidence (Hemmi and AOYAGI 1941). This resistance is attributed to the formation of silicified bulliform cells in the rice's leaf epidermis, which serves as a physical barrier to pathogen entry.

Subsequent research has further elucidated Si′s role, revealing its integration with hemicellulose within the cell wall, which not only creates a physical barrier but also enhances the wall's overall strength (He et al. 2015). Seebold et al. (2001) studied the impact of Si on rice blast disease on susceptible and resistant rice cultivars. The authors reported that silicon-supplemented rice plants, both resistant and susceptible cultivars, had fewer blast lesions compared to non-supplemented plants, clearly corroborating the physical barrier theory. The study by Kim et al. (2002) showed that Si deposition in rice leaves forms a double layer acting as a physical barrier that delays penetration by the rice blast fungus, *Pyricularia grisea*. Further research is needed to explore how Si interacts with other physical barriers, such as the wax layer, which offers an additional layer of protection.

In addition to its role in resistance to pathogens, Si deposition has significantly improved crop resistance against insect pests by increasing tissue stiffness, strengthening cell walls, and reducing pliability (Kvedaras et al. 2007; Massey et al. 2009; Massey and Hartley 2006, 2009). This phenomenon occurs in a diverse range of crops, including coffee (Yang et al. 2022), strawberry (Xiao et al. 2022), barley (Pavlovic et al. 2021; Sakr 2021), cucumbers (Sun et al. 2022), rice (Kim et al. 2002; Sun et al. 2010; Shivaraj et al. 2022), sugarcane (Nicchio et al. 2024), and wheat (Cruz et al. 2015; Dallagnol et al. 2020). This variability in crops underscores the broad applicability of Si in enhancing disease and resistance to pests through structural reinforcement. Recently, Muneer et al. (2018) studied intraspecific proteome changes in aphid-resistant and sensitive rose (*Rosa hybrida*) cultivars at two developmental stages.

18.4 Silicon and Activities of Antioxidant Enzymes

Research evidence has supported the physical barrier hypothesis (Sun et al. 2010; Pang et al. 2024; Nunez et al. 2021). However, recent studies have broadened its understanding, revealing that Si's influence extends to the biochemical and molecular domains, significantly improving plant disease tolerance (Muneer et al. 2018). The interaction of Si with organic compounds in epidermal cell walls improves resistance against degradation by hydrolytic enzymes. This element is instrumental in activating several defense-related enzymes, such as polyphenoloxidase, glucanase, and peroxidase, which play critical roles in plant defense mechanisms (Dorjee and Wangmu 2023; Gulzar et al. 2023; Gulzar et al. 2024).

The improvement in enzyme activity, the production of antimicrobial compounds, and the regulation of signaling molecules are key aspects of silicon-enhanced biochemical resistance in plants (Fortunato et al. 2014; Van Bockhaven et al. 2013; Wang et al. 2017). Notably, Si activates enzymes such as chitinase and peroxidases, essential for breaking fungal cell walls and generating pathogen-fighting compounds. Several studies have highlighted the role of Si in activating various defense-related enzymes, including chitinase, peroxidases, polyphenoloxidases, and superoxide dismutase, in addition to other plant mechanisms. Silicon strengthens plant defenses by activating various enzymes such as chitinase, which can break fungal cell walls, and peroxidases, which are involved in producing pathogen-fighting compounds (Waewthongrak et al. 2015; Tavallali and Karimi 2023; Shanmugaiah et al. 2023). Lata-Tenesaca et al. (2024) investigated the effects of leaf-applied Si against MLB and reported that defense enzymes such as chitinase, β-1,3-glucanase, polyphenoloxidase, peroxidase, lipoxygenase, and antioxidative enzymes were higher for infected Si-applied plants compared to infected plants without Si application. Similarly, researchers investigated leaf application of Si and magnesium against blast at the physiological and biochemical levels. Results showed an increased resistance against blast epidemics caused by the fungus *P. oryzae* (Araujo et al. 2023). For instance, silicon-mediated biochemical and molecular responses reduce the severity of diseases in a broad spectrum of plant species (Islam et al. 2020; Bhat et al. 2022).

Moreover, Si plays an important role in enhancing plant defense mechanisms, particularly by modulating immune responses through the activation of key signaling pathways involving salicylic acid (SA), jasmonic acid (JA), and ethylene (ET). This complex interaction underscores the role of Si in plant defense, emphasizing the need of understanding the functions of SA, JA, and ET to fully explain the impacts of Si on plant defense. Several studies have shown that Si can trigger these signaling pathways, significantly enhancing stress tolerance in plants (Ghareeb et al. 2011; Reynolds et al. 2016; Khan et al. 2021). The work of Fortunato et al.

(2014) explained the role of Si on the phenylpropanoid pathway, which in turn explains the resistance of banana plants to *F. oxysporum* f. sp. *cubense*. Similarly, the upregulation of *AJAZ1*-similar gene in tomatoes infected with *Ralstonia solanacearum* suggests that Si supplementation can activate pathways related to jasmonic acid (JA) and/or reactive oxygen species, further enhancing plant defenses (Song et al. 2021; Kim et al. 2014; Khan et al. 2023). Many studies have shown silicon's strength in enhancing plant defenses against various microbial pathogens (Ye et al. 2013). Studies have also explored the role of Si in plant-microbe interactions. Despite these findings, the specific mechanisms and the way Si regulates disease resistance remain poorly understood, and further studies are needed to understand this.

18.5 Silicon-Mediated Stress Mitigation

Silicon plays a pivotal role in mitigating stress in plants and in combating the perennial threat of pathogens such as bacteria, fungi, and viruses that lead to low crop yields. Vulnerable to a variety of pathogens present in the soil, seeds, and foliage in different growth stages, plants face significant challenges in disease management due to environmental variables and limited treatment options. Research has shown that Si supplementation can markedly decrease the severity of diseases caused by several pathogens, thereby enhancing plant resistance by strengthening physical, biochemical, and molecular defense mechanisms (Chérif et al. 1994; Ye et al. 2013; Rodrigues et al. 2015; Gulzar et al. 2023).

Silicon also helps plants to become more resistant to attacks by various biotic stresses by enhancing physical, biochemical, and molecular defense mechanisms (Hemmi and AOYAGI 1941; Islam et al. 2020; Ahammed and Yang 2021; Dorneles et al. 2017). This enhancement in plant resilience is evidenced through an improved resistance to environmental stressors, leading to notable benefits in stress mitigation and yield enhancement (Maghsoudi et al. 2015; Ning et al. 2023). For example, Si significantly alleviates drought stresses through better water retention (Liang et al. 2006, 2015), enhances salinity tolerance by reinforcing cell walls and reducing the uptake of harmful sodium ions (Ma 2004), and reduces damage from high temperatures by enhancing antioxidant activity and minimizing membrane damage (Chen et al. 2011) and nutrient uptake (Epstein 1999; Rafi and Epstein 1999).

Additionally, Si application offers protection against UV-B radiation (Mavrič Čermelj et al. 2021), thus improving development, photosynthesis, and antioxidant activities in seedlings of hydroponically grown crops such as lettuce (Bacchus 2010), soybean (Shen et al. 2010), and wheat (Yao et al. 2011; Shen et al. 2014a). Shen et al. (2014b) studied peanut (*Arachis hypogaea* L.) and reported the potential of Si to mitigate aluminum stress, highlighting its versatility in enhancing photosynthesis, biomass, and root-to-shoot ratio. Silicon supplementation in hydroponic systems is a powerful strategy for mitigating various environmental stresses, showing its significant contribution to cultivating more resilient and productive crops in controlled environments.

18.6 Conclusions

Silicon application in hydroponic systems offers significant potential benefits that underscore its importance in modern agriculture. These advantages include (1) silicon deposition on cell walls enhancing their mechanical strength and rigidity and improving structural support and resistance to mechanical stresses such as lodging or wind damage; (2) Si enhancing plants' tolerance to both abiotic and biotic stresses, contributing to improved resilience; and (3) Si easing an increased nutrient uptake efficiency, positively affecting overall plant health. The investigation of Si use in hydroponic farming is a growing field of research. Its optimal application may be reached as our understanding of Si activity deepens.

Although significant progress has been made in understanding the biochemical and molecular responses caused by silicon, further research is needed to fully elucidate the complex mechanisms of phytohormone signal transduction and transcriptomic regulations. The function of Si transporters, in particular, offers insights into the mechanisms of Si-mediated defense against biotic stressors. Prioritizing research on Si transporters, especially in contexts of biotic stress, may offer valuable knowledge about how plants use Si for defense and potentially develop novel strategies to improve plant health and crop protection.

References

Ahammed GJ, Yang Y (2021) Mechanisms of silicon-induced fungal disease resistance in plants. Plant Physiol Biochem 165:200–206

Alam A, Hariyanto B, Ullah H, Salin KR, Datta A (2021) Effects of silicon on growth, yield and fruit quality of cantaloupe under drought stress. SILICON 13:3153–3162

Altuntas O, Dasgan HY, Akhoundnejad Y (2018) Silicon-induced salinity tolerance improves photosynthesis, leaf water status, membrane stability, and growth in pepper (Capsicum annuum L.). HortScience 53(12):1820–1826

Araujo MU, Oliveira LM, Silva LC, Pinto LF, Cacique IS, Rodrigues FA (2023) Silicon, magnesium, and their interaction on wheat resistance against blast. Plant Soil 490(1):401–421

Asimov I (1954) The elementary composition of the earth's crust. J Chem Educ 31(2):70

Bacchus G (2010) An evaluation of the influence of biodynamic practices including foliar-applied silica spray on nutrient quality of organic and conventionally fertilised lettuce (Lactuca sativa L.). J Org Syst 5(1)

Bélanger R, Benhamou N, Menzies J (2003) Cytological evidence of an active role of silicon in wheat resistance to powdery mildew (Blumeria graminis f. sp. tritici). Phytopathology 93(4):402–412

Benevides Filho PRR, Blaskowski AE, Ramos MN, Lessa LGF, Filho ANZ, Abreu-Junior CH, Jani AD, Capra GF, Nogueira TAR (2023) Potential soil remineralizers from silicate rock powders (SRP) as alternative sources of nutrients for agricultural production (Amazon region). Fortschr Mineral 13(10):1255

Bhardwaj S, Kapoor D (2021) Fascinating regulatory mechanism of silicon for alleviating drought stress in plants. Plant Physiol Biochem 166:1044–1053

Bhat KA, Batool A, Mansoor M, Manzoor M, Bashir Z, Nazir M, Zargar SM (2022) Silicon-and nanosilicon-mediated disease resistance in crop plants. In: Silicon and nano-silicon in environmental stress management and crop quality improvement. Elsevier, pp 193–205

Broadley M, Brown P, Cakmak I, Ma JF, Rengel Z, Zhao F (2012) Beneficial elements. In: Marschner's mineral nutrition of higher plants. Elsevier, pp 249–269

Chaiwong N, Prom-U-Thai C (2022) Significant roles of silicon for improving crop productivity and factors affecting silicon uptake and accumulation in rice: a review. J Soil Sci Plant Nutr 22(2):1970–1982

Chen W, Yao X, Cai K, Chen J (2011) Silicon alleviates drought stress of rice plants by improving plant water status, photosynthesis and mineral nutrient absorption. Biol Trace Elem Res 142:67–76

Chérif M, Asselin A, Bélanger R (1994) Defense responses induced by soluble silicon in cucumber roots infected by Pythium spp. Phytopathology 84(3):236–242

Collin B, Doelsch E, Keller C, Panfili F, Meunier J-D (2013) Effects of silicon and copper on bamboo grown hydroponically. Environ Sci Pollut Res 20:6482–6495

Cornelis J-T, Delvaux B, Georg R, Lucas Y, Ranger J, Opfergelt S (2011) Tracing the origin of dissolved silicon transferred from various soil-plant systems towards rivers: a review. Biogeosciences 8(1):89–112

Cruz MF, Debona D, Rios JA, Barros EG, Rodrigues FA (2015) Potentiation of defense-related gene expression by silicon increases wheat resistance to leaf blast. Trop Plant Pathol 40:394–400

Currie HA, Perry CC (2007) Silica in plants: biological, biochemical and chemical studies. Ann Bot 100(7):1383–1389

Dallagnol LJ, Ramos AER, da Rosa DK (2020) Silicon use in the integrated disease management of wheat: current knowledge. In: Current trends in wheat research. IntechOpen, London, pp 5–26

Dorjee L, Wangmu T (2023) Role of silicon in inducing fungal disease resistance in Rice

Dorneles KR, Dallagnol LJ, Pazdiora PC, Rodrigues FA, Deuner S (2017) Silicon potentiates biochemical defense responses of wheat against tan spot. Physiol Mol Plant Pathol 97:69–78

Epstein E (1994) The anomaly of silicon in plant biology. Proc Natl Acad Sci 91(1):11–17

Epstein E (1999) Silicon. Annu Rev Plant Biol 50(1):641–664

Epstein E (2009) Silicon: its manifold roles in plants. Ann Appl Biol 155(2):155–160

Etesami H (2018) Can interaction between silicon and plant growth promoting rhizobacteria benefit in alleviating abiotic and biotic stresses in crop plants? Agric Ecosyst Environ 253:98–112

Etesami H, Jeong BR (2018) Silicon (Si): review and future prospects on the action mechanisms in alleviating biotic and abiotic stresses in plants. Ecotoxicol Environ Saf 147:881–896

Etesami H, Jeong BR (2020) Importance of silicon in fruit nutrition: agronomic and physiological implications. Fruit crops Diagnosis and Management of Nutrient Constraints. Elsevier, In, pp 255–277

Etesami H, Jeong BR (2022) Biodissolution of silica by rhizospheric silicate-solubilizing bacteria. In: Silicon and nano-silicon in environmental stress management and crop quality improvement. Elsevier, pp 265–276

Fortunato AA, da Silva WL, Rodrigues FÁ (2014) Phenylpropanoid pathway is potentiated by silicon in the roots of banana plants during the infection process of Fusarium oxysporum f. sp. cubense. Phytopathology 104(6):597–603

Ghareeb H, Bozsó Z, Ott PG, Repenning C, Stahl F, Wydra K (2011) Transcriptome of silicon-induced resistance against Ralstonia solanacearum in the silicon non-accumulator tomato implicates priming effect. Physiol Mol Plant Pathol 75(3):83–89

Gocke M, Liang W, Sommer M, Kuzyakov Y (2013) Silicon uptake by wheat: effects of Si pools and pH. J Plant Nutr Soil Sci 176(4):551–560

Gottardi S, Iacuzzo F, Tomasi N, Cortella G, Manzocco L, Pinton R, Römheld V, Mimmo T, Scampicchio M, Dalla Costa L (2012) Beneficial effects of silicon on hydroponically grown corn salad (Valerianella locusta (L.) Laterr) plants. Plant Physiol Biochem 56:14–23

Gulzar N, Lone R, Kamili AN, Shah MA, Alaklabi A (2023) Expression of defense genes and free radical scavenging on pre supplementation of silicon to Alternaria solani inoculated Lycopersicon esculentum plants. Physiol Mol Plant Pathol 123:101929

Gulzar N, Lone R, Al-Sadi AM, Azeez A (2024) Interactive role of silicon and Phenolics in biotic stress regulation in plants and expression of phenylpropanoid pathway genes. In: Plant phenolics in biotic stress management. Springer, pp 305–324

Guntzer F, Keller C, Meunier J-D (2012) Benefits of plant silicon for crops: a review. Agron Sustain Dev 32:201–213

He C, Wang L, Liu J, Liu X, Li X, Ma J, Lin Y, Xu F (2013) Evidence for 'silicon' within the cell walls of suspension-cultured rice cells. New Phytol 200(3):700–709

He C, Ma J, Wang L (2015) A hemicellulose-bound form of silicon with potential to improve the mechanical properties and regeneration of the cell wall of rice. New Phytol 206(3):1051–1062

Hemmi T, AOYAGI Z (1941) Germination of the macroconidia of Gibberella Fujikuroi in relation to some environmental factors ecological studies on important fungi pathogenic to the crops in the Far East. I. Jpn J Phytopathol 11(2):66–80

Hu AY, Xu SN, Qin DN, Li W, Zhao XQ (2020) Role of silicon in mediating phosphorus imbalance in plants. Plan Theory 10(1):51

Irfan M, Maqsood MA, Rehman H, Mahboob W, Sarwar N, Hafeez OBA, Hussain S, Ercisli S, Akhtar M, Aziz T (2023) Silicon nutrition in plants under water-deficit conditions: overview and prospects. Water 15(4):739

Islam W, Tayyab M, Khalil F, Hua Z, Huang Z, Chen HY (2020) Silicon-mediated plant defense against pathogens and insect pests. Pestic Biochem Physiol 168:104641

Iwasaki K, Maier P, Fecht M, Horst WJ (2002) Effects of silicon supply on apoplastic manganese concentrations in leaves and their relation to manganese tolerance in cowpea (Vigna unguiculata (L.) Walp.). Plant Soil 238:281–288

Jana S, Jeong BR (2014) Silicon: the most under-appreciated element in horticultural crops. Trends Hortic 4(1):1–19

Jones L, Handreck K (1965) Studies of silica in the oat plant: III. Uptake of silica from soils by the plant. Plant Soil 23:79–96

Karagiannis E, Michailidis M, Skodra C, Molassiotis A, Tanou G (2021) Silicon influenced ripening metabolism and improved fruit quality traits in apples. Plant Physiol Biochem 166:270–277

Khan MIR, Ashfaque F, Chhillar H, Irfan M, Khan NA (2021) The intricacy of silicon, plant growth regulators and other signaling molecules for abiotic stress tolerance: an entrancing crosstalk between stress alleviators. Plant Physiol Biochem 162:36–47

Khan I, Awan SA, Rizwan M, Brestic M, Xie W (2023) Silicon: an essential element for plant nutrition and phytohormones signaling mechanism under stressful conditions. Plant Growth Regul 100(2):301–319

Kim SG, Kim KW, Park EW, Choi D (2002) Silicon-induced cell wall fortification of rice leaves: a possible cellular mechanism of enhanced host resistance to blast. Phytopathology 92(10):1095–1103

Kim Y-H, Khan AL, Waqas M, Jeong H-J, Kim D-H, Shin JS, Kim J-G, Yeon M-H, Lee I-J (2014) Regulation of jasmonic acid biosynthesis by silicon application during physical injury to Oryza sativa L. J Plant Res 127:525–532

Kvedaras OL, Keeping MG, Goebel F-R, Byrne MJ (2007) Water stress augments silicon-mediated resistance of susceptible sugarcane cultivars to the stalk borer Eldana saccharina (Lepidoptera: Pyralidae). Bull Entomol Res 97(2):175–183

Lata-Tenesaca LF, Oliveira MJB, Barros AV, Picanço BBM, Rodrigues FÁ (2024) Physiological and biochemical aspects of silicon-mediated resistance in maize against Maydis leaf blight. Plan Theory 13(4):531

Li Y, Xiao J, Hu J, Jeong BR (2020) Method of silicon application affects quality of strawberry daughter plants during cutting propagation in hydroponic substrate system. Agronomy 10(11):1753

Liang Y, Hua H, Zhu YG, Zhang J, Cheng C, Römheld V (2006) Importance of plant species and external silicon concentration to active silicon uptake and transport. New Phytol 172(1):63–72

Liang Y, Nikolic M, Bélanger R, Gong H, Song A, Liang Y, Nikolic M, Bélanger R, Gong H, Song A (2015) Silicon-mediated tolerance to salt stress. In: Silicon in agriculture: from theory to practice. Springer, Dordrecht, pp 123–142

López-Pérez MC, Pérez-Labrada F, Ramírez-Pérez LJ, Juárez-Maldonado A, Morales-Díaz AB, González-Morales S, García-Dávila LR, García-Mata J, Benavides-Mendoza A (2018) Dynamic modeling of silicon bioavailability, uptake, transport, and accumulation: applicability in improving the nutritional quality of tomato. Front Plant Sci 9:647

Lux A, Luxová M, Abe J, Tanimoto E, Hattori T, Inanaga S (2003) The dynamics of silicon deposition in the sorghum root endodermis. New Phytol 158(3):437–441

Lux A, Lukačová Z, Vaculík M, Švubová R, Kohanová J, Soukup M, Martinka M, Bokor B (2020) Silicification of root tissues. Plants 9(1):111

Ma JF (2004) Role of silicon in enhancing the resistance of plants to biotic and abiotic stresses. Soil Sci Plant Nutr 50(1):11–18

Ma JF, Miyake Y, Takahashi E (2001) Silicon as a beneficial element for crop plants. In: Studies in plant science, vol 8, pp 17–39

Ma J, Ali S, Saleem MH, Mumtaz S, Yasin G, Ali B, Al-Ghamdi AA, Elshikh MS, Vodnar DC, Marc RA (2022) Short-term responses of spinach (Spinacia oleracea L.) to the individual and combinatorial effects of nitrogen, phosphorus and potassium and silicon in the soil contaminated by boron. Front Plant Sci 13:983156

Maghsoudi K, Emam Y, Ashraf M (2015) Influence of foliar application of silicon on chlorophyll fluorescence, photosynthetic pigments, and growth in water-stressed wheat cultivars differing in drought tolerance. Turk J Bot 39(4):625–634

Mandlik R, Thakral V, Raturi G, Shinde S, Nikolić M, Tripathi DK, Sonah H, Deshmukh R (2020) Significance of silicon uptake, transport, and deposition in plants. J Exp Bot 71(21):6703–6718

Massey FP, Hartley SE (2006) Experimental demonstration of the antiherbivore effects of silica in grasses: impacts on foliage digestibility and vole growth rates. Proc R Soc B Biol Sci 273(1599):2299–2304

Massey FP, Hartley SE (2009) Physical defences wear you down: progressive and irreversible impacts of silica on insect herbivores. J Anim Ecol 78(1):281–291

Massey FP, Massey K, Ennos AR, Hartley SE (2009) Impacts of silica-based defences in grasses on the feeding preferences of sheep. Basic Appl Ecol 10(7):622–630

Mavrič Čermelj A, Golob A, Vogel-Mikuš K, Germ M (2021) Silicon mitigates negative impacts of drought and UV-b radiation in plants. Plan Theory 11(1):91

McGinnity P (2015) Silicon and its role in crop production. A literature review 27

Meena V, Dotaniya M, Coumar V, Rajendiran S, Ajay KS, Subba Rao A (2014) A case for silicon fertilization to improve crop yields in tropical soils. Proc Natl Acad Sci India Sect B Biol Sci 84:505–518

Mir RA, Bhat BA, Yousuf H, Islam ST, Raza A, Rizvi MA, Charagh S, Albaqami M, Sofi PA, Zargar SM (2022) Multidimensional role of silicon to activate resilient plant growth and to mitigate abiotic stress. Front Plant Sci 13:819658

Mitani N, Ma JF (2005) Uptake system of silicon in different plant species. J Exp Bot 56(414):1255–1261

Mohaghegh P, Shirvani M, Ghasemi S (2010) Silicon application effects on yield and growth of two cucumber genotypes in hydroponics system. J Soil Plant Interact-Isfahan Univ Technol 1(1):35–40

Muneer S, Jeong BR (2015) Silicon decreases Fe deficiency responses by improving photosynthesis and maintaining composition of thylakoid multiprotein complex proteins in soybean plants (Glycine max L.). J Plant Growth Regul 34:485–498

Muneer S, Jeong HK, Park YG, Jeong BR (2018) Proteomic analysis of aphid-resistant and-sensitive rose (*Rosa Hybrida*) cultivars at two developmental stages. Proteomes 6(2):25

Muneer S, Park YG, Kim S, Jeong BR (2017) Foliar or subirrigation silicon supply mitigates high temperature stress in strawberry by maintaining photosynthetic and stress-responsive proteins. J Plant Growth Regul 36:836–845

Muneer S, Park YG, Manivannan A, Soundararajan P, Jeong BR (2014) Physiological and proteomic analysis in chloroplasts of *Solanum lycopersicum* L. under silicon efficiency and salinity stress. Int J Mol Sci 15(12):21803–21824
Nicchio B, Juliatti FC, Pereira HS (2024) Preventive application of Si to control orange rust on varieties of sugarcane seedlings cultivated under greenhouse conditions. J Plant Nutr 47:1214–1229
Ning D, Zhang Y, Li X, Qin A, Huang C, Fu Y, Gao Y, Duan A (2023) The effects of foliar supplementation of silicon on physiological and biochemical responses of winter wheat to drought stress during different growth stages. Plan Theory 12(12):2386
Noh K, Jeong BR (2022) Silicon supplementation alleviates adverse effects of ammonium on ssamchoo grown in home cultivation system. Plants 11(21):2882
Nunez P, Cabán-Acevedo M, Yu W, Richter MH, Kennedy K, Villarino AM, Brunschwig BS, Lewis NS (2021) Origin of the electrical barrier in electrolessly deposited platinum nanoparticles on p-si surfaces. J Phys Chem C 125(32):17660–17670
Pang Z, Peng H, Lin S, Liang Y (2024) Theory and application of a Si-based defense barrier for plants: implications for soil-plant-atmosphere system health. Crit Rev Environ Sci Technol 54(9):722–746
Pavlovic J, Kostic L, Bosnic P, Kirkby EA, Nikolic M (2021) Interactions of silicon with essential and beneficial elements in plants. Front Plant Sci 12:697592
Petkowski JJ, Bains W, Seager S (2020) On the potential of silicon as a building block for life. Life 10(6):84
Pontigo S, Ribera A, Gianfreda L, de la Luz MM, Nikolic M, Cartes P (2015) Silicon in vascular plants: uptake, transport and its influence on mineral stress under acidic conditions. Planta 242:23–37
Park YG, Muneer S, Kim S, Hwang SJ, Jeong BR (2018a) Foliar or subirrigational silicon supply modulates salt stress in strawberry during vegetative propagation. Horti Environ Biotechnol 59:11–18
Park YG, Muneer S, Kim S, Hwang SJ, Jeong BR (2018b) Silicon application during vegetative propagation affects photosynthetic protein expression in strawberry. Horti Environ Biotechnol 59:167–177
Park YG, Sivanesan I, Jeong BR (2013) Effect of silicon source and application method on growth and development, and incidence of powdery mildew (*Sphaerotheca pannosa* var. *rosae*) in potted Rosa hybrida'Apollo'and'Remata'. 한국원예학회 학 술발표요지:145–145
Rafi MM, Epstein E (1999) Silicon absorption by wheat (Triticum aestivum L.). Plant Soil 211.223–230
Rafi MM, Epstein E, Falk RH (1997) Silicon deprivation causes physical abnormalities in wheat (Triticum aestivum L). J Plant Physiol 151(4):497–501
Reynolds OL, Padula MP, Zeng R, Gurr GM (2016) Silicon: potential to promote direct and indirect effects on plant defense against arthropod pests in agriculture. Front Plant Sci 7:188037
Rizwan M, Ali S, Ibrahim M, Farid M, Adrees M, Bharwana SA, Zia-ur-Rehman M, Qayyum MF, Abbas F (2015) Mechanisms of silicon-mediated alleviation of drought and salt stress in plants: a review. Environ Sci Pollut Res 22:15416–15431
Rizwan M, Meunier J-D, Davidian J-C, Pokrovsky O, Bovet N, Keller C (2016) Silicon alleviates cd stress of wheat seedlings (Triticum turgidum L. cv. Claudio) grown in hydroponics. Environ Sci Pollut Res 23:1414–1427
Rodrigues F, Dallagnol LJ, Duarte HSS, Datnoff LE (2015) Silicon control of foliar diseases in monocots and dicots. In: Silicon and plant diseases. Springer, Cham, pp 67–108
Sakr N (2021) Silicon reduces Fusarium head blight development in barley. Open Agric J 15(1):54
Sambo P, Nicoletto C, Giro A, Pii Y, Valentinuzzi F, Mimmo T, Lugli P, Orzes G, Mazzetto F, Astolfi S (2019) Hydroponic solutions for soilless production systems: issues and opportunities in a smart agriculture perspective. Front Plant Sci 10:923
Savant NK, Datnoff LE, Snyder GH (1997) Depletion of plant-available silicon in soils: a possible cause of declining rice yields. Commun Soil Sci Plant Anal 28(13–14):1245–1252

Savvas D, Ntatsi G (2015) Biostimulant activity of silicon in horticulture. Sci Hortic 196:66–81
Savvas D, Manos G, Kotsiras A, Souvaliotis S (2002) Effects of silicon and nutrient-induced salinity on yield, flower quality and nutrient uptake of gerbera grown in a closed hydroponic system. J Appl Bot 76(5/6):153–158
Schuerger AC, Amaradasa BS, Dufault NS, Hummerick ME, Richards JT, Khodadad CL, Smith TM, Massa GD (2021) Fusarium oxysporum as an opportunistic fungal pathogen on Zinnia hybrida plants grown on board the international Space Station. Astrobiology 21(9):1029–1048
Seebold K, Kucharek T, La D, Correa-Victoria FJ, Marchetti M (2001) The influence of silicon on components of resistance to blast in susceptible, partially resistant, and resistant cultivars of rice. Phytopathology 91(1):63–69
Shanmugaiah V, Gauba A, Hari SK, Prasad R, Ramamoorthy V, Sharma MP (2023) Effect of silicon micronutrient on plant's cellular signaling cascades in stimulating plant growth by mitigating the environmental stressors. Plant Growth Regul 100(2):391–408
Shen X, Zhou Y, Duan L, Li Z, Eneji AE, Li J (2010) Silicon effects on photosynthesis and antioxidant parameters of soybean seedlings under drought and ultraviolet-B radiation. J Plant Physiol 167(15):1248–1252
Shen X, Li Z, Duan L, Eneji AE, Li J (2014a) Silicon mitigates ultraviolet-B radiation stress on soybean by enhancing chlorophyll and photosynthesis and reducing transpiration. J Plant Nutr 37(6):837–849
Shen X, Xiao X, Dong Z, Chen Y (2014b) Silicon effects on antioxidative enzymes and lipid peroxidation in leaves and roots of peanut under aluminum stress. Acta Physiol Plant 36:3063–3069
Shilpha J, Manivannan A, Soundararajan P, Jeong BR (2023) Heat stress mitigation by silicon nutrition in plants: a comprehensive overview. In: de Mello Prado R (ed) Benefits of silicon in the nutrition of plants. Springer, pp 329–346
Shivaraj S, Mandlik R, Bhat JA, Raturi G, Elbaum R, Alexander L, Tripathi DK, Deshmukh R, Sonah H (2022) Outstanding questions on the beneficial role of silicon in crop plants. Plant Cell Physiol 63(1):4–18
Singh VP, Su Q (2022) Water-environment-energy-food nexus: challenges and opportunities under climate change. Indian J Soil Conserv 50(3):177–189
Son M-S, Oh H-J, Song J-Y, Lim M-Y, Sivanesan I, Jeong B-R (2012) Effect of silicon source and application method on growth of kalanchoe'Peperu'. Korean J Horti Sci Technol 30(3):250–255
Son MS, Sivanesan I, Ko CH, Jeong BR (2013) Effect of silicon source and application method on growth of Spathiphyllum patinii'Chopin'. 한국원예학회 **학술발표요지**:146–146
Song X-P, Verma KK, Tian D-D, Zhang X-Q, Liang Y-J, Huang X, Li C-N, Li Y-R (2021) Exploration of silicon functions to integrate with biotic stress tolerance and crop improvement. Biol Res 54:19
Song J, Yang J, Jeong BR (2022a) Decreased solution pH and increased K+ uptake are related to ammonium tolerance in hydroponically cultured plants. Horticulturae 8(3):228
Song J, Yang J, Jeong BR (2022b) Silicon mitigates ammonium toxicity in cabbage (*Brassica campestris* L. ssp. pekinensis)'Ssamchu'. Front Sustain Food Syst 6:922666
Soundararajan P, Sivanesan I, Jana S, Jeong BR (2014) Influence of silicon supplementation on the growth and tolerance to high temperature in Salvia splendens. Horti Environ Biotechnol 55(4):271–279. https://doi.org/10.1007/s13580-014-0023-8
Souri Z, Khanna K, Karimi N, Ahmad P (2021) Silicon and plants: current knowledge and future prospects. J Plant Growth Regul 40:906–925
Sun W, Zhang J, Fan Q, Xue G, Li Z, Liang Y (2010) Silicon-enhanced resistance to rice blast is attributed to silicon mediated defence resistance and its role as physical barrier. Eur J Plant Pathol 128:39–49
Sun S, Yang Z, Song Z, Wang N, Guo N, Niu J, Liu A, Bai B, Ahammed GJ, Chen S (2022) Silicon enhances plant resistance to Fusarium wilt by promoting antioxidant potential and photosynthetic capacity in cucumber (Cucumis sativus L.). Front Plant Sci 13:1011859
Tamai K, Ma JF (2003) Characterization of silicon uptake by rice roots. New Phytol 158(3):431–436

Tavallali V, Karimi S (2023) Developing nano silicon-salicylic acid complex for inducing root-knot nematode resistance in tomato. SILICON 15(17):7335–7343
Trejo-Téllez LI, García-Jiménez A, Escobar-Sepúlveda HF, Ramírez-Olvera SM, Bello-Bello JJ, Gómez-Merino FC (2020) Silicon induces hormetic dose-response effects on growth and concentrations of chlorophylls, amino acids and sugars in pepper plants during the early developmental stage. PeerJ 8:e9224
Van Bockhaven J, De Vleesschauwer D, Höfte M (2013) Towards establishing broad-spectrum disease resistance in plants: silicon leads the way. J Exp Bot 64(5):1281–1293
Waewthongrak W, Pisuchpen S, Leelasuphakul W (2015) Effect of Bacillus subtilis and chitosan applications on green mold (Penicillium digitatum Sacc.) decay in citrus fruit. Postharvest Biol Technol 99:44–49
Wang M, Gao L, Dong S, Sun Y, Shen Q, Guo S (2017) Role of silicon on plant–pathogen interactions. Front Plant Sci 8:255703
Xiao J, Li Y, Jeong BR (2022) Foliar silicon spray to strawberry plants during summer cutting propagation enhances resistance of transplants to high temperature stresses. Front Sustain Food Syst 6:938128
Xuebin Q, Yatao X, Ahmad MI, Shehzad M, Zain M (2020) Silicon and its application methods improve physiological traits and antioxidants in Triticum aestivum (L.) under cadmium stress. J Soil Sci Plant Nutr 20:1110–1121
Yang J, Song J, Jeong BR (2022) Drenched silicon suppresses disease and insect pests in coffee plant grown in controlled environment by improving physiology and upregulating defense genes. Int J Mol Sci 23(7):3543
Yao X, Chu J, Cai K, Liu L, Shi J, Geng W (2011) Silicon improves the tolerance of wheat seedlings to ultraviolet-B stress. Biol Trace Elem Res 143:507–517
Ye M, Song Y, Long J, Wang R, Baerson SR, Pan Z, Zhu-Salzman K, Xie J, Cai K, Luo S (2013) Priming of jasmonate-mediated antiherbivore defense responses in rice by silicon. Proc Natl Acad Sci 110(38):E3631–E3639
Yilmaz Y, Korkmaz A (2023) Effects of different types of irrigation water quality and silicon doses on fruit yield, chlorophyll and carotenoid contents of tomato (Lycopersicon esculentum L.) under soilless culture technique. J Agric Sci 29(3)
Yoshida S, Ohnishi Y, Kitagishi K (1962) Histochemistry of silicon in rice plant: I. A new method for determining the localization of silicon within plant tissues. Soil Sci Plant Nutr 8(1):30–35
Zargar SM, Mahajan R, Bhat JA, Nazir M, Deshmukh R (2019) Role of silicon in plant stress tolerance: opportunities to achieve a sustainable cropping system. 3. Biotech 9(3):73
Zhu Y, Gong H (2014) Beneficial effects of silicon on salt and drought tolerance in plants. Agron Sustain Dev 34:455–472
Zia Z, Bakhat HF, Saqib ZA, Shah GM, Fahad S, Ashraf MR, Hammad HM, Naseem W, Shahid M (2017) Effect of water management and silicon on germination, growth, phosphorus and arsenic uptake in rice. Ecotoxicol Environ Saf 144:11–18

Chapter 19
Silicon and Plant Disease Control

Fabrício Ávila Rodrigues and Leandro José Dallagnol

Abstract Silicon (Si) helps to reduce the intensities of several seedborne, soilborne, and foliar diseases caused by biotrophic, hemibiotrophic, and necrotrophic plant pathogens in many crops. Plants displaying high levels of Si in their tissues are less prone to be severely affected by the infection process of the pathogen. The reduction in disease symptoms is linked to the effect of Si on some components of host resistance (e.g., incubation period, lesion size, lesion number, frequency of infection, and infectious period). The polymerization of Si beneath the cuticle and in the cell walls stops the penetration by the pathogens. On top of that, Si may potentiate the phenylpropanoid and terpenoid pathways and help plants to have a faster and stronger transcription of defense-related genes and higher activities of defense enzymes. When infected by pathogens, the plants supplied with Si show improved photosynthesis and a more robust antioxidant system.

Keywords Silicon · Nanosilica · Disease · Pathogen · Host resistance mechanisms

19.1 Introduction

Silicon (Si), a tetravalent metalloid, plays a fundamental role in shaping the composition of various minerals and characteristics of the Earth's crust, influencing geological processes, the formation of several types of soil, and the ecosystem dynamics overall (Datnoff et al. 1997; Debona et al. 2017, 2023). Silica (SiO_2), the primary constituent of many minerals, including quartz, chalcedony, and opal, results from

F. Á. Rodrigues (✉)
Department of Plant Pathology, Viçosa Federal University, Viçosa, Minas Gerais State, Brazil
e-mail: fabricio@ufv.br

L. J. Dallagnol
Department of Crop Protection, Faculty of Agronomy Eliseu Maciel, Federal University of Pelotas, Pelotas, Rio Grande do Sul State, Brazil

R. de Mello Prado et al. (eds.), *Silicon Advances for Sustainable Agriculture and Human Health*, Sustainable Plant Nutrition in a Changing World,
https://doi.org/10.1007/978-3-031-69876-7_19

the combination of Si and oxygen. Silicon also forms the backbone of silicate minerals, which comprise most of the Earth's crust. Silicate minerals are classified based on their structure, with groups including framework silicates (e.g., feldspar and quartz), sheet silicates (e.g., mica and clay minerals), chain silicates (e.g., pyroxene and amphibole), and isolated tetrahedra (e.g., olivine) (Debona et al. 2017, 2023).

The Si plays a significant role in the physiology of diatoms, yellow-brown or golden algae, and the Chrysophyceae and Equisetaceae (Debona et al. 2017, 2023). On the other hand, Si is considered to be quasi-essential for the growth and development of many plant species (Debona et al. 2017, 2023). The monoculture of profitable crops linked to the slow liberation of Si from minerals in the soil, along with its adsorption or resorption by the soil particles, is the leading cause of the depletion of this element in many types of soil, such as acidic alfisols, acidic inceptisols, organic soils (histosols), oxisols, spodosols, and ultisols (Datnoff et al. 1997; Zellner et al. 2021). In this scenario, soil amendment with soluble sources of Si is needed. The monosilicic acid (H_4SiO_4), an uncharged monomeric molecule, is the form of Si that plants uptake from the soil solution (Rodrigues and Datnoff 2015; Zellner et al. 2021; Debona et al. 2017, 2023). Following uptake, the H_4SiO_4 is transported to shoots either passively or actively (Rodrigues and Datnoff 2015; Zellner et al. 2021; Debona et al. 2017, 2023). Specific proteins found in barley, maize, rice, and wheat plants are involved in the active Si transport from roots to shoots (Zellner et al. 2021; Debona et al. 2017, 2023). Several families of transport proteins have been identified and are involved in the active uptake of Si in plants. The nodulin 26-like intrinsic protein aquaporins are integral membrane proteins that facilitate the movement of silicic acid across cell membranes in plant species known to uptake Si actively. For instance, two different types of transporters encoded by genes Low Silicon 1 (*Lsi1*) and Low Silicon 2 (*Lsi2*) were involved in both active uptake and distribution of Si in rice plant tissues (Rodrigues and Datnoff 2015; Zellner et al. 2021; Debona et al. 2017, 2023). The *Lsi1*, a Si permeable channel, belongs to the nodulin 26-like major intrinsic protein III subgroup of the aquaporin membrane protein family and facilitates the Si transport across the plasma membrane between the external solution (e.g., apoplast) and the plant cells (Zellner et al. 2021; Debona et al. 2017, 2023). Homologs of *Lsi1* were found in rice (*Lsi6*), barley (*HvLsi1* and *HvLsi6*), wheat (*TaLsi1*), maize (*ZmLsi1* and *ZmLsi6*), cucumber (*CSiT-1* and *CSiT-2*), pumpkin (*CmLsi1*), and soybean (*GmLsi1*) (Rodrigues and Datnoff 2015). A specific foliar concentration of Si that may range from 0.1 to 10% on a dry-weight basis is found on plants belonging to more than 44 angiosperm clades that represent over 100 orders or families (Zellner et al. 2021; Debona et al. 2017, 2023). However, the efficiency of Si uptake by roots is affected by several factors such as the crop, cultivar, plant growth stage, the source of Si amended to the soil, and the physical-chemical reactions in the soil (Rodrigues and Datnoff 2015; Zellner et al. 2021; Debona et al. 2017, 2023). Interestingly, crops such as coffee, pepper, and tomato are known to keep high levels of Si on the roots (greater than 0.5%) rather than in the shoots (less than 0.5%) (Zellner et al. 2021; Debona et al. 2017, 2023). Conversely, monocots (e.g., rice, sugarcane, and wheat) efficiently transport Si from the soil solution to the shoots.

A very efficient Si uptake by the root system ends up creating a physical barrier due to the polymerization of the H_4SiO_4 into insoluble silica (phytoliths) either in the root cell walls or in the cell wall of different tissues in the aerial parts of the plants especially beneath the cuticle (Rodrigues and Datnoff 2015; Zellner et al. 2021; Debona et al. 2017, 2023). Plants facing different types of abiotic (e.g., cold, drought, heat, salinity, and mineral deficiency or toxicity) and biotic (e.g., plant diseases and herbivores, and other pests) stress are more prone to be resilient against them due to a high concentration of Si found on their tissues (Rodrigues and Datnoff 2015; Zellner et al. 2021; Debona et al. 2017, 2023). The abundance of Si in nature and the lack of visible symptoms of deficiency in dicots such as rice and sugarcane contributed to overlooking its importance for more sustainable agriculture.

This chapter aims to provide the reader with the most important scientific information available in the literature regarding using Si to control diseases affecting profitable crops and the mechanisms that explain how this element helps plants defend themselves against infection by pathogens.

19.2 Diseases Controlled by Silicon

Many pathogens cause several diseases in different profitable crops, which are responsible for great yield losses. Some pathogens only cause mild symptoms, while others cause severe epidemics affecting large cultivated areas. Changes in the pathogen population in time and space, as well as the organ of the plant affected, make disease control a very difficult task (Strange and Scott 2005). Furthermore, pathogens may overcome the resistance of a certain cultivar and/or develop resistance to fungicides, requiring, therefore, increases in the frequency of sprays or the number of active ingredients simultaneously applied to achieve an acceptable level of disease control (Jones and Dangl 2006). In order to avoid the losses caused by pathogens, it is necessary to seek more unspecific and durable strategies. In this sense, Si emerges as an important measure that can be employed. Silicon has been reported as an element that increases the defense of different crops against biotrophic, hemibiotrophic, and necrotrophic pathogens belonging to different groups such as fungi, oomycetes, bacteria, and nematodes in both monocot and dicot plants.

In monocot plants, Si was considered an alternative for controlling diseases affecting rice, wheat, barley, sorghum, oat, sugarcane, maize, grass, ryegrass, banana, and finger millet (Table 19.1). The diseases on different crops reported to be negatively affected by Si in terms of reducing the development of symptoms and inoculum production are briefly discussed thereafter.

Table 19.1 Effect of silicon on the development of diseases occurring in monocots

Crops	Diseases	Pathogens
Rice	Blast	*Pyricularia oryzae*
	Brown spot	*Bipolaris oryzae*
	Sheath blight	*Rhizoctonia solani*
	Leaf scald	*Monographella albescens*
	Stem rot	*Magnaporthe salvinii* and *Exserohilum rostratum*
	Bacterial blight	*Xanthomonas oryzae* pv. *oryzae*
	Root-knot nematode	*Meloidogyne* spp. and *Meloidogyne graminicola*
Wheat	Powdery mildew	*Blumeria graminis* f. sp. *tritici*
	Blast	*Magnaporthe oryzae*
	Spot blotch	*Bipolaris sorokiniana*
	Tan spot	*Pyrenophora tritici-repentis*
	Leaf blotch	*Parastagonospora nodorum*
	Septoria leaf blotch	*Zymoseptoria tritici*
	Eyespot	*Oculimacula yallundae*
	Bacterial leaf streak	*Xanthomonas translucens* pv. *undulosa*
	Damping-off and root rot	*Rhizoctonia solani*
	Leaf rust	*Puccinia triticina*
Barley	Powdery mildew	*Blumeria graminis* f. sp. *hordei*
	Spot blotch	*Bipolaris sorokiniana*
	Fusarium head blight	*Fusarium* spp.
Sorghum	Anthracnose	*Colletotrichum sublineolum*
Oat	Stem rust	*Puccinia graminis* f. sp. *avenae*
	Root rot	*Rhizoctonia solani*
Sugarcane	Ringspot	*Leptosphaeria sacchari*
	Red rot	*Colletotrichum falcatum*
	Orange rust	*Puccinia kuehnii*
	Brown rust	*Puccinia melanocephala*
	Smut	*Sporisorium scitamineum*
	Leaf scald	*Xanthomonas albilineans*
Maize	Macrospora leaf spot	*Stenocarpella macrospora*
	Northern leaf blight	*Exserohilum turcicum*
	Maydis leaf blight	*Bipolaris maydis*
Creeping bentgrass	Large brown patch	*Rhizoctonia solani*
Kentucky bluegrass	Root rot	*Pythium aphanidermatum*
	Dollar spot	*Sclerotinia homoeocarpa*
	Brown patch	*Rhizoctonia solani*
	Powdery mildew	*Erysiphe graminis*

(continued)

Table 19.1 (continued)

Crops	Diseases	Pathogens
Perennial ryegrass	Gray leaf spot	*Pyricularia oryzae*
Annual ryegrass	Gray leaf spot	*Pyricularia oryzae*
	Brown leaf spot	*Bipolaris sorokiniana*
	Parrot's eye leaf spot	*Cercospora* sp.
Finger millet	Blast	*Pyricularia grisea*
Bermuda grass	Leaf spot	*Bipolaris cynodontis*
St. Augustine grass	Gray leaf spot	*Pyricularia oryzae*
Banana	Root necrosis	*Cylindrocladium spathiphylli*
	Fusarium wilt	*Fusarium oxysporum* f. sp. *cubense*
	Black Sigatoka	*Mycosphaerella fijiensis*
	Xanthomonas wilt	*Xanthomonas campestris* pv. *musacearum*

*19.2.1 Rice (***Oryza sativa***)*

Rice is considered a model plant for studying the mechanisms involved in potentiating host resistance by Si (Rodrigues and Datnoff 2015). The Japanese researchers performed pioneering studies demonstrating that application of various Si sources to Si-deficient paddy soils reduced blast, caused by *Pyricularia oryzae*, intensity (Rodrigues and Datnoff 2015). A linear and negative relationship was reported to occur between the number of blast lesions on the leaves of rice plants and the foliar Si concentration (Rodrigues and Datnoff 2015). The same trend was verified for several rice cultivars grown in similar environmental conditions but differing in their capacity to accumulate Si in both leaves and panicles. These cultivars showed differences in their level of resistance against leaf and panicle (neck) blast with lower incidence on plants from cultivars that had higher foliar Si concentration (Rodrigues and Datnoff 2015). Accordingly, blast intensity on some rice cultivars grown in soil that received different Si rates showed a negative correlation between disease and foliar Si concentration (Rodrigues and Datnoff 2015). However, for a given level of Si in the soil, a cultivar with the greatest foliar Si concentration was not necessarily more resistant to diseases than a cultivar that accumulated less Si in its tissues (Rodrigues and Datnoff 2015).

The source, rate, and particle size of the Si fertilizers are among the major factors that need to be taken into account to obtain a great control of diseases in rice. Various organic and inorganic Si sources reduced blast incidence by nearly 50% in rice (Rodrigues and Datnoff 2015). In Nigeria, the severity of neck blast was reduced in different rice cultivars by 40 and 50% due to the application of sodium silicate or sodium metasilicate, respectively, to Si-deficient soils (Rodrigues and Datnoff 2015). In Florida, the application of calcium silicate slag (22 dag Si kg^{-1}), at different rates (5, 10, and 15 t ha^{-1}), to a Si-deficient histosol resulted in negative linear and quadratic equations for the relationship of slag rates with neck blast and yield, respectively (Datnoff et al. 1997). In another study evaluating three grades of calcium silicate, namely, fine (100% < 0.15 mm), standard (90% < 2.4 mm), and pellets made from the fine material (100% > 1 mm and < 3.4 mm), for the control of diseases in rice showed that the fine grade was more efficient than the coarse grade for increasing foliar Si concentration and grain yield and for decreasing blast and brown spot (*Bipolaris oryzae*) severities (Datnoff et al. 1997). For several rice cultivars, blast severity decreased as the Si rates (0.25, 1.0, and 4.0 t ha^{-1}) increased in the soil (Vu et al. 2022). The Si nanoparticles applied as foliar spray decreased blast severity by nearly 70% (Du et al. 2022).

One important task to consider regarding using Si for plant disease control is using it in an integrated disease management program alongside other control measures. In this sense, some studies have demonstrated that applying Si as calcium silicate slag reduced neck blast severity as effectively as foliar application of a mercurial fungicide (Datnoff et al. 1997). In Florida, blast control was obtained at the same degree by using the fungicide benomyl or Si alone (0.4 t ha^{-1}), and the greatest blast reduction was obtained by combining Si in the soil and the foliar spray of fungicide (Datnoff et al. 1997). In Colombia, under low blast pressure, applying Si (1 t ha^{-1}) decreased leaf and neck blast severities at the same and even better than the full rate of the fungicides edifenphos or tricyclazole sprayed alone. Furthermore, combining the application of Si in the soil with the foliar fungicide spray at 10–25% of the labeled rate provided control equal to the fungicide applied at the full rate on plants without Si amendment under conducive conditions to leaf and neck blast development (Seebold et al. 2004). Therefore, the number of fungicide applications or even the rates used may be reduced by Si supply. Furthermore, Si appears to be an effective strategy to increase the defense of some rice cultivars against blast. For example, at two locations in eastern Colombia, application of Si to the soil (0.5 or 1.0 t ha^{-1}) decreased leaf and neck blast severities on partially resistant cultivars to levels compared to those observed on resistant cultivars non-supplied with Si (Seebold et al. 2000).

Another important issue is the interval between Si application. In this sense, a minimum 1-year residual effect on disease suppression was observed for soil Si applications, a result that indicates an amortizing of the costs for growers for both Si fertilizer and its application (Rodrigues and Datnoff 2015). In Florida, calcium silicate slag (5, 10, and 15 t ha^{-1}) applied to a histosol in 1987 or 1988, complemented or not with 5 t ha^{-1} (to assess the residual activity of Si), demonstrated that although in 1988 rice crop season, the decrease in the neck blast incidence was higher in soil receiving complementation of Si, the 1987 residual effect still were very effective (Datnoff et al. 1997). Experiments performed on an inceptisol and an oxisol for 2 years showed that a first-year residual effect of Si application had a significant effect in decreasing blast severity and increasing yield in the second year, at the same proportion to the Si application in the second year (Seebold et al. 2004).

For sheath blight (*Rhizoctonia solani*), disease intensity was reduced by Si regardless of the rates used (Rodrigues and Datnoff 2015). Initially, the reduction in sheath blight severity was believed to be restricted to the indica upland rice genotypes, because the Si effect was greater compared to japonica and intermediate genotypes (Rodrigues and Datnoff 2015). Another study showed that the accumulation of Si in rice tissues contributed to a decrease in sheath blight in both tropical Japonicus cultivars and an Indica-type (Rodrigues et al. 2001). In addition, authors reported that susceptible and moderately susceptible cultivars supplied with Si displayed similar severity of sheath blight compared to cultivars with high levels of partial resistance but non-supplied with Si. Furthermore, the total number of sheath blight lesions and the total area under the relative lesion extension progress curve decreased as the Si rates increased (from 0 to 8 t ha^{-1}) in the soil in different rice cultivars regardless of the plant growth stage (Rodrigues et al. 2003b). Another experiment using six rice cultivars grown in a Brazilian Si-deficient typic Acrustox red-yellow latosol amended with Si (8 t ha^{-1}) demonstrated that Si concentration in straw increased by 60%, while the total number of sheath blight lesions, total area under the relative lesion extension progress curve, severity of sheath blight, and the highest relative lesion height on the main tiller were reduced by 37, 40, 52, and 24%, respectively, in comparison to plants non-supplied with Si (Rodrigues et al. 2003c).

Studies performed under greenhouse demonstrated that Si supply (2 mM) increased rice defense by affecting several components of rice resistance (incubation period, infection efficiency, lesion size, and lesion expansion rate) resulting in a significant reduction in brown spot severity (Dallagnol et al. 2009, 2011). There was a linear and negative relationship between Si rates and brown spot severity in eight rice genotypes (Prabhu et al. 2012). Grain discoloration, caused by a complex of insects and fungal species, is another important constraint for irrigated and upland rice production worldwide that can be ameliorated by Si (Rodrigues and Datnoff 2015). An experiment performed with 48 genotypes grown in upland conditions showed a quadratic and negative relationship between Si rates and grain discoloration (Prabhu et al. 2012). Furthermore, plants supplied with Si and inoculated with *B. oryzae*, one of the main pathogens associated with grain discoloration, displayed a higher number of grains per panicle, weight of 1000 grains, and percentage of filled grains relative to their inoculated counterparts demonstrating that Si benefited both grain yield and quality (Dallagnol et al. 2014). Regarding seed quality, seeds from plants supplied with Si (2 mM) showed a higher percentage of seedling emergence and a lower percentage of seedlings infected by *B. oryzae* (Dallagnol et al. 2013). According to the authors, high Si concentration in the husk resulted in lower brown spot severity with an increase in the fungicide efficiency when seeds were treated, especially under low disease severity. However, a 2-year study conducted in lowland rice field demonstrated that the plants supplied with calcium silicate (4.9 t ha^{-1}) showed higher foliar Si concentration and lower intensities of brown spot and leaf scald (*Monographella albescens*), but without significant effect on grain yield and quality variables (brown rice content, head rice yield, and translucent rice content) (Rivera et al. 2021). For leaf scald, greenhouse experiments demonstrated that Si supply increased foliar Si concentration, reducing lesion length and disease severity (Tatagiba et al. 2014). In a field study, leaf scald severity decreased by 40% by applying Si (1 t ha^{-1}) (Seebold et al. 2000). Stem rot (*Magnaporthe salvinii*) has also been efficiently suppressed by Si (Rodrigues and

Datnoff 2015). In Peninsular Malaysia, a rice disease caused by *Exserohilum rostratum* was reduced by 24% due to soil amendment with calcium silicate (1.5 t ha^{-1}) (Mahmad-Toher et al. 2022).

Using Si also affects diseases caused by bacteria and nematodes in rice. The lesion length of bacterial blight (*Xanthomonas oryzae* pv. *oryzae*) decreased by 5–22% depending on the rice genotype (Chang et al. 2002), while the disease severity was reduced by 24% for Si-amended plants than for non-Si-amended ones (Song et al. 2016). Interestingly, plants that were switched from a nutrient solution without Si to a nutrient solution containing Si and simultaneously inoculated with *X. oryzae* pv. *oryzae* exhibited the same level of resistance to bacterial blight as the plants that continuously received Si; the first and second groups of plants exhibited reductions in disease severity by 52% and 62%, respectively (Song et al. 2016). Using Si reduced the damage caused by a virulent strain of *X. oryzae* pv. *oryzae* but had little effect on a less virulent strain in susceptible and resistant (*Xa4*, *Xa7*, and *Xa4* + *Xa7* genes) cultivars (Vu et al. 2022). According to the authors, Si had an additive effect on the level of host resistance in avoiding biomass reduction by 20% of infected plants. Regarding nematodes, higher resistance to the root-knot nematode (*Meloidogyne* spp.) was observed in rice cultivars with high Si concentrations in roots (Rodrigues and Datnoff 2015). The use of Si reduced the penetration of J2 from *Meloidogyne graminicola* in rice roots and delayed their development due to intense callose deposition and accumulation of phenolics, but no effect on the development of giant cells was observed (Zhan et al. 2018).

*19.2.2 Wheat (*Triticum aestivum*)*

Several wheat diseases have been controlled by Si (Debona et al. 2023). A 3-year study showed that incorporating rice straw in the soil reduced several wheat diseases including powdery mildew (*B. graminis* f. sp. *tritici*) (Rodgers-Gray and Shaw 2000). In the leaves of wheat plants supplied with Si (soil amended with calcium silicate at the rate of 3 t ha^{-1} or nutrient solution containing 2 mM Si), the colonies of *B. graminis* f. sp. *tritici* were reduced after 5 weeks after plant inoculation, and disease severity reached lesser than 5% in contrast to plants non-supplied with Si. On these plants, the first signs of *B. graminis* f. sp. *tritici* were noticed 5 days after plant inoculation and developed rapidly, and disease severity was up to 40% after 5 weeks (Bélanger et al. 2003). Root applications of Si (1.7 mM) reduced powdery mildew severity by 80% (Guével et al. 2007). A 3-year field experiment showed that calcium silicate, applied as a steel slag by-product at the rate of 4.8 t ha^{-1}, reduced powdery mildew severity to 44% (Provance-Bowley et al. 2010). For blast (*P. oryzae*), wheat plants grown in the presence of Si (2 mM) showed a reduction of leaf blast severity up to 70% (Aucique-Pérez et al. 2014; Debona et al. 2014). An increase in the incubation period of blast by 28% and reduction up to

45% on the number of lesions per cm^2 of leaf were reported by Xavier Filha et al. (2011). Foliar spray of Si, mainly using potassium silicate, reduced blast severity, but the positive effect changed according to the basal level of resistance of the cultivars used (Cruz et al. 2011) A 2-year field experiment showed that Si, applied in the soil as calcium and magnesium silicate in the furrow (1 t ha^{-1}) and as potassium silicate (30 g l^{-1}) applied to the leaves, reduced the incidence and severity of blast in the spikes, but effects were variable according to the year and cultivar (Pagani et al. 2014). The amendment of Si to soil reduced the intensity of spot blotch (*Bipolaris sorokiniana*) and tan spot (*Pyrenophora tritici-repentis*). For spot blotch, soil fertilization with calcium silicate (1.25 g wollastonite kg^{-1} soil) increased in the incubation period and decreased the number of lesions per cm^2 of leaf area and disease severity up to 59% (Rodrigues and Datnoff 2015). For tan spot, calcium silicate, a steel slag by-product at the rate of 8.2 t ha^{-1}, increased both foliar Si concentration and the incubation period but reduced the infection efficiency, the final number of lesions per cm^2, the rate of lesion expansion, and lesion size (Dorneles et al. 2017, 2018; Pazdiora et al. 2018). Under field conditions, a 3-year study showed that calcium silicate (4.0 t ha^{-1}) reduced tan spot severity and increased yield (Pazdiora et al. 2021). The effect on disease reduction and grain yield in Si-amended plants that received one fungicide spray was similar to the plants growing in soil amended with limestone and sprayed with fungicide twice indicating that Si may be an option to reduce fungicide spray (Pazdiora et al. 2021). Furthermore, the damage to wheat technological quality measured by flour color, gluten levels, and the alveograph and farinograph parameters caused by tan spot was minimized in Si-supplied plants (Dallagnol et al. 2023). Calcium silicate (4 t ha^{-1}) incorporated in the soil reduced fusarium head blight (FHB) (a complex of *Fusarium graminearum* species) on spikes index up to 68%, and the reduction was up to 85% when combined with fungicide spray (Pazdiora et al. 2023). However, wheat plants supplied only with Si displayed no effect or a slight increase in the concentration of the mycotoxin deoxynivalenol (DON) (Pazdiora et al. 2023). However, when plants were supplied with Si and sprayed with fungicide, the reduction in DON concentration was maximum (Pazdiora et al. 2023). An interesting point that should be highlighted is that for FHB control with fungicide, sprays should be done at the beginning of the anthesis. However, if the anthesis does not occur at the same time in the same plant, Si showed the potential to increase the period of fungicide spray while still providing good control of the disease by affecting the fungal colonization on spikes tissues (Pazdiora et al. 2023; Dallagnol et al. 2020). Other diseases on wheat plants positively affected by Si were leaf blotch (*Parastagonospora nodorum*), septoria leaf blotch (*Zymoseptoria tritici*), eyespot (*Oculimacula yallundae*), and bacterial leaf streak (*Xanthomonas translucens* pv. *undulosa*) (Rodrigues and Datnoff 2015). Wheat seeds treated with Si dioxide nanoparticles (100 μg ml^{-1}) showed reductions up to 83% and 78% in the post-emergence damping-off and root rot, respectively, due to infection by *R. solani* (Abdelrhim et al. 2021). The incidence of leaf rust (*Puccinia triticina*) was reduced by Si at the rate of 6 mM (Naz et al. 2021).

*19.2.3 Barley (*Hordeum vulgare*)*

Barley plants supplied with Si (3.6 mM) displayed a reduction of up to 35% in the number of colonies formed by *Blumeria graminis* f. sp. *hordei*, which was associated with an increase in papillae deposition in the attempted sites of fungal penetration and less powdery mildew severity (Wiese et al. 2005). The spot blotch (*Bipolaris sorokiniana*) severity was reduced to 50% when calcium silicate was added to the soil (Holz et al. 2022, 2023). According to the authors, the effect of Si was cultivar-dependent, but plants supplied with Si showed lower lesion expansion rate and lower values for both disease severity and area under disease progress curve compared to plants non-supplied with Si. Fungicide spray for spot blotch control was maximized in the presence of Si resulting in less disease severity values (Holz et al. 2022, 2023). Based on greenhouse and field studies, the supply of Si reduced FHB (*Fusarium* spp.) on spikes up to 20% and was associated with lower disease incidence and less colonization of spike tissues by fungal hyphae (Sakr 2021; Sakr and Kurdali 2023).

*19.2.4 Sorghum (*Sorghum bicolor*)*

Anthracnose (*Colletotrichum sublineolum*) severity and the area under disease index progress curve showed a negative and quadratic relationship with the calcium silicate rates ranging from 0.06 to 0.30 g Si kg^{-1} applied in a Si-deficient soil (Resende et al. 2009). The authors verified that by increasing the rate of calcium silicate in the soil, the foliar Si concentration increased, resulting in longer incubation and latent periods as well as lower values for area under relative infection efficiency progress curve, percentage of pigmented leaf area, and percentage of necrotic leaf area. A field experiment showed that calcium silicate (6 t ha^{-1}) application reduced the area under anthracnose progress curve by 39 and 42% in resistant and susceptible sorghum lines, respectively (Resende et al. 2013). Roots of sorghum plants supplied with Si (1 mM) were less prone to be infected by *Alternaria alternata* considering that the fungal hyphae penetration through root cell walls into the protoplasts was arrested due to intense depositions of suberin and phenolics in the exodermis (Bathoova et al. 2018, 2021).

*19.2.5 Oat (*Avena sativa*)*

Oat plants receiving Si (1.5 mM) showed reduced severity of stem rust (*Puccinia graminis* f. sp. *avenae*), and the pustules showed a less intense orange color (Li et al. 2022). Root rot (*R. solani*) was reduced by 44% on oat plants supplied with Si nanoparticles under greenhouse conditions (Ahmad et al. 2023).

*19.2.6 Sugarcane (*Saccharum officinarum*)*

Soil amendment with calcium silicate (6.7 t ha^{-1}) reduced the severity of ringspot (*Leptosphaeria sacchari*) across five cultivars by 67% (Rodrigues and Datnoff 2015). A field experiment using wollastonite and calcium silicate as the Si sources in the soil (600 kg of Si ha^{-1}) and foliar spray of potassium silicate and silicic acid indicated that only silicic acid (100 and 300 g ha^{-1}) reduced the severity of ringspot while none of the Si sources affected red rot (*Colletotrichum falcatum*) or orange rust (*Puccinia kuehnii*) (Nicchio et al. 2023). Orange rust severity was reduced on plants receiving Si as potassium silicate, calcium silicate, and silicic acid in a greenhouse experiment (Nicchio et al. 2023). For brown rust (*Puccinia melanocephala*), Si (555 kg Si ha^{-1}) applied to the Quartzipsamment, Rhodic Hapludox, and Rhodic Acrudox soil types decreased disease incidence by 29, 41, and 47%, respectively (Camargo et al. 2013). In South Africa, the severity and area under the brown rust progress curve were reduced by foliar spray of potassium silicate (> 400 mg Si l^{-1}) or soil amended with calcium silicate (1016 mg Si l^{-1}) (Ramouthar et al. 2016). The incidence of smut (*Sporisorium scitamineum*) was reduced due to Si supply to the soil, but the effect was variable among cultivars (Deng et al. 2020). According to the authors, disease incidence decreased as Si rates increased reaching zero by applying 45 g of exogenous Si $barrel^{-1}$. Bacterial diseases were also reduced by Si. The incidence of leaf scald (*Xanthomonas albilineans*) was lower, up to 63%, and the disease index was reduced by up to 60% on plants supplied with 7.70 g Si kg^{-1} of soil (Hong et al. 2021). Greenhouse and field experiments showed that Si applied in the form of organosilicon plus copper hydroxide spray reduced leaf scald index even in the lowest doses of copper hydroxide indicating, therefore, that Si may be an option to reduce the use of copper for disease control (Zhang et al. 2022).

*19.2.7 Maize (*Zea mays*)*

For macrospora leaf spot (*Stenocarpella macrospora*), the area under disease progress curve on maize leaves was reduced by 42% due to Si supply to the roots (Hawerroth et al. 2019). The severity of Northern leaf blight (*Exserohilum turcicum*) was reduced by 52% by Si amendment through soil irrigation with a solution containing monosilicic acid (2 mM) (Silveira et al. 2021). Furthermore, maydis leaf blight (*Bipolaris maydis*) showed disease symptoms (number and size of the necrotic lesions, lesions coalescence, and chlorosis) more expressive than in the leaves of plants supplied with Si (2 mM) (Fig. 19.1), and disease severity was significantly reduced by 47% (Lata-Tenesaca et al. 2024).

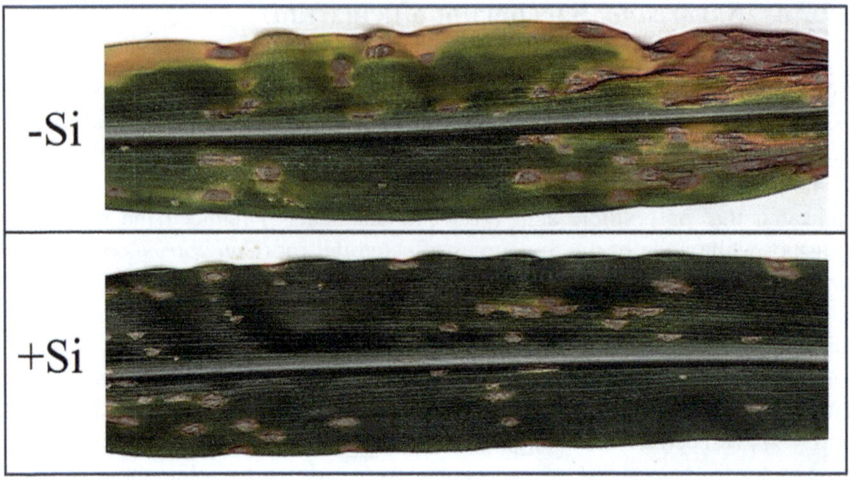

Fig. 19.1 Symptoms of maydis leaf blight on the leaves of maize plants non-supplied (−Si) or supplied (+Si) with silicon (Si)

*19.2.8 Banana (*Musa *spp.)*

The intensity of root necrosis caused by the fungus *Cylindrocladium spathiphylli* was significantly reduced by 50% due to Si supply (2 mM) (Vermeire et al. 2011). The incidence of *Fusarium* wilt (*Fusarium oxysporum* f. sp. *cubense*) was reduced by supplying Si (0.39 g Si kg^{-1}) to banana plants grown in Si-deficient soil (Fortunato et al. 2012). According to the authors, in the roots of plants supplied with Si, there was less colonization of root tissues by fungus hyphae, and the area under reflex leaf symptoms progress curve, area under root symptoms progress curve, area under disease progress curve, and area under asymptomatic fungal colonization of tissue progress curve were significantly reduced by 12, 50, 52, and 20%, respectively. Combining Si with biological control agents to reduce *Fusarium* wilt incidence is also interesting. In Malaysia, the application of sodium silicate in the soil associated with *Bacillus subtilis* reduced the *Fusarium* wilt incidence by 56% (Zakaria et al. 2023). The severity of black sigatoka (*Mycosphaerella fijiensis*) was reduced on the leaves of banana plants grown in the presence of Si (Kablan et al. 2012). Banana plants supplied with Si and in the presence of *Rhizophagus irregularis* (an arbuscular mycorrhizal fungus) showed reduction in the area under disease progress curve of up to 48% (Gbongue et al. 2019). In the presence of Si, banana plants showed increased resistance against *Xanthomonas* wilt (*Xanthomonas campestris* pv. *musacearum*), but the effect of Si was cultivar-dependent (Mburu et al. 2016).

19.2.9 Grass

On creeping bentgrass (*Agrostis stolonifera*), the severity of large brown patch (*R. solani*) decreased on plants growing in a soil amended with porous hydrate calcium silicate (Saigusa et al. 2000). Similarly, Si also reduced the intensities of root rot (*Pythium aphanidermatum*), dollar spot (*Sclerotinia homoeocarpa*), and brown patch (*R. solani*) on creeping bentgrass as well as powdery mildew (*Erysiphe graminis*) on Kentucky bluegrass (*Poa pratensis*) (Rodrigues and Datnoff 2015). On bermudagrass (*Cynodon dactylon*), Si reduced leaf spot (*Bipolaris cynodontis*) severity by 39% (Rodrigues and Datnoff 2015). Gray leaf spot (*Pyricularia oryzae*) severity was reduced by 19–78% on several cultivars of St. Augustine grass (*Stenotaphrum secundatum*) due to soil amendment with calcium silicate (Rodrigues and Datnoff 2015). Under field conditions, soil amendment with calcium silicate amendment (1 t ha^{-1}) reduced the area under gray leaf spot progress curve by 7–41%, fungicide treatment (chlorothalonil) reduced it by 34–65%, while the highest control (reduction by 59–68%) was achieved with calcium silicate + chlorothalonil (Rodrigues and Datnoff 2015).

19.2.10 Ryegrass (Lolium *spp.*)

For perennial ryegrass (*Lolium perenne*), gray leaf spot (*Pyricularia oryzae*) incidence and severity were significantly reduced by wollastonite and calcium silicate slag applied to the soil at rates ranging from 0.6 to 12 t ha^{-1} (Rodrigues and Datnoff 2015). Soil amendment with calcium silicate (5 t ha^{-1}) decreased gray leaf spot incidence and severity by 39 and 47%, respectively, on the leaves of perennial ryegrass plants (Rodrigues and Datnoff 2015). The area under gray leaf spot progress curve was reduced by 65–77% on perennial ryegrass plants supplied with Si (Rodrigues and Datnoff 2015). In annual ryegrass (*Lolium multiflorum*), soil amendment with calcium silicate (7.8 t ha^{-1}) reduced gray leaf spot severity up to 58%, and the combination of Si in the soil and seed treatment with the biological control agent *Trichoderma atroviride* reduced disease severity up to 42% (Rodrigues and Datnoff 2015). A 2-year field study showed that Si or *T. atroviride* reduced gray leaf spot severity by 40–80% as well as the brown leaf spot (*Bipolaris sorokiniana*) and Parrot's eye leaf spot (*Cercospora* sp.) severities (Rodrigues and Datnoff 2015).

19.2.11 Finger Millet (Eleusine coracana)

Finger millet grown in soil amended with calcium silicate showed reduced blast (*Pyricularia grisea*) severity on leaves due to higher foliar Si concentration (Jadhao and Rout 2020).

On dicots such as beans, coffee, cucurbits, cotton, grape, pea, potato, soybean, tomato, and strawberries, Si showed to be efficient in reducing the intensities of several diseases (Table 19.2).

Table 19.2 Effect of silicon on the development of diseases occurring in dicots

Crops	Diseases	Pathogens
Cucumber	Powdery mildew	*Podosphaera xanthii*
	Gray mold	*Botrytis cinerea*
	Target spot	*Corynespora cassiicola*
	Root decay	*Pythium ultimum* and *Pythium aphanidermatum*
	Root rot	*Phytophthora melonis*
	Fusarium wilt	*Fusarium oxysporum* f. sp. *cucumerinum*
Muskmelon and melon	Powdery mildew	*Podosphaera xanthii*
	Melon decay	*Alternaria alternata*, *Fusarium semitectum*, and *Trichothecium roseum*
	Bacterial fruit blotch	*Acidovorax citrulli*
Zucchini squash	Powdery mildew	*Podosphaera xanthii*
Watermelon	Fusarium wilt	*Fusarium oxysporum* f. sp. *niveum*
Soybean	Soybean rust	*Phakopsora pachyrhizi*
	Powdery mildew	*Erysiphe diffusa*
	Root rot	*Phytophthora sojae*
	Charcoal rot disease	*Macrophomina phaseolina*
	Frog eye leaf spot	*Cercospora sojina*
Common beans	Anthracnose	*Colletotrichum lindemuthianum*
	Alternaria leaf spot	*Alternaria alternata*
	Gray molt	*Botrytis cinerea*
	Cottony rot	*Pythium aphanidermatum*
Faba beans	Chocolate spot	*Botrytis fabae*
	Gray molt	*Botrytis cinerea*
Pea	Mycosphaerella blight	*Mycosphaerella pinodes*
	Blight disease	*Pseudomonas syringae* pv. *pisi*
	Root-knot nematode	*Meloidogyne incognita*
Coffee	Coffee leaf rust	*Hemileia vastatrix*
	Brown eye spot	*Cercospora coffeicola*
	Root-knot nematode	*Meloidogyne exigua*
Strawberry	Powdery mildew	*Sphaerotheca aphanis*
	Pestalotia leaf spot	*Pestalotia longisetula*
	Black root rot	*Rhizoctonia solani* and *Fusarium solani*
	Leaf blight	*Phomopsis obscurans*
Nectarine	Nectarine decay	*Rhizopus stolonifer*
Rose	Powdery mildew	*Podosphaera pannosa*
	Black spot	*Diplocarpon rosae*

(continued)

Table 19.2 (continued)

Crops	Diseases	Pathogens
Hemp	Powdery mildew	*Golovinomyces spadiceus*
Cotton	Angular leaf spot	*Xanthomonas citri* subsp. *malvacearum*
	Tropical rust	*Phakopsora gossypii*
	Ramulosis	*Colletotrichum gossypii* var. *cephalosporioides*
	Ramularia leaf spot	*Ramularia areola*
	Fusarium wilt	*Fusarium oxysporum* f. sp. *vasinfectum*
	Root-knot nematode	*Meloidogyne incognita*
Grape	Powdery mildew	*Uncinula necator*
	Downy mildew	*Plasmopara viticola*
	Bacterial speck	*Pseudomonas syringae* pv. *tomato*
	Bacterial wilt	*Ralstonia solanacearum*
	Powdery mildew	*Oidium neolycopersici*
	Late blight	*Phytophthora infestans*
Potato	Late blight	*Phytophthora infestans*
	Blackleg	*Pectobacterium* spp.
Eggplant	Root-knot nematode	*Meloidogyne incognita*
	Phomopsis blight	*Phomopsis vexans*
	Bacterial wilt	*Ralstonia solanacearum*
Asparagus	Stem blight	*Phomopsis asparagi*
Lettuce	Downy mildew	*Bremia lactucae*
	Root-knot nematode	*Meloidogyne incognita*
Beet	Root-knot nematode	*Meloidogyne incognita*
Carrot	Root-knot nematode	*Meloidogyne incognita*
Avocado	Root necrosis	*Calonectria ilicicola*
	Phytophthora root rot	*Phytophthora cinnamomi*
Passion fruit	Bacterial spot	*Xanthomonas axonopodis* pv. *passiflorae*

19.2.12 Cucurbits

The supply of Si through the roots or the foliar spray of products containing Si effectively reduced the severity of powdery mildew (*Podosphaera xanthii*) in several cucurbits. Plants of cucumber (*Cucumis sativus*) grown in a recirculated solution containing Si (1 mM) showed lower number of colonies formed by *P. xanthii* on leaves (Rodrigues and Datnoff 2015). Powdery mildew severity was also reduced in plants of muskmelon (*Cucumis melo*) and zucchini squash (*Cucurbita pepo*) either by foliar applications of potassium silicate (17 or 34 mM Si) or Si supplied through the roots (1.7 mM Si) (Rodrigues and Datnoff 2015). In zucchini, foliar Si application (250–1000 mg l^{-1}) reduced powdery mildew severity (Rodrigues and Datnoff 2015). For melon (*Cucumis melo*) plants, the area under powdery mildew progress curve was significantly reduced by 65 and 73% due to foliar (30 ml of potassium silicate l^{-1}) and root (0.34 ml of potassium silicate l^{-1}) supplies of Si (Dallagnol et al. 2012). The reduction in powdery mildew severity on melon plants was associated with a longer incubation period and a reduction in the number and size of the fungal colony as well as in the number of conidia produced by each colony (Dallagnol et al. 2012, 2015). The delay in the appearance of fungal colonies on melon leaves was due to a more robust defense response when Si was provided through roots, while foliar spray led to an alteration in the osmotic potential on the leaf surface (Dallagnol et al. 2012, 2015, 2020). A study evaluating several Si sources amended to the soil showed that powdery mildew severity was inversely related to the foliar Si concentration, and the wollastonite was the more efficient Si source to reduce disease severity (Torlon et al. 2016). Furthermore, the supply of Si to the roots of zucchini plants and their foliar spray with biocontrol agents (fungal or bacterial) resulted in better control of powdery mildew (Tesfagiorgis et al. 2014). According to the authors, Si drenched into the roots (250 ml at 100 mg l^{-1} applied weakly) decreased powdery mildew severity by 22–35%, while the foliar spray of biocontrol agents reduced disease severity by 29–94% (Tesfagiorgis et al. 2014). Using *Clonostachys rosea* combined with Si reduced the severity of gray mold (*Botrytis cinerea*) on the leaves of cucumber plants (Silva et al. 2020).

Furthermore, Si has been effective against root, vascular, and postharvest pathogens affecting cucurbits. On melon, the severity of bacterial fruit blotch (*Acidovorax citrulli*) and infection of seedlings by the bacterium was reduced by calcium silicate (1.41 g Si kg^{-1}) incorporated into the substrate along with the foliar spray of either yeast and potassium silicate (17 mM Si) or the yeasts, separately or in combination (Conceição et al. 2014). The calcium silicate (1.41 g Si kg^{-1}) increased the incubation period as well as reduced disease incidence, disease index, and area under disease progress curve (Ferreira et al. 2015). The incidence of melon decay (caused by *Alternaria alternata*, *Fusarium semitectum*, or *Trichothecium roseum*) was reduced by Si at a rate of 100 mM (Bi et al. 2006). In a recirculating solution, potassium silicate (1.7 mM Si) amendment reduced plant mortality, root decay, and yield loss caused by both *Pythium ultimum* and *P. aphanidermatum* on cucumber plants (Rodrigues and Datnoff 2015). The incidence of root rot (*Phytophthora melonis*) was also reduced by Si applied as sodium silicate to plants (1 or 1.7 mM Si) (Mohaghegh et al. 2011). Sodium silicate (2 mM Si) also enhanced cucumber seedling growth and decreased the index values for Fusarium wilt (*Fusarium oxysporum* f. sp. *cucumerinum*) disease (Zhou et al. 2018). Sodium silicate (3 mM) decreased Fusarium wilt up to 69% (Sun et al. 2022). In watermelon (*Citrullus lanatus*), mesoporous silica nanoparticles (MSN), with or without a chitosan coating, increased host resistance against *Fusarium oxysporum* f. sp. *niveum* infection, with MSN and chitosan-coated MSN reducing disease severity by 40% and 27%, respectively (Buchman et al. 2019). Furthermore, a tuneable form of MSN that released silicic acid decreased the incidence of Fusarium wilt more efficiently (Kang et al. 2021).

19.2.13 Soybean (Glycine max*)*

The severity of soybean rust (*Phakopsora pachyrhizi*) was reduced by foliar and root supply of Si to soybean plants. Field experiments evaluating the foliar spray of potassium silicate at rates ranging from 8 to 60 g l^{-1} (pH 5.5 and 10.5) resulted in a significative decrease in soybean rust severity (Rodrigues et al. 2009). According to the authors, at the highest potassium silicate rate (pH 5.5), soybean rust severity decreased by 70%. In a greenhouse experiment, the number of pustules produced by *P. pachyrhizi* per cm^2 of leaf tissue and soybean rust severity were reduced by potassium silicate (40 g l^{-1}) to the same as obtained for the mixture of the fungicides epoxiconazole and pyraclostrobin (Rodrigues et al. 2009). Another study also demonstrated that Si, sprayed on the leaves as Si dioxide, reduced soybean rust severity (Otolakoski et al. 2023). The amendment of soil with wollastonite (Si rates of 0, 0.96, and 1.92 t ha^{-1}) delayed the onset of soybean rust symptoms for up to 3 days, and the area under disease progress curve was reduced up to 43% (Lemes et al. 2011). Soybean plants supplied with Si (1.7 mM) showed increased foliar Si concentration, but the ability to uptake Si from the nutrient solution and translocate it to the shoots varied among genotypes (Arsenault-Labrecque et al. 2012). According to the authors, plants from the cultivar Hikmok sorip absorbed nearly four times more Si than those from Williams 82 and displayed less soybean rust symptoms when supplied with Si (Arsenault-Labrecque et al. 2012). Smaller and less developed uredia were observed in the leaflets of soybean plants supplied with Si (2 mM) along with significant reductions of 27, 23, and 60% for the number of lesions formed per cm^2 of leaf tissue, closed uredia, and opened uredia, respectively (Cruz et al. 2012). A 2-year field experiment showed that soybean cultivated after barley in a soil amended with calcium silicate displayed reductions in both soybean rust and powdery mildew (*Erysiphe diffusa*) severities indicating the positive effect of residual Si in the soil (Bruneto et al. 2022). Root rot (*Phytophthora sojae*) incidence was reduced by Si (1.7 mM) especially for a more resistant cultivar (Guérin et al. 2014). The severity of charcoal rot (*Macrophomina phaseolina*) was significantly reduced on the roots of soybean plants that received applications of potassium silicate and sodium silicate (El-Abdean et al. 2020).

19.2.14 Beans

On common beans (*Phaseolus vulgaris*), the severity of anthracnose (*Colletotrichum lindemuthianum*) was significantly reduced by Si. Plants grown in a hydroponic culture containing Si (2 mM) displayed an increase in the incubation period by 14%, while the area under anthracnose progress curve and disease severity significantly decreased by 33% and 27%, respectively (Polanco et al. 2012). Interestingly, observations performed under the scanning electron microscope showed that the veins in the leaves of common beans plants supplied with Si did not show drastic morphological changes in response to infection by *C. lindemuthianum* (Cruz et al. 2014). According to the authors, the concentrations of sulfur, potassium, and Si, quantified by X-ray microanalysis, increased in plants supplied with Si, contributing, therefore, to reducing anthracnose symptoms. Anthracnose severity was also reduced by 32% on common bean plants after foliar spray of potassium silicate (35 g l^{-1}) resulting in an increase in grain yield by 17% (Rodrigues et al. 2015). Nanoparticles of Si (200 ppm) reduced Alternaria leaf spot (*Alternaria alternata*) severity by 66% resulting in higher concentration of photosynthetic pigments and great yield (Taha et al. 2023). Nanoparticles of Si (1.5 and 3 mM) significantly reduced chocolate spot (*Botrytis fabae*) incidence and severity by 60% and 76%, respectively, for faba beans (*Vicia faba*) plants (Hasan et al. 2020). Phenolic concentration increased, and peroxidase and polyphenoloxidase activities were higher in the leaves of faba beans exposed to silver+Si dioxide nanocomposites in response to infection by *B. cinerea* (Baka and El-Zahed 2022). In snap beans (*Phaseolus vulgaris*) growth under field conditions, application of potassium silicate (2000 and 4000 ppm) reduced the incidence of cottony rot (*P. aphanidermatum*) and the severity of gray mold (*B. cinerea*) (El-Garhy et al. 2016).

*19.2.15 Coffee (*Coffea *spp.)*

The incidence and severity of coffee leaf rust (*Hemileia vastatrix*) in the leaves of coffee plants from experiments performed in both greenhouse and field conditions were not impacted by Si (Carré-Missio et al. 2009; Lopes et al. 2014). In this scenario, coffee plants were inefficient in the uptake of Si from the soil solution and translocating it to shoots (Carré-Missio et al. 2009; Lopes et al. 2014). The severity of coffee leaf rust did not decrease on coffee leaves of plants growing in soil amended with diatomaceous earth (Gokavi et al. 2021). On the other hand, the intensity of sporulation by *Hemileia vastatrix* on coffee leaves and the total number of pustules of coffee leaf rust per cm^2 of leaf tissue decreased as the potassium silicate rates ranged from 8 to 60 g l^{-1} at pH 5.5 (Carré-Missio et al. 2012b). The effect of foliar potassium silicate spray on reducing coffee leaf rust severity occurred only locally (Carré-Missio et al. 2012a). However, field experiments under high pressure of coffee leaf rust and potassium silicate alone or combined with copper hydroxide did not reduce disease incidence and severity (Lopes et al. 2013). In another study, Si was found to be uniformly distributed in the abaxial leaf surface of three coffee cultivars grown in a substrate amended with Si based on X-ray microanalysis (Pozza et al. 2004). Only plants from the coffee cultivar Catuaí benefited from Si supply aiming to control brown eye spot (*Cercospora coffeicola*) (Pozza et al. 2004). Brown eye spot severity and the total number of lesions per cm^2 of leaf tissue were significantly reduced by 63% and 43%, respectively, in plants supplied with Si (Pozza et al. 2004). Interestingly, higher Si concentration on the roots of coffee plants (cultivar Catuai 44) growing in a Si-deficient soil amended with calcium silicate increased their resistance (less number of galls and mass of eggs in the root system) against the root-knot nematode *Meloidogyne exigua* (Silva et al. 2012).

*19.2.16 Strawberries (*Fragaria *spp.)*

The use of Si has been demonstrated to be effective in reducing powdery mildew (*Sphaerotheca aphanis*) severity in strawberries regardless of root or foliar treatment. The use of potassium silicate (25 mg l^{-1}) added in the nutrient solution reduced powdery mildew severity by 88%, and disease symptoms were almost absent at the highest rate (50 and 100 mg potassium silicate l^{-1}) (Rodrigues and Datnoff 2015). At the highest dose (100 mg l^{-1}), the foliar Si concentration increased about 24 times compared to the control treatment, and the severity of powdery mildew significantly decreased when the foliar Si concentration was above 1.5 dag kg^{-1} (Rodrigues and Datnoff 2015). The soil drenching with potassium silicate decreased powdery mildew severity by 41–86% (Rodrigues and Datnoff 2015). A study carried out over 2 years demonstrated that supplying Si (0.017% v v^{-1}) via fertirrigation reduced powdery mildew severity from 29% to 39% (Liu et al. 2020). Seven years of field experiments on a commercial strawberry farm showed that weekly application of Si (0.017% v v^{-1}) through spray or root via fertirrigation reduced the severity of powdery mildew regardless of cultivar used (Liu et al. 2021). According to the authors, Si also delayed the epidemic rate for up to 14 days when compared to the control treatment. For pestalotia leaf spot (*Pestalotia longisetula*), the foliar spray of potassium silicate (30 g l^{-1}) reduced disease severity by 61% (Carré-Missio et al. 2010). The severity of leaf blight (*Phomopsis obscurans*) was significantly reduced by 86% on plants supplied with Si (Abd-El-Kareem et al. 2019a). The intensity of black root rot (*R. solani* and *Fusarium solani*) was reduced on plants supplied with Si (soil amendment with calcium silicate and foliar spray of potassium silicate) grown under field conditions (Abd-El-Kareem et al. 2019b).

*19.2.17 Cotton (*Gossypium hirsutum*)*

Cotton plants grown in soil amended with calcium silicate (1.5 g kg^{-1}) showed reduction in the severity of angular leaf spot (*Xanthomonas citri* subsp. *malvacearum*) by 55% (Oliveira et al. 2012). For tropical rust (*Phakopsora gossypii*), cotton plants grown in a hydroponic solution containing Si (2 mM) showed an increase in the incubation period and latent period and reduction in the area of pustule, area under disease progress curve (Debona et al. 2023). An increase in the incubation period and decreases in both incidence of ramulosis and area under disease progress curve were reported for cotton plants supplied with Si and infected by *Colletotrichum gossypii* var. *cephalosporioides* (Debona et al. 2023). For ramularia leaf spot (*Ramularia areola*), Si increased the incubation period and latent period by 10% and 15%, respectively, while the number of lesions per leaf was reduced by 39–62% and the lesion size by 17–27% depending on the cotton cultivar (Debona et al. 2023). Cotton plants supplied with Si displayed cellular reactions in response to infection by *Fusarium oxysporum* f. sp. *vasinfectum* (Whan et al. 2016). The Si supply (2.0 mM) efficiently affected the development of *Meloidogyne incognita* on the roots of cotton plants (Santos et al. 2022).

*19.2.18 Grape (*Vitis vinifera*)*

The severity of powdery mildew (*Uncinula necator*) on grape leaves was reduced by spraying Si at the rate of 1.7 mM (Rodrigues and Datnoff 2015). Grape plants of six cultivars growing in recirculation solution containing Si (10 and 112 mg l^{-1}) showed slightly increased resistance against powdery mildew, but only up to 2 weeks after exposure to Si (Rodrigues and Datnoff 2015). According to the authors, a minimal Si concentration was needed to achieve the resistance of grape plants against powdery mildew, but host susceptibility against infection by *U. necator* was not overcame by supplementary Si supply. In a 3-year field experiment, the effect of foliar and soil application of colloidal Si on the control of powdery mildew showed that foliar Si spray reduced disease incidence and severity by 40% and 50%, respectively (Schabl et al. 2020). For downy mildew (*Plasmopara viticola*), application of Si nanoparticles (150 ppm) to plants under field conditions reduced disease severity by 82% and improved the quality of berries (Rashad et al. 2021).

*19.2.19 Tomato (*Solanum lycopersicum*)*

The symptoms of bacterial speck, caused by *Pseudomonas syringae* pv. *tomato*, were reduced on tomato leaves sprayed with potassium silicate (2 ml l^{-1}), but no significant effect was obtained on plants grown in soil amended with calcium silicate at the rate of 0.16 g Si kg^{-1} (Andrade et al. 2013). The incidence of bacterial wilt, caused by *Ralstonia solanacearum*, was significantly reduced from 19 to 53% due to Si treatment (2 mM) compared to non-Si supplied plants (Wang et al. 2013). The reduction of tomato bacterial wilt index ranged from 46% to 72% when Si was supplied through irrigation at several rates that ranged from 0.5 to 2 mM (Wang et al. 2022a). The spray of Si nanoparticles (650 mg l^{-1}) increased both foliar and root Si concentrations by 5.4-fold and 1.8-fold, respectively, and reduced bacterial wilt severity from 21% to 88% (Wang et al. 2022b). Tomato plants grown in nutrient solution supplied with Si (100 mg l^{-1} using potassium silicate) showed less incidence and severity of powdery mildew (*Oidium neolycopersici*) (Garibaldi et al. 2011). The foliar spray of potassium silicate with intervals of 12 days significantly decreased the severity of powdery mildew (Yanar et al. 2011). The spray of tomato plants with potassium silicate was not efficient in reducing late blight (*Phytophthora infestans*) severity (Rodrigues and Datnoff 2015), but decreased disease severity in potato leaves by 25–44% (Soratto et al. 2012).

19.2.20 Other Edible Vegetables

On potato (*Solanum tuberosum*), sodium silicate (100 mM) applied foliarly significantly reduced the severity of late blight (*P. infestans*) (Xue et al. 2021). The foliar spray of Si reduced the incidence of blackleg (*Pectobacterium* spp.) by 47–51% and increased potato tuber yield (Soratto et al. 2012). On eggplant (*Solanum melongena*), Si nanoparticles (100 and 200 ppm) resulted in mortality of J_2 from *Meloidogyne incognita* up to 98%, and when Si nanoparticles were mixed with 50% of the rate of a synthetic nematicide, gall formation, the number of egg masses on roots, and final J2 population were significantly reduced (El-Ashry et al. 2022). Nanoparticles of Si applied foliarly to eggplant reduced the symptoms caused by *Phomopsis vexans*, *R. solanacearum*, and *M. incognita* (Khan et al. 2022). On asparagus (*Asparagus officinalis*), stem blight (*Phomopsis asparagi*) severity was significantly reduced by 32 and 14% for susceptible and resistant cultivars, respectively, using sodium silicate at the rate of 1.5 mM (Lu et al. 2008). Plants of lettuce (*Lactuca sativa*) showed a significant reduction of downy mildew (*Bremia lactucae*) severity when grown in a nutrient solution containing Si (Garibaldi et al. 2012). Lettuce seeds exposed to calcium silicate (100 μmol of Si) showed a reduction of 30% in the incidence of plants with downy mildew (Alves et al. 2022). The number of eggs and galls produced by *M. incognita* on the roots of lettuce plants grown in nutrient solution containing 3 mM Si were significantly reduced (Alonso et al. 2022). Seedlings of pea (*Lathyrus oleraceus*) grown in substrate amended with potassium silicate for 5 weeks showed fewer lesions of Mycosphaerella blight (*Mycosphaerella pinodes*) (Dann and Muir 2002). The severity of blight disease (*Pseudomonas syringae* pv. *pisi*) as well as the number of galls and the reproduction of *M. incognita* was reduced on pea plants originated from seeds soaked with a solution of Si nanoparticles (0.10 mg Si ml^{-1}) for 12 h (Kashyap and Siddiqui 2021). The Si nanoparticles (100 and 200 mg l^{-1}) supplied to seeds or via foliar spray significantly reduced the number of galls formed by *M. incognita* and its reproduction on the roots of beetroot (*Beta vulgaris*) plants (Khan and Siddiqui 2020). The reproduction of *M. incognita* was also affected in the roots of carrot (*Daucus carota*) plants supplied with Si nanoparticles (Ahamad and Siddiqui 2021).

On orchards of avocado (*Persea americana*), soil drench with potassium silicate improved yield and quality of the fruits, but did not affect the visual health of trees suffering from declined from Phytophthora root rot caused by *Phytophthora cinnamomi* (Dann and Le 2017). However, mineral mulch containing 5% available Si supplied to old avocado trees with decline by Phytophthora root rot had a beneficial effect on the visual health of trees (Dann and Le 2017). The supply of Si to avocado seedlings under greenhouse conditions decreased the symptoms of root necrosis caused by *Calonectria ilicicola* (Dann and Le 2017). The use of potassium silicate (0.2 mol Si) supplied to Hass avocado plants through an irrigation system for 10 days and before inoculation with *Phytophthora cinnamomi* reduced Phytophthora root rot symptoms (Álvarez et al. 2023). On nectarine (*Prunus persica* var. *nucipersica*), potassium silicate applied after fruit harvesting reduced the incidence of postharvest Rhizopus soft rot (*Rhizopus stolonifer*) from 20% to 85% (Dallagnol et al. 2021). According to the authors, the effectiveness of potassium silicate in reducing the incidence of *Rhizopus* soft rot was associated with less conidia germination due to the alkalinization on the fruit surface. On yellow passion fruit (*Passiflora edulis*), foliar sprays of Si clay (1–2%) inhibited the growth of *Xanthomonas axonopodis* pv. *passiflorae* in the leaf tissues resulting in less expressive symptoms of bacterial spot (Brancaglione et al. 2009). Plants of pedunculate oak (*Quercus robur*) supplied with Si showed an increase in the total root surface, total root volume, and length of thin roots by 34%, 58%, and 63%, respectively, and, consequently, reduced symptoms originating from infection by *Phytophthora plurivora* (Kostic et al. 2023). The supply of Si (300–900 kg wollastonite ha^{-1}) to the roots of hemp (*Cannabis sativa*) plants reduced powdery mildew (*Golovinomyces spadiceus*) severity up to 83% (Dixon et al. 2022).

19.2.21 Ornamentals

Rose plants growing in substrate supplied with 3.6 mM Si showed an increase in the foliar Si concentration by two- to fourfold and, consequently, a delay in the onset of powdery mildew (*Podosphaera pannosa*) symptoms by 1 to 2 days and a decrease of 49% in disease severity (Shetty et al. 2012). According to the authors, the most resistant rose genotypes showed the highest foliar Si concentration resulting in the best disease control. Rose plants that received potassium silicate solution (100 or 150 mg l^{-1}) daily showed reductions in the number of lesions of black spot (*Diplocarpon rosae*) per cm^2 of leaf tissue as well as less number of leaves with disease symptoms (Gillman et al. 2003). The use of Si was also efficient in reducing disease symptoms in the morning glory *Colletotrichum gloeosporioides*, paper daisies *C. gloeosporioides*, sunflower and zinnia *Golovinomyces cichoracearum*, and tiger grass *Exserohilum rostratum* (Debona et al. 2023) interactions. However, for gerbera daisy, neither calcium silicate nor potassium silicate was efficient in reducing powdery mildew severity due to the inability of the roots to uptake Si from the soil solution and translocate it to shoots (Moyer et al. 2008).

19.3 Silicon in the Potentiation of Host Defense Reactions

The physical or mechanical barrier hypothesis explains the role of Si in delaying the appearance of disease symptoms or even the lesion expansion. The rice *P. oryzae* pathosystem illustrates very well this hypothesis. Higher density of silicified bulliform in the epidermis of rice leaves impeded or delayed *P. oryzae* penetration resulting in smaller lesions due to the formation of complexes of Si with organic compounds especially hemicellulose and lignin (Debona et al. 2017). According to Kim et al. (2002), blast lesions in rice leaves did not expand well due to forming an outer electron-dense layer of Si and a thin electron-dense of Si layers embedded in cellulose microfibrils. Interestingly, the physical barrier hypothesis is one of the facets to explain how Si might reduce the intensities of plant diseases considering the availability of physiological, biochemical, and molecular data showing evidence of Si involvement in the potentiation of host defense reactions against pathogen infection.

The size of blast lesions on the leaves of rice plants supplied with Si was smaller than those formed on the leaves of plants non-supplied with Si (Rodrigues et al. 2003a). Ultrastructural observations of leaf samples containing smaller blast lesions obtained from Si-supplied plants revealed less fungal colonization and intense osmiophilic material deposition surrounding the empty fungal cells (Rodrigues et al. 2003a). Leaf cells of rice plants supplied with Si reacted against *M. albescences* infection by accumulating phenolic-like compounds or amorphous granular material (Araujo et al. 2016). The terpenoid pathway in the leaves of rice plants supplied with Si in response to infection by *P. oryzae* was stimulated as noticed by an increase in the accumulation of the phytoalexins momilactone A and momilactone B (Rodrigues et al. 2004). In the roots of banana plants supplied with Si, the accumulation of flavonoids in the sclerenchyma and metaxylem vessels increased in response to infection by *F. oxysporum* f. sp. *cubense* (Fortunato et al. 2014). Moreover, denser depositions of both lignin and dopamine were closely associated with Si-increasing resistance of banana roots against fungal infection (Fortunato et al. 2014). Great deposition of flavonoids in the leaves of wheat plants supplied with Si helped to reduce the colonization of cells by hyphae of *P. oryzae* resulting in lower disease severity values (da Silva et al. 2015). Great activity of phenylalanine ammonia-lyase following higher accumulation of antifungal phenolics was noticed in the leaves of wheat plants supplied with Si and infected by *P. tritici-repentis* (Dorneles et al. 2018). In the leaves of wheat plants supplied with Si, the formation of papilla containing callose and deposition of electron-dense osmiophilic material (e.g., glycosylated phenolics) were enhanced in response to infection by *B. graminis* f. sp. *tritici* (Bélanger et al. 2003). According to the authors, the phenolics greatly affected the integrity of haustoria formed by *B. graminis* f. sp. *tritici* in the cytosol of wheat leaf cells (Bélanger et al. 2003).

An earlier accumulation of phenolics occurred in the leaf cells of cucumber plants supplied with Si infected by *P. xanthii* (Debona et al. 2023). The phytoalexin flavonol aglycone rhamnetin was involved in the increased resistance of leaves from cucumber plants supplied with Si and infected by *P. xanthii* (Debona et al. 2023). An electron-dense phenolic-like material in the roots of cucumber plants supplied with Si resulted in hyphae of *P. ultimum* of empty appearance (Debona et al. 2023). The roots of cucumber plants supplied with Si reacted very efficiently against infection by *Pythium* spp. due to great chitinase, peroxidase, and polyphenoloxidase activities as well as the fungistatic action of glycosidically bound phenolics (Debona et al. 2023). Pea seedlings supplied with Si displayed great activities of chitinase and β-1,3-glucanase during the infection process of *M. pinodes* (Debona et al. 2023). Phenolics and flavonoids were involved in the increased resistance of leaves from rose plants against infection by *P. pannosa* along with a great transcription of genes coding for phenylalanine ammonia-lyase, cinnamyl alcohol dehydrogenase, and chalcone synthase (Rodrigues and Datnoff 2015). Higher production of transcripts for glucanase, peroxidase, and PR-1 were noticed in the leaf cells of rice plants supplied with Si and infected by *P. oryzae* (Rodrigues et al. 2005). An increase in the activities of peroxidase, polyphenoloxidase, and phenylalanine ammonia-lyase as well as higher lignin concentration was associated with the increased resistance of rice plants supplied with Si against blast disease (Debona et al. 2023). Leaves from perennial ryegrass plants supplied with Si and infected by *P. oryzae* showed greater activities of peroxidase and polyphenoloxidase, higher levels of phenolics (e.g., chlorogenic acid) and flavonoids, and higher transcript levels of genes encoding for phenylalanine ammonia-lyase and lipoxygenase (Rahman et al. 2015). The great activity of phenylalanine ammonia-lyase was reported to affect the infection process of pathogens in the banana-*F. oxysporum* f. sp. *cubense*, coffee-*M. exigua*, cotton-*C. gossypii* var. *cephalosporioides* and cotton-*R. areola*, rice-*M. albescens* and rice-*R. solani*, soybean-*P. pachyrhizi*, and tomato-*P. syringae* pv. *tomato* interactions (Debona et al. 2023). For the banana-*F. oxysporum* f. sp. *cubense*, cotton-*C. gossypii* var. *cephalosporioides* and cotton-*R. areola*, and coffee-*M. exigua*, the concentrations of phenolics and lignin increased due to higher phenylalanine ammonia-lyase activity (Debona et al. 2023). Increased activities of chitinase, β-1,3-glucanase, peroxidase, and polyphenoloxidase gained by Si reduced the disease symptoms in the banana-*F. oxysporum* f. sp. *cubense*; cotton-*C. gossypii* var. *cephalosporioides* and cotton-*P. gossypii*; coffee-*M. exigua*; rice-*B. oryzae*, rice-*M. albescens*, and rice-*R. solani*; sorghum-*C. sublineolum*; soybean-*P. pachyrhizi*; and tomato-*P. syringae* pv. *tomato* interactions (Debona et al. 2023). Higher deposition of anthocyanins in leaves of sorghum plants supplied with Si reduced anthracnose symptoms (Resende et al. 2013). Activities of phenylalanine ammonia-lyase and lipoxygenase in tomato leaves infected with *P. syringae* pv. *tomato* increased in the presence of Si (Andrade et al. 2013). In the presence of Si, lower production of malondialdehyde, a cellular metabolite resulted from lipid peroxidation, was reported to occur in the banana-*F. oxysporum* f. sp. *cubense*, cotton-*R. areola*, rice-*P. oryzae*, sorghum-*C. sublineolum*, and wheat-*P. oryzae* interactions due to higher activities of antioxidant enzymes such as ascorbate peroxidase, catalase, glutathione peroxidase, glutathione reductase, glutathione-*S*-transferase, peroxidase, and superoxide dismutase (Debona et al. 2023). The activities of ascorbate peroxidase, catalase, peroxidase, glutathione-*S*-transferase, and superoxide dismutase were greater in the leaves of wheat plants supplied with Si and infected by *P. oryzae* (Debona et al. 2014).

Brunings et al. (2009) conducted a microarray study to determine if rice plants supplied with Si could boost their resistance against blast. According to the authors, genes involved in the ethylene signaling pathway, genes encoding for a thaumatin/pathogenesis-related protein and a class III peroxidase, as well as a number of transcription factors and protein kinases were differently regulated in the Si-supplied and infected plants. Interestingly, less than half the number of differentially expressed genes in Si-supplied plants compared to non-supplied plants (298 vs. 738) occurred in response to infection by *P. oryzae* (Brunings et al. 2009).

The transcript levels for genes coding for PR-1, peroxidase, β-1,3-glucanase, chitinase, phenylalanine ammonia-lyase, and lipoxygenase were higher for wheat plants supplied with Si and infected by *P. oryzae* (Cruz et al. 2015). On top of that, the Si-supplied plants showed less colonization of epidermal and parenchyma cells by hyphae of *P. oryzae* and great deposition of phenolics (Cruz et al. 2015). Rice plants supplied with Si and infected by *B. oryzae* displayed increased photorespiration rates, which helped counteract fungal-induced leaf senescence and cell death that could favor its infection process (Van Bockhaven et al. 2015a). Potentiation of host defense responses mediated by ethylene, jasmonic acid, and salicylic acid with the participation of Si has been reported in the Arabidopsis-*E. cichoracearum*, rice-*P. oryzae* and rice-*B. oryzae*, and tomato-*R. solanacearum* interactions (Debona et al. 2023). In the tomato-*R. solanacearum* interaction, pathways regulated by auxin, abscisic acid, cytokinin, and brassinosteroid were greatly regulated (Jiang et al. 2019).

Considering that orthosilicic acid is a biochemically inert substance in plant tissues, the hypothesis that Si could act directly in activating plant defense reactions seems very difficult to accept. In the rice-*B. oryzae* interaction, the reduction of brown spot symptoms gained by Si did not involve the participation of both jasmonic acid and salicylic acid (Van Bockhaven et al. 2015b). According to the authors, Si was able to neutralize the ethylene produced by *B. oryzae* compromising its ability to suppress plant immunity and resulting in a faster and stronger activation of host defense reactions. Using comparative transcriptomic analysis, Rasoolizadeh et al. (2018) showed that soybean plants non-supplied with Si displayed early (4 days post inoculation) and higher expression of defense-related genes as well as plant receptors in response to infection by *P. sojae*. It is known that *P. sojae* can secrete many effectors that contribute to its increased aggressiveness in the root tissues of soybean plants. On the other hand, the transcriptomic response of Si-supplied plants did not change upon infection by *P. sojae*, and the expression of effector-coding genes by the pathogen was significantly reduced. Considering the interaction between effectors released by *P. sojae* and the receptors in the roots of soybean plants as well as Si accumulation in the apoplast, Si most likely abolished or reduced the interaction of effectors produced by *P. sojae* with the receptors in the root cells (Rasoolizadeh et al. 2018). Based on these findings, the apoplastic obstruction hypothesis was proposed to explain the Si-mediated resistance in plants against infection by the pathogen (Coskun et al. 2019).

19.4 Concluding Remarks

The role of the quasi-essential element Si in the resilience of several plant species against different types of stress is indisputable, with infection by pathogens of different lifestyles being the most notable. However, the scientific community is still struggling to recognize that Si can greatly contribute to a more sustainable agriculture by reducing the application of pesticides while improving the overall production cost. Using fertilizers containing soluble Si as a soil amendment or foliar spray may become an outstanding strategy for integrated disease management, especially for soils that are low or limiting in Si. Plants benefiting from the effect of Si display reductions in disease symptoms because several components of host resistance (e.g., incubation period, latent period, lesion size, lesion number, and inoculum production per infection site) are greatly impacted, therefore contributing to reducing the disease epidemic rate and number of fungicide sprays. Intensities of some crop diseases that have a high Si content in their plant tissues are well-known to be lowered to the same level gained by fungicide sprays. On top of that, cultivars with important agronomic traits, but susceptible to the major root or foliar diseases, display a phenotype similar to cultivars with high levels of partial resistance (non-race-specific resistance). Lower intensities of diseases in profitable crops result from the polymerized Si beneath the cuticle and in the cell wall of different plant tissues. In this scenario, the pathogen needs more time to penetrate the plant cell, allowing it to defend more rapidly against its arrival in the cytosol. Interestingly, Si does not only act passively to confer plant protection against infection by pathogens. Plants supplied with Si during their developmental stages exhibit earlier and more efficient defense reactions that will extend the appearance of disease symptoms and their expression (e.g., size, format, and coloration of lesions) and the amount of inoculum produced per infection site. This spectacular cellular response is the result of a joint effort of some biochemical pathways (e.g., phenylpropanoid and terpenoid) that are promptly activated toward a more abundant production of antimicrobial compounds (e.g., different types of phenolics and phytoalexins), enzymes indirectly (e.g., catalase, peroxidase, phenylalanine ammonia-liase, polyphenoloxidase, and superoxide dismutase) or directly (e.g., chitinase and β-1,3-glucanase) involved in host defense, and proteins that are hallmarks for increased resistance (e.g., PR-1). In a nutshell, Si plays a multifaceted role in sustainable agriculture by enhancing plant resilience to diseases, improving nutrient uptake and utilization by plants toward better growth and higher yields, and reducing the environmental impacts of pesticides. Adopting Si-based management strategies into the current agricultural practices in the most profitable crops can greatly contribute to more sustainable and resilient farming systems worldwide.

References

Abd-El-Kareem F, Elshahawy IE, Abd-Elgawad MMM (2019a) Management of strawberry leaf blight disease caused by *Phomopsis obscurans* using silicate salts under field conditions. Bull Natl Res Cent 43:1

Abd-El-Kareem F, Elshahawy IE, Abd-Elgawad MMM (2019b) Effectiveness of silicon and silicate salts for controlling black root rot and induced pathogenesis-related protein of strawberry plants. Bull Natl Res Cent 43:91

Abdelrhim AS, Mazrou YSA, Nehela Y et al (2021) Silicon dioxide nanoparticles induce innate immune responses and activate antioxidant machinery in wheat against *Rhizoctonia solani*. Plan Theory 10:2758

Ahamad L, Siddiqui ZA (2021) Effects of silicon dioxide, zinc oxide and titanium dioxide nanoparticles on *Meloidogyne incognita*, *Alternaria dauci* and *Rhizoctonia solani* disease complex of carrot. Exp Parasitol 230:108176

Ahmad F, Jabeen K, Iqbal S et al (2023) Influence of silicon nano-particles on *Avena sativa* L. to alleviate the biotic stress of *Rhizoctonia solani*. Sci Rep 13:15191

Alonso TAS, Silva DL, Prado RM et al (2022) Silicon promotes the control of *Meloidogyne incognita* in lettuce by increasing ascorbic acid and phenolic compounds. J Pest Sci 95:1453–1466

Álvarez A, Oliveros D, Ávila YC et al (2023) Resistance induction with silicon in Hass avocado plants inoculated with *Phytophthora cinnamomi* Rands. Plant Signal Behav 31:2178362

Alves RC, Santos Zucco MF, Oliveira KR et al (2022) Seed priming with silicon improves plant resistance to downy mildew (*Bremia lactucae*) in lettuce seedlings by intensifying antioxidant defense systems. Silicon 14:12721–12731

Andrade CCL, Resende RS, Rodrigues FA et al (2013) Silicon reduces bacterial speck development on tomato leaves. Trop Plant Pathol 38:436–442

Araujo L, Paschoalino RS, Rodrigues FA (2016) Microscopic aspects of silicon-mediated rice resistance to leaf scald. Phytopathology 106:132–141

Arsenault-Labrecque G, Menzies JG, Bélanger RR (2012) Effect of silicon absorption on soybean resistance to *Phakopsora pachyrhizi* in different cultivars. Plant Dis 96:37–42

Aucique-Pérez CE, Rodrigues FA, Moreira WR et al (2014) Leaf gas exchange and chlorophyll *a* fluorescence in wheat plants supplied with silicon and infected with *Pyricularia oryzae*. Phytopathology 104:143–149

Baka ZA, El-Zahed MM (2022) Antifungal activity of silver/silicon dioxide nanocomposite on the response of faba bean plants (*Vicia faba* L.) infected by *Botrytis cinerea*. Bioresour Bioprocess 9:102

Bathoova M, Švubová R, Bokor B et al (2021) Silicon triggers sorghum root enzyme activities and inhibits the root cell colonization by *Alternaria alternata*. Planta 253:29

Bathoovaa M, Bokorab B, Soukupac M et al (2018) Silicon-mediated cell wall modifications of sorghum root exodermis and suppression of invasion by fungus *Alternaria alternata*. Plant Pathol 67:1891–1900

Bélanger RR, Benhamou N, Menzies JG (2003) Cytological evidence of an active role of silicon in wheat resistance to powdery mildew (*Blumeria graminis* f. sp. *tritici*). Phytopathology 93:402–412

Bi Y, Tian SP, Guo YR et al (2006) Sodium silicate reduces postharvest decay on Hami melons: induced resistance and fungistatic effects. Plant Dis 90:279–283

Brancaglione P, Sampaio AC, Fischer IH et al (2009) Eficiência de argila silicatada no controle de *Xanthomonas axonopodis* pv. *passiflorae*, in vitro e em mudas de maracujazeiro-amarelo. Rev Bras Frutic 31:718–724

Brunetto AE, Louzada FD, Dorneles KR et al (2022) Residual effect of calcium silicate on the management of soybean diseases. Australas Plant Pathol 51:559–563

Brunings AM, Datnoff LE, Ma JF et al (2009) Differential gene expression of rice in response to silicon and rice blast fungus *Magnaporthe oryzae*. Ann App Biol 155:161–170

Buchman JT, Elmer WH, Ma C et al (2019) Chitosan-coated meso-porous silica nanoparticle treatment of *Citrullus lanatus* (watermelon): enhanced fungal disease suppression and modulated expression of stress-related genes. ACS Sustam Chem Eng 7:19649–19659

Camargo MS, Amorim L, Júnior ARG (2013) Silicon fertilisation decreases brown rust incidence in sugarcane. Crop Prot 53:72–79

Carré-Missio V, Rodrigues FA, Schurt DA et al (2009) Ineficiência do silício no controle da ferrugem do cafeeiro em solução nutritiva. Trop Plant Pathol 34:416–421

Carré-Missio V, Rodrigues FA, Schurt DA et al (2010) Aplicação foliar de silicato de potássio, acibenzolar-S-metil e fungicidas na redução da mancha de Pestalotia em morango. Trop Plant Pathol 35:182–185

Carré-Missio V, Rodrigues FA, Schurt DA et al (2012a) Proteção local, não sistêmica, do silicato de potássio reduz os sintomas da ferrugem do cafeeiro. Trop Plant Pathol 37:275–280

Carré-Missio V, Rodrigues FA, Schurt DA et al (2012b) Componentes epidemiológicos da ferrugem do cafeeiro afetados pela aplicação foliar de silicato de potássio. Trop Plant Pathol 37:50–56

Chang SJ, Tzeng DDS, Li CC (2002) Effect of silicon nutrient on bacterial blight resistance of rice (*Oryza sativa* L.). In: Matoh T (ed) Abstracts of the 2nd silicon in agriculture conference. Press-Net, Kyoto, pp 31–33

Conceição CS, Felix KCS, Mariano RLR et al (2014) Combined effect of yeast and silicon on the control of bacterial fruit blotch in melon. Sci Hortic 174:164–170

Coskun D, Deshmukh R, Sonah H et al (2019) The controversies of silicon's role in plant biology. New Phytol 221:67–85

Cruz MFA, Diniz APC, Rodrigues FA et al (2011) Aplicação foliar de produtos na redução da severidade da brusone do trigo. Trop Plant Pathol 36:424–428

Cruz MFA, Silva LF, Rodrigues FA et al (2012) Silício no processo infeccioso de *Phakopsora pachyrhizi* em folíolos de plantas de soja. Pesq Agropec Bras 47:142–145

Cruz MFA, Rodrigues FA, Diniz APC et al (2014) Soybean resistance to *Phakopsora pachyrhizi* as affected by acibenzolar-S-methyl, jasmonic acid and silicon. J Phytopathol 162:133–136

Cruz MFA, Debona D, Rios JA et al (2015) Potentiation of defense-related gene expression by silicon increases wheat resistance to leaf blast. Trop Plant Pathol 40:394–400

da Silva WL, Cruz MFA, Fortunato AA et al (2015) Histochemical aspects of wheat resistance to leaf blast mediated by silicon. Sci Agric 72:322–327

Dallagnol LJ, Rodrigues FA, Mielli MVB et al (2009) Defective active silicon uptake affects some components of rice resistance to brown spot. Phytopathology 99:116–121

Dallagnol LJ, Rodrigues FA, DaMatta FM et al (2011) Deficiency in silicon uptake affects cytological, physiological, and biochemical events in the rice-*Bipolaris oryzae* interaction. Phytopathology 101:92–104

Dallagnol LJ, Rodrigues FA, Tanaka FAO et al (2012) Effect of potassium silicate on epidemic components of powdery mildew on melon. Plant Pathol 61:323–330

Dallagnol LJ, Rodrigues FA, Mielli MVB (2013) Silicon improves the emergence and sanity of rice seedlings obtained from seeds infected with *Bipolaris oryzae*. Trop Plant Pathol 38:478–484

Dallagnol LJ, Rodrigues FA, Mielli MVB et al (2014) Rice grain resistance to brown spot and yield are increased by silicon. Trop Plant Pathol 39:56–63

Dallagnol LJ, Rodrigues FA, Pascholati SF et al (2015) Comparison of root and foliar applications of potassium silicate in potentiating post-infection defences of melon against powdery mildew. Plant Pathol 64:1085–1093

Dallagnol LJ, Ramos AR, Dorneles KR (2020) Silicon use in the integrated disease management of wheat: current knowledge. In: Ansari MR (Org) current trends in wheat research, 1st edn. IntechOpen, London, pp 1–21

Dallagnol LJ, Silva-Junior GJ, Moreira AS et al (2021) Potassium silicate reduces postharvest nectarine decay caused by. Acta Hort 1:385–390

Dallagnol LJ, Pazdiora PC, Miranda MZ et al (2023) Silicon ameliorates wheat technological quality under biotic stress. Trop Plant Pathol 48:226–235

Dann EK, Le DP (2017) Effects of silicon amendment on soilborne and fruit diseases of avocado. Plan Theory 20:51

Dann EK, Muir S (2002) Peas grown in media with elevated plant-available silicon levels have higher activities of chitinase and β-1,3-glucanase, are less susceptible to a fungal leaf spot pathogen and accumulate more foliar silicon. Aust Plant Pathol 31:9–13

Datnoff LE, Deren CW, Snyder GH (1997) Silicon fertilization for disease management of rice in Florida. Crop Prot 16:525–531

Debona D, Rodrigues FA, Rios JA et al (2014) The effect of silicon on antioxidant metabolism of wheat leaves infected by *Pyricularia oryzae*. Plant Pathol 63:581–589

Debona D, Rodrigues FA, Datnoff LE (2017) Silicon's role in abiotic and biotic plant stresses. Annu Rev Phytopathol 55:85–107

Debona D, Datnoff LE, Rodrigues FA (2023) Silicon and plant disease. In: Datnoff LE, Elmer WH, Rodrigues FA (eds) Mineral nutrition and plant disease, 2nd edn. The American Phytopathological Society, Saint Paul, pp 381–424

Deng Q, Wu J, Chen J et al (2020) Physiological mechanisms of improved smut resistance in sugarcane through application of silicon. Front Plant Sci 11:1587

Dixon E, Leonberger K, Amsden B et al (2022) Suppression of hemp powdery mildew using root-applied silicon. Plant Health Prog 23:250–264

Dorneles KR, Dallagnol LJ, Pazdiora PC et al (2017) Silicon potentiates biochemical defense responses of wheat against tan spot. Physiol Mol Plant Pathol 97:69–78

Dorneles KR, Pazdiora PC, Hoffmann JF et al (2018) Wheat leaf resistance to *Pyrenophora tritici-repentis* induced by silicon activation of phenylpropanoid metabolism. Plant Pathol 67:1713–1724

Du J, Liu B, Zhao T et al (2022) Silica nanoparticles protect rice against biotic and abiotic stresses. J Nanobiotech 20:197

El-Abdean WZ, Abo-Elyousr KAM, Hassan MHA et al (2020) Effect of silicon compounds against *Macrophomina phaseolina* the causal agent of soybean charcoal rot disease. Arch. Phytopathol Pflanzenschutz. ID:83494584

El-Ashry RM, El-Saadony MT, El-Sobki AEA et al (2022) Biological silicon nanoparticles maximize the efficiency of nematicides against biotic stress induced by *Meloidogyne incognita* in eggplant. Saudi J Biol Sci 29:920–932

El-Garhy HA, Rashid IA, Abou-Ali RM et al (2016) Field application of safe chemical elicitors induced the expression of some resistance genes against grey mold and cottony rot diseases during snap bean pods storage. Gene 576:358–365

Ferreira HA, Nascimento CWA, Datnoff LE et al (2015) Effects of silicon on resistance to bacterial fruit blotch and growth of melon. Crop Prot 78:277–283

Fortunato AA, Rodrigues FA, Baroni JCP et al (2012) Silicon suppresses Fusarium wilt development in banana plants. J Phytopathol 160:674–679

Fortunato AA, da Silva WL, Rodrigues FA (2014) Phenylpropanoid pathway is potentiated by silicon in the roots of banana plants during the infection process of *Fusarium oxysporum* f. sp. *cubense*. Phytopathology 104:597–603

Garibaldi A, Gilardi G, Gullino ML (2011) Effect of potassium silicate and electrical conductivity in reducing powdery mildew of hydroponically grown tomato. Phytopathol Mediterr 50:192–202

Garibaldi A, Gilardi G, Cogliati EE et al (2012) Silicon and increased electrical conductivity reduce downy mildew of soilless grown lettuce. Eur J Plant Pathol 132:123–132

Gbongue LR, Lalaymia I, Zeze A et al (2019) Increased silicon acquisition in bananas colonized by *Rhizophagus irregularis* MUCL 41833 reduces the incidence of *Pseudocercospora fijiensis*. Front Plant Sci 9:1977

Gillman JH, Zlesak DC, Smith JA (2003) Applications of potassium silicate decrease black spot infection in Rosa hybrida 'Meipelta' (Fuschia Meidiland™). HortSci 38:1144–1147

Gokavi N, Jayakumar M, Mote K et al (2021) Diatomaceous earth as a source of silicon and its impact on soil physical and chemical properties, yield and quality, pests and disease incidence of *Arabica Coffee* cv. Chandragiri. Silicon 13:4583–4600

Guérin V, Lebreton A, Cogliati EE et al (2014) A zoospore inoculation method with *Phytophthora sojae* to assess the prophylactic role of silicon on soybean cultivars. Plant Dis 98:1632–1638

Guével MH, Menzies JG, Bélanger RR (2007) Effect of root and foliar applications of soluble silicon on powdery mildew control and growth of wheat plants. Eur J Plant Pathol 119:429–436

Hasan KA, Soliman H, Baka Z et al (2020) Efficacy of nano-silicon in the control of chocolate spot disease of *Vicia faba* L. caused by *Botrytis fabae*. Egypt J Basic Appl Sci 7:53–66

Hawerroth C, Araujo L, Bermúdez-Cardona MB et al (2019) Silicon-mediated maize resistance to macrospora leaf spot. Trop Plant Pathol 44:192–196

Holz TM, Dorneles KR, Brunetto AE et al (2022) Effect of silicon and fungicide on photosynthetic responses in barley leaves challenged by *Bipolaris sorokiniana*. Physiol Mol Plant Pathol 120:101849

Holz TM, Dorneles KR, Brunetto AE et al (2023) Silicon complements disease control by fungicides by enhancing the biochemical defense responses of barley against *Bipolaris sorokiniana*. J Plant Pathol 105:1593–1607

Hong D-K, Talha J, Yao Y et al (2021) Silicon enhancement for endorsement of *Xanthomonas albilineans* infection in sugarcane. Ecotoxicol Environ Saf 220:112380

Jadhao KR, Rout GR (2020) Silicon (Si) enhances the resistance in finger millet genotypes against blast disease. J Plant Pathol 102:985–1006

Jiang N, Fan X, Lin W et al (2019) Transcriptome analysis reveals new insights into the bacterial wilt resistance mechanism mediated by silicon in tomato. Int J Mol Sci 20:761

Jones J, Dangl J (2006) The plant immune system. Nature 444:323–329

Kablan L, Lagauche A, Delvaux B et al (2012) Silicon reduces black Sigatoka development in banana. Plant Dis 96:273–278

Kang H, Elmer W, Shen Y et al (2021) Silica nanoparticle dissolution rate controls the suppression of Fusarium wilt of watermelon (*Citrullus lanatus*). Environ Sci Technol 55:13513–13522

Kashyap D, Siddiqui ZA (2021) Effect of silicon dioxide nanoparticles and *Rhizobium leguminosarum* alone and in combination on the growth and bacterial blight disease complex of pea caused by *Meloidogyne incognita* and *Pseudomonas syringae* pv. *pisi*. Arch Phytopathol Pflanzenschutz 54:499–515

Khan M, Siddiqui ZA (2020) Use of silicon dioxide nanoparticles for the management of *Meloidogyne incognita*, *Pectobacterium betavasculorum* and *Rhizoctonia solani* disease complex of beetroot (*Beta vulgaris* L.). Sci Hortic 265:109211

Khan M, Siddiqui ZA, Parveen A et al (2022) Elucidating the role of silicon dioxide and titanium dioxide nanoparticles in mitigating the disease of the eggplant caused by *Phomopsis vexans*, *Ralstonia solanacearum*, and root-knot nematode *Meloidogyne incognita*. Nanotechnol Rev 11:1606–1619

Kim SG, Kim KW, Park EW et al (2002) Silicon-induced cell wall fortification of rice leaves: a possible cellular mechanism of enhanced host resistance to blast. Phytopathology 92:1095–1103

Kostic I, Nikolic N, Milanovic S et al (2023) Silicon modifies leaf nutriome and improves growth of oak seedlings exposed to phosphorus deficiency and *Phytophthora plurivora* infection. Front Plant Sci 14:1265782

Lata-Tenesaca LF, Oliveira JB, Barros AV et al (2024) Physiological and biochemical aspects of silicon-mediated resistance in maize against Maydis leaf blight. Plan Theory 13:531

Lemes EM, Mackowiak CL, Blount A et al (2011) Effects of silicon applications on soybean rust development under greenhouse and field conditions. Plant Dis 95:317–324

Li Y, Liu J, Lv P et al (2022) Silicon improves the photosynthetic performance of oat leaves infected with *Puccinia graminis* f. sp. *avenae*. Front. Plant Sci 13:1037136

Liu B, Davies K, Hall A (2020) Silicon builds resilience in strawberry plants against both strawberry powdery mildew *Podosphaera aphanis* and two-spotted spider mites *Tetranychus urticae*. PLoS One 15:e0241151

Liu B, Asiana C, Wileman H et al (2021) Benefits of using silicon as a nutrient in sustainable strawberry production. Acta Hortic 1309:413–420

Lopes UP, Zambolim L, Neto PNS et al (2013) Effect of foliar application of potassium silicate on the progress of coffee leaf rust. Trop Plant Pathol 38:547–551

Lopes UP, Zambolim L, Neto PNS et al (2014) Silicon and triadimenol for the management of coffee leaf rust. J Phytopathol 162:124–128

Lu G, Jian W, Zhang J et al (2008) Suppressive effect of silicon nutrient on Phomopsis stem blight development in asparagus. HortScience 43:811–817

Mahmad-Toher AS, Govender N, Dorairaj D et al (2022) Effects of silica soil amendment against *Exserohilum rostratum*, the fungal pathogen of rice brown spot disease in peninsular Malaysia. Sci Rep 12:15690

Mburu K, Oduor R, Mgutu A et al (2016) Silicon application enhances resistance to xanthomonas wilt disease in banana. Plant Pathol 65:807–818

Mohaghegh P, Khoshgoftarmanesh AH, Shirvani M et al (2011) Effect of silicon nutrition on oxidative stress induced by *Phytophthora melonis* infection in cucumber. Plant Dis 95:455–460

Moyer C, Peres NA, Datnoff LE et al (2008) Evaluation of silicon for managing powdery mildew on gerbera daisy. J Plant Nutr 31:2131–2144

Naz R, Batool S, Shahid M et al (2021) Exogenous silicon and hydrogen sulfide alleviates the simultaneously occurring drought stress and leaf rust infection in wheat. Plant Physiol Biochem 166:558–571

Nicchio B, Juliatti FC, Pereira HS et al (2023) Effect of silicon sources on sugarcane orange rust, ring spot and red rot in Brazil. J Agric Sci 15:9

Oliveira JC, Albuquerque GMR, Mariano RLR et al (2012) Reduction of the severity of angular leaf spot of cotton mediated by silicon. J Plant Pathol 94:297–304

Otolakoski MG, Viegas BG, Bagio BZ et al (2023) Reduction of the severity of Asian soybean rust with foliar application of silicon dioxide. Crop Protect 173:106387

Pagani APS, Dianese AC, Café-Filho AC (2014) Management of wheat blast with synthetic fungicides, partial resistance and silicate and phosphite minerals. Phytoparasitica 42:609–617

Pazdiora PC, Dorneles KR, Forcelini CA et al (2018) Silicon suppresses tan spot development on wheat infected by *Pyrenophora tritici-repentis*. Eur J Plant Pathol 150:49–56

Pazdiora PC, Dorneles KR, Morello TN et al (2021) Silicon soil amendment as a complement to manage Tan spot and Fusarium head blight in wheat. Agron Sustain Dev 41:21

Pazdiora PC, Crizel RL, Dorneles KR et al (2023) Silicon amendment improves wheat defence against and complements the control by fungicide of Fusarium head blight. Plant Pathol 72:53–66

Polanco LR, Rodrigues FA, Nascimento KJT et al (2012) Biochemical aspects of bean resistance to anthracnose mediated by silicon. Ann Appl Biol 161:140–150

Pozza AAA, Alves E, Pozza EA et al (2004) Efeito do silício no controle da cercosporiose em três variedades de cafeeiro. Fitopatol Bras 29:185–188

Prabhu AS, Barbosa Filho MP, Datnoff LE et al (2012) Silicon reduces brown spot severity and grain discoloration on several rice genotypes. Trop Plant Pathol 37:409–414

Provance-Bowley MC, Heckman JR, Durner EF (2010) Calcium silicate suppresses powdery mildew and increases yield of field grown wheat. Soil Sci Soc Amer J 74:1652–1661

Rahman A, Wallis CM, Uddin W (2015) Silicon-induced systemic defense responses in perennial ryegrass against infection by *Magnaporthe oryzae*. Phytopathology 105:748–757

Ramouthar PV, Caldwell PM, McFarlane SA (2016) Effect of silicon on the severity of brown rust of sugarcane in South Africa. Eur J Plant Pathol 145:53–60

Rashad YM, El-Sharkawy HHA, Belal BEA et al (2021) Silica nanoparticles as a probable anti-oomycete compound against downy mildew, and yield and quality enhancer in grapevines: field evaluation, molecular, physiological, ultrastructural, and toxicity investigations. Front Plant Sci 12:763365

Rasoolizadeh A, Labbé C, Sonah H et al (2018) Silicon protects soybean plants against *Phytophthora sojae* by interfering with effector-receptor expression. BMC Plant Biol 18:97

Resende RS, Rodrigues FA, Soares JM et al (2009) Influence of silicon on some components of resistance to anthracnose in susceptible and resistant sorghum lines. Eur J Plant Pathol 124:533–541

Resende RS, Rodrigues FA, Gomes RJ et al (2013) Microscopic and biochemical aspects of sorghum resistance to anthracnose mediated by silicon. Ann Appl Biol 163:114–123

Rivera JF, Duarte HSS, Dorneles KR et al (2021) Effects of calcium silicate fertilization on disease intensity and yield and quality of rice grain. Mod Concep Dev Agrono 8:000688

Rodgers-Gray BS, Shaw MW (2000) Substantial reductions in winter wheat diseases caused by addition of rice straw but not manure to soil. Plant Pathol 49:590–599

Rodrigues FA, Datnoff LE (eds) (2015) Silicon and Plant Diseases. 1st Edition, Springer International Publishing Switzerland. 148 p.

Rodrigues FA, Datnoff LE, Korndörfer GH et al (2001) Effect of silicon and host resistance on sheath blight development in rice. Plant Dis 85:827–832

Rodrigues FA, Benhamou N, Datnoff LE et al (2003a) Ultrastructural and cytochemical aspects of silicon-mediated rice blast resistance. Phytopathology 93:535–546

Rodrigues FA, Vale FXR, Datnoff LE et al (2003b) Effect of rice growth stages and silicon on sheath blight development. Phytopathology 93:256–261

Rodrigues FA, Vale FXR, Korndörfer GH et al (2003c) Influence of silicon on sheath blight of rice in Brazil. Crop Prot 22:23–29

Rodrigues FA, McNally DJ, Datnoff LE et al (2004) Silicon enhances the accumulation of diterpenoid phytoalexins in rice: a potential mechanism for blast resistance. Phytopathology 94:177–183

Rodrigues FA, Jurick WM, Datnoff LE et al (2005) Silicon influences cytological and molecular events in compatible and incompatible rice-*Magnaporthe grisea* interactions. Physiol Mol Plant Pathol 66:144–159

Rodrigues FA, Duarte HSS, Domiciano GP et al (2009) Foliar application of potassium silicate reduces the intensity of soybean rust. Australas Plant Pathol 38:366–372

Rodrigues FA, Polanco LR, Duarte HSS et al (2015) Photosynthetic gas exchange in common bean submitted to foliar sprays of potassium silicate, sodium molybdate and fungicide and infected with *Colletotrichum lindemuthianum*. J Phytopathol 163:554–559

Saigusa M, Onozawa K, Watanabe H et al (2000) Effects of porous hydrate calcium silicate on the wear resistance, insect resistance, and disease tolerance of turf grass "Miyako". Grassl Sci 45:416–420

Sakr N (2021) Silicon reduces fusarium head blight development in barley. Open Agric J 15:54–65

Sakr N, Kurdali F (2023) Root silicon amendment enhances barley's resistance to fusarium head blight in the field. Open Agric J 17:e187433152309120

Santos LB, Souza Júnior JP, Mello Prado R et al (2022) Silicon allows halving cadusafos dose to control *Meloidogyne incognita* and increase cotton development. Silicon 14:3809–3816

Schabl P, Gabler C, Kührer E et al (2020) Effects of silicon amendments on grapevine, soil and wine. Plant Soil Environ 66:403–414

Seebold KW, Datnoff LE, Correa-Victoria FJ et al (2000) Effect of silicon rate and host resistance on blast, scald, and yield of upland rice. Plant Dis 84:871–876

Seebold KW, Datnoff LE, Correa-Victoria FJ et al (2004) Effects of silicon and fungicides on the control of leaf and neck blast in upland rice. Plant Dis 88:253–258

Shetty R, Jensen B, Shetty NP et al (2012) Silicon induced resistance against powdery mildew of roses caused by *Podosphaera pannosa*. Plant Pathol 61:120–131

Silva RV, Oliveira RDL, Nascimento KJT et al (2012) Biochemical responses of coffee resistance against *Meloidogyne exigua* mediated by silicon. Plant Pathol 59:586–593

Silva PRA, Araujo L, Resende RS et al (2020) Silicon, *Clonostachys rosea*, and their interaction for gray mold management in cucumber. J Plant Pathol 102:1257–1262

Silveira PR, Aucique-Pérez CE, Cruz MFA et al (2021) Biochemical and physiological changes in maize plants supplied with silicon and infected by *Exserohilum turcicum*. J Phytopathol 69:393–408

Song A, Xue G, Cui P et al (2016) The role of silicon in enhancing resistance to bacterial blight of hydroponic- and soil-cultured rice. Sci Rep 6:24640

Soratto RP, Fernandes AM, Crusciol CAC et al (2012) Produtividade, qualidade de tubérculos e incidência de doenças em batata, influenciados pela aplicação foliar de silício. Pesq Agrop Bras 47:1000–1006

Strange RN, Scott PR (2005) Plant disease: a threat to global food security. Annu Rev Phytopathol 43:3.1–3.34

Sun S, Yang Z, Song Z et al (2022) Silicon enhances plant resistance to Fusarium wilt by promoting antioxidant potential and photosynthetic capacity in cucumber (Cucumis sativus L.) front. Plant Sci 13:1011859

Taha NA, Hamden S, Bayoumi YA et al (2023) Nanofungicides with selenium and silicon can boost the growth and yield of common bean (*Phaseolus vulgaris* L.) and control alternaria leaf spot disease. Microorganisms 11:728

Tatagiba SD, Rodrigues FA, Filippi MCC et al (2014) Physiological responses of rice plants supplied with silicon to *Monographella albescens* infection. J Phytopathol 162:596–606

Tesfagiorgis HB, Laing MD, Annegarn HJ (2014) Evaluation of biocontrol agents and potassium silicate for the management of powdery mildew of zucchini. Biol Control 73:8–15

Torlon JL, Heckman JR, Simon JE et al (2016) Silicon soil amendments for suppressing powdery mildew on pumpkin. Sustain For 8:293

Van Bockhaven J, Spíchal L, Novák O et al (2015a) Silicon induces resistance to the brown spot fungus *Cochliobolus miyabeanus* by preventing the pathogen from hijacking the rice ethylene pathway. New Phytol 206:761–773

Van Bockhaven J, Steppe K, Bauweraerts I et al (2015b) Primary metabolism plays a central role in moulding silicon-inducible brown spot resistance in rice. Mol Plant Pathol 16:811–824

Vermeire ML, Kablan L, Dorel M et al (2011) Protective role of silicon in the banana-*Cylindrocladium spathiphylli* pathosystem. Eur J Plant Pathol 131:621–630

Vu Q, Dossa GS, Mundaca EA et al (2022) Combined effects of soil silicon and host plant resistance on planthoppers, blast and bacterial blight in tropical rice. Insects 13:604

Wang L, Cai K, Chen Y et al (2013) Silicon-mediated tomato resistance against *Ralstonia solanacearum* is associated with modification of soil microbial community structure and activity. Biol Trace Elem Res 152:275–283

Wang L, Gao Y, Jiang N et al (2022a) Silicon controls bacterial wilt disease in tomato plants and inhibits the virulence-related gene expression of *Ralstonia solanacearum*. Int J Mol Sci 23:6965

Wang L, Pan T, Gao X et al (2022b) Silica nanoparticles activate defense responses by reducing reactive oxygen species under *Ralstonia solanacearum* infection in tomato plants. NanoImpact 28:100418

Whan JA, Dann EK, Aitken EAB (2016) Effects of silicon treatment and inoculation with *Fusarium oxysporum* f. sp. *vasinfectum* on cellular defences in root tissues of two cotton cultivars. Ann Bot 118:219–226

Wiese J, Wiese H, Schwartz J et al (2005) Osmotic stress and silicon act additively in enhancing pathogen resistance in barley against barley powdery mildew. J Plant Nutr Soil Sci 168:269–274

Xavier Filha MS, Rodrigues FA, Domiciano GP et al (2011) Wheat resistance to leaf blast mediated by silicon. Australas Plant Pathol 40:28–38

Xue X, Geng T, Liu H et al (2021) Foliar application of silicon enhances resistance against *Phytophthora infestans* through the ET/JA- and NPR1-dependent signaling pathways in potato. Front Plant Sci 12:609870

Yanar Y, Yanar D, Gebologlu N (2011) Control of powdery mildew (*Leveillula taurica*) on tomato by foliar sprays of liquid potassium silicate (K_2SiO_3). Afr J Biotechnol 10:3121–3123

Zakaria MAT, Sakimin SZ, Ismail MR et al (2023) Biostimulant activity of silicate compounds and antagonistic bacteria on physiological growth enhancement and resistance of banana to fusarium wilt disease. Plan Theory 12:1124

Zellner W, Tubaña B, Rodrigues FA et al (2021) Silicon's role in plant stress reduction and why this element is not used routinely for managing plant health. Plant Dis 105:2033–2049

Zhan LP, Peng DL, Wang XL et al (2018) Priming effect of root-applied silicon on the enhancement of induced resistance to the root-knot nematode *Meloidogyne graminicola* in rice. BMC Plant Biol 18:50

Zhang M-Y, Hong D-K, Chen Y-H et al (2022) Synergistic effects of organosilicon and cu(OH)$_2$ in controlling sugarcane leaf scald disease. Int J Mol Sci 23:13532

Zhou X, Shen Y, Fu X et al (2018) Application of sodium silicate enhances cucumber resistance to fusarium wilt and alters soil microbial communities. Front Plant Sci 9:624

Chapter 20
Optimized Plant Nutrition for Precision Biofortification to Meet Nutrition Custom Human

Massimiliano D'Imperio, Davide Onofrio Palmitessa, and Francesco Serio

Abstract Biofortification, the enhancement of plant nutritional quality, involves increasing organic and mineral factors or reducing antinutritional elements. Market-available successes include selenium-enriched potatoes and iodine-enriched carrots. The evolving challenge focuses on tailored plant nutrition, producing foods for specific populations, known as "tailored food." Soilless cultivation systems prove ideal, allowing precise mineral modulation for human health benefits or, if required, reducing harmful elements for individuals with metabolic disorders. Examples include high-silicon vegetables for osteoporosis patients and low-potassium plant foods for kidney disease patients. Modularity in plant nutrition, achievable through soilless systems, offers promising prospects for tailored foods addressing nutritional needs. This innovation, easily integrated into existing production facilities, aligns with global health initiatives, promoting nutrient-rich diets and countering malnutrition. The simplicity, sustainability, and cost-effectiveness of this approach, alongside emerging technologies, position customized plant-based foods as versatile ingredients for fortified products and daily consumption.

Keywords Personalized nutrition · Tailored vegetables · Silicon biofortification · Soilless system · Therapeutic diet

M. D'Imperio (✉) · F. Serio
Institute of Sciences of Food Production, CNR – National Research Council of Italy, Bari, Italy
e-mail: massimiliano.dimperio@ispa.cnr.it

D. O. Palmitessa
Department of Soil, Plant and Food Sciences, University of Bari "Aldo Moro", Bari, Italy

R. de Mello Prado et al. (eds.), *Silicon Advances for Sustainable Agriculture and Human Health*, Sustainable Plant Nutrition in a Changing World,
https://doi.org/10.1007/978-3-031-69876-7_20

20.1 Introduction

Biofortification is the process of enhancing the nutritional quality of a plant or a portion of it. It can be achieved by increasing nutritional factors—organic and/or minerals—introduced into the plant through various techniques or by reducing naturally occurring antinutritional factors in vegetables, such as phytates and oxalates that can reduce the absorption of calcium, iron, and zinc in the intestines. Regardless of the method used, only with a multidisciplinary approach is it possible to "measure" the nutritional value of biofortified products. This is done through the application of in vitro and/or in vivo protocols that assess fundamental nutritional parameters such as bioaccessibility and bioavailability (D'Imperio et al. 2016a, b, 2017; Piątkowska et al. 2016). Successful examples currently available on the market include selenium-enriched potatoes, carrots, and onions, iodine-enriched potatoes, as well as "nickel-free" and/or low-histamine and heavy metal-free products. The new challenge of biofortification is represented by tailored plant nutrition, aimed at obtaining products for specific population groups. In this case, we refer to "tailored food." Soilless cultivation systems are ideal for achieving this result because they allow for precise and appropriate plant nutrition management by modulating the accumulation of essential minerals (calcium, iodine, zinc, selenium, silicon, iron, etc.) beneficial for human health (D'Imperio et al. 2016a, b; Ferrarese et al. 2012; Giordano et al. 2019; Gonnella et al. 2019; Montesano et al. 2016; Pannico et al. 2020; White et al. 2018) or, if necessary, harmful to individuals with specific metabolic disorders. A typical example is the reduction of potassium in vegetables intended for individuals with chronic kidney diseases who must follow a low-potassium diet (a nutrient typically abundant in vegetables and fruits). Another example is the reduction of nickel concentration in vegetables for people allergic to this metal. The modularity of plant nutrition, achievable with soilless cultivation systems, opens up significant prospects for obtaining plant-based foods tailored to specific nutritional needs (D'Imperio et al. 2019; Rouphael and Kyriacou 2018). The potential for introducing prototypes of tailored vegetables is promising, considering the state of malnutrition not only in underdeveloped and developing countries but also in industrialized nations, where there is an exponential increase in "malnourished" individuals due to excessive and "incorrect" eating habits (junk food). It should not be forgotten that numerous public health campaigns promoting healthy eating suggest increasing the consumption of fruits and vegetables. The application of tailoring production to fruits and vegetables, therefore, allows for the dual result of promoting the consumption of foods beneficial to human health and increasing the intake of micronutrients with specific functions in the human body. A typical example is iodine-enriched vegetables, which can replace iodized salt and avoid the negative health effects associated with table salt consumption (Gonnella et al. 2019; Gonzali et al. 2017). Furthermore, the potential and spread of this innovation are based on the relative simplicity of the production process and the ability to obtain tailored foods without modifying soilless cultivation systems' production facilities, especially without reducing production per unit area (Somma et al. 2024; D'Imperio et al. 2023). In recent years, further innovations have emerged in the field of customized foods with the goal of improving sustainability and reducing production costs (biostimulants, nanocarriers, calibrated light inputs, and organic matrices, the latter used as a natural source of mineral elements). Finally, enriched plant-based foods can easily be used in the formulation of new fortified food products for daily consumption (Klopsch et al. 2018).

20.2 Silicon, Nutrition, and Human Health

Silicon, the second most abundant element on Earth after oxygen, is unlikely to cause nutritional deficiencies in humans due to its abundance. Scientific research has shown that silicon is essential for human health (Van Dyck et al. 1999). It plays a crucial role in bone mineralization and soft tissue development. Consequently, silicon is believed to have positive effects on various human conditions, such as osteoporosis (Eisinger and Clairet 1993), aging of the skin, hair, and nails (Lassus 1993), and arteriosclerosis (Lassus 1993; Schwarz et al. 1977). While its essentiality for humans hasn't been definitively established, supplements are recommended for specific groups. The main dietary sources of silicon are cereals like barley, oats, wheat, rice, and certain herbaceous plants, providing 30% of silicon intake. Fruits, vegetables, beverages, and nuts follow (Jugdaohsingh 2007; Montesano et al. 2016). Cereal-based products such as bread, rice, pasta, breakfast cereals, and flour also offer significant silicon. Other sources include fish, meat, eggs, milk, juices, drinking water, alcoholic beverages, and pharmaceutical supplements (Powell et al. 2005). However, food processing can result in up to 50% silicon loss, so unprocessed whole foods have higher amounts. Leafy and fruit vegetables provide a limited silicon source, but innovative agricultural techniques can increase silicon content in the edible parts (D'Imperio et al. 2016a, 2017; Alves et al. 2024; Gonzalez-Porras et al. 2024). Additionally, vegetable consumption not only offers silicon but also bioactive compounds that work in synergy with silicon to promote bone health (D'Imperio et al. 2018). Silicon is a common dietary component, with its most bioavailable form being orthosilicic acid (H_4SiO_4). After ingestion, most absorbed silicon is excreted in the urine within 4–8 h, likely as orthosilicic acid and/or magnesium orthosilicate (Nielsen 2014). Soluble silicon is present in liquids like mineral water (2–5 mg/L silicon) and beer (9–39 mg of silicon/L) but not in solid foods (Rodella et al. 2014). Silicon absorption depends on the food source, with green beans having a high absorption rate (>50% of the total content), while bananas have a low absorption rate (2% of the total content) (Sripanyakorn et al. 2009). Silicon supplements are commonly used, with intestinal availability varying from less than 1% to 50%, likely depending on its chemical form (Calomme et al. 1998; Van Dyck et al. 1999). Research by Jugdaohsingh et al. (2000) showed that 53% of administered H_4SiO_4 (in monomeric and oligomeric forms) was excreted in urine, while polymeric silicic acid [SiOx(OH)4-2x]n had only a slight impact on urinary silicon levels (Jugdaohsingh et al. 2000). In addition to H_4SiO_4, water-soluble silicates also contribute to bioavailable silicon forms. For example, pharmaceutically acceptable alkali metal silicates (Na_2SiO_3 and K_2SiO_3) release H_4SiO_4 when appropriately diluted in aqueous solutions, coming into contact with stomach hydrochloric acid (Popplewell et al. 1998). The same authors used radiolabeled ammonium silicate to measure total absorption and urinary excretion. Silicon's significance as a mineralization initiator is seen in its high concentration in immature osteoids, decreasing as bone maturity increases (Jugdaohsingh et al. 2004). Beneficial effects of silicon on bones and cartilage have been confirmed in various in vivo, ex vivo, or in vitro studies. A study by Jugdaohsingh (2007) showed that a 28 mg/day silicon supplement over 12 weeks increased spinal bone mineral density by 2.5% in women with low bone density. Numerous studies on cell cultures and tissues have been conducted to understand the mechanisms of silicon's effect on bones, including its role in cell proliferation, alkaline phosphatase activity, osteocalcin production, and increased type I collagen in osteosarcoma and osteoblastic cell lines (Brady et al. 1991; Reffitt et al. 2003). From a bone perspective, silicon helps initiate the mineralization process. It is highly concentrated in immature osteoid and decreases as the calcium content increases in mature bones (Jugdaohsingh et al. 2004). Silicon may also modulate immune or inflammatory responses and has been associated with mental health (Nielsen 2014). Given silicon's numerous beneficial properties, introducing silicon-rich foods into the diet may enhance human well-being.

20.3 From Biofortification to Tailored Vegetables

Plants intended for human consumption, regardless of the species, represent a source of essential nutrients for consumers, both organic (such as vitamins, sugars, and proteins) and inorganic (such as minerals). The intake of these nutrients is crucial for the health of consumers. While a varied diet is recommended, it is important to note that in many parts of the world, especially in low-income countries, diets are often lacking in diversity. This situation can lead to the development of nutritional deficiencies, which can be addressed through biofortification approaches.

Over the years, the concept of biofortification has evolved to include the production of plant-based foods designed to meet specific nutritional needs (Fig. 20.1). These foods are known as "tailored vegetables."

20.3.1 Biofortification, History, Concepts, and Approaches

Biofortification is the scientific response to a significant social problem—malnutrition—a challenge exacerbated by the widespread adoption of unhealthy lifestyles and a high consumption of unhealthy foods worldwide, regardless of socio-economic context. Over the past 24 years, the number of scientific articles published on

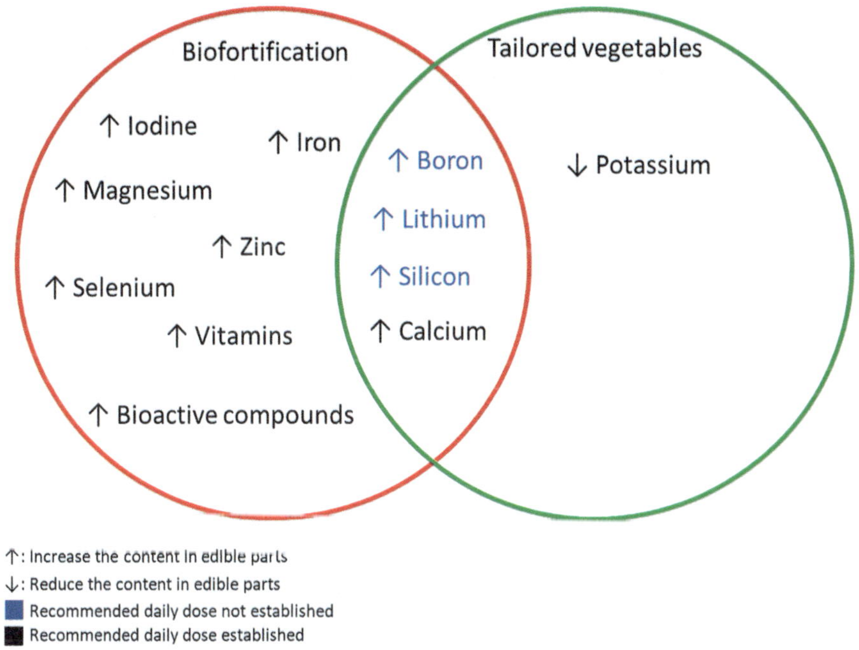

Fig. 20.1 Evolution of the biofortification concept in the development of tailored vegetables to specific nutritional needs. In this perspective, mineral elements without a recommended daily dose are highlighted in blue, as they are not considered essential

Scopus, a renowned platform for indexing peer-reviewed scientific content, has surged from a single paper in 2000 to over 731 in 2024 alone. This growing interest is tied to the potential to enhance the nutritional quality of commonly consumed foods by the population. Vegetables are rich in essential nutrients such as polyphenols, carotenoids, vitamins, folate, and various minerals and compounds influencing human health and potentially bone health (Macdonald et al. 2008). A healthy eating strategy, emphasizing increased consumption of plant-based foods in quantities recommended by FAO and the World Health Organization (WHO), is crucial for a well-balanced diet and the prevention of non-communicable diseases like heart disease, cancer, stroke, diabetes, Alzheimer's disease, cataracts, age-related function decline, malnutrition, and food-borne diseases (food safety). Consumer demands are evolving, with a preference for food that aligns with their specific life situations, including health status, age, sex, employment status, cultural background, etc. This poses a challenge for the food industry to better understand and meet these diverse needs, ensuring that new foods, as part of balanced diets, are accepted by consumers. In recent years, the scientific community has published numerous studies exploring different approaches to improve the nutritional status of target populations, including supplementation, fortification, and biofortification. Supplementation, a short-term strategy to enhance nutritional health, involves distributing pills or mineral solutions for immediate consumption (Gomez-Galera et al. 2010). While chemical supplements are used to prevent chronic diseases related to micronutrient deficiency, they are less acceptable, as consumers generally prefer food over "chemical supplements." Fortification is another method to increase nutrient concentration in food, allowing the addition of nutrients beyond natural levels (e.g., adding vitamin C, folic acid to breakfast cereals) (Berner et al. 2001). This practice has been successful in the developed world since 1980 for preventing various diseases. Moreover, biofortification involves deliberately increasing the content of essential micronutrients, such as vitamins and minerals, in plant foods. This practice aims to improve the nutritional quality and provide public health benefits with minimal risk to health.

20.4 Strategies for Biofortification of Food Crops

20.4.1 Genetic Engineering

Genetic engineering allows to get biofortified crop production by increasing the micronutrient content in the edible parts of plants. In recent years, studies of applied genomics combined with new technologies in molecular biology have made it possible to sequence entire genomes of the major basic crops such as rice, sorghum, and soybeans (Paterson et al. 2009; Schmutz et al. 2010). Biotechnological approaches have different points of weakness linked to reduced consent by the consumer" (Winkler 2011). However, the strength of genetic engineering is the ability to act upstream of the biochemical processes by inhibiting or inducing the synthesis of anti-nutritional or nutritional factors, respectively. An example is the gold-rice, a special type of rice, obtained by engineering genetic, constituted by a high content of carotenoids (Ye et al. 2000) used to combat the deficiencies of vitamin A in the developing countries (Paine et al. 2005). More recently, the genetic engineering has moved into a new phase that aims at adding several micronutrients to the same plant (i.e., "multigene transfer") (Winkler 2011). An example of success is the "multivitamin corn," which has achieved high levels of β-carotene, ascorbate (vitamin C), and folate (vitamin B9) in the plant (Naqvi et al. 2009).

20.4.2 Conventional Plant Breeding

Considering that there are many varieties of the same species, it is possible to develop, through conventional techniques, new variants of crops with better nutritional profiles. Thus, more recently, an effort has been put into conventional plant breeding for enhancing the micronutrient concentrations of both minerals and vitamins in the edible parts of the plant (Carvalho and Vasconcelos 2013). The most important program is the *HarvestPlus* biofortification, aimed to increase the nutritional levels of iron, zinc, and pro-vitamin A in seven basic crops (beans, cassava, maize, rice, wheat, sweet potatoes, and pearl millet) (Winkler 2011). This strategy is a widespread public acceptance. Moreover, this approach presents a reduced cost, as the seeds can be further collected and reused in subsequent years (Winkler 2011). For these reasons, this strategy, turns out to be convenient to find a long-term application, is considered the most cost-effective solution to increase the micronutrient density in edible plant tissues (Hirschi 2009). The main weaknesses are represented by the high time required to obtain crops biofortified and high dependence on phyto-availability of mineral nutrients in the soil as well as sufficient genetic variation of a particular phenotypic trait within the same species (Carvalho and Vasconcelos 2013).

20.4.3 Agronomic Techniques

In the soil, mineral elements can be present as free ions, as surface-adsorbed ions, as dissolved compounds or precipitates (White and Broadley 2009). A common limitation for biofortification, by using agronomic approaches, is the generally low phyto-availability of mineral micronutrients in the soil. Thus, agronomic efforts have been directed toward the application of mineral fertilizers by using foliar or soil fertilization, obtaining immediate results by using relatively simple method. The weaknesses of these approaches are represented by short-term strategy, recurrent expensive cost (regular application is needed), and negative environmental impact, such as increase of chemical fertilization in environment and/or change of chemical/physical parameters of the soil (Carvalho and Vasconcelos 2013). Considering the difficulty to obtain biofortified vegetables by using common agronomic procedures, one of the easy methods for the biofortification is the floating hydroponic system approach (Malorgio et al. 2009; Voogt et al. 2010; Tomasi et al. 2016). This system—a closed soilless system utilized to produce leafy vegetables—allows a quickly nutrient uptake in plant (Malorgio et al. 2009; Voogt et al. 2010; D'Imperio et al. 2016a), by using management of growing conditions. More particularly, the concentration of the nutrient solutions is one of the most important aspects for successful leafy vegetable production (Tomasi et al. 2016). Changes on nutrient solution composition can also have a great impact on the shelf life of the fresh cut product (Tomasi et al. 2016) and increase of nutritional quality, such as silicon, iodine, iron, zinc, and other enrichment (Table 20.1). The strengths of floating hydroponic system are reducing the time of cultivation, increasing the shelf life of the fresh cut product, and enhancing the nutritional value for animals and humans (Tomasi et al. 2016). From an application standpoint, vegetable production is more straightforward using soilless cultivation systems. The simple addition or increase of a nutrient in the nutrient solution facilitates root absorption and subsequent translocation to the edible portions of the vegetables. Furthermore, the addition of micronutrients does not result in any alterations to visual quality or productive performance, as long as excessive amounts are avoided to prevent phytotoxicity phenomena. On

Table 20.1 Some instances of biofortified vegetables that have been enriched with various nutrients, cultivated through soilless systems

Species	Elements and level	Info cultivation system	Authors
Basil, chicory, tatsoi, mizuna, purslane, and Swiss chard	Silicon (50 and 100 mg/L)	Floating system	D'Imperio et al. (2016)
Green beans	Silicon (100 mg/L)	Pots	Montesano et al. (2016)
Chicory	Silicon (four combinations of Si and NaCl levels in the nutrient solution)	Floating system	D'Imperio et al. (2017)
Tomato and unripe melon seedlings	Silicon (50 and 100 mg/L)	Floating system	Somma et al. (2024)
Strawberry	Silicon (50 and 100 mg/L)	Growing media	Valentinuzzi et al. (2017)
Purslane	Boron (3 and 6 mg/L)	Floating system	D'Imperio et al. (2020)
Purslane	Zinc (1.3, 2.6, and 5.2 mg/L Zn)	Floating system	D'Imperio et al. (2022)
Pea and sunflower microgreens	Zinc (nutri-priming: 25, 50, 100, and 200 mg/L)	Growing media	Poudel et al. (2023)
Basil	Zinc (12.5, 25.0, 37.5, and 50 μM)	Floating system	Ciriello et al. (2022)
Rocket	Zinc and copper (25, 50, or 100 μM)	Floating system	Chrysargyris et al. (2023)
Basil	Zinc (0, 12.5, 25, 37.5, and 50 μM)	Floating system	Ciriello et al. (2023)
Tomato	Iron (foliar and nutrient solution application)	Pots	Buturi et al. (2022a)
Lettuce	Iron (1.02 and 2.02 mmol/L)	Pots	Buturi et al. (2022b)
Basil, endive, tatsoi, mizuna	Calcium (200 mg/L)	Floating system	D'Imperio et al. (2016)
Lettuce	Calcium (100–800 mg/L)	Floating system	Borghesi et al. (2013)
Tomato	Iodine (foliar application 1 and 5 mM)	Growing media	Somma et al. (2024)
Lettuce	Iodine (different form of iodoquinolines)	Nutrient film technique	Dyląg et al. (2023)
Kale	Iodine (different form of iodoquinolines)	Nutrient film technique	Krawczyk et al. (2024)
Broccoli raab, curly kale, mizuna, red mustard	Iodine (0.75 and 1.5 mg/L)	Pots	Gonnella et al. (2019)

(continued)

Table 20.1 (continued)

Species	Elements and level	Info cultivation system	Authors
Radish, peat, rocket and Swiss chard microgreens	Iodine (1.5 and 3 mg/L)	Growing media	D'Imperio et al. (2023)
Tomato	Selenium (1.0, 2.0, and 4.0 μmol/L)	Growing media	Sabatino et al. (2021)
Kale microgreens	Selenium (10, 20, and 40 μM)	Growing media	Tavan et al. (2024)
Lamb's lettuce, lettuce, wild rocket, and spinach	Selenium (2.6, 3.9, and 5.2 μmol/L)	Floating system	Francini et al. (2023)
Rocket	Selenium (10, 20, 30, 40, 50, and 60 μM)	Nutrient film technique	Nascimento et al. (2022)

the other hand, when the goal is to increase a macronutrient, such as calcium or magnesium, it is advisable not to excessively raise the concentration in the nutrient solution. Such a practice could lead to an excessive increase in electrical conductivity, with the potential for reduced yield and commercial quality. The optimal concentration of a nutrient capable of producing biofortified vegetables can depend on various factors, including genotype (species-specific absorption), cultivation system (floating system, aeroponics, substrate cultivation), application method (root, foliar spray), growing-media type (organic, inorganic, substrate mix), nutrient solution management (open or closed cycle), and climatic conditions (temperature, relative humidity, light).

Table 20.1 reports some examples of research developed with the aim of producing biofortified vegetables with different nutrients, such as silicon, boron, zinc, iron, iodine, and selenium, using various soilless cultivation systems.

In recent years, with the increasingly widespread adoption of lighting systems, new approaches have emerged that harness LED lights to provide specific wavelengths and/or extend the photoperiod. The goal of these approaches is to enhance nutritional quality by "encouraging" plants to synthesize bioactive compounds and/or vitamins beneficial for human health.

20.5 Production of Tailored Vegetables

The production of vegetables tailored to specific nutritional needs poses an increasingly relevant challenge in modern agriculture, especially with the advent of personalized nutrition. Growing concerns related to food safety and individual health have spurred research in the development of innovative agronomic techniques to produce vegetables capable of meeting nutritional requirements, even for small groups of individuals.

Among these groups are individuals with chronic kidney disease, mental disorders, and osteoporosis. For individuals with chronic kidney disease, the primary nutritional recommendation involves reducing the consumption of potassium-rich foods, such as vegetables (Fig. 20.1). However, since these foods often contain bioactive compounds with anti-inflammatory activity, reducing their intake can have significant impacts on consumer health, particularly considering that such products are rich in bioactive compounds with anti-inflammatory properties, and chronic kidney disease is an inflammatory-based chronic condition.

For individuals with mental disorders, such as bipolar disorder, it is necessary to supplement lithium, typically taken through supplements. Lithium compounds appear to be among the most promising and effective drugs for treating this disorder, especially concerning bipolar affective disorder. Lithium, a metal naturally present in the Earth's crust, is used in the form of carbonate (Li_2CO_3) for psychiatric disorder treatment. Commonly prescribed doses for such disorders range from 600 to 1200 mg of Li_2CO_3 per day. There is evidence that a daily intake of 1 mg of lithium could effectively prevent mood disorders, reducing impulsivity and nervousness, especially in at-risk individuals. Some lithium-enriched foods, such as nuts, cereals, fish, and vegetables, could contribute to mood stabilization. There is also evidence that lithium could be useful in preventing neurodegenerative diseases such as Alzheimer's, in treating depression, and in mood stabilization, thanks to its antimanic, antisocial, and prophylactic properties.

Bone tissue health is influenced by factors such as gender, hormones, genetic and environmental variables, and, not least, nutrition. From a nutritional standpoint, it is well-known that optimal intake of calcium and vitamins D and potassium is crucial for both primary and secondary prevention of osteoporosis. The recommended daily intake of calcium for individuals aged 19 to 50 is 1 g, while it is 1.2 g for those over 50 (Del Valle et al. 2011). Silicon is another element associated with promoting bone formation and increasing bone mineral density in premenopausal men and women (Macdonald et al. 2012; Reffitt et al. 2003). An intake of 10–25 mg per day seems to improve bone health (Hunt 2012; Nielsen 2008). Furthermore, some data indicate that an intake of 1.0–3.0 mg per day of boron has beneficial effects on bone health (Hunt 2012; Nielsen 2008). It is important to remember, however, that no single food or nutrient can guarantee bone health on its own. Instead, a balanced diet with adequate amounts of fruits and vegetables containing vitamins, minerals, and alkalizing substrates is considered the best approach (Prentice 2004; Stránský and Ryšavá 2009). In light of these observations, consuming vegetables with high content of calcium, silicon, and boron appears to be an effective way to promote bone health, especially in individuals susceptible to osteoporosis.

The production of vegetables tailored to specific nutritional needs requires an integrated approach that combines genetic selection, cultivation techniques, and the use of advanced technologies. In general, in the soilless system, the winning strategy is represented by the innovative management of the composition of the nutrient solution, as shown in Fig. 20.2. The reduction of an element, typically present in the nutrient solution, such as potassium, allows the production of different types of vegetables with reduced potassium content. Meanwhile, the increase in other

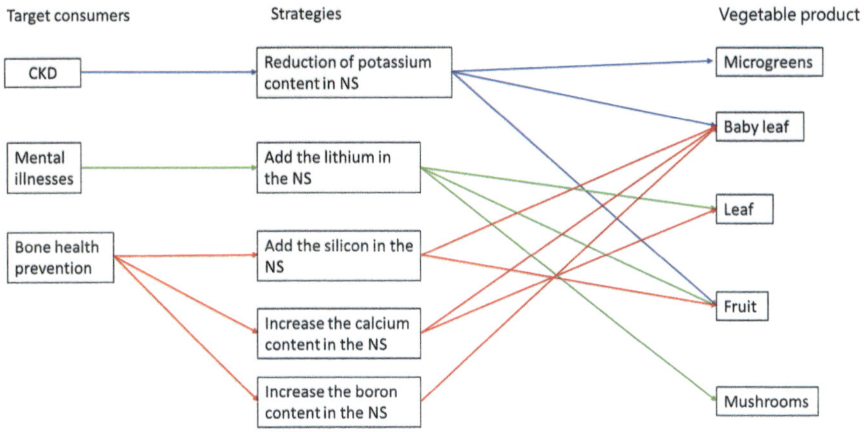

Fig. 20.2 Overview of the production of different types of tailored vegetables to specific nutritional needs as chronic kidney disease (CKD), mental illnesses, and bone health prevention. These vegetables are cultivated using soilless cultivation techniques through innovative nutrient solution (NS) management strategies. (Extracted from Renna et al. 2022)

elements such as calcium and boron, or the addition of elements normally absent in the nutrient solution, such as silicon and lithium, enables the creation of tailored vegetables for individuals with mental illnesses and for those who aim to maintain good bone health.

20.6 Silicon Biofortification Through Foliar or Root Applications

The positive impacts of silicon are significantly hindered by the absorption capabilities of plants. Higher plants exhibit three potential types of silicon uptake (Fig. 20.3) concerning water absorption: active (faster/higher than water uptake), passive (similar speed to water uptake), and rejective (slower/lower than water uptake) (Kaur and Greger 2019; Yan et al. 2018). Working with soilless systems, plants with an active silicon uptake system lead to substantial silicon depletion, resulting in a reduction of silicon concentration in the nutrient solution (Kaur and Greger 2019; Yan et al. 2018). Conversely, plants utilizing passive silicon uptake maintain an unchanged silicon concentration in the nutrient solution (Kaur and Greger 2019; Yan et al. 2018). In contrast, plants employing rejective silicon uptake tend to elevate silicon concentration in the external medium. Moreover, active accumulators have a shoot silicon content ranging from 1.5% to 10%, passive accumulators have a shoot silicon content of 0.5–1.5%, and rejective classification applies to plants with a silicon content of <0.2% (Coskun et al. 2019). Passive transport through diffusion and/or facilitated diffusion via proteinaceous channels constitutes a concentration-dependent aspect of silicon uptake present in all plant species, irrespective of their

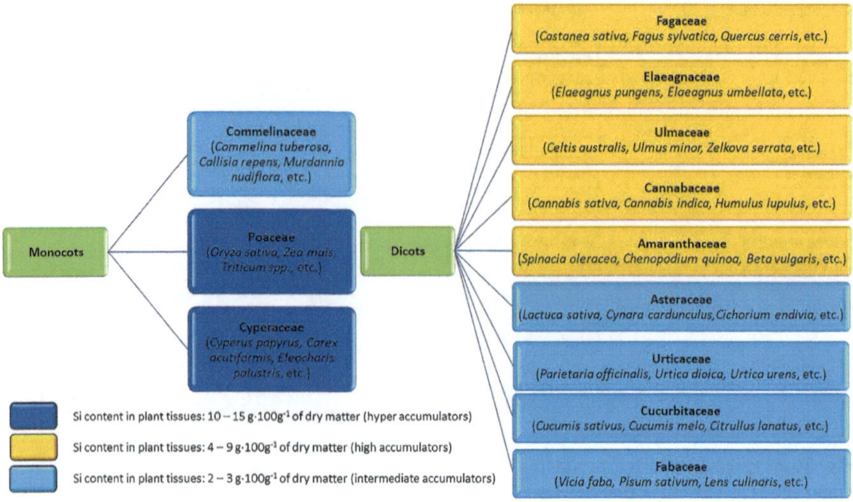

Fig. 20.3 Classification of plants based on silicon content

silicon accumulation abilities. Conversely, active uptake, an independent concentration component of silicon transport, has been identified in various silicon-accumulating plant species, encompassing both high and intermediate accumulators (Kaur and Greger 2019). Recent molecular studies on silicon uptake and transport have revealed a diverse group of proteins and genes with distinct characteristics responsible for silicon transport from the soil solution into the roots and from the roots to the shoots in different plant species. Variations in the existence, density, and localization of the complex silicon transport protein system at least partially explain the differences in silicon accumulation levels among species and tissues (Yan et al. 2018).

The concentration of available silicon is predominantly influenced by soil parent materials and factors affecting silicon adsorption-desorption processes in soil, such as pH, water status, temperature, and accompanying ions. Variances in available silicon between soil types indicate potential silicon deficiency in specific soils. However, measured plant-available SiO_2 content in calcareous paddy soils can reach as high as 250–380 mg/kg (Yan et al. 2018). To address silicon deficiency and enhance crop production, various silicon sources, including slag-based silicate, soluble potassium, or sodium silicates, have been utilized as silicon fertilizers. Silicon-based fertilizers have been documented to enhance the growth, yield, and quality of a wide range of agriculturally and horticulturally important crops. Different forms of silicon-based fertilizers, derived from chemical products, natural minerals, and metallurgic waste by-products, are generally effective in promoting crop performance, with variations in efficacy and economic costs (Yan et al. 2018). One of the most used silicon-based fertilizers is potassium metasilicate (K_2SiO_3) that provided by root application at silicon concentration by 3.6 mM, in soilless system, increased the nutritional value and quality of the obtained chicory

(D'Imperio et al. 2018). However, the main limitation to the use of potassium metasilicate in soilless systems is the clogging of the nozzles of drip systems. To avoid salt precipitation phenomena, careful management of the pH of the nutrient solution is essential, which must be maintained between 5.0 and 6.0.

Recent research has focused on foliar nutrition using silicon, which offers unequivocal production benefits (Basirat and Mousavi 2022), is more cost-effective compared to soil fertilization (Artyszak 2018). The use of silicon for foliar nutrition may become a crucial element of integrated plant protection, mandatory for all farmers in the European Union. Importantly, foliar application of silicon is environmentally safe and applicable in organic farming. Studies have utilized Si concentrations of 17 mM in the form of potassium silicate for foliar nutrition, surpassing the standard 1.7 mM used in root applications (Guével et al. 2007). Despite this, in certain species (e.g., wheat), Si deposition was most pronounced in root applications at the recommended concentrations of 1.7 mM, while foliar applications resulted in negligible or inconsistent absorption (Guével et al. 2007). In some cases, it was demonstrated that mitigating potassium deficiency in plants is more apparent with Si supply via nutrient solution (roots) compared to foliar applications, although foliar spray can also reduce deficiency. Foliar application of silicon effectively provides this beneficial element to plants (Dos Santos et al. 2021). The exploration of new potential sources for foliar application necessitates studies assessing silicon supply efficacy and its effects on biological indicators to maximize benefits, especially for plants that do not naturally accumulate the beneficial element (Dos Santos et al. 2021).

Furthermore, soil salinity poses an abiotic stress limiting the yield of agronomic crops crucial for food and industrial production. Silicon reduces sodium uptake by decreasing bypass flow, inhibiting transpiration, regulating polyamine metabolism, and reducing sodium accumulation in both root apex and cortex, facilitating sodium exclusion. A study on sunflowers and sorghum plants demonstrated that the adverse effects of salt stress were more prominently attenuated when silicon was applied via nutrient solution in sorghum plants (Calero Hurtado et al. 2020). Additionally, foliar application of silicon alone may serve as a potential strategy to alleviate detrimental effects of salt stress in both sorghum and sunflower plants when silicon is not available in the growth medium (soil or nutrient solution) (Calero Hurtado et al. 2020). These findings suggest that the method of silicon application plays a crucial role in sodium detoxification by modifying the antioxidative defense mechanism, actively mediating essential metabolic processes, and enhancing salt stress tolerance in Si accumulator and intermediate-Si accumulator plants.

The ability of silicon to reduce the uptake of harmful non-nutrient ions is not restricted merely to NaCl but is extended also to other toxic metals (Savvas and Ntatsi 2015). It was studied that silicon addition was capable of alleviating Al toxicity in some crops by drastically restricting the aluminum uptake by roots (Savvas and Ntatsi 2015). When concentrations of nutrients or harmful non-nutrient elements in the root zone are excessive, application of Si, mainly by roots, may reduce their translocation to the shoot, thereby mitigating their toxicity, by reducing the rates of transpiration (Savvas and Ntatsi 2015).

Fig. 20.3 shows the silicon storage capacity of the main botanical families of agricultural interest based on the silicon concentration in the dry matter of plant tissues (Kovács et al. 2022).

20.7 Bioavailability and Bioaccessibility: Valuable Tools to Enhance Precision Biofortification

In the field of food science, in vitro systems play a fundamental role in assessing nutritional efficiency, and these systems can be applied to various food matrices, including fresh plant products, cooked items, and processed and ultra-processed products. They can be used to evaluate the bioaccessibility, bioavailability, and the biological effect induced by the food on the target tissue.

Bioaccessibility represents the percentage or concentration of a specific nutrient released from the food matrix, such as a biofortified food, in the gastrointestinal tract during an in vitro digestion process (Fig. 20.4). Some in vitro models have been developed to simulate the physiological conditions of human gastrointestinal digestion. The biochemical, physiological, and dynamic conditions of the gastrointestinal system (mouth, stomach, and intestines), including temperature, mechanical forces, pH, concentrations of enzymes (pepsin, pancreatin, and lipase), and the presence of bile (Fernández-García et al. 2009), have been artificially reproduced in the in vitro gastrointestinal digestion model. This model provides information on the amount of nutrients released from the food matrix (bioaccessibility), their biochemical transformation and chemical degradation, nutrient-nutrient and nutrient-antinutrient interactions, and the effects of the matrix and food processing (Fernández-García et al. 2009).

Bioaccessibility depends on various factors, such as the species and type of plant materials and nutrient localization, and is influenced by multiple variables,

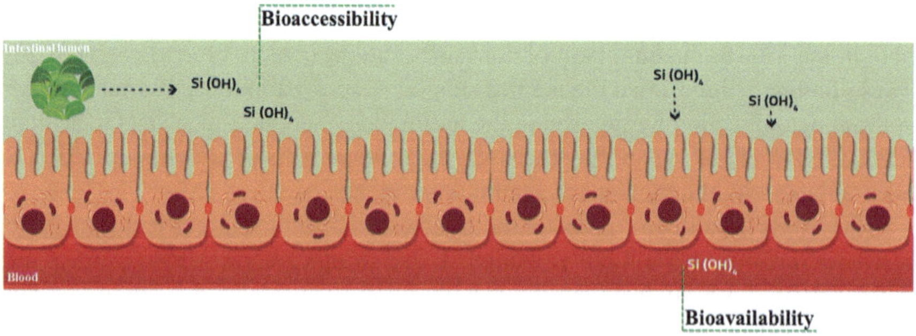

Fig. 20.4 Representation of an in vitro model for assessing the silicon bioaccessibility (percentage of nutrient release from the food matrix during the gastrointestinal digestion process) and silicon bioavailability (percentage of nutrients absorbed in the intestinal tract after the gastrointestinal digestion process) of organic and inorganic nutrients

including concentrations of other nutrients and/or antinutritional compounds, cooking methods used, if and how the food has been processed, and the interaction of other nutrients. Nutrient concentration in the edible parts is crucial, but the biofortification process itself can also modify nutrient release, as reported in numerous studies on different mineral elements (D'Imperio et al. 2016a, 2017, 2022).

The gastrointestinal digestion model has been utilized to identify biofortified species or treatments capable of releasing high quantities of nutrients in the gastrointestinal tract (Zou et al. 2014; Dong et al. 2021; Cunha Coelho et al. 2021). Bioaccessibility assessment methodologies can be used to select species, cultivars, and/or genotypes capable of releasing high quantities of nutrients during the digestive process to maximize health effects.

In the in vitro digestion process, bioaccessibility assessment is often coupled with the evaluation of the amount of a nutrient absorbed in the gastrointestinal tract, potentially capable of inducing a beneficial effect on the consumer (Fig. 20.4). Currently, several in vitro models can simulate histological, physiological, and biochemical conditions of the gastrointestinal tract with characteristics comparable to in vivo conditions. One of the most widely used cellular models is represented by Caco-2, with applications in the study of active transport of mineral nutrients (D'Imperio et al. 2017; Koronowicz et al. 2016).

Normally, this information is obtained from in vivo studies on laboratory animals, which are more expensive and time-consuming. Considering the principles of the 3Rs (replacement, reduction, and refinement), developed over 50 years ago and supported by the European Community, alternative approaches, such as in vitro methods, are increasing in both academic and industrial sectors. These methods are useful for screening to reduce costs and the number of animals, offering benefits in reproducing human physiological conditions and understanding the mechanism of action. Through a multidisciplinary approach, it is possible to optimize both the process and the effectiveness of biofortification protocols, aiming to refine the calibration process.

References

Alves DMR, de Mello PR, Barreto RF (2024) Silicon and sodium attenuate potassium deficiency in Eruca sativa Mill. Food Chem 432:137225

Artyszak A (2018) Effect of silicon fertilization on crop yield quantity and quality—a literature review in Europe. Plan Theory 7:54

Basirat M, Mousavi SM (2022) Effect of foliar application of silicon and salicylic acid on regulation of yield and nutritional responses of greenhouse cucumber under high temperature. J Plant Growth Regul 41:1978–1988

Berner LA, Clydesdale FM, Douglass JS (2001) Fortification contributed greatly to vitamin and mineral intakes in the United States, 1989–1991. J Nutr 131:2177–2183

Borghesi E, Carmassi G, Uguccioni MC, et al (2013) Effects of calcium and salinity stress on quality of lettuce in soilless culture. J plant nut, 36(5):677–690

Brady MC, Dobson PRM, Thavarajah M et al (1991) Zeolite a stimulates proliferation and protein synthesis in human osteoblast-like cells and osteosarcoma cell line MG-63. J Bone Miner Res:S139

Buturi CV, Coelho SRM, Cannata C et al (2022a) Iron biofortification of greenhouse cherry tomatoes grown in a soilless system. Horticulturae 8(10):858

Buturi CV, Sabatino L, Mauro RP et al (2022b) Iron biofortification of greenhouse soilless lettuce: an effective agronomic tool to improve the dietary mineral intake. Agronomy 12(8):1793

Calero Hurtado A, Chiconato DA, Prado R et al (2020) Different methods of silicon application attenuate salt stress in sorghum and sunflower by modifying the antioxidative defense mechanism. Ecotoxicol Environ Saf 203:110964

Calomme MR, Cos P, D'Haese PC et al (1998) Absorption of silicon in healthy subjects. In: Collery P, Bratter P, Negretti de Bratter V, Khassanova L, Etienne JC (eds) Metal ions in biology and medicine, vol 5. John Libbey Eurotext, Paris, pp 228–232

Carvalho SM, Vasconcelos MW (2013) Producing more with less: strategies and novel technologies for plant-based food biofortification. Food Res Int 54(1):961–971

Chrysargyris A, Maggini R, Incrocci L et al (2023) Copper and Zinc Accumulation in Young Leaves of Eruca sativa (L.) Grown in Soilless Culture. Horticulturae, 9(9):976

Ciriello M, Formisano L, Kyriacou M et al (2022) Zinc biofortification of hydroponically grown basil: stress physiological responses and impact on antioxidant secondary metabolites of genotypic variants. Front Plant Sci 13:1049004

Ciriello M, Formisano L, Zarrelli A et al (2023) Zinc biofortification of Genovese basil: influence on mineral profile and estimated daily intake in adults and children. Food Res Int 164:112374

Coskun D, Deshmukh R, Sonah H et al (2019) The controversies of silicon's role in plant biology. New Phytol 221:67–85

Cunha Coelho R, Fernandes Barsotti RC, França Maltez H et al (2021) Expanding information on the bioaccessibility and bioavailability of iron and zinc in biofortified cowpea seeds. Food Chem 347:129027

D'Imperio M, Renna M, Cardinali A et al (2016a) Silicon biofortification of leafy vegetables and its bioaccessibility in the edible parts. J Sci Food Agr 96(3):751–756

D'Imperio M, Renna M, Cardinali A et al (2016b) Calcium biofortification and bioaccessibility in soilless "baby leaf" vegetable production. Food Chem 213:149–156

D'Imperio M, Brunetti G, Gigante I et al (2017) Integrated in vitro approaches to assess the bioaccessibility and bioavailability of silicon-biofortified leafy vegetables and preliminary effects on bone. In Vitro Cell Dev Biol Anim 53(3):217–224

D'Imperio M, Montesano FF, Renna M et al (2018) NaCl stress enhances silicon tissue enrichment of hydroponic "baby leaf" chicory under biofortification process. Sci Hortic 235:258–263

D'Imperio M, Montesano FF, Renna M et al (2019) Hydroponic production of reduced-potassium Swiss chard and spinach: a feasible agronomic approach for chronic kidney disease patients tailored vegetables. Agronomy 9:627

D'Imperio M, Parente A, Montesano FF et al (2020) Boron biofortification of Portulaca oleracea L. through soilless cultivation for a new tailored crop. Agronomy 10(7):999

D'Imperio M, Montesano FF, Santovito E et al (2022) Mineral composition and bioaccessibility in rocket and purslane after Zn biofortification process. Food Secur 11(3):484

D'Imperio M, Bonelli L, Mininni C et al (2023) Soilless cultivation systems to produce tailored microgreens for specific nutritional needs. J Sci Food Agric. https://doi.org/10.1002/jsfa.13222

Del Valle HB, Yaktine AL, Taylor CL, Ross AC (eds) (2011) Dietary reference intakes for calcium and vitamin D. National Academies Press, Washington, DC

Dong Z, Liu Y, Dong G et al (2021) Effect of boiling and frying on the selenium content, speciation, and in vitro bioaccessibility of selenium-biofortified potato (*Solanum tuberosum* L.). Food Chem 348:129150

Dos Santos SMM, de Mello PR, de Souza JJP et al (2021) Silicon supplied via foliar application and root to attenuate potassium deficiency in common bean plants. Sci Rep 11:1–13

Dyląg A, Smoleń S, Wisła-Świder A et al (2023) Evaluation of the chemical composition and nutritional value of lettuce (*Lactuca sativa* L.) biofortified in hydroponics with iodine in the form of iodoquinolines. Front Plant Sci 14

Eisinger J, Clairet D (1993) Effects of silicon, fluoride, etidronate and magnesium on bone mineral density: a retrospective study. Magnes Res 3:247–249

Fernández-García E, Carvajal-Lérida I, Pérez-Gálvez A (2009) In vitro bioaccessibility assessment as a prediction tool of nutritional efficiency. Nutr Res 29(11):751–760

Ferrarese M, Sourestani M, Quattrini E et al (2012) Biofortification of spinach plants applying selenium in the nutrient solution of floating system. Veg Crop Res Bull 76:127–136

Francini A, Quattrini E, Giuffrida F et al (2023) Biofortification of baby leafy vegetables using nutrient solution containing selenium. J Sci Food Agric. https://doi.org/10.1002/jsfa.12622

Giordano M, El-Nakhel C, Pannico A et al (2019) Iron biofortification of red and green pigmented lettuce in closed soilless cultivation impacts crop performance and modulates mineral and bioactive composition. Agronomy 9(6):290

Gomez-Galera S, Rojas E, Sudhakar D et al (2010) Critical evaluation of strategies for mineral fortification of staple food crops. Transgenic Res 19:165–180

Gonnella M, Renna M, D'Imperio M et al (2019) Iodine biofortification of four brassica genotypes is effective already at low rates of potassium iodate. Nutrients 11(2):451

Gonzalez-Porras CV, Teixeira GCM, Prado RDM et al (2024) Silicon via fertigation with and without potassium application, improve physiological aspects of common beans cultivated under three water regimes in field. Sci Rep 14(1):2051

Gonzali S, Kiferle C, Perata P (2017) Iodine biofortification of crops: agronomic biofortification, metabolic engineering and iodine bioavailability. Curr Opin Biotechnol 44:16–26

Guével MH, Menzies JG, Bélanger RR (2007) Effect of root and foliar applications of soluble silicon on powdery mildew control and growth of wheat plants. Eur J Plant Pathol 119:429–436

Hirschi KD (2009) Nutrient biofortification of food crops. Annu Rev Nutr 29:401–421

Hunt CD (2012) Dietary boron: progress in establishing essential roles in human physiology. J Trace Elem Med Biol 26:157–160

Jugdaohsingh R (2007) Silicon and bone health. J Nutr Health Aging 11(2):99–110

Jugdaohsingh R, Anderson SH, Tucker KL et al (2002) Dietary silicon intake and absorption. Am J Clin Nutr 75:887–893

Jugdaohsingh R, Tucker KL, Qiao N et al (2004) Dietary silicon intake is positively associated with bone mineral density in men and premenopausal women of the Framingham offspring cohort. J Bone Miner Res 19(2):297–307

Kaur H, Greger M (2019) A review on Si uptake and transport system. Plan Theory 8:1–8

Klopsch R, Baldermann S, Voss A et al (2018) Bread enriched with legume microgreens and leaves-ontogenetic and baking-driven changes in the profile of secondary plant metabolites. Front Chem 6:322

Koronowicz AA, Kopeć A, Master A et al (2016) Transcriptome profiling of Caco-2 cancer cell line following treatment with extracts from iodine-biofortified lettuce (*Lactuca sativa* L.). PLoS One 11(1):e0147336

Kovács S, Kutasy E, Csajbók J (2022) The multiple role of silicon nutrition in alleviating environmental stresses in sustainable crop production. Plan Theory 11

Krawczyk KK, Smoleń S, Wisła-Świder A et al (2024) Kale (*Brassica oleracea* L. var. sabellica) biofortified with iodoquinolines: effectiveness of enriching with iodine and influence on chemical composition. Sci Hort 323:112519

Lassus A (1993) Colloidal silicic acid for oral and topical treatment of aged skin, fragile hair and brittle nails in females. J Int Med Res 4:209–215

Macdonald HM, Black AJ, Aucott L, Duthie G et al (2008) Effect of potassium citrate supplementation or increased fruit and vegetable intake on bone metabolism in healthy postmenopausal women: a randomized controlled trial. Am J Clin Nutr 88:465–474

Macdonald HM, Hardcastle AC, Jugdaohsingh R et al (2012) Dietary silicon interacts with oestrogen to influence bone health: evidence from the Aberdeen prospective osteoporosis screening study. Bone 50(3):681–687

Malorgio F, Diaz KE, Ferrante A et al (2009) Effects of selenium addition on minimally processed leafy vegetables grown in floating system. J Sci Food Agric 89:2243–2251

Montesano FF, D'Imperio M, Parente A, et al (2016) Green bean biofortification for Si through soilless cultivation: plant response and Si bioaccessibility in pods. Scie Rep 6(1). https://doi.org/10.1038/srep31662

Naqvi S, Zhu C, Farre G et al (2009) Transgenic multivitamin corn through biofortification of endosperm with three vitamins representing three distinct metabolic pathways. Proc Natl Acad Sci 106:7762–7767

Nascimento CS, Nascimento CS, Lopes G et al (2022) Biofortified rocket (*Eruca sativa*) with selenium by using the nutrient film technique. Horticulturae 8(11):1088

Nielsen FH (2008) Is boron nutritionally relevant? Nutr Rev 66(4):183–191

Nielsen FH (2014) Update on the possible nutritional importance of silicon. J Trace Elem Med Biol 28(4):379–382

Paine JA, Shipton CA, Chaggar S et al (2005) Improving the nutritional value of Golden Rice through increased pro-vitamin a content. Nat Biotechnol 23:482–487

Pannico A, El-Nakhel C, Graziani G et al (2020) Selenium biofortification impacts the nutritive value, polyphenolic content, and bioactive constitution of variable microgreens genotypes. Antioxidants 9(4):272

Paterson AH, Bowers JE, Bruggman R et al (2009) The Sorghum bicolor genome and the diversification of grasses. Nature 457:551–556

Piątkowska E, Kopeć A, Bieżanowska-Kopeć R et al (2016) The impact of carrot enriched in iodine through soil fertilization on iodine concentration and selected biochemical parameters in wistar rats. PLoS One 11(4)

Popplewell JF, King SJ, Day JP et al (1998) Kinetics of uptake and elimination of silicic acid by a human subject: a novel application of 32Si and accelerator mass spectrometry. J Inorg Biochem 69(3):177–180

Poudel P, Di Gioia F, Lambert JD et al (2023) Zinc biofortification through seed nutri-priming using alternative zinc sources and concentration levels in pea and sunflower microgreens. Front Plant Sci 14:1177844

Powell JJ, McNaughton SA, Jugdaosingh R et al (2005) A provisional database for the silicon content of foods in the United Kingdom. Br J Nutr 94:804–812

Prentice A (2004) Diet, nutrition and the prevention of osteoporosis. Public Health Nutr 7(1a):227–243

Reffitt DM, Ogston N, Jugdaohsingh R et al (2003) Orthosilicic acid stimulates collagen type I synthesis and osteoblastic differentiation in human osteoblast-like cells in vitro. Bone 32:127–135

Renna M, D'Imperio, M, Maggi, et al (2022) Soilless biofortification, bioaccessibility, and bioavailability: Signposts on the path to personalized nutrition. Front Nut, 9, 966018.

Rodella LF, Bonazza V, Labanca M et al (2014) A review of the effects of dietary silicon intake on bone homeostasis and regeneration. J Nutr Health Aging 18:820–826

Rouphael Y, Kyriacou MC (2018) Enhancing quality of fresh vegetables through salinity eustress and biofortification applications facilitated by soilless cultivation. Front Plant Sci 9:1254

Sabatino L, La Bella S, Ntatsi G et al (2021) Selenium biofortification and grafting modulate plant performance and functional features of cherry tomato grown in a soilless system. Sci Hort 285:110095

Savvas D, Ntatsi G (2015) Biostimulant activity of silicon in horticulture. Sci Hortic 196:66–81

Schmutz J, Cannon SB, Schlueter J et al (2010) Genome sequence of the palaeopolyploid soybean. Nature 463:178–183

Schwarz K, Ricci BA, Punsar S et al (1977) Inverse relation of silicon in drinking water and atherosclerosis in Finland. Lancet 8010(538–539):6

Somma A, Palmitessa OD, Conversa G et al (2024) Localized foliar application of iodine on tomato: an effective approach for targeted agronomic biofortification. Sci Hort 327:112807

Sripanyakorn S, Jugdaohsingh R, Dissayabutr W et al (2009) The comparative absorption of silicon from different foods and food supplements. Br J Nutr 102:825–834

Stránský M, Ryšavá L (2009) Nutrition as prevention and treatment of osteoporosis. Physiol Res 58:S7–S11

Tavan M, Wee B, Fuentes S et al (2024) Biofortification of kale microgreens with selenate-selenium using two delivery methods: selenium-rich soilless medium and foliar application. Sci Hort 323:112522

Tomasi N, Pinton R, Gottardi S et al (2016) Selenium fortification of hydroponically grown corn salad (*Valerianella locusta*). Crop Pasture Sci 66:1128–1136

Valentinuzzi F, Cologna K, Pii Y, et al (2017) Assessment of silicon biofortification and its effect on the content of bioactive compounds in strawberry (Fragaria× ananassa'Elsanta') fruits. In VIII International Symposium on Mineral Nutrition of Fruit Crops 1217 (pp. 307–312)

Van Dyck K, Van Cauwenbergh R, Robberecht H et al (1999) Bioavailability of silicon from food and food supplements. Fresenius J Anal Chem 363:541–544

Voogt W, Harmen T, Holwerda HT et al (2010) Biofortification of lettuce (*Lactuca sativa* L.) with iodine: the effect of iodine form and concentration in the nutrient solution on growth, development and iodine uptake of lettuce grown in water culture. J Sci Food Agric 90:906–913

White PJ, Broadley MR (2009) Biofortification of crops with seven mineral elements often lacking in human diets—iron, zinc, copper, calcium, magnesium, selenium and iodine. New Phytol 182:49–84

White PJ, Pongrac P, Sneddon CC et al (2018) Limits to the biofortification of leafy brassicas with zinc. Agriculture 8:32

Winkler JT (2011) Biofortification: improving the nutritional quality of staple crops. In: Pasternak C (ed) Access not excess. Smith-Gordon, Cambs, pp 100–112

Yan G, Nikolic M, Ye M et al (2018) Silicon acquisition and accumulation in plant and its significance for agriculture. J Integr Agric 17:2138–2150

Ye X, Salim A, Klati A et al (2000) Engineering the provitamin A (β-carotene) biosynthetic pathway into (carotenoid-free) rice endosperm. Science. https://doi.org/10.1126/science.287.5451.303

Zou T, Xu N, Hu G et al (2014) Biofortification of soybean sprouts with zinc and bioaccessibility of zinc in the sprouts. J Sci Food Agric 94(14):3053–3060

Chapter 21
Evidence of the Benefits of Silicon for Human Health

Forrest H. Nielsen

Abstract Silicon is not generally recognized as essential for humans because it has not been found to be required for growth, reproduction, or sustenance for life. However, nutritional amounts of silicon prevent bone, hexosamine, and collagen abnormalities in experimental animals fed silicon-deprived diets. In vitro studies have shown that silicon promotes bone formation and maintenance. Epidemiological and supplementation studies indicate that nutritional intakes of silicon are beneficial for bone growth and maintenance and normal connective tissue properties in humans. Support for being beneficial for human health is that the absorption, tissue uptake, distribution, transport, and excretion of silicon have the characteristics of essential mineral elements. Silicon moves through the body mainly as orthosilicic acid. A likely mechanism of action of silicon in bone development is that in the bone formation or remodeling front, orthosilicic acid is transformed into the silicate ion, which is substituted by phosphate. This substitution promotes amorphous bone to become crystalline hydroxyapatite bone. The mechanism of action in connective tissue may be through the ability of silicon to form complexes with the polyols hexosamine and ascorbate. These polyols are used to form glycosaminoglycans, mucopolysaccharides, and collagen, which are involved in connective tissue formation and stabilization.

Keywords Silicon · Bone · Connective tissue · Metabolism · Beneficial intake

21.1 Introduction

Silicon is essential for some lower forms of life (Carlisle 1997). Silicon has a structural role in diatoms, radiolarians, and some sponges. Diatoms, which are unicellular microscopic plants, have an absolute requirement for silicon as monomeric

F. H. Nielsen (✉)
Research Nutritionist Consultant, Grand Forks, ND, USA

R. de Mello Prado et al. (eds.), *Silicon Advances for Sustainable Agriculture and Human Health*, Sustainable Plant Nutrition in a Changing World,
https://doi.org/10.1007/978-3-031-69876-7_21

silicic acid for normal cell growth. Silicon apparently is required by some higher plants such as rice. Because silicon deprivation has not been shown to interrupt the life cycle in mammals or to have a clearly defined biochemical function, silicon is not generally accepted as an essential nutrient for higher animals and humans. However, since the early 1970s, numerous animal, in vitro, and human studies have indicated that silicon has beneficial effects on human health.

21.2 Evidence from Animal and In Vitro Studies

The first evidence that silicon might have beneficial effects was reported in 1972 when it was reported that supra nutritional amounts of silicon (100 and 500 mg/kg diet) alleviated bone abnormalities in chicks and improved growth in rats fed low-silicon diets (Carlisle 1972; Schwarz and Milne 1972). These and studies performed shortly thereafter found that silicon alleviated bone abnormalities in silicon-deprived chicks that included abnormal skull structure with depressed collagen content in bone; long bones with small, poorly formed joints and defective endochondral bone growth; and tibias with depressed contents of articular cartilage, water, hexosamine, and collagen (Carlisle 1997).

Reports after these initial studies also have found that supra nutritional or pharmacological amounts of silicon have bone promoting effects in ovariectomized experimental animals. A silicon supplement of 500 mg/kg diet for 30 days increased longitudinal growth and mineral content in ovariectomized rats (Rico et al. 2000). Ovariectomized rats given a 500 mg/kg diet silicon supplement as sodium metasilicate in a calcium-deficient AIN-93M diet exhibited increased bone density of the femur and tibia and C-telopeptide type 1 collagen in serum (Kim et al. 2009). Choline-stabilized orthosilicic acid orally supplemented at a dose of 1.0 mg/kg body weight daily for 30 weeks partially prevented femoral bone loss in aged ovariectomized rats (Calomme et al. 2006). A study with aged (17 weeks) ovariectomized rats found that daily oral administration of 20 mg silicon/kg body weight significantly increased femur and tibia bone mineral density (Bae et al. 2008).

Experiments performed since 2000 have shown that only nutritional amounts of silicon (e.g., 10–50 mg/kg diet) are needed to prevent bone, hexosamine, and collagen abnormalities in animals fed silicon-low diets (e.g., <2.0 mg/kg diet). The nutritional amounts prevented decreased collagen formation (Seaborn and Nielsen 2002a); decreased femur hydroxyproline concentration (Seaborn and Nielsen 2002b); increased urinary helical peptide excretion (Nielsen and Poellot 2004); and bone growth plate thickness and increased chondrocyte density occurring in silicon-deprived rats (Jugdaohsingh et al. 2008). In mice, the silicon supplement increased femur strength and stiffness, dry and ash weights, calcium and hydroxyproline contents, and alkaline phosphatase activity and decreased femur tartrate-resistant acid phosphatase and urinary excretion of hydroxy proline (Maehira et al. 2009). Silicon supplementation also increased bone marrow mRNA expression related to osteoclastogenesis.

In vitro studies also have shown that silicon promotes bone formation and maintenance. Orthosilicic acid at physiological concentrations was found to stimulate collagen type 1 synthesis in human osteoblast-like cells and enhance osteoblastic differentiation in culture (Refitt et al. 2003). The efficiency of bone engineering implants silica-based bioactive glass and ceramics has been attributed to silicon being involved in osteoblast proliferation and differentiation, type 1 collagen synthesis, and apatite formation (Hench et al. 2004).

21.3 Human Studies Indicating Health Benefits of Silicon

A few epidemiological studies have indicated that nutritional intakes of silicon are beneficial for bone health in humans. Dietary silicon was positively associated with bone mineral density in four hip sites of men and postmenopausal women in the Framingham Offspring Cohort of 1251 men and 1596 women (Jugdaohsingh et al. 2004). Large differences of up to 10% in density were found between the highest (>40 mg/day) and lowest (<14 mg/day) quintiles of silicon intake. Also, in the Framingham Offspring Cohort, the beneficial effect of moderate beer consumption was associated with the silicon in the beer (Tucker et al. 2009). Silicon intake also was positively associated with bone mineral density at the femur neck in late premenopausal women and postmenopausal women using hormone replacement therapy (Macdonald et al. 2012). Energy-adjusted silicon intakes were negatively associated with urinary markers of bone resorption (pyridinoline and deoxypyridinoline cross-links) and positively associated with serum N-terminal propeptide of type 1 collagen, a marker of bone formation. The lowest quartile of silicon intake was 16 mg/day, and the highest quartile was 31.5 mg/day.

A few supplementation studies provide further evidence that silicon may have beneficial effects on bone and connective tissue. Spine bone mineral density was increased by a 28 mg/day silicon supplement for 12 weeks in women with low bone mass (Jugdaohsingh 2007). In a double-blind, placebo-controlled study, 6 and 12 mg/day supplementation with silicon as choline-stabilized orthosilicic acid for 12 months increased the bone formation marker serum procollagen type I N-terminal propeptide at 12 months (Spector et al. 2008). A supplement of 10 mg/day silicon as choline-stabilized orthosilicic acid for 20 weeks improved photodamaged skin surface and mechanical properties and decreased hair and nail brittleness (Barel et al. 2005).

Further evidence that silicon is beneficial to bone growth and maintenance are findings showing that including silicon in regenerative bone materials or scaffolds enhances biomineralization, osteogenesis, and hard tissue formation (Göltz et al. 2019).

In addition to being beneficial to bone health, silicon may have a beneficial effect on cognitive function. In an 8-year follow-up of subjects in the Paquid cohort, the exposure to drinking water with silica concentrations of 11.25 mg/L or higher was associated with a lower risk for developing Alzheimer's disease (Rondeau et al.

2000). Higher silica in drinking water was also found to be associated with a reduced risk of developing Alzheimer's disease in the 7-year follow-up of 1462 women (aged ≥75 years) participating in the Epidemiology of Osteoporosis study (Gillette-Guyonnet et al. 2005).

21.4 Silicon Metabolism

The absorption, tissue uptake, distribution, transport, and excretion of silicon have the characteristics of essential mineral elements. This supports the contention that silicon has health benefits for humans.

Silicon absorption is strongly influenced by the solubility of silicon compounds consumed and formed during digestion. Orthosilicic acid [$Si(OH)_4$] is the major source of bioavailable silicon for both humans and animals. Orthosilicic acid and soluble silicates in beverages are easily absorbed, whereas forms of silica present in solid plant foods are less absorbable. One study found that silicon absorption was highest from monomethylsilanetriol and beer (64% of dose), followed by green beans (44%), orthosilicic acid solution (43%), choline-stabilized orthosilicic acid (17%), and bananas (4%) (Sripanyakorn et al. 2009). High dietary calcium and magnesium, through the formation of insoluble silicates, and high fiber through binding silicon to make it unavailable may decrease the absorption of silicon. Silicon in polymeric and colloidal forms that are not readily broken down in the gastrointestinal tract is excreted in the feces.

Although not definitively established, simple uncharged silicon species such as orthosilicic acid that are readily and rapidly absorbed predominantly in the proximal small intestine likely do so through a paracellular or small-pore transcellular pathway. Charged polymeric silica species are less mobile and/or are too large to easily cross the gastrointestinal tract.

Once absorbed, silicon circulates mainly as a neutral orthosilicic acid species in plasma. Orthosilicic acid entering the bloodstream is not bound to proteins and seems to be efficiently transferred to tissues or excreted in urine because the silicon concentration in blood remains relatively constant over a range of dietary intakes.

The transport of silicon into cells apparently can occur though aquaporins, especially the aquaglyceroporin (AQGP) subfamily AQP 7, AQP 9, and AQP 10 (Garneau et al. 2015). These AQGPs are present in tissues where silicon has bioactive activity, and they transcriptionally respond to changes in dietary silicon.

The efflux of silicon from cells may depend on SLC34A2, which is also known as sodium-phosphate cotransporter 2B (Ratcliffe et al. 2017). SLC34A2 was upregulated in rat kidney following chronic dietary silicon deprivation. SLC34A2 transports orthosilicic acid out of cells in the reverse direction of phosphate transport, which is consistent with the observation that the upregulation of SLC34A2 in silicon-depleted rats was accompanied by increased urinary phosphate excretion and the observation of conserved bone silicon but depleted bone phosphate in animals on silicon-deficient diets (Jugdaohsingh et al. 2008).

Connective tissues, including aorta, immature bone, skin, tendon, trachea, and fingernails, contain much of the silicon retained in the body. The retention in these tissues may reflect where silicon has beneficial effects.

Absorbed silicon is rapidly excreted in the urine. A human balance study indicated that in 24 h the amount of silicon excreted in the urine was equal to a soluble dose of dietary silicon (Pruksa et al. 2014). Silicon is readily filtered by the renal glomerulus. Reabsorption or silicon efflux as orthosilicic acid back into the circulation by the kidney occurs through the transporter Slc34a2 (also called NaPiIIb) (Ratcliffe et al. 2017). With this efflux, phosphate is moved in the opposite direction. Animal experiments indicate that silicon deprivation increases the efflux to maintain circulating silicon levels. Silicon exists as orthosilicic acid or magnesium orthosilicate in urine. Intraperitoneal injection of silicon can elevate urinary excretion above the upper limit achieved by dietary intake.

21.5 Possible Mechanisms of Action of Silicon in Humans

One of the most likely mechanisms of the beneficial action of silicon is an involvement in bone development. There is no evidence for a structural role for silicon in bone; thus it is found in low amounts in mineralized bone. However, in the bone formation or remodeling, front silicon is found in measurable amounts. At these sites orthosilicic acid apparently can be transformed into the silicate ion, which is substituted by phosphate. This substitution promotes amorphous bone to become crystalline hydroxyapatite bone (Chappell et al. 2020).

Cell culture studies have demonstrated that silicon promotes osteoblastic differentiation and inhibits osteoclastogenesis. Silicon can upregulate BMP-2 that activates the BMP-2/Smad1/5//Runx2 signaling that increases osteocalcin in human osteoblast-like cells (Dong et al. 2016) and type l collagen expression by MC3T3-E1 cells (Kim et al. 2013). The inhibition of osteoclastogenesis may occur through the suppression of M-CSF and RANKL expression (Mladenović et al. 2014). In RANKL-stimulated RAW264.7 cells, silicon inhibited the expression of osteoclast phenotypic genes (Magnusson et al. 2021).

Silicon also may have a beneficial mechanism of action in connective tissue formation and function. The basis for silicon being present in high amounts in connective tissue may involve the formation or utilization of polyols. These polyols include hexosamines and ascorbate used to form glycosaminoglycans, mucopolysaccharides, and collagen, which are involved in connective tissue formation and stabilization. Support for this mechanism is that silicate-carbohydrate complexes have been identified. Silicate forms 4-, 5-, and 6-coordinate complexes with aliphatic polyols with at least two hydroxy groups in the *threo* (opposite side) configuration or furanoidic molecules with vicinal *cis*-diol functionality (Vis et al. 2020).

Some reported beneficial effects of silicon, especially supra nutritional or pharmacological amounts, might have occurred through altering the absorption and utilization of other mineral elements involved in bone metabolism, immune or inflammatory response, oxidative stress, or cognitive function. For example, it has been suggested that silicon has a protective effect in Alzheimer's disease through preventing the absorption and retention of aluminum, which has been associated with some cases of that disease. The formation of aluminosilicates may prevent aluminum from competing for iron-binding sites that result in decreased functions related to iron. These functions include that of prolyl hydroxylase, which has a role in collagen formation.

21.6 Toxicity of Silicon

There is little concern about silicon toxicity through food and water intake. No ULs for silicon intake were set by the Institute of Medicine (2001). Long-term use of silicate-containing drugs and chronic hemodialysis may cause silicon accumulation in the kidney that induces nephropathy (Jugdaohsingh 2007). However, there are no reports of silicon toxicity through excessive dietary intake for people with normal renal function.

21.7 Food Sources of Silicon

Daily dietary silicon intake is estimated at 15–50 mg in the Western world. It largely comes in the form of digestible phytolytic silica from plants (primarily cereals) and as orthosilicic acid (OSA), a weak acid with a pKa of 9.8, in fluids such as drinking water. Various studies have shown that it is the soluble OSA component either from direct ingestion or following digestion in the gut that is readily absorbed.

The best sources of silicon are cereal and cereal products, especially less refined barley, millet, oats, and rice products (Powell et al. 2005; Prescha et al. 2019). Substantial amounts of silicon are found in Kenyan, French, and runner beans, spinach, dried fruits, bananas, and red lentils. Undetectable amounts of silicon occur in tomatoes, oranges, and onions. The silicon in barley and hops is solubilized during the beer-making process, which makes this beverage a rich source of silicon. Drinking water can be a significant source of silicon, and the levels of silicon in water may vary with the geographic region and source. The silicon content in lowland Britain and European mineral waters was found to be 2.8–16 mg/L. Higher concentrations of silicon were found in Malaysian spritzer and Fiji mineral waters. Silicon is added to processed foods in various forms for anticaking, thickening, and stabilizing purposes. These forms include calcium silicate, sodium aluminosilicate, magnesium hydrogen metasilicate (talc), magnesium trisilicate, calcium aluminum silicate, bentonite, and kaolin, all of which are inert and not readily absorbed from the gastrointestinal tract.

21.8 Dietary Considerations for Health Benefits of Silicon

In the United States, Dietary Reference Intakes (DRIs) have not been established for silicon because it was judged that animal and human data were too limited to justify doing so (Institute of Medicine 2001). However, there are some findings that support a suggested beneficial intake. In the Framingham Offspring Cohort study, the highest bone mineral density occurred in the highest quintiles with intakes ranging from 30.2 to 63.2 mg/day and 34.4 to 118.0 mg/day compared with the lowest quintile with intakes ranging from 7.1 to 16.7 mg/day and 7.6 to 18.8 mg/day, respectively, for premenopausal women and men (Jugdaohsingh et al. 2004). Based on these findings, extrapolations from animal data and questionable balance data, a beneficial intake for silicon may be near 25 mg/day (Nielsen 2020).

Mean intakes of silicon in individuals consuming Western style diets range from 18 to 33 mg silicon per day, and healthy adult subjects have fasting serum concentrations of silicon in the range of 100–310 μg/L (Jugdaohsingh et al. 2002; Robberecht et al. 2009; Prescha et al. 2019). A fasting serum silicon concentration near or below 100 μg/L and a dietary food record indicating a silicon intake less than 25 mg/day could be an indication that a person would benefit from an increased intake of silicon.

References

Bae Y-J, Kim J-Y, Choi M-K et al (2008) Short-term administration of water-soluble silicon improves mineral density of the femur and tibia in ovariectomized rats. Biol Trace Elem Res 124:157–163

Barel A, Calomme M, Timchenko A et al (2005) Effect of oral intake of choline-stabilized orthosilicic acid on skin, nails, and hair in women with photodamaged skin. Arch Dermatol Res 297:147–153

Calomme M, Geusens P, Demeester N et al (2006) Partial prevention of long-term femoral bone loss in aged ovariectomized rats supplemented with choline-stabilized orthosilicic acid. Calcif Tissue Int 78:227–232

Carlisle EM (1972) Silicon an essential element for the chick. Science 178:619–621

Carlisle EM (1997) Silicon. In: O'Dell BL, Sunde RA (eds) Handbook of nutritionally essential minerals. Marcel Dekker, New York, pp 603–618

Chappell HF, Jugdaohsingh R, Powell JJ (2020) Physiological silicon incorporation into bone mineral requires orthosilicic acid metabolism to SiO_4^{4}. J R Soc Interface 17:20200145

Dong M, Jiao G, Liu H et al (2016) Biological silicon stimulates collagen type 1 and osteocalcin synthesis in human osteoblast-like cells through the BPM-2/Smad/RUNX2 signaling pathway. Biol Trace Elem Res 173:306–315

Garneau AP, Carpentier GA, Marcoux AA et al (2015) Aquaporins mediate silicon transport in humans. PLoS One 10(8):e0136149

Gillette-Guyonnet S, Andrieu S, Nourhashemi F et al (2005) Cognitive impairment and composition of drinking water in women: findings of the EPIDOS study. Am J Clin Nutr 81:897–902

Göltz W, Tobiasch E, Witzleben S et al (2019) Effects of silicon compounds on biomineralization, osteogenesis, and hard tissue formation. Pharmaceutics 11:117

Hench LL, Xynos ID, Polak JM (2004) Bioactive glass for in situ tissue regeneration. J Biomater Sci Polymer Ed 15:543–562
Institute of Medicine (2001) Dietary reference intakes for vitamin A, vitamin K, arsenic, boron, chromium, copper, iodine, iron, manganese, molybdenum, nickel, silicon, vanadium, and zinc. National Academy Press, Washington, DC
Jugdaohsingh R (2007) Silicon and bone health. J Nutr Health Aging 11:99–110
Jugdaohsingh R, Anderson SH, Tucker K et al (2002) Dietary silicon intake and absorption. Am J Clin Nutr 75:887–893
Jugdaohsingh R, Tucker KL, Qiao N et al (2004) Dietary silicon intake is positively associated with bone mineral density in men and pre-menopausal women of the Framingham offspring cohort. J Bone Miner Res 19:297–307
Jugdaohsingh R, Calomme MR, Robinson K et al (2008) Increased longitudinal growth in rats on a silicon-depleted diet. Bone 43:596–606
Kim M-H, Bae Y-J, Choi M-K (2009) Silicon supplementation improves the bone mineral density of calcium-deficient ovariectomized rats by reducing bone resorption. Biol Trace Elem Res 128:239–247
Kim E-J, Bu S-Y, Sung M-K et al (2013) Analysis of antioxidant and anti-inflammatory activity of silicon in murine macrophages. Biol Trace Elem Res 156:329–337
Macdonald HM, Hardcastle AC, Jugdaohsingh R et al (2012) Dietary silicon interacts with oestrogen to influence bone health: evidence from the Aberdeen Prospective Osteoporosis Screening Study. Bone 50:681–687
Maehira F, Miyagi I, Eguchi Y (2009) Effects of calcium sources and soluble silicate on bone metabolism and the related gene expression in mice. Nutrition 25:581–589
Magnusson C, Uribe P, Jugdaohsingh R et al (2021) Inhibitory effects of orthosilicic acid on osteoclastogenesis in RANKL-stimulated RAW264.7 cells. J Biomed Mater Res Part A 109:1967–1978
Mladenović Ž, Johansson A, Willman B et al (2014) Soluble silica inhibits osteoclast formation and bone resorption in vitro. Acta Biomater 10:406–418
Nielsen FH (2020) Manganese, molybdenum, boron, chromium, and other trace elements. In: Marriott BP, Birt DF, Stallings VA, Yates AA (eds) Present knowledge in nutrition, vol 1, 11th edn. Elsevier Academic Press, London, pp 485–500
Nielsen FH, Poellot R (2004) Dietary silicon affects bone turnover differently in ovariectomized and sham-operated growing rats. J Trace Elem Exp Med 17:137–149
Powell JJ, McNaughton SA, Jugdaohsingh R et al (2005) A provisional database for the silicon content of foods in the United Kingdom. Br J Nutr 94:804–812
Prescha A, Zablocka-Slowińska K, Grajeta H (2019) Dietary silicon and its impact on plasma silicon levels in the Polish population. Nutrients 11:980
Pruksa S, Siripinyanond A, Powell JJ et al (2014) Silicon balance in human volunteers; a pilot study to establish the variance in silicon excretion versus intake. Nutr Metab (Lond) 11:4
Ratcliffe S, Jugdaohsingh R, Vivancos J et al (2017) Identification of a mammalian silicon transporter. Am J Physiol Cell Physiol 312:C550–C561
Refitt DM, Ogston N, Jugdaohsingh R et al (2003) Orthosilicic acid stimulates collagen type 1 synthesis and osteoblastic differentiation in human osteoblast-like cells in vitro. Bone 32:127–135
Rico H, Gallego-Lago JL, Hernández ER et al (2000) Effect of silicon supplement on osteopenia induced by ovariectomy in rats. Calcif Tissue Int 66:53–55
Robberecht H, Van Cauwenbergh R, Van Vlaslaer V et al (2009) Dietary silicon intake in Belgium: sources, availability from foods, and human serum levels. Sci Total Environ 407:4777–4782
Rondeau V, Commenges D, Jacqmin-Gadda H et al (2000) Relation between aluminum concentrations in drinking water and Alzheimer's disease: an 8-year follow-up study. Am J Epidemiol 152:59–66
Schwarz K, Milne DB (1972) Growth-promoting effects of silicon. Nature 239:333–334
Seaborn CD, Nielsen FH (2002a) Silicon deprivation decreases collagen formation in wounds and bone, and ornithine transaminase enzyme activity in liver. Biol Trace Elem Res 89:251–261

Seaborn CD, Nielsen FH (2002b) Silicon deprivation and arginine and cystine supplementation affect bone collagen and bone and plasma trace mineral concentrations in rats. J Trace Elem Exp Med 15:113–122
Spector TD, Calomme M, Timchenko A et al (2008) Choline-stabilized orthosilicic acid supplementation as an adjunct to calcium/vitamin D_3 stimulates markers of bone formation in osteopenic females: a randomized, placebo-controlled trial. BMC Muscloskelet Disord 9:85
Sripanyakorn S, Jugdaohsingh R, Dissayabutr W et al (2009) The comparative absorption of silicon from different foods and food supplements. Br J Nutr 102:825–834
Tucker KL, Jugdaohsingh R, Powell JJ et al (2009) Effects of beer, wine, and liquor intakes on bone mineral density in older men and women. Am J Clin Nutr 89:1188–1196
Vis BM, Wen J, Mellerup SK et al (2020) Silicon forms a rich diversity of aliphatic polyol complexes in aqueous solution. J Am Chem Soc 142:9188–9202

Chapter 22
Role of Silicon Nanoparticles in Bone Regeneration

Fanrui Zeng and Kai Zheng

Abstract This chapter delves into the significant role of silicon nanoparticles (SiNPs) in bone regeneration. Bone tissue possesses intrinsic regenerative capabilities crucial for healing and remodeling. However, challenges arise in cases of delayed union or non-union fractures and in addressing large bone defects from various causes. SiNPs, due to their unique properties such as high surface area-to-volume ratio, biocompatibility, and ability to stimulate cellular activities, are promising for bone regeneration. The chapter summarizes various synthesis approaches for SiNPs, including top-down and bottom-up methods, each with its advantages and challenges. SiNPs have been shown to promote osteogenesis and angiogenesis, which are essential for bone regeneration. The controlled release of bioactive ions and biomolecules from SiNPs can further promote cell differentiation and tissue regeneration. The chapter also discusses the use of SiNPs incorporated in hydrogels to improve mechanical properties and bioactivity, as well as their application in drug-delivery systems for bone repair. The potential benefits of using SiNPs in bone repair are thoroughly examined, with specific examples illustrating their practical implications in relevant contexts. In conclusion, SiNPs offer a promising solution for advancing bone tissue engineering and regenerative medicine.

Keywords Silicon (Si) · Silicon nanoparticles (SiNPs) · Nanoparticles (NPs) · Bone regeneration · Hydrogels · Bioactive materials · Chemical vapor deposition (CVD) · Porous SiNPs (pSiNPs)

F. Zeng · K. Zheng (✉)
Jiangsu Province Engineering Research Center of Stomatological Translational Medicine, College of Stomatology, Nanjing Medical University, Nanjing, China

Jiangsu Key Laboratory of Oral Diseases, Nanjing Medical University, Nanjing, China
e-mail: kaizheng@njmu.edu.cn

R. de Mello Prado et al. (eds.), *Silicon Advances for Sustainable Agriculture and Human Health*, Sustainable Plant Nutrition in a Changing World,
https://doi.org/10.1007/978-3-031-69876-7_22

22.1 Introduction

Bone tissue exhibits intrinsic regenerative properties essential for the healing process after injury, the development of the skeletal system, and the remodeling in adults. The regenerative capacity of bone is a complex physiological phenomenon indispensable for the normal progression of fracture healing and the continuous remodeling processes (Giannoudis et al. 2007; Duda et al. 2023). This multifaceted process encompasses a delicate orchestration of biological phenomena, cellular interactions, and molecular signaling pathways in a precise temporal and spatial sequence (Duda et al. 2023). Nonetheless, compromised bone regeneration cases are observed in scenarios characterized by delayed union or non-union fractures, with a notable prevalence in tibial fractures (Winkler et al. 2018; Wildemann et al. 2021). Furthermore, in the context of orthopedic and oral/maxillofacial surgical interventions, there is often a requisite for substantial bone regeneration to address large bone defects that arise from trauma, infection, oncological resection, or conditions that impede the regenerative potential, including avascular necrosis and osteoporosis (Wildemann et al. 2021).

Various strategies have been developed to address the challenges posed by inadequate or dysfunctional bone regeneration, a phenomenon that carries profound medical and socioeconomic ramifications (Wildemann et al. 2021). Surgeons utilize these methods either independently or in combination to tackle complex clinical scenarios. Prevalent methodologies encompass distraction osteogenesis, bone transport, and a spectrum of bone-grafting procedures, including autologous bone grafts, allografts, the application of biomaterials, and the utilization of growth factors (Winkler et al. 2018; Montoya et al. 2021). Among these bone-grafting strategies, there is a burgeoning interest in the evolution of bioactive materials that mimic the biomechanical attributes of native bone. Such advancements can potentially expedite the regenerative process and ameliorate systemic conditions, such as skeletal disorders and osteoporosis, thereby offering a more effective therapeutic intervention for bone regeneration (Winkler et al. 2018).

Diverse types of bioactive materials have been deployed to facilitate the process of osteogenesis, including inorganic non-metallic biomaterials (e.g., bioactive glasses, bioceramics) (El-Rashidy et al. 2017; Ginebra et al. 2018), polymeric matrices (e.g., collagen, chitosan, polycaprolactone) (Bai et al. 2018; Filippi et al. 2020), and degradable metals (e.g., magnesium, zinc, and their composites) (Shuai et al. 2019; Zhang et al. 2021). In particular, silicon (Si) has attracted considerable attention in bone regeneration applications. Si, the third most abundant element in the human body, is widely present in bone tissue (Rodella et al. 2014; Henstock et al. 2015). The natural presence of Si in bone tissue, coupled with its osteogenic properties, biocompatibility, enhancement of mineralization, gene regulation, and synergistic effects with other bone-regenerative elements, underscores its vital role in maintaining bone health and structure (Rodella et al. 2014; O'Neill et al. 2018). With their unique characteristics, silicon-based materials offer promising approaches for advancing bone tissue engineering and regenerative medicine (Henstock et al. 2015).

Bioactive materials can be engineered into various morphologies customized to address the nuanced requirements of specific bone regeneration applications (Ho-Shui-Ling et al. 2018) including granules (Murakami et al. 2010), three-dimensional scaffolds (Yan et al. 2019), fibrous membranes (Aprile et al. 2020), surface coatings (Pereira et al. 2020), and nanoparticles (Zheng et al. 2021). Notably, nanoparticles have emerged as a particularly promising modality in tissue regeneration, attributable to their unique physicochemical and biological characteristics facilitating an accelerated and more efficient bone healing process (Fathi-Achachelouei et al. 2019).

This chapter provides a comprehensive exploration of the role of silicon nanoparticles (SiNPs) in bone regeneration applications. It begins by delving into the properties of SiNPs that make them suitable for bone regeneration. Following this, various synthesis approaches for SiNPs are briefly summarized. Subsequently, the discussion shifts to strategies for employing SiNPs in bone regeneration, exploring different applications and approaches. Throughout this section, the potential bene fits and challenges associated with the use of SiNPs in bone repair are thoroughly examined. To illustrate the practical implications of SiNPs in bone regeneration, the chapter highlights specific examples of their application in relevant contexts. Concluding remarks provide a concise summary of the pivotal role played by SiNPs in bone regeneration, emphasizing their potential to revolutionize clinical treatments.

22.2 Properties of SiNPs Favorable for Bone Regeneration

Nanoparticles have emerged as promising materials in bone regeneration, leveraging their unique properties and advantages to enhance tissue repair and regeneration processes. A key attribute of nanoparticles is their high surface area-to-volume ratio, significantly amplifying their interaction with the biological milieu (Fathi-Achachelouei et al. 2019). This interaction is pivotal for cell adhesion, proliferation, and differentiation processes, all of which are indispensable for osteogenesis. Moreover, nanoparticles offer a sophisticated platform for encapsulating and controlled release of bioactive molecules, including growth factors and therapeutic drugs (Couvreur 2013). This targeted delivery mechanism ensures the sustained release of these agents at the site of bone injury or defect, thereby promoting tissue regeneration while concurrently minimizing systemic side effects. The biodegradability of many nanoparticles used in bone regeneration is a notable advantage, as they degrade over time into non-toxic byproducts that are readily metabolized and excreted from the body (Croissant et al. 2017). This feature circumvents the need for surgical implant removal and mitigates the risk of long-term complications. Furthermore, the surface chemistry of nanoparticles can be tailored and functionalized to augment their performance in specific biomedical applications, including the incorporation of targeting ligands to confer improved tissue specificity (Mout et al. 2012). Surface modification also allows nanoparticles to mimic the biochemical cues present in the native bone microenvironment, promoting cellular adhesion, migration, and differentiation. In addition to their therapeutic properties, certain nanoparticles possess intrinsic imaging properties, such as fluorescence or magnetic resonance, enabling real-time, non-invasive monitoring of the bone regeneration process (Han et al. 2019). These imaging modalities facilitate the tracking of nanoparticle distribution, degradation, and integration within bone tissue over time. Nanoparticles also exhibit the potential to synergize with other biomaterials, such as hydrogels, thereby amplifying their regenerative capacity. By integrating nanoparticles as fillers into composite materials, the distinct attributes of each component can be harnessed to fabricate multifunctional composites that promote tissue regeneration through diverse mechanisms (Thoniyot et al. 2015). In recent years, there has been considerable interest in the immunomodulatory properties of nanoparticles, which hold significant promise for enhancing tissue regeneration outcomes (Cui et al. 2022). These properties render nanoparticles indispensable tools in developing advanced therapies targeting bone defects, fractures, and various orthopedic conditions. However, the design and application of nanoparticles necessitate a careful consideration of critical factors such as biocompatibility, toxicity, and long-term effects. This careful evaluation is essential to ensure the safety and efficacy of these nanoscale entities in clinical treatments, thereby facilitating their translation from bench to bedside.

SiNPs can exist in nonporous or porous forms. They have emerged as a pivotal component within the field of nanotechnology, with their potential for diverse applications, including energy, catalysis, biotechnology, and medicine (Valtchev and Tosheva 2013; Santos et al. 2014). The versatility of SiNPs stems from their unique characteristics, such as biocompatibility, abundance, and distinctive electronic, optical, and mechanical properties (O'Farrell et al. 2006; Chang and Sun 2014). The biomedical applications of SiNPs have garnered increasing attention in recent years, particularly in drug delivery, tissue engineering, and immunotherapy. In addition to the intrinsic properties of Si, several factors contribute to the appeal of silicon nanomaterials in these applications, including their customizable size and porosity, as well as their amenable surface chemistry (Santos et al. 2014). Furthermore, Si is renowned for its biodegradability, biocompatibility, and non-toxicity, with its degradation product, silicic acid, easily eliminated from the body via urine and potentially offering therapeutic benefits (O'Neill et al. 2018; Bosch-Rué et al. 2023). Additionally, the intrinsic photoluminescence (PL) of SiNPs facilitates in vivo detection, a feature that is highly valuable for optical biosensors and bioimaging applications (Park et al. 2009, 2019). The abundance of Si, which is the second most prevalent element on Earth, following oxygen, also positions it as a cost-effective and plentiful resource (O'Farrell et al. 2006; Chang and Sun 2014). This abundance ensures the feasibility of large-scale production of SiNPs, which is a critical factor in their commercial viability and widespread adoption in various industries.

22.3 Effects of Silicon on Bone Regeneration

Silicon (Si), primarily present as silicic acid (Si ions), in living organisms, plays a crucial role in metabolic processes involved in bone formation (Wang et al. 2018). The biological mechanism of Si for bone regeneration involves its effects on osteogenesis, angiogenesis, and immunomodulation (O'Neill et al. 2018). Biomaterials that can deliver Si ions (e.g., bioactive glasses) have been widely used for bone regeneration (Zhu et al. 2021). Degradable SiNPs are also promising delivery platforms for Si ions that exhibit multifaceted effects on bone metabolism, including promoting osteogenesis and angiogenesis, as well as inhibiting osteoclastogenesis and regulating local immune response. To enhance the comprehension of silicon's contribution to bone regeneration, dissecting its impact on two critical biological processes, osteogenesis and angiogenesis, is imperative. Each process plays a distinct yet interrelated role in the complex cascade of events that facilitate bone regeneration.

Osteogenesis is fundamentally influenced by Si that has been demonstrated to stimulate osteoblast proliferation and differentiation (Hoppe et al. 2011; O'Neill et al. 2018; Wang et al. 2018). This osteogenic effect of Si is attributed to its ability to modulate the activity of these bone-forming cells, thereby promoting the deposition of extracellular matrix and the mineralization that lead to new bone tissue formation (Hoppe et al. 2011). Si has been implicated in the fundamental stage of bone calcification, where its concentration peaks to facilitate the crystallization of hydroxyapatite into the bone matrix. This process is critical for the establishment of bone mineralization and the subsequent development of bone tissue. Clinical investigations have substantiated that silicon supplementation can positively influence bone health. Specifically, it has been shown to enhance bone density, which is a significant indicator of bone strength and resistance to fracture (Hoppe et al. 2011). Furthermore, si supplementation has been linked to a reduction in bone resorption, a process that is often exacerbated by calcium deficiency, thereby highlighting its potential role in attenuating bone loss (O'Neill et al. 2018). In addition, silicon contributes to maintaining the equilibrium of collagen and other extracellular matrix (ECM) proteins within the bone matrix. Si has been demonstrated to promote osteoblast differentiation, a key event in bone formation, and to stimulate the production of collagen I, the most abundant protein in bone tissue. Si ions have also been observed to activate critical signaling pathways in bone regeneration, such as the bone morphogenetic protein 2 (BMP-2) pathway and the AMP-activated protein kinase (AMPK) pathway. These pathways are integral to osteogenesis, promoting osteoblast differentiation and new bone formation (Wang et al. 2013). Moreover, the interaction of si ions with prolyl hydroxylase, an enzyme involved in collagen stabilization and maturation, promotes the synthesis of collagen type I (Col-I) in the bone matrix (Wang et al. 2013). Col-I is the most abundant protein in bone and is essential for the structural integrity of bone tissue. Si has been demonstrated to exert inhibitory effects on osteoclast differentiation, a critical process in bone resorption (Ma et al. 2021). Understanding the mechanisms by which si modulates osteoclastogenesis is essential for its potential application in the treatment of bone diseases characterized by excessive bone resorption. One of the primary mechanisms through which silicon inhibits osteoclast differentiation involves the modulation of the receptor activator of nuclear factor ≌B (RANK), its ligand (RANKL), and osteoprotegerin (OPG) signaling axis (Ma et al. 2021). This axis is central to the regulation of osteoclastogenesis, with RANKL binding to RANK to promote osteoclast differentiation and OPG acting as a decoy receptor to inhibit this interaction. In addition to its influence on the RANK/RANKL/OPG axis, Si ions have been shown to downregulate the expression of connexin 43 (Cx43), a gap junction protein that plays a crucial role in osteoclast precursor fusion and osteoclast function (Uribe et al. 2020). By reducing Cx43 expression, si ions inhibit the formation and activation of osteoclasts, further contributing to its inhibitory effect on osteoclastogenesis. The regulation of the RANK/RANKL/OPG axis and downregulation of Cx43 expression both contribute to the suppression of osteoclast formation and activation (Uribe et al. 2020). These mechanisms highlight si's potential as a therapeutic agent for conditions associated with increased bone resorption, such as osteoporosis and inflammatory bone diseases.

In addition to osteogenesis, si has emerged as a promising agent for promoting angiogenesis—the formation of new blood vessels (Bosch-Rué et al. 2023). Although research into the mechanisms of si-induced angiogenesis has been less extensive compared to other ions such as Cu and Ce, recent studies have begun to elucidate the pathways involved. Current evidence suggests that the proangiogenic effects of si involve the upregulation of hypoxia-inducible factor-1≌ (HIF-1≌), a key regulator of oxygen homeostasis and angiogenesis. This upregulation is mediated through the inhibition of prolyl hydroxylase 2 (PHD2), an enzyme responsible for the degradation of HIF-1≌ under normoxic conditions. The inhibition of PHD2 by si leads to the stabilization and accumulation of HIF-1≌, which in turn facilitates the expression of proangiogenic growth factors such as vascular endothelial growth factor (VEGF) and basic fibroblast growth factor (bFGF) (Li and Chang 2013b; Dashnyam et al. 2017). The influence of si on angiogenesis is further exemplified by its role in cell-cell interactions, where it amplifies proangiogenic effects through paracrine signaling mechanisms. In co-culture systems involving human dermal fibroblasts (HDFs) and human umbilical vein endothelial cells (HUVECs), si has been shown to significantly augment the gene expression of VEGF in HDFs (Li and Chang 2013a). This increase in VEGF expression subsequently leads to the upregulation of the kinase insert domain receptor (KDR), a receptor for VEGF, in HUVECs through paracrine signaling pathways (Li et al. 2014). The upregulation of the KDR receptor in HUVECs triggers a cascade of events pivotal for angiogenesis. This activation improves the process of blood vessel formation. Si, through its ability to modulate paracrine signaling between HDFs and HUVECs, thus facilitates the development of a vascular network necessary for tissue regeneration. Further research has demonstrated that si enhances the expression of angiogenic markers and the activity of antioxidant enzymes. Specifically, si has been shown to upregulate the expression of superoxide dismutase-1 (SOD-1), catalase-1, and nitric oxide synthase-3 (NOS3) in HUVECs. This upregulation occurs under both normal and oxidative stress conditions, suggesting a role for si in modulating cellular redox balance (Monte et al. 2018). The modulation of antioxidant enzymes by si contributes to decreased cell death and increased proliferation rates in HUVECs and tubular network structure formation.

22.4 Synthesis of Silicon Nanoparticles

The synthesis of SiNPs is a multifaceted field with various methods offering distinct advantages and challenges. These methods can be broadly categorized into top-down and bottom-up approaches, each yielding SiNPs with specific characteristics (Kim and Lee 2023). The top-down approach to fabricating SiNPs is distinguished by its ability to produce high-purity and uniform particles. This method entails systematically reducing bulk si material into nanoscale particles, leveraging various techniques to achieve this transformation. The bottom-up fabrication method is advantageous for the mass production of SiNPs due to its scalability and potential for large-scale synthesis. This method involves the assembly of molecular precursors into nanostructures, allowing for producing a substantial quantity of SiNPs (Chang and Sun 2014; Kim and Lee 2023).

Electrochemical etching of si wafers is a top-down approach that generates a colloidal suspension of SiNPs with diameters around 150 nm. In this process, a mixture of hydrofluoric acid (HF) and water acts as the etchant, facilitating the generation of colloidal si suspensions (Kolasinski 2005; Qin et al. 2014). This method is notable for its simplicity and the ability to produce a relatively monodisperse distribution of particle sizes (Qin et al. 2014). However, the particles may exhibit reduced sphericity. The synthesis of porous SiNPs (pSiNPs) is achieved through a specialized variant of electrochemical etching. This involves the anodization of boron-doped p-type single crystalline si wafers in cells containing HF. The porous si layer is detached from the wafer through a lift-off etching process. The ensuing free-standing porous si film is then subjected to fragmentation into nano-sized particles via external stimulation techniques such as ultrasonication, ball milling, or micro-fluidization (Qin et al. 2014). Mechanical grinding, either through traditional grinding or blending, is another top-down technique that can produce SiNPs ranging from 100 to 1400 nm in diameter (Russo et al. 2011). While this method is straightforward, it often results in a wide distribution of particle sizes and may not achieve the same level of size stratification as electrochemical etching (Bukovsky et al. 2016). Laser ablation is a versatile method for producing spherical SiNPs (Amoruso et al. 2004). By irradiating a si target in a liquid medium, a colloidal solution is formed. The use of nanosecond, picosecond, and femtosecond lasers allows for the control of particle size through pulse duration, with shorter pulses minimizing thermal effects and leading to the formation of smaller nanoparticles (Amoruso et al. 2004). Laser printing on a substrate is a specialized technique where annular femtosecond laser intensities are used to transfer molten si nanodroplets onto a glass surface, creating SiNPs exceeding 400 nm in diameter. Similarly, the application of Gaussian laser pulses to the top of a silicon-on-insulator (SOI) wafer results in the formation of SiNPs over 160 nm in diameter through rapid heating and local melting. While these laser-based methods offer precise control over particle positioning, they may be limited in terms of production volume. Plasma-based techniques can also be employed to produce SiNPs, typically yielding particles larger than 10 nm (Kim and Lee 2023). However, these methods face challenges in precisely controlling particle size.

Bottom-up synthesis is a chemical approach to fabricating SiNPs that involves building up materials from smaller molecular units. This strategy is highly advantageous for the controlled synthesis of nanoparticles with specific properties and is particularly suitable for large-scale production (Kim and Lee 2023). One of the most common bottom-up methods is chemical vapor deposition (CVD). In the CVD process, gaseous precursors such as silane (SiH_4) are introduced into a reaction chamber where they decompose at high temperatures, leading to the formation of SiNPs on a substrate (Leach et al. 2002). By carefully controlling the reaction parameters, including temperature, pressure, and gas flow rates, it is possible to regulate the size, shape, and crystallinity of the resulting nanoparticles (Lumen et al. 2021). The selection of appropriate precursors is crucial in bottom-up synthesis. Silane is a popular choice due to its high reactivity and the ability to produce high-purity SiNPs (Lumen et al. 2021). Pyrolysis, which involves the thermal decomposition of these precursors, can be induced by various heating methods, including laser or plasma heating. For instance, CO_2 lasers are frequently used due to the efficient absorption of their radiation by silane. Plasma-enhanced chemical vapor deposition (PECVD) is another bottom-up technique that utilizes plasma to activate the reaction between gaseous precursors and a substrate. This method can lead to the formation of SiNPs at lower temperatures than conventional CVD and can provide additional control over particle properties through plasma parameters (Hartel et al. 2011).

The reduction of silicon halides, such as silicon tetrachloride ($SiCl_4$), using reducing agents like sodium (Na) or potassium (K) is a well-established bottom-up process (Zou and Kauzlarich 2008). This method can be performed at high temperatures and pressures or, with recent advancements, at room temperature using sodium silicide (NaSi) or potassium silicide (KSi) as reducing agents. Carbothermal reduction involves the reaction of silica (SiO_2) with carbon at high temperatures, typically in an inert atmosphere. This process is energetically demanding due to the strong Si-O bond, but it can produce SiNPs with controlled sizes and properties. The high temperatures facilitate the reduction of silica to silicon, while the carbon source acts as a reductant (Moshtaghioun et al. 2011). A related approach is magnesiothermic reduction, where magnesium is used as a reducing agent for silica. This reaction occurs at lower temperatures than carbothermal reduction and can produce nanosilicon with a narrow size distribution. However, challenges such as incomplete reduction and the formation of byproducts like magnesium silicide need to be addressed (Curtis et al. 2021). Bottom-up methods offer a versatile and scalable approach to producing SiNPs, with the ability to fine-tune particle characteristics to meet specific application requirements. Despite challenges related to size control and byproduct management, ongoing research is focused on optimizing these processes for higher yield, purity, and efficiency of SiNP production.

22.5 Applications of Silicon Nanoparticles in Bone Regeneration

Silicon nanoparticles (SiNPs) have emerged as a promising material in the biomedical field, particularly for bone tissue regeneration. Their unique properties make them ideal candidates for use in composites, functional coatings, and drug delivery systems (Sun et al. 2007; Rodella et al. 2014). The integration of SiNPs in bone regeneration strategies is driven by their ability to stimulate the body's natural healing processes. SiNPs provide a conducive environment for cell proliferation and differentiation, thereby promoting the formation of new bone tissue. Moreover, the controlled release of bioactive ions and biomolecules from SiNPs can further promote the regenerative process (Rodella et al. 2014). These released components can stimulate cellular activities, modulate the immune response, and inhibit infection, leading to improved clinical outcomes. Furthermore, the synergistic effects of SiNPs and other bioactive components can lead to a more comprehensive therapeutic approach. By combining SiNPs with growth factors, stem cells, or other biomaterials, researchers can create multifunctional systems that address the complex requirements of bone regeneration (Anglin et al. 2008). This synergistic approach can result in enhanced osteogenesis and angiogenesis, ultimately leading to more effective bone repair and regeneration (Mao et al. 2017).

Hydrogels have garnered significant attention in the field of bone regeneration due to their unique properties. These three-dimensional networks of crosslinked hydrophilic polymers offer several advantages for bone tissue engineering. Hydrogels are generally biocompatible and can mimic the natural extracellular matrix, providing a suitable environment for cell adhesion, proliferation, and differentiation (Hunt et al. 2014). The mechanical strength and stiffness of hydrogels can be tailored to match that of the surrounding tissue, providing support during the healing process and gradually degrading as new tissue forms. Many hydrogels possess a porous structure, which facilitates nutrient and oxygen diffusion, as well as waste removal, essential for cell survival and tissue regeneration (Catoira et al. 2019). However, many hydrogels, particularly natural polymer-derived hydrogels, have limited mechanical strength, which may not provide sufficient support in load-bearing applications or large bone defects without additional reinforcement. Some hydrogels degrade too quickly, potentially compromising the structural integrity before sufficient tissue regeneration has occurred. Achieving precise control over the degradation rate of hydrogels can be challenging, leading to a potential mismatch between scaffold degradation and tissue formation (Vedadghavami et al. 2017).

The integration of inorganic nanoparticles, such as PSiNPs, into hydrogels has emerged as a promising strategy for improving the mechanical properties and bioactivity of these composite materials (Thoniyot et al. 2015; Andisheh Motealleh 2017). This approach leverages the unique characteristics of inorganic nanoparticles to create advanced hydrogel systems with enhanced performance in biological applications. For example, incorporating PSiNPs into hydrogels represents a significant advancement in the development of bioactive materials with enhanced mechanical properties. By harnessing the bioactivity of PSiNPs and their ability to stimulate extracellular matrix production, regulate cell activity, and facilitate bone mineralization, these composite hydrogel systems hold great promise for applications in bone tissue engineering and regenerative medicine. The capacity to tailor the properties of PSiNPs and load them with biomolecules further expands their potential, offering a versatile platform for the design of advanced biomaterials with targeted functionalities. Aghajanzadeh et al. (2024) developed self-crosslinked nanocomposite hydrogel based on gelatin/aldehyde-modified xanthan (Gel-AXG), which was successfully created by adjusting the concentrations of PSiNPs. They investigated impact of PSiNPs integration on the physical, mechanical, and biological performance of the nanocomposite hydrogels. Morphological analysis revealed the formation of highly porous 3D microstructures with interconnected pores in all nanocomposite hydrogels. Higher PSiNPs content led to a lower swelling ratio, reduced porosity, and pore size, which hindered media penetration and slowed the degradation process. Additionally, notable improvements in dynamic mechanical properties were observed in Gel-AXG-8%Si (compressive strength, 0.6223 MPa at 90% strain, and compressive modulus, 0.054 MPa), along with enhanced biomineralization ability through hydroxyapatite formation after immersion in simulated body fluid (SBF). This optimized nanocomposite hydrogel enabled a sustained release of Si ions at safe dose levels. Furthermore, in vitro cytocompatibility studies using MG-63 cells demonstrated remarkable performance in terms of cell attachment, proliferation, and alkaline phosphatase (ALP) activity for Gel-AXG-8%Si. These findings indicate that the developed nanocomposite hydrogel holds significant potential as a scaffold for bone tissue engineering.

While the degradation products of Si ions have been well-recognized as beneficial for osteogenesis and angiogenesis toward enhanced bone regeneration (O'Neill et al. 2018), SiNPs alone have not been widely used in bone regeneration applications. Other nanoparticles able to deliver Si ions, such as silica nanoparticles and bioactive glass nanoparticles, are more often selected for the delivery of Si ions (Zheng et al. 2021). Alternatively, pSiNPs are applied as delivery platforms of biomolecules (e.g., antibiotics, growth factors) for bone repair and regeneration applications (Santos et al. 2014). pSiNPs can serve as drug delivery systems, targeting the direct delivery of growth factors, cytokines, or other drugs that promote bone regeneration directly to damaged bone sites. pSiNPs, known for their controllable particle size, surface applicability, and good biocompatibility, are often chosen as carriers for drug delivery, controlled release, and targeted imaging in bone repair therapy. In nanomedicine, pSiNPs have emerged as a promising delivery system due to their outstanding biocompatibility, biodegradability, high payload capacity, sustained release kinetics, and efficient transfection capabilities (Santos et al. 2014). pSiNPs can undergo in vivo biodegradation, resulting in the rapid clearance of non-toxic byproducts Si ions. Furthermore, pSiNPs demonstrate high biocompatibility and can serve as multifunctional platforms for applications ranging from bio-imaging to theranostics. Various therapeutic agents, including drugs, growth factors, and miRNAs, have been combined with pSiNPs for bone regeneration applications (Santos et al. 2014). For example, in the context of rheumatoid arthritis (RA) management, the lack of drugs targeting both disease progression and tissue preservation presents a significant challenge. To address this issue, Jeong et al. (2022) developed a novel porous silicon-based nanomedicine capable of efficiently delivering an antirheumatic drug to inflamed synovium while simultaneously degrading into bone-remodeling products. They loaded methotrexate (MTX) into pSiNPs using a calcium silicate-based condenser chemistry to form calcium silicate-pSiNPs constructs (pCaSiNPs). This system exhibited preferential degradation and drug release in inflammatory environments. Notably, the biodegradation products of Si and Ca ions possess immunomodulatory and antiresorptive effects. In a mouse model of collagen-induced arthritis, systemically administered MTX-loaded pCaSiNPs could effectively accumulate in inflamed joints, attenuating RA progression at both early and established stages of the disease. Importantly, combining the nanoparticle carrier with MTX proves more effective than monotherapies, as evidenced by disease state readouts. The efficient accumulation and dissolution of pCaSiNPs in inflamed joints, along with the release of antiresorptive substances, mitigate immunogenic responses and manage bone erosion in the RA model. The high loading capacity of pCaSiNPs suggests the potential for delivering other anti-arthritic agents, highlighting the translational promise of this pSiNPs for RA treatment.

Combining mesenchymal stromal cells (MSC), hydrogels, and osteogenic signals is an effective strategy for bone regeneration. However, osteogenic signals such as growth factors face challenges in clinical use due to cost, stability issues, and immunogenicity. This has led to a growing interest in microRNAs (miRNAs) as a more practical alternative (Leng et al. 2020). Various pro-osteogenic miRNAs have been employed to promote bone regeneration. pSiNPs have been used to deliver various bone regenerating molecucles, including miRNAs. Shrestha et al. (Shrestha et al. 2024) modified pSiNPs with polyamidoamine dendrimers (PAMAM-pSiNP) to deliver pro-osteogenic miRNAs efficiently to MSC embedded within gelatin-PEG hydrogels. Among the miRNAs evaluated, miR-29b-3p, miR-101-3p, and miR-125b-5p emerged as particularly potent pro-osteogenic factors. They were found to target FASN and ELOVL4 in the fatty acid biosynthesis pathway, thereby regulating MSC osteogenesis. Importantly, hydrogel delivery of miRNA:PAMAM-pSiNP complexes showed enhanced transfection efficiency compared to traditional 2D methods. Notably, a dual effect on osteogenesis was observed, with miRNAs increasing the expression of ALP and Runt-related transcription factor 2 (RUNX2), while the pSiNPs enhanced mineralization, likely through degradation into silicic acid. This work sheds light on the role of miRNAs and fatty acid signaling in osteogenesis, offering potential targets for improving bone formation and presenting a promising avenue for enhancing bone tissue engineering.

Insufficient angiogenesis can inhibit bone regeneration and pose a significant challenge in tissue engineering. Bioactive materials capable of effectively maintaining and delivering pro-angiogenic cells and factors during transplantation are demanded to overcome this barrier. Silicon nanoparticles have been used to deliver various therapeutic agents to promote angiogenesis. Duan et al. (2021) developed biodegradable, therapeutic, and self-reporting luminescent porous silicon (PSi) microparticles to deliver VEGF and Si ions to synergistically promote angiogenesis. These PSi microparticles exhibited rapid release of therapeutic Si ions, promoting cell migration, tube formation, and angiogenic gene expression of HUVECs in vitro. VEGF can be electrostatically adsorbed onto the PSi microparticles. The simultaneous release of Si ions and VEGF occurs with the dissolution of the Si skeleton, leading to a decrease in photoluminescence (PL) intensity and a blue shift of the maximum PL wavelength. This allows for real-time self-reporting and assessment of drug release using the two-dimensional PL signal. The simultaneous release of Si ions and VEGF also demonstrated a synergistic effect, resulting in highly efficient angiogenesis. The VEGF and Si ion co-delivery system holds great promise in tissue engineering as a biodegradable and self-reporting theranostic platform for promoting angiogenesis for bone regeneration.

SiNPs can serve as coating precursors for drug delivery systems, biosensors, and medical implants. For instance, Zhao et al. (2020) utilized the magnetron sputtering method to apply nanoscale Si coating onto the surface of carbon fiber (CF) to enhance the properties of CF-reinforced hydroxyapatite (HA) bioceramics as load-bearing bone substitutes. Si-coated CF (Si-CF) was then used to create HA composites with controllable alignment by pressureless and hot-press sintering. The results revealed that the CF surface was uniformly covered with continuous Si coating (approximately 0.2 μm in thickness). The mechanical strength of three-layer Si-CF reinforced HA (3Si-CF/HA) was significantly enhanced compared to that of three-layer CF reinforced HA (3CF/HA) bioceramics. Notably, the introduction of Si-coated CF effectively minimized the contact between CF and the oxygen produced during the decomposition of HA, probably due to the chemical reaction between Si and O_2, resulting in the formation of SiO_2. The protective barrier layer provided by the Si and SiO_2 coatings effectively prevented oxidative damage of CF during sintering, thereby preserving the original reinforcing and toughening effects of CF. Consequently, the bending strength and fracture toughness of the composites were significantly enhanced, meeting the requirements for load-bearing bone defect repair.

The integration of SiNPs into bone regeneration applications represents a significant advancement in the biomedical field. Their multifaceted utility, coupled with the synergistic effects of released biomolecules and ions, positions SiNPs as a valuable tool for improving bone regeneration. With the development of nanotechnology and medicine, we can expect to see more innovative strategies and therapies that harness the power of SiNPs to promote healing and restore function in patients with bone injuries or diseases.

22.6 Conclusions

Silicon nanoparticles (SiNPs) have emerged as a promising biomaterial for bone regeneration due to their biocompatibility, pro-osteogenic nature, and ability to stimulate cellular activities. The controlled release of bioactive ions from SiNPs can enhance osteogenesis and promote cell differentiation, crucial factors for successful bone healing. Future advancements in SiNP-based therapies include tailored properties for specific therapeutic needs; hybrid systems with other biomaterials, growth factors, and stem cells; individualized patient-specific bone grafts and scaffolds; and a deeper understanding of the mechanistic pathways by which SiNPs influence cellular processes. As research continues to unravel the complexities of bone healing and the role of SiNPs, we can expect to see continued innovation and the eventual widespread use of these nanoparticles in clinical practice, offering new hope for patients suffering from bone injuries and diseases. We expect that the present chapter will serve as a useful starting point for researchers who are entering the field of silicon technology and tissue engineering, giving a comprehensive overview of the synthesis and applications of this very promising biomaterial.

References

Aghajanzadeh MS, Imani R, Nazarpak MH, McInnes SJP (2024) Augmented physical, mechanical, and cellular responsiveness of gelatin-aldehyde modified xanthan hydrogel through incorporation of silicon nanoparticles for bone tissue engineering. Int J Biol Macromol 259. https://doi.org/10.1016/j.ijbiomac.2024.129231

Amoruso S, Bruzzese R, Spinelli N et al (2004) Generation of silicon nanoparticles via femtosecond laser ablation in vacuum. Appl Phys Lett 84:4502–4504

Andisheh Motealleh NSK (2017) Nanocomposite hydrogels and their applications in tissue engineering. Adv Healthc Mater 6:1600938. https://doi.org/10.1002/adhm.201600938

Anglin EJ, Cheng L, Freeman WR, Sailor MJ (2008) Porous silicon in drug delivery devices and materials. Adv Drug Deliv Rev 60:1266–1277

Aprile P, Letourneur D, Simon-Yarza T (2020) Membranes for guided bone regeneration: a road from bench to bedside. Adv Healthc Mater 9:2000707. https://doi.org/10.1002/adhm.202000707

Bai X, Gao M, Syed S et al (2018) Bioactive hydrogels for bone regeneration. Bioact Mater 3:401–417

Bosch-Rué È, Díez-Tercero L, Buitrago JO et al (2023) Angiogenic and immunomodulation role of ions for initial stages of bone tissue regeneration. Acta Biomater 166:14–41. https://doi.org/10.1016/j.actbio.2023.06.001

Bukovsky EV, Castro KP, Wyatt BM et al (2016) Anaerobic vs. aerobic preparation of silicon nanoparticles by stirred media milling. The effects of dioxygen, milling solvent, and milling time on particle size, surface area, crystallinity, surface/near-surface composition, and reactivity. RSC Adv 6:112370–112380

Catoira MC, Fusaro L, Di Francesco D et al (2019) Overview of natural hydrogels for regenerative medicine applications. J Mater Sci Mater Med 30:1–10

Chang H, Sun S-Q (2014) Silicon nanoparticles: preparation, properties, and applications. Chinese Phys B 23:88102

Couvreur P (2013) Nanoparticles in drug delivery: past, present and future. Adv Drug Deliv Rev 65:21–23

Croissant JG, Fatieiev Y, Khashab NM (2017) Degradability and clearance of silicon, organosilica, silsesquioxane, silica mixed oxide, and mesoporous silica nanoparticles. Adv Mater 29:1604634. https://doi.org/10.1002/adma.201604634

Cui Y, Li H, Li Y, Mao L (2022) Novel insights into nanomaterials for immunomodulatory bone regeneration. Nanoscale Adv 4:334–352

Curtis IS, Wills RJ, Dasog M (2021) Photocatalytic hydrogen generation using mesoporous silicon nanoparticles: influence of magnesiothermic reduction conditions and nanoparticle aging on the catalytic activity. Nanoscale 13:2685–2692

Dashnyam K, Jin G-Z, Kim J-H et al (2017) Promoting angiogenesis with mesoporous microcarriers through a synergistic action of delivered silicon ion and VEGF. Biomaterials 116:145–157. https://doi.org/10.1016/j.biomaterials.2016.11.053

Duan W, Jin Y, Cui Y et al (2021) A co-delivery platform for synergistic promotion of angiogenesis based on biodegradable, therapeutic and self-reporting luminescent porous silicon microparticles. Biomaterials 272:120772. https://doi.org/10.1016/j.biomaterials.2021.120772

Duda GN, Geissler S, Checa S, Tsitsilonis S, Petersen A, Schmidt-Bleek K (2023) The decisive early phase of bone regeneration. Nat Rev Rheumatol 19:78–95. https://doi.org/10.1038/s41584-022-00887-0

El-Rashidy AA, Roether JA, Harhaus L et al (2017) Regenerating bone with bioactive glass scaffolds: a review of in vivo studies in bone defect models. Acta Biomater 62:1–28. https://doi.org/10.1016/j.actbio.2017.08.030

Fathi-Achachelouei M, Knopf-Marques H, da Silva CE et al (2019) Use of nanoparticles in tissue engineering and regenerative medicine. Front Bioeng Biotechnol 7:113

Filippi M, Born G, Chaaban M, Scherberich A (2020) Natural polymeric scaffolds in bone regeneration. Front Bioeng Biotechnol 8:474

Giannoudis PV, Einhorn TA, Marsh D (2007) Fracture healing: the diamond concept. Injury 38:S3–S6

Ginebra M-P, Espanol M, Maazouz Y et al (2018) Bioceramics and bone healing. EFORT Open Rev 3:173–183

Han X, Xu K, Taratula O, Farsad K (2019) Applications of nanoparticles in biomedical imaging. Nanoscale 11:799–819

Hartel AM, Hiller D, Gutsch S et al (2011) Formation of size-controlled silicon nanocrystals in plasma enhanced chemical vapor deposition grown SiOxNy/SiO2 superlattices. Thin Solid Films 520:121–125

Henstock JR, Canham LT, Anderson SI (2015) Silicon: the evolution of its use in biomaterials. Acta Biomater 11:17–26

Hoppe A, Güldal NS, Boccaccini AR (2011) A review of the biological response to ionic dissolution products from bioactive glasses and glass-ceramics. Biomaterials 32:2757–2774. https://doi.org/10.1016/j.biomaterials.2011.01.004

Ho-Shui-Ling A, Bolander J, Rustom LE et al (2018) Bone regeneration strategies: engineered scaffolds, bioactive molecules and stem cells current stage and future perspectives. Biomaterials 180:143–162. https://doi.org/10.1016/j.biomaterials.2018.07.017

Hunt JA, Chen R, van Veen T, Bryan N (2014) Hydrogels for tissue engineering and regenerative medicine. J Mater Chem B 2:5319–5338

Jeong M, Jung Y, Yoon J et al (2022) Porous silicon-based nanomedicine for simultaneous management of joint inflammation and bone erosion in rheumatoid arthritis. ACS Nano 16:16118–16132. https://doi.org/10.1021/acsnano.2c04491

Kim T, Lee J (2023) Silicon nanoparticles: fabrication, characterization, application and perspectives. Micro Nano Syst Lett 11:18. https://doi.org/10.1186/s40486-023-00184-9

Kolasinski KW (2005) Silicon nanostructures from electroless electrochemical etching. Curr Opin Solid State Mater Sci 9:73–83

Leach WT, Zhu J, Ekerdt JG (2002) Thermal desorption effects in chemical vapor deposition of silicon nanoparticles. J Cryst Growth 243:30–40

Leng Q, Chen L, Lv Y (2020) RNA-based scaffolds for bone regeneration: application and mechanisms of mRNA, miRNA and siRNA. Theranostics 10:3190

Li H, Chang J (2013a) Bioactive silicate materials stimulate angiogenesis in fibroblast and endothelial cell co-culture system through paracrine effect. Acta Biomater 9:6981–6991. https://doi.org/10.1016/j.actbio.2013.02.014

Li H, Chang J (2013b) Stimulation of proangiogenesis by calcium silicate bioactive ceramic. Acta Biomater 9:5379–5389

Li H, Xue K, Kong N et al (2014) Silicate bioceramics enhanced vascularization and osteogenesis through stimulating interactions between endothelia cells and bone marrow stromal cells. Biomaterials 35:3803–3818. https://doi.org/10.1016/j.biomaterials.2014.01.039

Lumen D, Wang S, Mäkilä E et al (2021) Investigation of silicon nanoparticles produced by centrifuge chemical vapor deposition for applications in therapy and diagnostics. Eur J Pharm Biopharm 158:254–265

Ma W, Wang F, You Y et al (2021) Ortho-silicic acid inhibits RANKL-induced osteoclastogenesis and reverses ovariectomy-induced bone loss in vivo. Biol Trace Elem Res 199:1864–1876

Mao L, Xia L, Chang J et al (2017) The synergistic effects of Sr and Si bioactive ions on osteogenesis, osteoclastogenesis and angiogenesis for osteoporotic bone regeneration. Acta Biomater 61:217–232. https://doi.org/10.1016/j.actbio.2017.08.015

Monte F, Cebe T, Ripperger D et al (2018) Ionic silicon improves endothelial cells' survival under toxic oxidative stress by overexpressing angiogenic markers and antioxidant enzymes. J Tissue Eng Regen Med 12:2203–2220

Montoya C, Du Y, Gianforcaro AL et al (2021) On the road to smart biomaterials for bone research: definitions, concepts, advances, and outlook. Bone Res 9:12. https://doi.org/10.1038/s41413-020-00131-z

Moshtaghioun BM, Monshi A, Abbasi MH, Karimzadeh F (2011) A study on the effects of silica particle size and milling time on synthesis of silicon carbide nanoparticles by carbothermic reduction. Int J Refract Met Hard Mater 29:645–650

Mout R, Moyano DF, Rana S, Rotello VM (2012) Surface functionalization of nanoparticles for nanomedicine. Chem Soc Rev 41:2539–2544

Murakami Y, Honda Y, Anada T et al (2010) Comparative study on bone regeneration by synthetic octacalcium phosphate with various granule sizes. Acta Biomater 6:1542–1548

O'Farrell N, Houlton A, Horrocks BR (2006) Silicon nanoparticles: applications in cell biology and medicine. Int J Nanomedicine 1:451–472

O'Neill E, Awale G, Daneshmandi L et al (2018) The roles of ions on bone regeneration. Drug Discov Today 23:879–890. https://doi.org/10.1016/j.drudis.2018.01.049

Park J-H, Gu L, Von Maltzahn G et al (2009) Biodegradable luminescent porous silicon nanoparticles for in vivo applications. Nat Mater 8:331–336

Park Y, Yoo J, Kang M-H et al (2019) Photoluminescent and biodegradable porous silicon nanoparticles for biomedical imaging. J Mater Chem B 7:6271–6292

Pereira HF, Cengiz IF, Silva FS et al (2020) Scaffolds and coatings for bone regeneration. J Mater Sci Mater Med 31:1–16

Qin Z, Joo J, Gu L, Sailor MJ (2014) Size control of porous silicon nanoparticles by electrochemical perforation etching. Part Part Syst Charact 31:252–256

Rodella LF, Bonazza V, Labanca M et al (2014) A review of the effects of dietary silicon intake on bone homeostasis and regeneration. J Nutr Health Aging 18:820–826

Russo L, Colangelo F, Cioffi R et al (2011) A mechanochemical approach to porous silicon nanoparticles fabrication. Materials (Basel) 4:1023–1033

Santos HA, Mäkilä E, Airaksinen AJ et al (2014) Porous silicon nanoparticles for nanomedicine: preparation and biomedical applications. Nanomedicine 9:535–554

Shrestha S, Tieu T, Wojnilowicz M et al (2024) Delivery of miRNAs using porous silicon nanoparticles incorporated into 3D hydrogels enhances MSC osteogenesis by modulation of fatty acid signaling and silicon degradation. Adv Healthc Mater 2400171. https://doi.org/10.1002/adhm.202400171

Shuai C, Li S, Peng S et al (2019) Biodegradable metallic bone implants. Mater Chem Front 3:544–562

Sun W, Puzas JE, Sheu T-J et al (2007) Nano-to microscale porous silicon as a cell interface for bone-tissue engineering. Adv Mater 19:921–924

Thoniyot P, Tan MJ, Karim AA et al (2015) Nanoparticle–Hydrogel composites: concept, design, and applications of these promising, multi-functional materials. Adv Sci 2:1–13

Uribe P, Johansson A, Jugdaohsingh R et al (2020) Soluble silica stimulates osteogenic differentiation and gap junction communication in human dental follicle cells. Sci Rep 10:9923

Valtchev V, Tosheva L (2013) Porous nanosized particles: preparation, properties, and applications. Chem Rev 113:6734–6760. https://doi.org/10.1021/cr300439k

Vedadghavami A, Minooei F, Mohammadi MH et al (2017) Manufacturing of hydrogel biomaterials with controlled mechanical properties for tissue engineering applications. Acta Biomater 62:42–63

Wang C, Lin K, Chang J, Sun J (2013) Osteogenesis and angiogenesis induced by porous $β$-CaSiO3/PDLGA composite scaffold via activation of AMPK/ERK1/2 and PI3K/Akt pathways. Biomaterials 34:64–77

Wang X, Gao L, Han Y et al (2018) Silicon-enhanced adipogenesis and angiogenesis for vascularized adipose tissue engineering. Adv Sci 5. https://doi.org/10.1002/advs.201800776

Wildemann B, Ignatius A, Leung F et al (2021) Non-union bone fractures. Nat Rev Dis Prim 7:57

Winkler T, Sass FA, Duda GN, Schmidt-Bleek K (2018) A review of biomaterials in bone defect healing, remaining shortcomings and future opportunities for bone tissue engineering: the unsolved challenge. Bone Jt Res 7:232–243. https://doi.org/10.1302/2046-3758.73.BJR-2017-0270.R1

Yan Y, Chen H, Zhang H et al (2019) Vascularized 3D printed scaffolds for promoting bone regeneration. Biomaterials 190:97–110

Zhang J, Jiang Y, Shang Z et al (2021) Biodegradable metals for bone defect repair: a systematic review and meta-analysis based on animal studies. Bioact Mater 6:4027–4052

Zhao X, Zheng J, Zhang W et al (2020) Preparation of silicon coated-carbon fiber reinforced HA bio-ceramics for application of load-bearing bone. Ceram Int 46:7903–7911. https://doi.org/10.1016/j.ceramint.2019.12.010

Zheng K, Sui B, Ilyas K, Boccaccini AR (2021) Porous bioactive glass micro- and nanospheres with controlled morphology: developments, properties and emerging biomedical applications. Mater Horizons 8:300–335. https://doi.org/10.1039/d0mh01498b

Zhu H, Zheng K, Boccaccini AR (2021) Multi-functional silica-based mesoporous materials for simultaneous delivery of biologically active ions and therapeutic biomolecules. Acta Biomater 129:1–17. https://doi.org/10.1016/j.actbio.2021.05.007

Zou J, Kauzlarich SM (2008) Functionalization of silicon nanoparticles via silanization: alkyl, halide and ester. J Clust Sci 19:341–355

Index

R. de Mello Prado et al. (eds.), *Silicon advances for sustainable agriculture and human health*, Sustainable Plant Nutrition in a Changing World, https://doi.org/10.1007/978-3-031-69876-7

Printed by Printforce, the Netherlands